Bioactive Compounds in Leafy Vegetables

Editor:
Hosakatte Niranjana Murthy
Department of Life Sciences
Christ University
Bengaluru, India

CRC Press
Taylor & Francis Group
Boca Raton London New York

CRC Press is an imprint of the
Taylor & Francis Group, an **informa** business
A SCIENCE PUBLISHERS BOOK

First edition published 2026
by CRC Press
2385 NW Executive Center Drive, Suite 320, Boca Raton FL 33431

and by CRC Press
4 Park Square, Milton Park, Abingdon, Oxon, OX14 4RN

CRC Press is an imprint of Taylor & Francis Group, LLC

Library of Congress Cataloging-in-Publication Data (applied for)

ISBN: 978-1-041-12489-4 (hbk)
ISBN: 978-1-041-17080-8 (pbk)
ISBN: 978-1-003-68783-2 (ebk)

DOI: 10.1201/9781003687832

Typeset in Times New Roman
by Prime Publishing Services

Preface

Leafy green vegetables are essential components of a balanced diet. Leafy greens contain minerals, proteins, fatty acids, and carbohydrates. They are also rich in vitamins and fiber. Leafy greens also contain specialized metabolites with significant nutraceutical advantages, including phenolics, flavonoids, terpenoids, alkaloids, and glucosinolates. Leafy greens are therefore considered a nutrient-dense and beneficial addition to a diet that can help prevent and improve risk factors for various diseases, including diabetes, cancer, obesity, metabolic syndrome, and cardiovascular diseases.

The nutritional value, bioactive components, and biological activity of green leafy vegetables from throughout the world are covered in this book. I wish to thank and express my deepest gratitude to all the contributors and all the officials of CRC Press who helped us to complete this book.

Hosakatte Niranjana Murthy

Contents

List of Contributors

Aayush Sinha
Department of Life Sciences, School of Sciences, Christ University, Bengaluru 560 029, India.

Adeyemi O. Aremu
Indigenous Knowledge Systems Centre, Faculty of Natural and Agricultural Sciences, North-West University, Private Bag X2046, Mmabatho 2790, South Africa.
School of Life Sciences, College of Agriculture, Engineering and Science, University of KwaZulu-Natal, Private Bag X54001, Durban 4000, KwaZulu-Natal, South Africa.

Ahmed Olatunde
Department of Medical Biochemistry, Abubakar Tafawa Balewa University, Bauchi 740272, Nigeria.

Alberto Martín
Nutrición y Bromatología, Escuela de Ingenierías Agrarias, Universidad de Extremadura, Avd. Adolfo Suárez s/n, 06007 Badajoz, Spain.
Instituto Universitario de Investigación en Recursos Agrarios (INURA), Universidad de Extremadura, Avd. de la Investigación, 06006 Badajoz, Spain.

Bonta Ramesh Kumar
Center for Process Research and Innovation (CPRI), Dr.Reddy's Institute of Life Sciences, University of Hyderabad Campus, Gachibowli, Hyderabad 500046, Telangana, India.

Clarissa Marcelle Naidoo
Department of Biology and Environmental Sciences, Sefako Makgatho Health Sciences University, Pretoria 0204, South Africa.

Damaji S. Bhise
Department of Botany, Shri Shivaji Mahavidyalaya, Barshi 413 411 Maharashtra, India.

Dhanasekaran Sneha
Department of Life Sciences, School of Sciences, Christ University, Bengaluru 560 029, India.

Gayane A. Atazhanova
School of Pharmacy, Karaganda Medical University, Gogol Street, 40, Karaganda 100017, Kazakhstan.

Guggalada Govardhana Yadav
Department of Botany, ADB First Grade College, Harapanhalli 583131, India.
Department of Botany, Karnatak University, Dharwad 580003, India.

Hosakatte Niranjana Murthy
Department of Life Sciences, CHRIST (Deemed to be University), Bengaluru 560029, India.
Department of Horticultural Science, Chungbuk National University, Cheongju 28644, Republic of Korea.
Department of Botany, Karnatak University, Dharwad 580003, India.

Iris Gudiño
Nutrición y Bromatología, Escuela de Ingenierías Agrarias, Universidad de Extremadura, Avd. Adolfo Suárez s/n, 06007 Badajoz, Spain.
Instituto Universitario de Investigación en Recursos Agrarios (INURA), Universidad de Extremadura, Avd. de la Investigación, 06006 Badajoz, Spain.

Jagdish Vishnu Dalavi
Botanical Survey of India, Western Regional Centre, Pune, Maharashtra 411001, India.

Jameel M. Al-Khayri
Department of Agriculture Biotechnology, College of Agriculture and Food Sciences, King Faisal University, Al-Ahsa 31982, Saudi Arabia.

Jaya Gangwar
Department of Life Sciences, Christ University, Bangalore, 560029, India.

Kadanthottu Sebastian Joseph
Department of Life Sciences, Christ University, Bangalore, 560029, India.

Karabo Maselepe Makoana
Department of Biology and Environmental Sciences, Sefako Makgatho Health Sciences University, Pretoria 0204, South Africa.

Karakoz Zh. Badekova
School of Pharmacy, Karaganda Medical University, Gogol Street, 40, Karaganda 100017, Kazakhstan.

Keamogela I. Nko
Indigenous Knowledge Systems Centre, Faculty of Natural and Agricultural Sciences, North-West University, Private Bag X2046, Mmabatho 2790, South Africa.

Kee Yoeup Paek
Department of Horticultural Science, Chungbuk National University, Cheongju 28644, Republic of Korea.

Lecholocholo N.
Department of Crop Science, Tshwane University of Technology, Pretoria, South Africa.

Makarand M. Aitawade
Department of Botany, Prof. Dr. N. D. Patil Mahavidyalaya, Malkapur-Perid 415 101 Maharashtra, India.

Makhotso Lekhooa
DSI/NWU Preclinical Drug Development Platform, Faculty of Health Sciences, North-West University, Private Bag X6001, Potchefstroom 2521, South Africa.

María José Benito
Nutrición y Bromatología, Escuela de Ingenierías Agrarias, Universidad de Extremadura, Avd. Adolfo Suárez s/n, 06007 Badajoz, Spain.
Instituto Universitario de Investigación en Recursos Agrarios (INURA), Universidad de Extremadura, Avd. de la Investigación, 06006 Badajoz, Spain.

Nilesh Appaso Madhav
Department of Botany, HPT Arts and RYK Science College, Nashik, M.S., India

Nqobile Monate Mkolo
Department of Biology and Environmental Sciences, Sefako Makgatho Health Sciences University, Pretoria 0204, South Africa.

Praveen Nagella
Department of Life Sciences, School of Sciences, Christ University, Bengaluru 560029, India.

Priyanka V. Patil
Department of Botany, The New College, Kolhapur 416 012 Maharashtra, India.

Rocío Casquete
Nutrición y Bromatología, Escuela de Ingenierías Agrarias, Universidad de Extremadura, Avd. Adolfo Suárez s/n, 06007 Badajoz, Spain.
Instituto Universitario de Investigación en Recursos Agrarios (INURA), Universidad de Extremadura, Avd. de la Investigación, 06006 Badajoz, Spain.

Sachin B. Patil
Department of Botany, Anandibai Raorane Arts, Commerce and Science College, Vaibhavwadi, Sindhudurg 416 810 Maharashtra, India.

Sagar A. Deshmukh
Department of Botany, The New College, Kolhapur 416 012 Maharashtra, India.

Saikat Sen
Center for Process Research and Innovation (CPRI), Dr.Reddy's Institute of Life Sciences, University of Hyderabad Campus, Gachibowli, Hyderabad 500046, Telangana, India.

Samsur Ali Dafadar
Center for Process Research and Innovation (CPRI), Dr.Reddy's Institute of Life Sciences, University of Hyderabad Campus, Gachibowli, Hyderabad 500046, Telangana, India.

Shrinivas Lamani
Department of Botany, DVS College of Arts and Science, Shimoga 577201, India.

Sivakumar D.
South African Research in Phytochemical Food Network, Tshwane University of Technology, Pretoria, South Africa.
Centre for Nutrition and Food Sciences, Queensland Alliance for Agriculture and Food Innovation, The University of Queensland, QLD 4108, Queensland, Australia.

Smita Shinde
Department of Botany, Karnatak University, Dharwad 580003, India.

Sowmya Viswanathan
Department of Life Sciences, School of Sciences, Christ University, Bengaluru 560029, India.

So-Young Park
Department of Horticultural Science, Chungbuk National University, Cheongju 28644, Republic of Korea.

Srinivas Oruganti
Center for Process Research and Innovation (CPRI), Dr.Reddy's Institute of Life Sciences, University of Hyderabad Campus, Gachibowli, Hyderabad 500046, Telangana, India.

Supreeth Holenarasipura Subramanya
Department of Life Sciences, School of Sciences, Christ University, Bengaluru 560 029, India.

Suraj P. Devkar
Department of Botany, The New College, Kolhapur 416 012 Maharashtra, India.

Tshilongo L.
Department of Crop Science, Tshwane University of Technology, Pretoria, South Africa.

Vasundhara Vilas Khot
Department of Botany, Balwant College Vita, (Sangli), M.S., India

Vijayalaxmi S. Dandin
Department of Biology, Alliance Pre-university College, Bengaluru 560099, India.

Yana K. Levaya
School of Pharmacy, Karaganda Medical University, Gogol Street, 40, Karaganda 100017, Kazakhstan.

Yuanfeng Wu
School of Biological and Chemical Engineering, Zhejiang University of Science and Technology, Hangzhou 310023, China.

1 Leafy Vegetables: Nutritional Value, Bioactive Compounds, and Health Benefits

Hosakatte Niranjana Murthy[1-3*], Guggalada Govardhana Yadav,[3,4] Smita Shinde[3], Kee Yoeup Paek[2] and So-Young Park[2]

Introduction

Leafy vegetables are defined by their edible leaves, which can be eaten raw, cooked, or processed (Putriani et al., 2022). In addition to their great nutritional value, these vegetables are key components of a balanced diet and are widely respected and popular since they provide a wide variety of plants with different features, such as color, flavor, and texture (Koukounaras et al., 2020). Swiss chard, kale, spinach, cabbage, and lettuce are some of the types that are frequently eaten. In addition to well-known options, there are lesser-known greens including fenugreek, parsley, fennel, and amaranth leaves. Furthermore, the variety of possibilities is further increased by coriander, curry leaf, moringa, and dandelion greens (García Herrera et al., 2020). Leafy vegetables include not just well-known types but also underutilized ones that have important nutritional value. Malabar spinach, sorrel, dill greens, and numerous other vegetables are a few examples. These underutilized types add to the nutritional and gastronomic diversity with their own flavors and textures.

Nutrients like proteins, carbohydrates, minerals, and dietary fibers can all be found in leafy vegetables (Das, 2022; Gupta et al., 2005). Leafy vegetables containing chemicals of economic and pharmacological significance, such as phenolics, tannins, flavonoids, saponins, alkaloids, and glucosinolates, have been identified through phytochemical investigation (Khude, 2023). Leafy vegetables have been shown in numerous studies to be great sources of vitamins, antioxidants, and amino acids (Gupta et al., 2005). These nutrient-dense, therapeutically useful vegetables are used to cure several chronic illnesses, including cancer, diabetes, rheumatism, constipation, and dysentery (Horo and Topno, 2015). In drought and famine conditions, wild leafy greens are essential for the food security of both

[1] Department of Life Sciences, CHRIST (Deemed to be University), Bengaluru 560029, India.
[2] Department of Horticultural Science, Chungbuk National University, Cheongju 28644, Republic of Korea.
[3] Department of Botany, Karnatak University, Dharwad 580003, India.
[4] Department of Botany, ADB First Grade College, Harapanahalli 583131, India.
* Corresponding author: hnmurthy60@gmail.com; hosakatte.niranjanamurthy@christuniversity.in

DOI: 10.1201/9781003687832-1

urban and rural environments, according to the Food and Agriculture Organization (WHO, 2000). Leafy vegetables generally can improve nutritional quality and food diversity.

Research on leafy greens has exploded in the last several years, particularly in the areas of nutritional analysis, bioactive chemical evaluation, toxicological investigations, and health benefits. Therefore, the main objective of the book is to give examples of leafy vegetables along with information about their nutritional value, bioactive compounds, and biological activities. Researchers have provided insight into the nutritional and bioactive elements of a broad range of leafy vegetables from all over the world. The scientific research on many leafy vegetables that are widely consumed in South Asia and Southeast Asian nations, including moringa (*Moringa oleifera* Lam., Family: Moringaceae, a fast-growing deciduous tree grows in tropical and subtropical regions, Figure 1), colocasia (*Colocasia esculenta* L. (Schott), Family: Araceae, perennial tropical plant grown for corms and leaves, Figure 1), purslane (*Portulaca oleraceae* L., Family: Portulacaceae, a succulent herbaceous plant grows in common in north Africa, middle east, Indian subcontinent, and Australia, Figure 1), and Indian sorrel (*Oxalis corniculata* L., Family: Oxalidaceae, a small creeping herb grows in moist climates, Figure 1), is summarized in this chapter. This covers details about their nutritional and phytochemical composition, safety profiles, health advantages, and valued-added products.

Nutrients

Numerous studies examine the nutritional components of leafy vegetables from diverse places/regions. According to these studies, leafy vegetables have the same nutrients as other plant portions, including carbohydrates, proteins, fat, fiber, ash, minerals, vitamins, and amino acids. The composition, however, differs depending on the plant species or cultivar and the geographical area from which they are sourced (Sivakumar et al., 2018; Mungofa et al., 2022; Masiala et al., 2024; Niko et al., 2024). This review compares and discusses the nutritional value of South Asia and Southeast Asian nations edible leafy vegetables, such as moringa (*Moringa oleifera*), colocasia (*Colocasia esculenta*), purslane (*Portulaca oleracea*), and Indian sorrel (*Oxalis corniculata*).

Moringa, colocasia, purslane, and Indian sorrel leaves are typically abundant in nutrients, including carbohydrates, proteins, lipids, and dietary fiber (Table 1). However, the content varies according to the plant species, age, and agroclimatic conditions. Moringa leaves had 8.28 g/100 g of carbohydrates (Haytowitz David, 2019). The carbohydrate content of colocasia leaves was reported to be 3.0 g/100 g (Haytowitz David, 2019). The carbohydrate content of purslane and Indian sorrel leaves was 3.39 g/100 g and 21.1 g/100 g, respectively (Haytowitz David, 2019; Radha et al., 2021). Moringa and Indian sorrel leaves have higher carbohydrate content than other regularly consumed green leafy vegetables like spinach (2.9 g/100 g) and mustard leaves (3.2 g/100 g). Proteins play a vital structural and

Fig. 1 Some examples of edible leafy vegetables. Moringa (A-B), Colocasia (C-D), Purslane (E-F), and Indian Sorrel (G-H).

functional role in the growth and maintenance of the human body (Sa et al., 2020). Moringa leaves had a protein level of 9.40 g/100 g (Table 1), which was significantly higher than colocasia (5.0 g/100 g), purslane (2.03 g/100 g), and Indian sorrel (4.3 g/100 g. Leafy vegetables are generally low in fat, resulting in less calories (Obeng et al., 2020). Moringa, purslane, and Indian sorrel leaves had fat contents of 1.40, 0.36, and 1.42 g/100 g, respectively (Table 1). According to Tshabalala et al. (2020), moringa leaves have less saturated fatty acids and more monounsaturated fatty acids. Furthermore, omega-3 and omega-6 fatty acids, which are important for cardiovascular function and body vitality, have been found in moringa leaves. Similarly, Palaniswamy et al. (2001) found omega-3 and omega-6 fatty acids in purslane leaf extract. Furthermore, it has been observed that moringa leaves contain important amino acids such as lysine, isoleucine, histidine, phenylalanine, methionine, tryptophan, threonine, tyrosine, and valine (Moyo et al., 2011). Similarly, Nemzer et al. (2020) found essential amino acids such as leucine, valine, lysine, phenylalanine, isoleucine, threonine, histidine, and methionine in purslane leaves, as well as non-essential amino acids including glutamic acid, aspartic acid, alanine, and arginine. Dietary fiber in vegetables functions as a prebiotic, controlling blood glucose, lowering cholesterol levels, improving gut health, and protecting against carcinogens (Neela and Fanta, 2019). The dietary fiber content of moringa leaves was 2.0 g/100 g, according to Haytowitz David (2019). Other researchers have reported crude fiber content ranging from 7.09 to 35 g/100 g (Moyo et al., 2011). Likewise, the crude fiber content of colocasia and Indian sorrel leaves was determined to be 3.7 g/100 g (Haytowitz David, 2019) and 29.96 g/100 g (Radha et al., 2021). Moringa leaves contain high levels of vitamins, including vitamin B3 (niacin, 2.22 mg/100 g), vitamin C (ascorbic acid, 51.7 mg/100 g), vitamin A (carotenoids, 0.37 mg/100 g), vitamin B9 (folic acid, 40 μg/100 g), vitamin B1 (thiamine, 257 μg/100 g), vitamin B2 (riboflavin, 660 μg/100 g), and vitamin B6 (1.2 mg/100 g) (Haytowitz David, 2019). Longvah et al. (2017) found that colocasia leaves contain high levels of vitamin B3 (0.29 mg/100 g), B5 (0.27 mg/100 g), B6 (0.29 mg/100 g), B7 (12.10 mg/100 g), B9 (159 μg/100 g), and C (40.71 mg/100 g). Longvah et al. (2017) observed higher quantities of vitamin B3 (0.33 mg/100 g), B5 (0.22 mg/100 g), B6 (0.15 mg/100 g), B7 (4.14 mg/100 g), B9 (142 μg/100 g), and C (30.28 mg/100 g) in colocasia leaves than spinach. Purslane leaves contain several carotenoids, such as α-carotene (0.009 mg/100 g), β-carotene (3.5 mg/100 g), lutein (5.4 mg/100 g), and zeaxanthin (0.19 mg/100 g) (Dias et al., 2009). Purslane also contains vitamins C, E, and K, as well as B-complex vitamins like riboflavin, niacin, and pyridoxine (Nemzer et al., 2020).

Leafy vegetables are high in micronutrients and minerals, which are essential for body metabolism and act as cofactors in enzyme activities. Moringa leaves are especially high in minerals including calcium (185 mg/100 g), iron (4.0 mg/100 g), sodium (9.0 mg/100 g), magnesium (147 mg/100 g), manganese (0.36 mg/100 g), potassium (337 mg/100 g), zinc (0.60 mg/100 g), and phosphorus (112 mg/100 g) (Haytowitz David, 2019; Table 1). Colocasia leaves were likewise mineral-rich, with 2.16 mg/100 g calcium, 3.41 mg/100 g iron, 57.88 g/100 g phosphorus, and 0.82 mg/100 g zinc (Longvah et al., 2017). Purslane leaves contained potassium (6400 mg/100 g DW), magnesium (1266.67 mg/100 g DW), calcium (914.33 mg/100 g DW), phosphorous (281.67 mg/100 g DW), sodium (35.07 mg/100 g DW), manganese (6.77 mg/100 g DW), copper (1.18 mg/100 g DW), iron (41.73 mg/100 g DW), and zinc (6.62 mg/100 g DW) (Nemzer et al., 2020). Similarly, Indian sorrel leaves contain 67.34 mg/100 g calcium, 0.16 mg/100 g iron, 0.02 g/100 g sodium, 17.86 mg/100 g magnesium, 0.05 mg/100 g manganese, 1.87 mg/100 g potassium, 0.21 mg/100 g phosphorus, 0.03 mg/100 g copper, and 2.39 mg/100 g sulfur (Table 1, Radha et al., 2021). The green leafy vegetables mentioned above had a variety of nutritious components, including carbohydrates, proteins, vital fatty acids, fiber, vitamins, and minerals. Consumption of leafy vegetable species may provide health benefits due to their nutritional content. These vegetables are an affordable source of nutrition, especially for rural people.

Table 1 Nutritional composition of selected leafy vegetables.

Composition (g/100 g)	*Moringa (Moringa oleifera Lam.)*	*Colocasia (Colocasia esculenta (L.) Schott)*	*Purslane (Portulaca oleracea L.)*	*Indian Sorrel (Oxalis corniculata L.)*
Reference/s	Haytowitz David (2019)	Haytowitz David (2019)	Haytowitz David (2019)	Radha et al. (2021)
Method of analysis	Fresh weight basis	Fresh weight basis	Fresh weight basis	Fresh weight basis
Proximate				
Moisture (g/100 g)	78.66	86.0	92.86	69.06
Calorific value (Kcal/100 g)	64	42	20	ND
Carbohydrates (g/100 g)	8.28	3.0	3.39	21.1
Protein (g/100 g)	9.40	5.0	2.03	4.3
Fat (g/100 g)	1.40	ND	0.36	1.42
Fiber (g/100 g)	2.0	3.7	ND	29.96
Ash (g/100 g)	ND	ND	ND	4.26
Niacin (mg/100 g)	2.22	1.5	0.48	ND
Ascorbic acid (mg/100 g)	51.7	52	21	ND
Carotenoids (mg/100 g)	0.37	2.8	1320 IU	ND

Contd...

Table 1 *Contd...*

Composition (g/100 g)	*Moringa (Moringa oleifera Lam.)*	*Colocasia (Colocasia esculenta (L.) Schott)*	*Purslane (Portulaca oleracea L.)*	*Indian Sorrel (Oxalis corniculata L.)*
Folic acid (μg/100 g)	40	126	12	ND
Thiamine (μg/100 g)	257	0.209	47	ND
Riboflavin (μg/100 g)	660	0.456	112	ND
Vitamin B6 (mg/100 g)	1.20	0.146	0.073	ND
Vitamin E (mg/100 g)	ND	2.02	12.2	ND
Vitamin K (μg/100 g)	ND	106.6	ND	ND
Minerals (mg/100 g)				
Calcium	185	107	65	67.34
Iron	4.00	2.25	1.99	0.16
Sodium	9.00	ND	ND	0.02
Magnesium	147	45	68	17.86
Manganese	0.36	0.714	0.303	0.05
Potassium	337	648	494	1.87
Zinc	0.60	0.41	0.17	ND
Phosphorous	112	ND	44	0.21
Copper	ND	ND	ND	0.03
Sulphur	ND	ND	ND	2.39

Notes: ND, not determined; IU, International units.

Bioactive Compounds

Metabolites produced by plants are categorized into two types: primary and secondary metabolites. Primary metabolites are substances that aid in growth and development, such as carbohydrates, amino acids, proteins, and lipids. Secondary metabolites are compounds that are thought to improve plants' overall ability to survive and overcome local challenges by allowing them to interact with their surroundings. The bulk of plant bioactive compounds are formed as secondary metabolites, which are responsible for plant color, flavor, and fragrance. The simplest meaning of a bioactive molecule is secondary metabolites from plants that have pharmacological or toxicological effects in humans and animals. Bioactive compounds are present in all plant materials. They are primarily responsible for herbs' medicinal properties and health benefits, but they can also have adverse and toxic effects on others (Tiaz and Zeiger, 2015).

Secondary metabolites are divided into three types: terpenes and terpenoids, alkaloids, and phenolic compounds. Bioactive compounds are divided into various families based on their unique structural features, which influence their biosynthesis and functionality. According to Tiaz and Zeiger (2015), secondary metabolites can be produced by four pathways: shikimic acid, malonic acid, mevalonic acid, and non-mevalonate. Aromatic amino acids (via the shikimic acid pathway) and aliphatic amino acids are the principal sources of alkaloids. Phenolics are produced in two ways: shikimic acid and malonic acid, whereas terpenes are formed by mevalonic and non-mevalonate acid routes (Tiaz and Zeiger, 2015).

Secondary metabolites present in green leafy vegetables are significant for humans because they provide a variety of health benefits, including antibacterial, antioxidant, antidiabetic, anti-inflammatory, and anticancer activity. The next section discusses the many substances present in the leafy crops moringa, colocasia, purslane, and Indian sorrel, as well as their medicinal properties.

Polyphenolics

Phenolics are classified into phenolic acids and derivatives, flavonoids, tannins, stilbenes, lignans, quinones, and others based on the number of phenolic rings and the structural elements that connect them. The biosynthesis of mono and polyphenolic chemicals from carbohydrates occurs via shikimic acid, phenylpropanoid, and flavonoid pathways (Tiaz and Zeiger, 2015). Generally, phenolic acids have two distinct constitutive carbon frameworks: hydroxycinnamic and hydroxybenzoic structures. Several flavonoids, including apigenin, astragalin, kaempferol, luteolin, and their glycosides, have been identified from moringa leaves (Table 2; Figure 2) (Yang et al., 2008; Nouman et al., 2016; Swamy et al., 2022). Similarly, phenolic acids such as sinapic acid, syringic acid, chlorogenic acid, and 4-O-caffeorly quinic acid have been isolated from moringa leaves (Zhang et al., 2011; Kashiwada et al., 2012). Niazinin A, niazinin B, marumoside A, and marumoside B were among the carbamates isolated from moringa leaves (Faizi et al., 1992; Cheenpracha et al., 2010; Sahakitpichan et al., 2011). Iwashina et al. (1999) assessed the flavonoid composition of Colocasia leaf extracts. The primary flavonoids found in leaf extracts were reported to be isovitexin 3′ -O-glucoside, isovitexin, leteolin-7-O-glucoside, vitexin X"-O-glucoside, vicenin-2, isoorientin, orientin-7-O-glucoside, and orientin (Table 3; Figure 3). A phytochemical analysis of purslane leaves indicated the presence of apigenin, oleracone D, portulacanonne B, oleracone F, and (E)-5-Hydroxy-7-methoxy-3-(2-hydroxybenzyl)-4-chromanone. Some lignans reported from purslane leaves include oleralignan C, monomethyl-3,3',4,4'-tetrahydroxy-δ-truxinate, and olerabenzofuran (Table 4; Figure 4) (Yan et al., 2012; Nayaka et al., 2014; Yang et al., 2018; Lee et al., 2019; Park et al., 2019). Polyphenolics derived from the Indian sorrel plant include rutin, isoorientin, swetisin, vitexin, and vanillic acid (Table 5, Figure 5) (Mizokami et al., 2008; Mukherjee et al., 2018; Pandey et al., 2020).

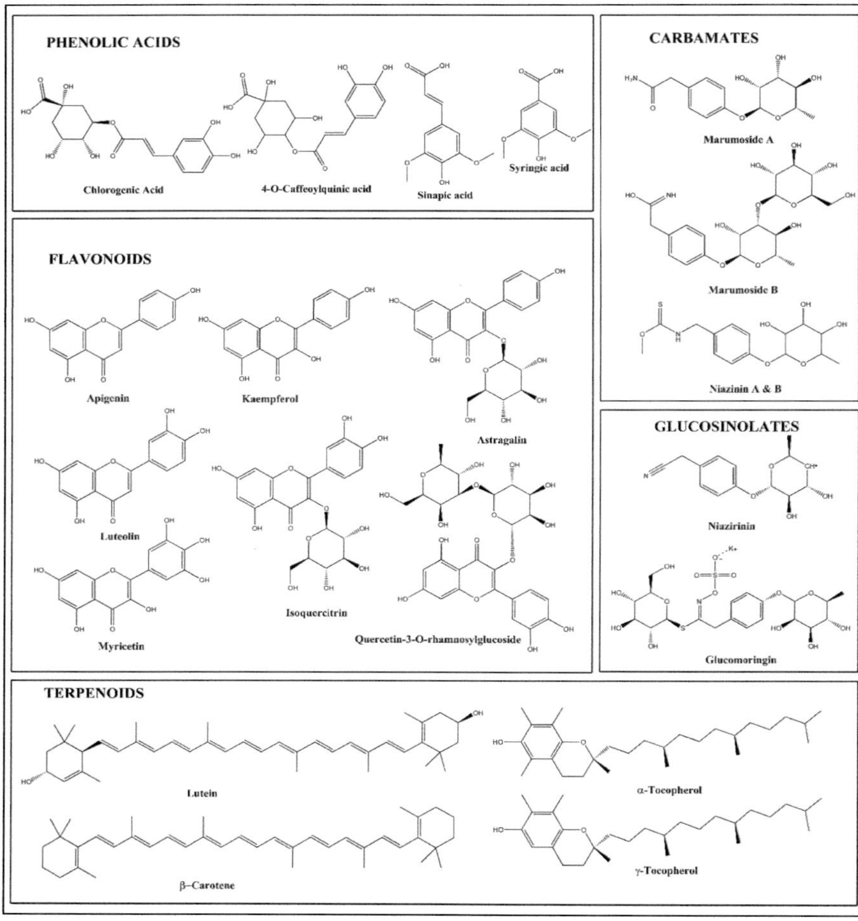

Fig. 2 Structures of bioactive compounds isolated from moringa leaves.

Table 2 Phytochemical constituents of moringa leaves.

Phytochemical Class	*Compound*	*References*
Phenolic acids	2-Hydroxylcoumaric acid	Zhang et al. (2011)
	4-O-(3′-O-α-D-glucopyranosyl)-caffeoyl quinic acid	Kashiwada et al. (2012)
	4-O-Caffeoyl quinic acid	Kashiwada et al. (2012)
	5-O-Caffeoyl quinic acid	Kashiwada et al. (2012)
	Benzaldehyde-4-O-β-glucoside	Manguro et al. (2007)

Contd...

Table 2 *Contd...*

Phytochemical Class	*Compound*	*References*
	Benzoic acid 4-O-α-rhamnosyl-(1→2)-β-glucoside	Manguro et al. (2007)
	Benzoic acid-4-O-β-glucoside	Manguro et al. (2007)
	Benzyl β-d-xylopyranosyl-(1→6)-β-d-glucopyranoside	Sahakitpichan et al. (2011)
	Benzyl-β-d-glucopyranoside	Sahakitpichan et al. (2011)
	Chlorogenic acid	Kashiwada et al. (2012)
	Cryptochlorogenic acid	Vongsak et al. (2014)
	Epicatechin	Bajpai et al. (2005)
	Gentistic acid	Zhang et al. (2011)
	Sinapic acid	Zhang et al. (2011)
	Syringic acid	Zhang et al. (2011)
Flavonoids	Apigenin-7-C-glucoside	Nouman et al. (2016)
	Apigenin-O-8-glucoside	Nouman et al. (2016)
	Astragalin	Swamy et al. (2022)
	Daidzein	Bajpai et al. (2005)
	Genistein	Bajpai et al. (2005)
	Isoquercitrin	Sahakitpichan et al. (2011)
	Isorhamnetin	Yang et al. (2008)
	Kaempferide-3-O-(2″-O-galloylrhamnoside)	Manguro et al. (2007)
	Kaempferide-3-O-(2″-O-galloylrutinoside)-7-O-α-rhamnoside	Manguro et al. (2007)
	Kaempferide-3-O-2″,3″-diacetylglucoside	Manguro et al. (2007)
	Kaempferol-3-O-(6″-malonylglucoside)	Bennett et al. (2003)
	Kaempferol-3-O-[α-rhamnosyl-(1→2)]-[α-rhamnosyl-(1→4)-β-glucoside-7-O-α-rhamnoside	Manguro et al. (2007)
	Kaempferol-3-O-[β-glucosyl-(1→2)]-[α-rhamnosyl-(1→6)-O-β-glucoside-7-O-rhamnoside	Manguro et al. (2007)
	Kaempferol-3-O-β-d-(6″-O-malonyl)-glucoside	Sahakitpichan et al. (2011)
	Kaempferol-3-O-β-d-glucopyranoside	Lim (2012)
	Kaempferol-O-3,7-diglucoside	Nouman et al. (2016)
	Kaempferol-O-3-glucoside	Nouman et al. (2016)

Contd...

Table 2 *Contd...*

Phytochemical Class	*Compound*	*References*
	Kaempferol-O-7-glucoside	Nouman et al. (2016)
	Luteolin	Yang et al. (2008)
	Myricetin	Sultana & Anwar (2008)
	Quercetin 3-O-glucoside	Amaglo et al. (2010)
	Quercetin-3-acetylglucoside	Nouman et al. (2016)
	Quercetin-3-O-(6″-malonylglucoside)	Bennett et al. (2003)
	Quercetin-3-O-rhamnosylglucoside	Bennett et al. (2003)
	Quercetin-3-O-β-d-(6″-O-3-hydroxy-3-methylglutaryl)-glucoside	Jiang et al. (2020)
	Quercetin-3-O-β-d-(6″-O-malonyl)-glucoside	Sahakitpichan et al. (2011)
	Quercetin-3-O-β-d-glucopyranoside	Lim (2012)
	Quercetin-O-3,7-diglucoside	Nouman et al. (2016)
	Quercetin-O-3-glucoside	Nouman et al. (2016)
Carbamates	Marumoside A	Sahakitpichan et al. (2011)
	Marumoside B	Sahakitpichan et al. (2011)
	Niazimicin	Faizi et al. (1992)
	Niazimimin A	Faizi et al. (1992)
	Niazimimin B	Faizi et al. (1992)
	Niazinin A	Cheenpracha et al. (2010)
	Niazinin B	Faizi et al. (1992)
	O-Ethyl,4-[(2′,3′,4′-tri-O-acetyl-α-l-rhamnosyloxy)benzyl] thiocarbamate (Z)	Faizi et al. (1995)
	O-Ethyl-4-[(2′,3′,4′-tri-O-acetyl-α-L-rhamnosyloxy) benzyl] carbamate (E)	Faizi et al. (1995)
	O-Ethyl-4-[(2′,3′,4′-tri-O-acetyl-α-L-rhamnosyloxy) benzyl] thiocarbamate (E)	Faizi et al. (1995)
	O-Ethyl-4-[α-l-rhamnosyloxy)benzyl] thiocarbamate(Z)	Faizi et al. (1995)
	O-Methyl-4-[(2′,3′,4′-tri-O-acetyl-α-L-rhamnosyloxy) benzyl] carbamate (E)	Faizi et al. (1995)
	O-Methyl-4-[(2′,3′,4′-tri-O-acetyl-α-L-rhamnosyloxy) benzyl] carbamate (Z)	Faizi et al. (1995)
	O-Methyl-4-[(2′,3′,4′-tri-O-acetyl-α-L-rhamnosyloxy) benzyl] thiocarbamate(E)	Faizi et al. (1995)

Contd...

Table 2 *Contd...*

Phytochemical Class	*Compound*	*References*
	O-Methyl-4-[(2′,3′,4′-tri-O-acetyl-α-L-rhamnosyloxy)benzyl] thiocarbamate (Z)	Faizi et al. (1995)
	O-Methyl-4-[(4′-O-acetyl-α-L-rhamnosyloxy) benzyl] carbamate (E)	Faizi et al. (1995)
	Pterygospermin	Horwath and Benin (2011)
	S-Methyl-N-{4-[(α-l-rhamnopyranosyloxy)benzyl]} thiocarbamate	Cheenpracha et al. (2010)
Terpenoids	Lutein	Foidl et al. (2001)
	α-Tocopherol	Liu et al. (2022)
	β-Carotene	Foidl et al. (2001)
	γ- Tocopherol	Liu et al. (2022)
Glucosinolates	4-[(2′,3′,4′-Tri-O-actyl-α-l-rhamnosyloxy) benzyl] nitrile	Faizi et al. (1995)
	Glucomoringin	Kashyap et al. (2022)
	Niazirinin	Tewari & Bhakuni (2006)

Terpenoids

Terpenoids are a diverse group of naturally occurring secondary plant metabolites. They are composed of isoprene units (C-5) and possess several pharmacological activities, including antibacterial, anti-inflammatory, antiallergic, anticancer, antimetastatic, anti-angiogenesis, and apoptotic induction (Camara et al., 2024). Moringa leaf terpenoids included β-carotene and lutein (Foidl et al., 2001; Table 2; Figure 2).

Alkaloids

Alkaloids are a class of nitrogen-containing chemicals, many of which are harmful, that plants use to defend themselves against other species. Many alkaloids have several biological effects, including antioxidant, anti-inflammatory, anticancer, antibacterial, antimalarial, immunomodulatory, hepatoprotective, antidiabetic, cardioprotective, and analgesic characteristics. However, in greater concentrations, they are hazardous to the human system (Debnath et al., 2018). Several oleraceins, cyclo-DOPA phenolic alkaloids characterized by a 5,6-dihydroxyindoline-2-carboxylic acid nucleus N-acylated with cinnamic acid derivatives and glycosylated with one or more glucose units, have been isolated from purslane (Table 4, Figure 4) (Yang et al., 2009; Liu et al., 2011; Jiao et al., 2015).

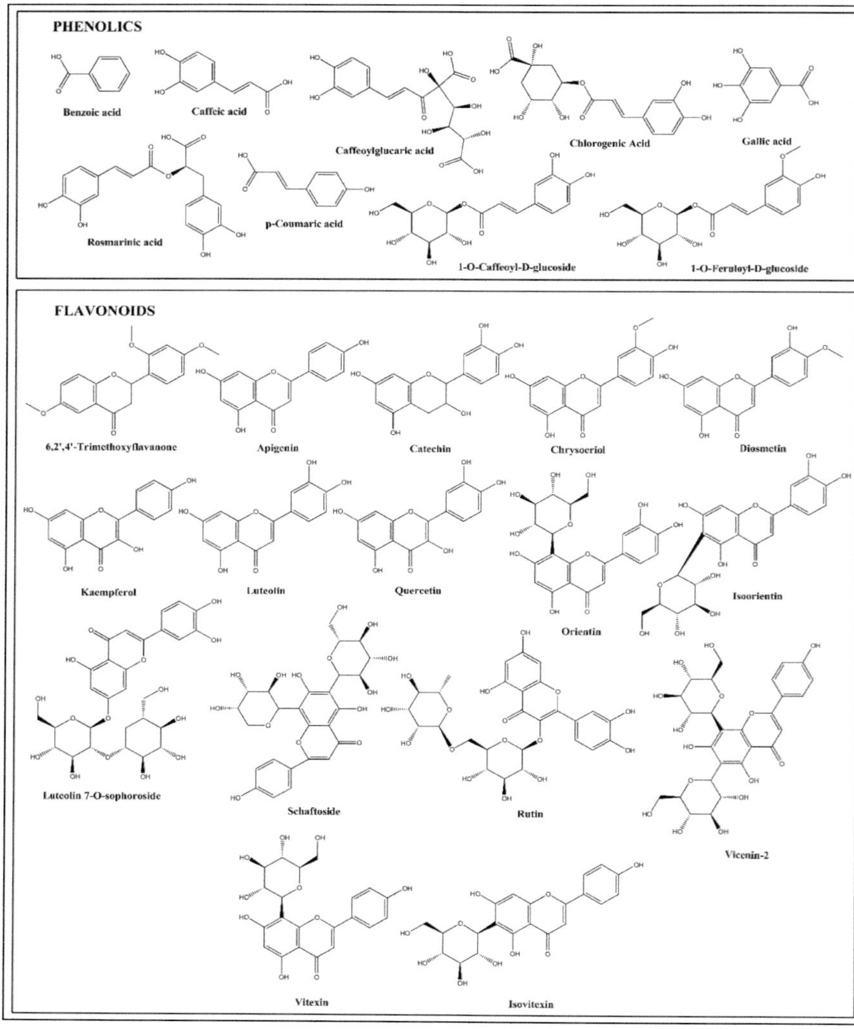

Fig. 3 Structures of bioactive compounds isolated from Colocasia leaves.

Table 3 Phytochemical constituents isolated from colocasia leaves.

Phytochemical class	*Compounds*	*References*
Phenolic acids	1-O-Caffeoyl-D-glucoside	Li et al. (2014)
	1-O-Feruloyl-D-glucoside	Li et al. (2014)
	Benzoic acid	El-Mesallamy et al. (2021)
	Caffeic acid	Ferreres et al. (2012); El-Mesallamy et al. (2021)

Contd...

Table 3 *Contd...*

Phytochemical class	*Compounds*	*References*
	Caffeic acid derivatives	Ferreres et al. (2012)
	Caffeoylglycaric acid	Gupta et al. (2019)
	Chlorogenic acid	El-Mesallamy et al. (2021)
	Dihydrocaffeoylquinic acid derivatives	Ferreres et al. (2012)
	Ellagic acid	El-Mesallamy et al. (2021)
	Gallic acid	El-Mesallamy et al. (2021)
	p-Coumaric acid	Ferreres et al. (2012)
	Rosmarinic acid	Li et al. (2014)
	Sinapoyl hexoside	Ferreres et al. (2012)
Flavonoids	6,2',4'-Trimethoxyflavanone	El-Mesallamy et al. (2021)
	Apigenin-6,8-di-C-hexoside	Ferreres et al. (2012); Gonçalves et al. (2013)
	Apigenin-6-C-(2-O-hexosyl) hexoside-8-C-pentoside	Ferreres et al. (2012)
	Apigenin-6-C-(6-O-hexosyl) hexoside	Ferreres et al. (2012)
	Apigenin-6-C-glucoside (isovitexin)	Gonçalves et al. (2013)
	Apigenin-6-C-hexoside	Ferreres et al. (2012)
	Apigenin-6-C-hexoside-8-C-pentoside	Ferreres et al. (2012); Gonçalves et al. (2013)
	Apigenin-6-C-pentoside-8-C-(2-O-hexosyl)hexoside	Ferreres et al. (2012)
	Apigenin-6-C-pentoside-8-C-hexoside	Gonçalves et al. (2013); Ferreres et al. (2012)
	Apigenin-6-C-pentoside-8-C-hexoside-7-O-hexoside	Ferreres et al. (2012)
	Apigenin-8-C-(2-O-pentosyl) hexoside	Ferreres et al. (2012); Gonçalves et al. (2013)
	Apigenin-8-C-hexoside	Ferreres et al. (2012)
	Astilbin	El-Mesallamy et al. (2021)
	Catechin	El-Mesallamy et al. (2021)
	C-Glucoglucosyl: vitexin X" – O -glucoside	Iwashina et al. (1999)
	C-Glucosyl :orientin	Iwashina et al. (1999)
	C-Glucosyl: orientin 7-0-glucoside	Iwashina et al. (1999)

Contd...

Table 3 *Contd...*

Phytochemical class	*Compounds*	*References*
	Chrysoeriol-6-C-hexoside	Ferreres et al. (2012)
	Chrysoeriol-6-C-hexoside-8-C-pentoside	Ferreres et al. (2012)
	Chrysoeriol-7-O-hexoside	Ferreres et al. (2012)
	Chrysoeriol-7-O-rhamnosyl(1→6) hexoside	Ferreres et al. (2012)
	Chrysoeriol-8-C-hexoside	Ferreres et al. (2012); Gonçalves et al. (2013)
	Diosmetin-6-C-hexoside-8-C-pentoside	Ferreres et al. (2012)
	Glucosyl: isovitexin 4'-O-glucoside	Iwashina et al. (1999)
	Isoorientin	Iwashina et al. (1999); Leong et al. (2010); Gonçalves et al. (2013); Li et al. (2014)
	Isoschaftoside	Leong et al. (2010)
	Isovitexin	Iwashina et al. (1999); Leong et al. (2010); Li et al. (2014)
	Kampferol	El-Mesallamy et al. (2021)
	Lucosyl:luteolin 7-0-glucoside	Iwashina et al. (1999)
	Luteolin-7-O-sophoroside	Leong et al. (2010)
	Luteolin-6,8-C-diglucoside	El-Mesallamy et al. (2021)
	Luteolin-6,8-di-C-hexoside	Ferreres et al. (2012)
	Luteolin-6-C-(2-O-pentosyl) hexoside	Ferreres et al. (2012)
	Luteolin-6-C-(3-O-hexosyl) hexoside-8-C-pentoside	Ferreres et al. (2012)
	Luteolin-6-C-(6-O-hexosyl) hexoside	Ferreres et al. (2012)
	Luteolin-6-C-hexoside	Ferreres et al. (2012)
	Luteolin-6-C-hexoside-8-C-pentoside	Ferreres et al. (2012); Gonçalves et al. (2013)
	Luteolin-6-C-pentoside-8-C-hexoside	Ferreres et al. (2012); Gonçalves et al. (2013)
	Luteolin-7-O-glucoside	Li et al. (2014)
	Luteolin-7-O-rhamnosyl(1→2) hexoside	Ferreres et al. (2012)

Contd...

Table 3 *Contd...*

Phytochemical class	*Compounds*	*References*
	Luteolin-7-O-rutinoside	Li et al. (2014)
	Luteolin-8-C-hexoside	Ferreres et al. (2012)
	Orientin	Leong et al. (2010); Gonçalves et al. (2013); Li et al. (2014)
	Quercetin	El-Mesallamy et al. (2021)
	Rutin	El-Mesallamy et al. (2021)
	Schaftoside	Leong et al. (2010)
	Vicenin-2	Iwashina et al. (1999)
	Vitexin	Leong et al. (2010); Gonçalves et al. (2013); Li et al. (2014); El-Mesallamy et al. (2021)

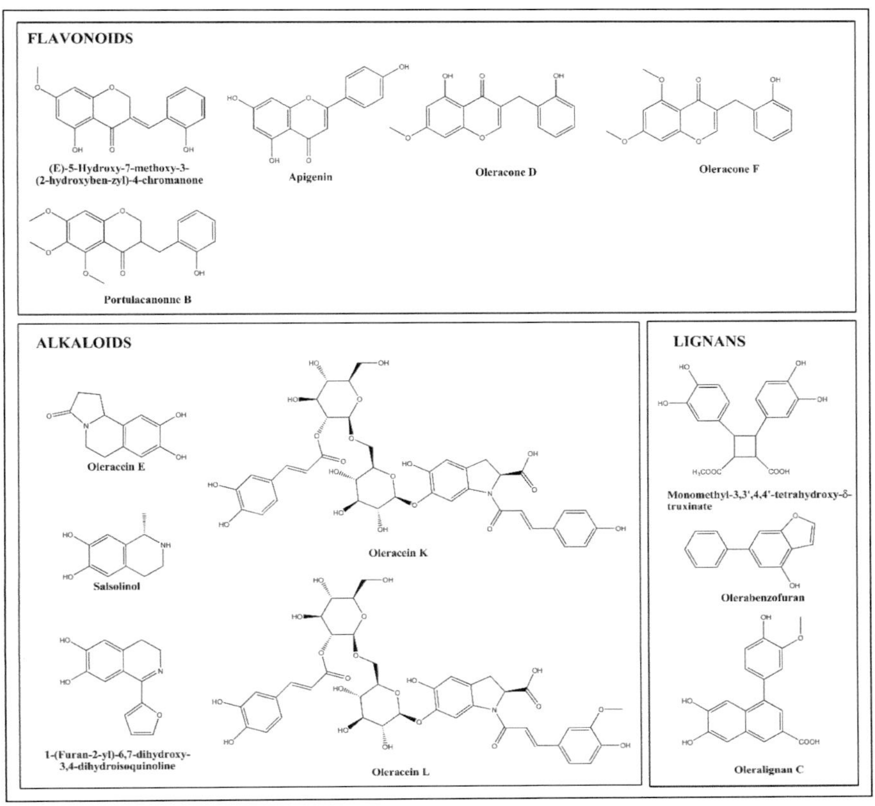

Fig. 4 Structures of bioactive compounds isolated from Purslane leaves.

Table 4 Phytochemical constituents of Purslane leaves.

Phytochemical Class	*Compound/s*	*References*
Flavonoids	(E)-5-Hydroxy-7-methoxy-3-(2-hydroxyben-zyl)-4-chromanone	Park et al. (2019a, b)
	2,2'-Dihydroxy-4',6'-dimethoxychalcone	Yan et al. (2012); Lee et al. (2019)
	Apigenin	Nayaka et al. (2014)
	Oleracone C	Yang et al. (2018)
	Oleracone D	Yang et al. (2018)
	Oleracone E	Yang et al. (2018)
	Oleracone F	Yang et al. (2018)
	Oleracone G	Daun et al. (2021)
	Oleracone J	Daun et al. (2022)
	Oleracone K	Daun et al. (2022)
	Portulacanone B	Yan et al. (2012); Lee et al. (2019)
	Portulacanone C	Yan et al. (2012); Lee et al. (2019)
Lignans	(+)-Lirioresinol A	Ma et al. (2019)
	(+)-Syringaresinol	Ma et al. (2019)
	Monomethyl-3,3',4,4'-tetrahydroxy-δ-truxinate	Ma et al. (2019)
	Olerabenzofuran	Tian et al. (2022)
	Oleraindenone	Tian et al. (2022)
	Oleralignan	Ma et al. (2019)
	Oleralignan A	Xu et al. (2021)
	Oleralignan B	Daun et al. (2021)
	Oleralignan C	Wang et al. (2022)
	Oleralignan D	Wang et al. (2022)
Alkaloids	(−)-(R)2-[1-(3,4-Dihydroxyphenethyl)-3-hydroxy-2,5-dioxopyrrolidin-3-yl]acetic acid	Jin et al. (2018)
	(−)-(S)-3-Hydroxy-1-(3,4-dihydroxy-phenethyl) pyrrolidine-	Jin et al. (2018)
	(+)-(S)-2-[1-(3,4-Dihydroxyphenethyl)-3-hydroxy-2,5-dioxopyrrolidin-3-yl]acetic acid	Jin et al. (2018)
	(2R,3S,3′S,5S,6R)-2-Ethoxy-5-hydroxy-6-(hydroxymethyl)-	Jin et al. (2018)
	(3R)-3,5-Bis(3-methoxy-4-hydroxyphenyl)-2,3-dihydro-2(1H)-pyridinone	Tian et al. (2014)

Contd...

Table 4 *Contd...*

Phytochemical Class	*Compound/s*	*References*
	(7′R)-N-Feruloyl normetanephrine	Tian et al. (2014)
	(E)-3-[4-(β-D-Glucopyranosyloxy)-3-methyoxyphenyl]-N-[2-(4-hydro-xy-3,5-dimet-hoxyphenyl)ethyl]-2-propenamide	Fu et al. (2021)
	(E)-Ferulamide	Xia et al. (2021)
	(E)-N-[3-Hydroxy-4-(β-D-glucopyranosyloxy) phenethyl]-3-(4-hydroxyphenyl)acrylamide	Fu et al. (2021)
	(E)-N-[4-(β-D-Glucopyranosyloxy)-phenethyl]-3-(4-hydroxyphenyl) acrylamide	Fu et al. (2021)
	(E)-p-Coumaramide	Xia et al. (2019)
	(R)-(+)-1-Benzyl-6,7-dihydroxy-1,2,3,4-tetrahydroisoquinoline	Jin et al. (2018)
	(R)-(+)-1-Isobutyl-6,7-dihydroxy-1,2,3,4-tetrahydroisoquinoline	Jin et al. (2018)
	(S)-(−)-Oleracein E	Yang et al. (2009); Jin et al. (2018)
	(S)-(−)-Salsolinol	Jin et al. (2018)
	1-(5′-Hydroxylmethylfuran-2-yl)-6,7-dihydrox-y-3,4-dihydroisoquinoline	Jin et al. (2018)
	1-(Furan-2-yl)-6,7-dihydroxy-3,4-dihydroisoquinoline	Jin et al. (2018)
	1,5-Dimethyl-6-phenyl-1,2-dihydro-1,2,4-triazin-3(2H)-one	Tian et al. (2014)
	2,3,4,9-Tetrahydro-1H-pyrido[3,4-b] indole-3-carboxylic acid	Xia et al. (2019)
	2-Sulfonic acid dopamine	Jin et al. (2018)
	4-Carboxy-6,7-dihydro-9,10-dihydroxy-benzo-[a]quinolizinium	Jin et al. (2018)
	6,7-Dihydro-1,10,11-trihydroxybenzo[a]-quinolizinium hydroxide	Jin et al. (2018)
	6,7-Dihydroxy-1-methyl-3,4-dihydroisoquinoline	Jin et al. (2018)
	6,7-Dihydroxy-3,4-dihydroisoquinoline	Jin et al. (2018)
	Benzamide	Xia et al. (2019)
	Carboxylate inner salt	Jin et al. (2018)
	Dopamine	Jin et al. (2018)
	Iseluxine	Xiu et. al. (2019)

Contd...

Table 4 *Contd...*

Phytochemical Class	*Compound/s*	*References*
	L(1S,3S)-1-Methyl-1,2,3,4-tetrahydro-β-carboline-3-carboxylic acid	Xia et al. (2019)
	N-trans-Feruloyl tyramine	Tian et al. (2014)
	Oleracein A	Yang et al. (2009)
	Oleracein B	Yang et al. (2009)
	Oleracein C	Jiao et al. (2015)
	Oleracein D	Jiao et al. (2015)
	Oleracein F	Liu et al. (2011)
	Oleracein G	Liu et al. (2011)
	Oleracein K	Jiao et al. (2015)
	Oleracein L	Jiao et al. (2015)
	Oleracein N	Jiao et al. (2015)
	Oleracein O	Jiao et al. (2015)
	Oleracein R	Jiao et al. (2015)
	Oleracein S	Jiao et al. (2015)
	Oleracein X	Fu et al. (2021)
	Oleracein Y	Fu et al. (2021)
	Oleracein Z	Fu et al. (2021)
	Oleracein ZA	Fu et al. (2021)
	Oleracein ZB	Fu et al. (2021)
	Oleraciamide D	Zhao et al. (2018)
	Oleraciamide E	Liu et al. (2021)
	Oleraciamide E	Liu et al. (2021)
	Oleraciamide G	Xu et al. (2021)
	Oleracone L	Cui et al. (2021)
	Oleracrylimide A	Song et al. (2022)
	Oleracrylimide B	Song et al. (2022)
	Oleracrylimide C	Song et al. (2022)
	Oleraindole A	Zhao et al. (2019)
	Oleraindole A	Zhao et al. (2019)
	Oleraindole B	Zhao et al. (2019)
	Oleraindole B	Zhao et al. (2019)
	Oleraindole D	Xu et al. (2021)
	Oleraisoindole A	Ma et al. (2019)

Contd...

Table 4 *Contd...*

Phytochemical Class	*Compound/s*	*References*
	Oleraurea	Xia et al. (2019)
	Portulacatal	Cui et al. (2021)
	Portulacatone	Yue et al. (2015)
	Portulacatone B	Cui et al. (2021)
	Soyalkaloid A	Xia et al. (2019)
	β-Carboline-3-carboxylic acid	Xia et al. (2019)

Glucosinolates

Glucosinolates are a naturally occurring class of chemical compounds composed of glucose and an amino acid. They contain sulphur and nitrogen. Glucosinolates have been connected to a variety of health advantages, including antibacterial, antidiabetic, cardiovascular, and central nervous system protection, as well as cancer prevention (Miekus et al., 2020). Moringa leaves contain 4-[(2′,3′,4′-tri-O-actyl-α-l-rhamnosyloxy) benzyl] nitrile and niazirinin (Table 2, Figure 2), according to Faizi et al. (1995) and Tiwari and Bhakuni (2006).

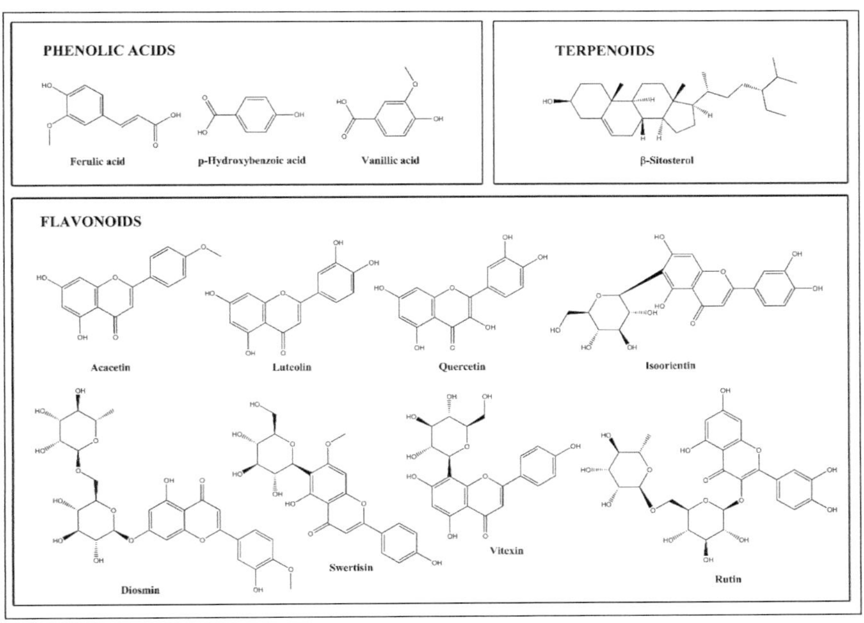

Fig. 5 Structures of bioactive compounds isolated from Indian Sorrel leaves.

Table 5 Phytochemical constituents of Indian Sorrel leaves.

Phytochemical Class	*Compound*	*References*
Phenolic acids	Ethyl gallate	Pandey et al. (2020)
	Ferulic acid	Mukherjee et al. (2018)
	p-Hydroxybenzoic acid	Mukherjee et al. (2018)
	Vanillic acid	Pandey et al. (2020)
Flavonoids	5,7,4′-Trihydroxy-6,8-dimethoxyflavone	Rehman et al. (2015)
	5,7-Dimethoxy-3,3′,4′- trihydroxyflavone	Pandey et al. (2020)
	5-Hydroxy-6,7,8,4'-tetra methoxy flavone	Rehman et al. (2015)
	Acacetin	Draz et al. (2022)
	Apigenin 7,4′-diglucoside	Pandey et al. (2020)
	Diosmetin-7-O-β-D-glucopyranoside	Pandey et al. (2020)
	Diosmin	Kiran et al. (2023)
	Isoorientin	Pandey et al. (2020)
	Isovitexin	Mizokami et al. (2008)
	Luteolin	Ibrahim et al. (2013)
	Quercetin	Pandey et al. (2020)
	Rutin	Mukherjee et al. (2018)
	Swertisin	Mizokami et al. (2008); Pandey et al. (2020)
	Vitexin	Pandey et al. (2020)
Terpenoids	β-Sitosterol	Gupta et al. (2011); Dighe et al. (2016)

Health Benefits

Leafy vegetables are a unique type of edible plant, well-known for their capacity to address nutritional deficiencies and dramatically improve health. They contain vitamins like ascorbic acid, tocopherol, and carotenoids (Edelman and Colt, 2016). Leafy vegetables are rich in minerals, as well as phytochemicals such as phenolic antioxidants, anthocyanins, and glucosinolates. For example, lettuce, kale, spinach, cabbage are significantly higher in calcium, magnesium, iron, manganese, zinc, and molybdenum than other vegetables (Natesh et al., 2017). Leafy vegetables improve the bioavailability and bioaccessibility of minerals such iron, potassium, magnesium, and zinc (Natesh et al., 2017). They outperform other vegetables in terms of nutritional value, containing higher amounts of minerals like iron, calcium, and phosphorus, as well as vitamins like folic acid, vitamin C, omega-6 and omega-3 fatty acids (Edelman and Colt, 2016).

Leafy vegetables are well-known for their health-promoting characteristics, which include anti cancer, antibacterial, antioxidant, anti-inflammatory, and anti diabetic benefits. They have demonstrated potential in the regulation of hormone-dependent malignancies such as breast and prostate cancer, as well as the prevention of cardiovascular disease (Mungofa et al., 2022). This section discusses the health advantages of bioactive chemicals found in moringa, colocasia, purslane, and Indian sorrel.

Antioxidant Activity

Teixeira et al. (2014) tested the antioxidant activity of moringa leaf extracts in methanol, acetone, dichloromethane, water, diethyl ether, chloroform, and ethyl acetate and found that the acetate extract had greater scavenging activity for super oxide anion radicals, indicating better antioxidant properties. In another study, Saleem et al. (2020) investigated the *in vitro* antioxidant activity of various moringa leaf extracts at concentrations ranging from 0.1563–5.0 mg/ml and discovered that the methanolic extract had higher 2,2-diphenyl-1-picrylhydrazyl (DPPH) scavenging activity at 0.1563 mg/ml. Similarly, Saleem et al. (2020) demonstrated the maximum hydrogen peroxide (H_2O_2) scavenging and reducing power activities at a concentration of 1 mg/ml. Aju et al. (2019) investigated the effects of methanol extract from moringa leaves on streptozotocin-treated rats *in vivo*. Moringa leaf extracts were administered orally to streptozotocin-treated rats at a dosage of 300 mg/kg body weight for 60 days. There were six groups: group 1: normal rats, group 2: normal rats treated with moringa extract, group 3: high energy diet control rats, group 4: diabetic control rats, group 5: diabetic rats treated with moringa extract, and group 6: diabetic rats treated with standard drugs (metformin and atorvastatin). Aju et al. (2019) found a substantial increase in antioxidant enzyme activity such as catalase, glutathione, glutathione peroxidase, and superoxide dismutase in rats' blood from groups 2, 5, and 6. However, they detected a significant drop in these enzymes in rats in groups 3 and 4. Moringa leaves' antioxidant action is attributed to a variety of bioactive chemicals, including quercetin, kaempferol, ascorbic acid, β-carotene, isothiocyanates, and rutin (Bajpai et al., 2005; Atawodi et al., 2010; Wang et al., 2017).

Gonçalves et al. (2013) investigated the antioxidative potential of colocasia leaf varieties (giant white, red, and white) and reported DPPH, superoxide radical, and nitric oxide scavenging activity in terms of IC_{50} values in the range of 0.226–1.140 mg/ml, 0.076–0.145 mg/ml, and 0.521–1.485 mg/ml, respectively, which was significantly ($p \leq 0.05$) lower than standard ascorbic acid (7.1 mg/ml DPPH activity; 0.425 mg/ml superoxide radical activity, and 0.250 mg/ml nitric oxide activity). Similarly, Singh et al. (2011) investigated the influence of several solvent extraction methods on the antioxidant activity of diverse indigenous vegetables including colocasia from the Andaman and Nicobar Islands in India. The DPPH radical scavenging activity of colocasia leaves (73.83%) in methanol extract was considerably ($p < 0.05$) higher than acetone (70.70%) and aqueous extracts (67.47%), most likely due to variations in antioxidative component solubility index and solvent polarity. Colocasia leaf extracts' antioxidant activity was related to their greater polyphenolic content.

Li et al. (2014) investigated the antioxidant activity of purslane-derived polysaccharide components (POP II and POP III). It was discovered that both of them have substantial antioxidant properties in both cell-free and cell-mediated radical generation systems. In cell-free radical producing systems, POP II and POP III had DPPH radical scavenging activities of 35.24% and 17.21% at 1.0 mg/ml, respectively. Both POP II and POP III prevented lipid peroxidation in a dose-dependent manner. Both POP II and POP III inhibited alloxan-induced reactive oxygen species (ROS) production and protected HIT-T15 cells (a beta cell line isolated from a Syrian golden hamster's islet of Langerhans) from alloxan-induced ROS stress, thereby contributing to anti-cytotoxic effects in cell-mediated radical generation systems. Jiao et al. (2015) demonstrated the DPPH scavenging activity of indoline amide glucosides derived from purslane leaves, specifically oleracein K and oleracein L. The EC_{50} values ranged from 15.30– 43.52 μmol/l, indicating effective DPPH radical scavenging action. Oleracein K and L had twice the DPPH radical scavenging activity of vitamin C (EC_{50} = 30.15 μmol/l), with EC50 values of 15.30 and 16.13 μmol/l, respectively.

Indian sorrel has been discovered to contain phytochemicals with antioxidant activity, including phytosterols and polyphenols (Handique et al., 2012). These chemicals have been found to scavenge free radicals and lower oxidative stress. Khan et al. (2012) found that treating rats with methanolic extract of Indian sorrel greatly reduced oxidative stress caused by carbon tetrachloride (CCl_4), as demonstrated by a drop in liver enzyme levels.

Antimicrobial Activity

Roccetti et al. (2020) tested methanolic leaf extracts against various bacterial strains and found a maximum zone of inhibition against *Listeria innocua* (10.21 ± 0.08 mm), *Salmonella enteritides* (5.67 ± 0.47 mm), *Salmonella typhimurium* (5.33 ± 0.47 mm), and *Bacillus cereus* (15 ± 0.0 mm) with a 50:50 v/v methanol-water extract. Similarly, Abalaka et al. (2012) studied the antifungal efficacy of steam distilled moringa leaves against *Aspergillus niger*, *A. oryzae*, *A. nidulans*, and *A. terreus*, and found that the leaf extract inhibited *A. niger* more effectively. The presence of different polyphenolics accounts for strong antimicrobial activity.

Chakraborty et al. (2015) tested the antibacterial efficacy of *Colocasia esculenta* leaf extract against *Pseudomonas aeruginosa*, *Serratia* sp., *Escherichia coli*, *Shigella* sp., *Streptococcus aures*, *Salmonella* sp., *Klebsiella* sp., *Proteus mirabilis*, and *Enterococcus* sp. The leaf extracts were found to be effective against both gram-positive and gram-negative bacteria, with the highest antimicrobial activity against *Proteus mirabilis* at 100 mg/ml and the lowest for *Klebsiella* sp. and *Proteus mirabilis* at 25 mg/ml. Similarly, Ehiobu and Ogu (2018) tested a *Colocasia esculenta* leaf aqueous extract at concentrations of 25, 50, and 75 g/l against *Aspergillus niger* and *Botryodiplodia theobromine*. The results indicated very good activity in terms of mycelial growth inhibition (92.2% for *A. niger* and 85.9% for *B. theobromine*) at 75 g/l. The antibacterial activity of extracts is attributed to polyphenols found in leaves.

Tleubayeva et al. (2021) investigated the carbon dioxide extract of purslane against *Bacillus subtilis* ATCC 6633, *Staphylococcus aureus* ATCC 6538-P, *Candida albicans* ATCC 10231, and *Escherichia coli* ATCC 8739, and found that it had higher activity against all of the species tested. They also conducted a phytochemical examination of the extract and concluded that the action is due to the presence of triterpenoids and phytosterols.

Manandhar et al. (2019) tested the methanolic extract of *Oxalis corniculata* against *Staphylococcus aureus*, MRSA and *E. coli*, *Salmonella typhi*, MDR *S. Typhi*, *Pseudonymous aeruginosa*, MDR *Klebsiella pneumoniae*, *Citrobacter koseri*, *Rhizopus* sp., *Aspergillus niger*, *Aspergillus flavus*, *Candida albicans*, and reference strains *E. coli* ATCC 25922 and *S. aureus* ATCC 25923. The extract of *O. corniculata* showed the strongest potential against *Escherichia coli*, *Salmonella typhi*, MDR *Salmonella typhi*, *Klebsiella pneumoniae*, and *Citrobacter koseri*, with zones of inhibition of 17 mm, 13 mm, 16 mm, 11 mm, and 12 mm, respectively.

Anticancer Activity

Jung et al. (2015) studied the antitumor efficacy of moringa leaf aqueous extract on hepatocellular carcinoma HepG2 cells. When the leaf extracts were administered orally, HepG2 cell growth was significantly reduced (44–52%), indicating that they are effective anticancer agents. Khan et al. (2020) studied the effect of a methanolic extract of moringa leaves on human prostate cancer DU145 cells. Cell survival and nuclear alteration were dose-dependently inhibited, and cell death was produced by simultaneous overexpression and downregulation of the Bax and Bcl-2 gene expressions. Furthermore, moringa extract reduced Notch-1 and Hes-1 expression, effectively suppressing the aberrant notch signaling pathway.

Barhoi et al. (2021) evaluated moringa leaves' anti-carcinogenic potential using both *in vitro* and *in vivo* experiments. *In vitro* research was conducted using Ehrlich ascites carcinoma (EAC) and human laryngeal carcinoma (Hep-2) cells, whereas *in vivo* studies were carried out on Balb/c animals. Mice were given Mitimycin C (a chemotherapeutic medication) at a dose of 2 mg/kg body weight and aqueous moringa extract at 200 and 400 mg/kg body weight. The tumor volume decreased by 78.69% with an administration of 400 mg/kg body weight, while a dose of 200 mg/kg body weight resulted in a 36.97% reduction after 50 days. Similarly, tumor weight was considerably lowered at dosages of 200 mg/kg (12.45 ± 1.20 g) and 400 mg/kg (8.43 ± 0.49 g) of moringa, as well as 2 mg/kg mitomycin C (14.42 ± 1.09 g), compared to the control (27.91 ± 1.50 g). *In vitro* tests revealed dose- and time-dependent harmful effects on both cell lines.

Antidiabetic Activity

Diabetes mellitus is a category of metabolic illnesses distinguished by an abnormal and prolonged increase in plasma glucose levels. Type-1 diabetes is characterized by an entire lack of insulin secretion, necessitating the use of insulin replacements. The most prevalent kind of diabetes is Type 2 (non-insulin dependent diabetes mellitus), which is caused by impaired insulin production and resistance. Several studies have shown that moringa leaves help treat both forms of diabetes. In a streptozotocin-induced rat study, an aqueous extract from *Moringa olieifera* leaves was found to be effective in treating Type I and insulin-resistant Type 2 diabetes (Khan et al., 2017). Streptozotocin-induced ATP dephosphorylation results in the generation of free radicals and superoxide in beta cells via xanthine oxidase. These ROS destroy beta cells and limit insulin production, resulting in hyperglycemia and Type 2 diabetes. The antioxidants found in moringa leaves reduce ROS, protect beta cells from injury, and keep hyperglycemia under control.

Jimoh et al. (2018) investigated the antioxidant and inhibitory effects of *Moringa oleifera* leaves on enzymes associated with Type 2 diabetes. Moringa leaves effectively inhibit α-amylase and α-glucosidase *in vitro*, with IC_{50} values of 6.49 μg/ml for α-amylase and 4.73 μg/mL for α-glucosidase. Moringa leaves have also shown increased antioxidant activity. Moringa leaves' higher phenolic content is responsible for their antioxidant activity and inhibition of both enzymes, indicating their potential for treating Type 2 diabetes.

In another study, Bamagous et al. (2018) investigated the effects of moringa leaf extracts on streptozotocin-induced diabetic mice. For 30 days, streptozotocin-induced diabetic rats were treated orally with the ethyl acetate fraction of moringa leaves at a dose of 200 mg/kg body weight. Moringa leaves dramatically raised body weight, food and water consumption, blood glucose, insulin, and glycosylated hemoglobin levels. Moringa leaf extract reduced hepatic marker enzymes (alanine transaminase, aspartate transaminase, lactate dehydrogenase, alkaline phosphatase), lipid profile levels (triglycerides, total cholesterol, low density lipoprotein-cholesterol), pancreatic tumor necrosis factor and interleukin-6, and serum interleukin-1β levels in streptozotocin-induced diabetic rats. In addition, there was a considerable rise in antioxidant enzymes such as catalase, glutathione S-transferases, superoxide dismutase, glutathione peroxidase, vitamin C, and vitamin E.

Akter et al. (2013) studied the antihyperglycemic effect of a methanol extract from colocasia leaves in Swiss albino mice. The results showed that colocasia leaf extract effectively lowered blood glucose levels in a dosage-dependent manner, with the highest reduction (35.9%) at a dose of 400 mg/kg body weight. The chlorogenic acid and other phytochemicals found in colocasia leaf have been demonstrated to reduce body weight and visceral fat, which alters lipid-lipoprotein fractions and regulates blood glucose while enhancing insulin sensitivity.

Cho et al. (2019) investigated the effect of an ethanolic extract of purslane on Glucagon-like peptide-1 (GLP-1, a target for diabetes treatment) release and found that the purslane extract had the highest activity in generating GLP-1 secretion. The suggested mechanism could be related to satiety enhancement and appetite reduction. Purslane has been demonstrated to have anti-hyperglycemic effects in numerous studies using a variety of biological processes. One of these methods includes the PI3K and AMPK pathways. Ethanolic extract from purslane increased glucose absorption by increasing glucose transporter type 4 (GLUT4) translocation to the plasma membrane by activating the PI3K and AMPK pathways in 3T3-L1 adipocytes. In a more recent work, Park et al. (2019b) showed that an HM-chromanone, called (E)-5-hydroxy-7-methoxy-3-(2-hydroxyben-zyl)-4-chromanone (a bioactive chemical derived from purslane), might improve glucose absorption in 3T3-L1 adipocytes via activating the PI3K/AKT and AMPK pathways.

Anti-Inflammatory Activity

Inflammation is an immune system reaction that ensures survival after infection or tissue injury. Several green vegetables have considerable anti-inflammatory properties, as demonstrated by *in-vivo* and *in-vitro* research. Oguntibeju et al. (2020) evaluated the effect of moringa leaves on inflammatory biomarkers in streptozotocin-treated male Wistar rats. Rats were divided into four groups: non-diabetic, non-diabetic moringa-treated, diabetic, and diabetic moringa-treated. Rats were given 55 mg/kg of streptozotocin and then treated with 250 mg/kg of methanolic moringa leaf extract. Moringa-treated diabetic rats had considerably lower serum levels of NF-κβ and IL-18 interleukin in the kidney, as well as IL-1 and IL-18 in the liver. Bcl-2 expression was also elevated in the kidneys and liver. Moringa treatment increases cellular antioxidant capability, which helps to reduce aberrant cell proliferation, and elevation of inflammatory markers demonstrates moringa leaves' anti-apoptotic and anti-inflammatory response.

In another investigation, ethanolic moringa leaf extract significantly reduced diclofenac sodium-induced liver damage in male albino rats. The acute toxicity of leaf extract was assessed using a dosage pattern ranging from 500 to 4000 mg/kg body weight, with 300 mg/kg being the maximum safe dose. Diclofenac sodium-induced rats (100 mg/kg) were given two doses, 150 and 300 mg/kg, via gastric gavage. Moringa administration dramatically decreased the concentrations of liver marker enzymes (alkaline transaminase, alkaline phosphatase, and aspartate transaminase), as well as bilirubin, creatinine, urea, and urease (El-Hadary and Ramadan, 2019).

Lee et al. (2012) found that an aqueous extract of purslane inhibited TNF-α-induced vascular inflammation in human umbilical vein endothelial cells (HUVEC) by inhibiting ROS production, activating NF-κB, and reducing adhesion molecule expression. According to Jin et al. (2018), isoquinoline alkaloids derived from purslane show anti-inflammatory activity by inhibiting nitric oxide production in lipopolysaccharide (LPS)-induced murine macrophage RAQ 264.7 cells. The isoquinoline alkaloids include (S)-(−)-oleracein E, (S)-(−)-salsolinol, (S)-(−)-oleracein E, and 1-(furan-2-yl)-6,7-dihydroxy-3,4-dihydroisoquinoline was more effective than positive control 3,4-dihydroxybenzohydroxamic acid (EC_{50} = 82.4 μmol/l), with EC_{50} values of 58.7, 35.4, 18.1, and 36.3 μmol/l, respectively.

In animal studies, ethanolic leaf extracts of Indian sorrel have shown anti-inflammatory properties (Mei-Xian et al., 2014). In a rat model of carrageenan-induced paw edema, β-sitosterol extracted from leaves using petroleum ether extracts showed dose-dependent anti-inflammatory action and reduced paw edema (Dighe et al., 2016).

Cardiovascular Activity

Cardiovascular diseases refer to a group of disorders that affect the heart and blood arteries. Moringa leaves had a significant effect on the circulatory system because they included gossypetin, quercetagenin, and proanthicyanadins, which lowered mortality from various coronary heart disorders. Other phytochemicals contained in leaves, such as niazinin and its derivatives, as well as glucomoringinin, have been shown to have a hypertensive bradycardic impact (Gupta et al., 2018).

Aekthammarat et al. (2019) studied the effect of an aqueous extract of leaves on hypertensive rats. N-nitro-L-arginine-methyl ester was administered to rats at a dose of 50 mg/kg/day for three weeks, increasing their blood pressure and heart rate, while moringa extract was given at doses of 30 and 60 mg/kg/day. The results showed a dose-dependent drop in blood pressure and tachycardia. Treatment with moringa leaf extract improved acetylcholine-induced relaxation and decreased hyper-reactivity of adrenergic-mediated contraction. It also demonstrated antioxidant activity and antihypertensive benefits by decreasing endothelial-dependent vasorelaxation.

Vasant et al. (2012) investigated the antihypertensive activity of aqueous extracts of Colocasia leaves in an animal model and found that extracts at doses of 200 and 400 mg/kg significantly reduced systolic and diastolic blood pressure in hypertensive rats not given adrenaline. Colocasia leaves' antihypertensive action is likely due to inhibition of phosphodiesterase or angiotensin-converting enzymes, as well as the release of endothelial-derived relaxing factors.

Neuroprotective Activity

Many diseases are linked to the central nervous system, including Parkinson's, Alzheimer's, Huntington's, epilepsy, and many others, which impact a variety of bodily functions such as concentration, movement, balance, and memory loss. *Moringa oleifera* leaves were traditionally used to treat disorders such as epilepsy and Alzheimer's of the central nervous system, causing anti-convulsant activity through the release of GABA (γ-amino butyric acid). Mahaman et al. (2018) investigated the effects of moringa leaves on hyperhomocysteinemia (HHcy)-induced Alzheimer's disease in mice. Rats were given 400 μg/kg/day of HHcy through the vena caudalis for 14 days. Rats were given a methanolic extract of moringa as a prophylactic and curative treatment at doses of 200 and 400 mg/kg/day, both simultaneously and after injection. SCR1693 (a non-competitive acetylcholinesterase inhibitor at 1 mg/kg/day) was utilized as a positive control. Moringa therapy reduced neurodegeneration while simultaneously restoring synaptic protein levels. Treatment with moringa leaf extract reduced HHcy-induced tau phosphorylation and Aβ pathogenesis. Moringa leaves inhibit calpain activity, which is critical in this process.

Several studies have demonstrated that purslane extracts, like moringa, have neuroprotective properties. For example, Abdel Moneim et al. (2013) demonstrated that dietary feeding of purslane extract to rats has the potential to reduce rotenone-induced oxidative stress due to its antioxidant nature and ability to modulate the activities of antioxidant defenses, thus possessing anti-neurotoxic activity. In another study, Al-Quraishy et al. (2012) demonstrated purslane's anti-apoptotic and antioxidant effects on rotenone-induced brain damage in rats. All this research suggests that the purslane has the potential to prevent brain damage and neurodegenerative disorders caused by oxidative stress.

Anti-urolithiatic Activity

Urolithiasis is defined as the production of stones in the kidneys or other parts of the urinary tract. Renal calculi have become more common as a result of dietary changes and a sedentary lifestyle. Dwivedi et al. (2016) tested the anti-urolithiatic efficacy of colocasia leaf extracts *in-vitro* on artificial kidney membranes. The results revealed that methanolic extracts had higher anti-urolithiatic action than chloroform extracts.

Other Activities

Wound healing activity: Wounds are physical injuries that compromise the anatomical and functional integrity of the skin. *Colocasia esculenta* leaves have long been used to cure wounds (Das et al., 2021). Gonçalves et al. (2013) studied the wound-healing properties of various *Colocasia esculenta* leaf extracts. Among the examined samples, the white variety had the highest hyaluronidase inhibitory activity (HIA), with IC_{50} values ranging from 0.33 to 4.002, while the big white and red variants had the lowest activities (IC_{50} values of 5.21 and 4.65, respectively). In another study, Sjamsudin (2021) found that Colocasia leaf extract at a 25% concentration significantly increased wound healing markers such as neutrophils, macrophages, fibroblast growth factor-2, and re-epithelization in wounds contaminated with Staphylococcus aureus in Sprague Dawley rats.

Anti-photoaging activity: Lee et al. (2014) revealed that purslane extract can protect human fibroblast and keratinocyte cells against UVB-induced damage *in vitro*. Specifically, the extract effectively reduced cell death and apoptotic DNA cleavage following UVB irradiation. These findings provided a helpful tip for the practical application of purslane in cosmetics.

Anti-mutagenic activity: Continuous efforts are underway to investigate the antimutagenic components of plants as biotherapeutics (Akram et al., 2020). Botting et al. (1999) studied the antimutagenic effect of Colocasia leaf extracts against the heterocyclic amine 2-amino-3-methylimidazo [4,5-f] quinolone (IQ).

The study indicated that the maximum antimutagenic activity was observed in the heptane fraction of Colocasia leaves, and the extract at a dosage of 0.02 g reduced mutagenicity by 50%.

Toxicity and Safety Profiles

Green leafy vegetables are a staple in the global diet, containing essential nutrients, metabolites, and antioxidants. In general, consuming safe and high-quality foods is essential for good health. However, antinutritional elements such as nitrates, oxalates, tannins, phytates, and cyanogenic glycosides are considered antinutrients, and their amounts vary depending on the species, cultivar, geographic area, and ecological location where the plants were grown or farmed (Natesth et al., 2017). Heat treatment is currently the most efficient way for reducing the antinutritional elements found in green leafy vegetables (Mosha and Gaga, 1999). Cooking and blanching aid to remove antinutrients (Yadav and Sehgal, 2003). The following section discusses the toxicity and safety evaluation literature for moringa, colocasia, purslane, and Indian sorrel.

Moringa leaves are highly recommended as a natural dietary supplement due to their high nutritional value and minimal anti-nutritional content. No negative effects of moringa leaves have been identified in human research thus far. Adedapo et al. (2009) found that 70 g of moringa leaf extract per day was safe to consume. The toxicity of the aqueous extract of moringa leaf extract was assessed in mice by administering 6400 mg/kg and 1500 mg/kg interperitoneally in the acute phase, and 250, 500, and 1500 mg/kg orally for 60 days in the subchronic research. For mice, the lethal dose was LD_{50} = 1585 mg/kg. Histopathological and biochemical parameters showed no significant alterations, and oral administration was recognized as safe consumption (Adedapo et al., 2009).

Camille et al. (2018) studied the acute and subacute toxicity of an aqueous extract of *Colocasia esculanta* leaves in rats. The toxicity of the extract was assessed in both male and female Wistar rats following a single injection (acute model) and over 21 days (subacute model). Acute administration of extract up to a dose of 4000 mg/kg did not cause any toxicity, hence the LD_{50} in the subacute treatment was determined to be 4000 mg/kg.

Musa et al. (2007) investigated the toxicity of the methanolic extract of *Portulaca oleracea* (Purslane) on mice. The extract was injected intraperitoneally at doses of 1000, 1250, 1500, 1750, 2000, and 2260 mg/kg, and the LD_{50} was reported as 1870 mg/kg. Similarly, Sakat et al. (2012) calculated the acute oral toxicity of a methanolic extract of Indian sorrel (*Oxalis corinculata*) in male and female mice. There were no signs of toxicity, mortality, or behavioral abnormalities with the treatment of 2000 mg/kg. As a result, toxicological investigations determined that moringa, colocasia, purslane, and Indian sorrel preparations were safe to consume at the prescribed doses and quantities.

Conclusions

This review primarily gives a comprehensive account, focusing on the nutritional value, phytochemistry, health benefits, and toxicity of leafy vegetables in Southeast Asia, including moringa, colocasia, purslane, and Indian sorrel. To summarize, green leafy vegetables include essential elements for human health and welfare. These include amino acids, vitamins, vital fatty acids, minerals, and fiber. Leafy vegetables are also rich in phytochemicals such as polyphenols, terpenes and terpenoids, alkaloids, and glucosinolates. The bioactive chemicals' antioxidative capability protects against a variety of degenerative illnesses, hence validating their medicinal potential.

There is room for research to identify and investigate the potential of edible leaves from indigenous and underutilized plants for use as food and medicine, as well as integration in mainstream agri-food systems. Research is also required to investigate alternative kinds and the possibilities of using agronomic measures that would minimize the concentration and influence of anti-nutritional elements in green leafy vegetables while increasing their nutritional content. Advanced processing technologies can increase the bioavailability of micronutrients and bioactive chemicals, allowing for more targeted therapeutic effects. Clinical trials are needed to validate the medicinal potential of leafy vegetables and their derivatives. Once clinically proven, leafy greens can be used to create medicinal and value-added food items.

Acknowledgments

Hosakatte Niranjana Murthy is thankful for the "Brain Pool" fellowship (Grant No. RS-2025-25425249) of the National Research Foundation, Republic of Korea.

Disclosure Statement

No potential conflict of interest was reported by the authors.

Author contributions: HNM: conceptualization; HNM: investigation, data curation, and formalization; SYP: resources and validation; HNM, GGY, KYP, and SYP: writing, review, and editing. All the authors have read and agreed to the published version of this manuscript.

Funding

This work was supported by the National Research Foundation of Korea (NRF) grant (MSIT) (No. NRF-2020R1A2C2102401), Korea, and the Technology Innovation Program (Grant number P0018148) funded by the Ministry of Trade, Industry, & Energy (MOTIE, Korea).

References

Abalaka, M.E., Daniyan, S.Y., Oyelekke, S.B. and Adeyemo, S.O. 2012. The antibacterial evaluation of *Moringa oleifera* leaf extracts on selected bacterial pathogens. *J. Microbiol. Res., 2*: 1–4. https://doi.org/10.5923/j.microbiology.20120202.01.

Abdel Moneim, A.E., Dkhil,M.A. and Al-Quraishy, S. 2013. The potential role of Portulaca oleracea as a neuroprotective agent in rotenone-induced neurotoxicity and apoptosis in the brain of rats. Pestic. *Biochem. Physiol.*, *105*: 203–212. https://doi.org/10.1016/j.pestbp.2013.02.004.

Adedapo, A.A., Mogbojuir, O.M.and Emikpe, B.O. 2009. Safety evaluation of the aqueous extract of *Moringa oleifera* in rats. *J. Med. Plants Res*., *3*: 586–591.

Aekhammrat, D., Pannagpetch, P. and Tangsucharit, P. 2019. *Moringa oleifera* leaf extract lowers high blood pressure by alleviating vascular dysfunction and decreasing oxidative stress in L-NAME hypertensive rats. *Phytomedicine*, *54*: 9–16. https://doi.org/10.1016/j.phymed.2018.10.023.

Aju, B.Y., Rajalakshmi, R.and Mani, S. 2019. Protective role of *Moringa oleifera* leaf extract on cardiac antioxidant status and lipid peroxidation in streptozotocin induced diabetic rats. *Heliyon, 5*: e02035. https://doi.org/10.1177/00368504211032079.

Akram, M., Riaz, M., Wadood, A.W.C., Hazrat, A., Mukhtiar, M., Zakki, S.A., Daniyal, M., Shariati, M.A., Khan, F.S. and Zainab, R.. 2020. Medicinal plants with anti-mutagenic potential. *Biotechnol. Biotechnol. Equipment*, *34*: 309–318. https://doi.org/10.1080/13102818.2020.1749527.

Akter, A., Rehamn, S., Morshed, M.T., Hossain, S., Jahan, S., Swarna, A. and Rahmatullah, A. 2013. Evaluation of antihyperglycemic and antinociceptive potential of *Colocasia esculenta* (L.) Schoot (Araceae) leaves. *Adv. Nat. Appl. Sci*., *7*: 143–148.

Al-Quraishy, M.A. Dkhil, and A.E. Abdel Moneim. 2012. Protective effects of *Portulaca oleracea* against rotenone mediated depletion of glutathione in the stratum of rats as an animal model of Parkinson's disease. *Pestic. Biochem. Physio*., *103*: 108–114. https://doi.org/10.1016/j.pestbp.2012.04.005.

Amaglo, N.K., Bennett, R.N., Lo Curto, R.B., Rosa, E.A.S., Turco, V.L., Giuffrida, A., Curto, A.L., Crea, F. and Timpo, G.M. 2010. Profiling selected phytochemicals and nutrients in different tissues of the multipurpose tree *Moringa oleifera* L. grown in Ghana. *Food Chem*., *122*: 1047–1054. https://doi.org/10.1016/j.foodchem.2010.03.073.

Atawodi, S.E., Atawodi, J.C., Idakwo, G.A., Pfundstein, B., Haubner, R., Wartele, G., Bartsch, H. and Owen, R.W. 2010. Polyphenol content and antioxidant properties of methanol extract of the leaves, stem, and root barks of *Moringa oleifera* Lam. *J. Med. Food, 13*: 710–716.

Bajpai, M., Pande, A., Tewari, S.K. and Prakash, D.. 2005. Phenolic contents and antioxidant activity of some food and medicinal plants. *Int. J. Food Sci Nutr*., *56*: 287–291. https://doi.org/10.1080/09637480500146606.

Bamagous, G.A., Al-Ghamdi, S.S., Ibrahim, I.A.A., Mahfoz, A.M., Afify, M.A., Alsugoor, M.H., Shamah, A.A., Arulselvan, P. and Rengarajan, T. 2018. Antidiabetic and antioxidant activity of ethyl acetate extract fraction of *Moringa oleifera* leaves in streptozotocin induced diabetic rats via inhibition of inflammatory mediators. *Asian Pac. J. Trop. Biomed*., *8*: 320. https://doi.org/10.4103/2221-1691.235327.

Barhoi, D., Upadhayay, P., Barbhiya, S.N., Giri, A. and Giri, S. 2021. Aqueous extract of *Moringa oleifera* exhibit potential anticancer activity and can be used as a possible cancer therapeutic agent: A study involving *in vitro* and *in vivo* approach. *J. Am. Cell Nutr.*, *40*: 70–85. https://doi.org/10.1080/07315724.2020.1735572.

Bennett, R.N., Mellon, F.A., Foidl, N., Pratt, J.H., Susan Dupont,M., Perkins, L. and Kroon, P.A. 2003. Profiling glucosinolates and phenolics in vegetative and reproductive tissues of the multi-purpose tree *Moringa oleifera* L. (Horseradish tree) and *Moringa stenopetala*. L. *J. Agric. Food Chem*., *51*: 3456–3553. https://doi.org/10.1021/jf0211480.

Botting, K.J., Young, M.M., Pearson, A.E., Harris, P.J. and Ferguson, L.R. 1999. Antimutagens in food plants eaten by Polynesians: Micronutrients, phytochemicals, and protection against mutagenicity of the heterocyclic amine 3-aminon 3-methylimidazo (4,5-f) quinoline. *Food Chem. Technol., 37*: 95–103. https://doi.org/10.1016/S0278-6915(98)00121-5.

Camara, J.S., Perestrelo, R., Ferreira, R., Berenguer, C.V., Pereira, J.A.M. and Casilho, P.C. 2024. Plant-derived terpenoids: A plethora of bioactive compounds with several health functions and industrial applications: A comprehensive overview. *Molecules*, *29*: 3861. https://doi.org/10.3390/molecules29163861.

Camille, N.N.D., Ide, N.N.M., Fortune, B.E., Claver, O.O.P., Landry, K.B., Calvin, B.Z., Philippe, B.E., Modeste, S. and Dieudonne, M.L. 2018. Evaluation of the toxicity *Colocasia esculenta* (Araceae): Preliminary study of leaves infected by *Phytophthora colocasia* on wistar albino rats. *Biomed. Pharmacother*., *99*: 1009–1013. https://doi.org/10.1016/j.biopha.2017.12.061.

Chakraborty, P., Deb, P., Chakraborty, S., Chatterjee, B. and Abraham, J. 2015. Cytotoxicity and antimicrobial activity of *Colocasia esculenta. J. Chem. Pharma. Res.*, *7*: 627–635.

Cheenpracha, S., Park, E.J., Yoshida, W.Y., Barit, C., Wall, M., Pezzuto, J.M. and Chang, L.C. 2010. Potential anti-inflammatory phenolic glycosides from the medicinal plant *Moringa oleifera* fruits. *Bioorg. Med. Chem.*, *18*: 6598–6602. https://doi.org/10.1016/j.bmc.2010.03.057.

Cho, H., Kim, K., Jang, D.J. and Kim, H.Y. 2019. Effect of six Korean plants on glucagon like peptide-1 release. *Food Sci. Biotechnol.*, *28*: 1571–1576. https://doi.org/10.1007/s10068-019-00603-x.

Cui, X., Ying, Z., Ying, X., Jia, L. and Yang, G. 2021. Three new alkaloids from *Portulaca oleracea* L. and their bioactivities. *Fitoterapia*, *154*: 105020. https://doi.org/10.1016/j.fitote.2021.105020.

Das, B.D., Paudel, N., Paudel, M., Khadka, M.K., Dhakal, S. and Amrit, K.C. 2021. Ethnobotanical knowledge of kewrat community of Morang district, Eastern Nepal. *Ethnobot. Res. Appl.*, *21*: 1–11.

Das, M.K.I. 2022. Documentation of wild leafy vegetable plants of Chitwan, Nepal. *BMC J. Sci. Res.*, *5*: 80–84. https://doi.org/10.3126/bmcjsr.v5i1.50677.

Debnath, B., Singh, W.S., Goswmi, S., Singh, M.K., Maiti, D.and Manna, K. 2018. Role of plant alkaloids on human health: A review of biological activities. *Mater. Today Chem.*, *9*: 56–72. https://doi.org/10.1016/j.mtchem.2018.05.001.

Dias, M.G., Comes, M.F.G.F.C., and Oliveira, L. 2009. Carotenoids in traditional Portuguese fruits and vegetables. *Food Chem.*, *113*: 808–815. https://doi.org/10.1016/j.foodchem.2008.08.002.

Dighe, S.B., Kuchekar, B.S., Wankhede, S.B. and Dighe, M. 2016. Analgesic and anti-inflammatory activity of β-sitosterol isolated from levees of *Oxalis corniculata. Int. J. Pharmcol. Res.*, *6*: 109–113. https://doi.org/10.7439/ijpr.

Draz, A.A., Kawashty, S.A., Shamso, E.M., Hosni, H.A. and Hussein, S.R. 2022. Chemical profiling of *Oxalis* species growing wild in Egypt using HRLC/MS spectroscopy. *Int. J. Second. Metab.*, *9*: 426–439. https://doi.org/10.21448/ijsm.1110807.

Duan, Y., Ying, Z., Zhang, M., Ying, X. and Yang, G. 2022. Two new homoisoflavones from *Portulaca oleracea* L. and their activities. *Nat. Prod. Res.*, *36*: 1765–1773. https://doi.org/10.1080/14786419.2020.1815742.

Duan, Y., Ying, Z., He, F., Ying, X., Jia, L. and Yang, G. 2021. A new skeleton flavonoid and a new lignan from *Portulaca oleracea* L. and their activities. *Fitoterapia*, *153*: 104993. https://doi.org/10.1016/j.fitote.2021.104993.

Dwivedi, P., Dwivedi, J., Patel, D., Desai, S. and Meshram, D. 2016. Phytochemical analysis and assessment of *in vitro* urolithiatic activity of *Colocasia* leaves. *J. Med. Plants Studies*, *4*: 43–47.

Edelman, M. and Colt, M. 2016. Nutrient value of leaf vs. seed. *Front. Chem.*, *4*: 32. https://doi.org/10.3389/fchem.2016.00032.

Ehiobu, J.M. and Ogu, G.I. 2018. Phytochemical content and *in vitro* antimycelial efficacy of *Colocasia esculenta* (L.), *Manihot esculenta* (Crantz) and *Dioscorea rotundata* (Poir) leaf extracts on

Aspergillus niger, and *Botryodiplodia theorbromane*. *J. Hort. Plant Res., 1*: 9–18. https://doi.org/10.18052/www.scipress.com/JHPR.1.9.

El-Hadary, A.E. and M.F. Ramadan. 2019. Antioxidant traits and protective impact of *Moringa oleifera* leaf extract against diclofenac sodium-induced liver toxicity in rats. *J. Food Biochem., 43*: e12704. https://doi.org/10.1111/jfbc.12704.

El-Mesallmy, A.M., El-Tawil, N.A., Ibrahim, S.A. and Hussein, S.A.M. 2021. Phenolic profile: Antimicrobial activity and antioxidant capacity of *Colocasia esculenta* (L.) Schoot. *Egypt J. Chem., 64*: 2165–2172. https://doi.org/10.21608/EJCHEM.2021.56495.3213.

Faizi, S., Siddiqui, B.S., Saleem, R., Siddiqui, S., Aftab, K. and Gilani, A.H. 1995. Fully acetylated carbamate and hypotensive thiocarbamate glycosides from *Moringa oleifera. Phytochemistry, 38*: 957–963. https://doi.org/10.1016/0031-9422(94)00729-d.

Faizi, S., Siddiqui,B.S., Saleem, R., Siddiqui, S., Aftab, K. and Gilani, A.H. 1992. Isolation and structure elucidation of novel hypotensive agents, niazinin A, niazinin B, niazimicin, and niazimminin A+B from *Moringa oleifera*: The first naturally occurring thiocarbamates. *J. Chem. Soc. Perkin Trans. 1, 23*: 3237–3241. https://doi.org/10.1039/p19920003237.

Ferreres, F., Goncalves, R.F., Gil-Izuierdo, A., Valentao, P., Silva, A.M.S., Silva, J.B., Santos, D. and Andrade, P.B. 2012. Further knowledge on the phenolic profile of *Colocasia esculenta* (L.) Schott. *J. Agric. Food Chem., 60*: 7005–7015. https://doi.org/10.1021/jf301739q.

Foidl, N., Makkar, H.P.S. and Becker, K. 2001. The potential of *Moringa oleifera* for agricultural and industrial uses. In: Lowell, J., C.T.A. Fuglie (Eds.), *The Miracle Tree: The Multiple Attributes of Moringa*. CTA Publication, The Netherlands, pp. 45–76.

Fu, J., Wang, H., Dong, C., Xi, C., Xie, J., Lai, S., Chen, R. and Kang, J. 2021. Water-soluble alkaloids isolated from *Portulaca oleracea* L. *Bioorg. Chem., 113*: 105023. https://doi.org/10.1016/j.bioorg.2021.105023.

Garcia-Herrera, P., Morales, P., Camera, M., Fernandez-Ruiz, V., Tardio, J. and Sanchez-Mata, M.C. 2020. Nutritional and phytochemical composition of Mediterranean wild vegetables after culinary treatment. *Food, 9*: 1761. https://doi.org/10.3390/foods9121761.

Goncalves, R.F., Silva, A.M., Silva, A.M., Valentao, P., Ferreres, F., Gil-Izquierdo, A., Silva, J.B., Santos, D. and Andrade, P.B. 2013. Influence of taro (*Colocasia eschulenta* L. Schoot) growth conditions on the phenolic composition and biological properties. *Food Chem., 141*: 3480–3485. https://doi.org/10.1016/j.foodchem.2013.06.009.

Gupta R., Sharma A.K., Dobhal M.P., Sharma M.C. and Gupta R.S. 2011. Antidiabetic and antioxidant potential of β-Sitosterol in streptozotocin induced experimental hyperglycemia. *J. Diabetes*, 3: 29–37.

Gupta, S., Jain, R., Kacchwaha, S. and Kothari, S.L. 2018. Nutritional and medicinal applications of *Moringa oleifera* Lam.: Review of current status and future possibilities. *J. Herb. Med., 11*: 23–25. https://doi.org/10.1016/j.hermed.2017.07.003.

Gupta, S., Lakshmi, J.A., Manjunath, M.N. and Prakash, J. 2005. Analysis of nutrient and antinutrient content of underutilized green leafy vegetables. *Food Sci. Technol., 38*: 339–345. https://doi.org/10.1016/j.lwt.2004.06.012.

Handique, J.G., Boruah, M.P., and Kalith, D. 2012. Antioxidant activities and total phenolic and flavonoid contents in three indigenous medicinal vegetables of north-east India. *Nat. Prod. Commun., 7*: 1021–1023. https://doi.org/10.1177/1934578X1200700815.

Haytowitz David, B., Ahuja Jaspreet, K.C., Xianli, Wu, Meena, S., Melissa, N., Nguyen Quyen, A., Roseland Janet, M., Williams Juhi, R., Patterson Kristine, Y., Yang, Li and Pehrsson Pamela, R. 2019. *USDA National Nutrient Database for Standard Reference, Legacy Release*. Nutrient Data Laboratory, Beltsville Human Nutrition Center, ARS, USDA. https://doi.org/10.15482/USDA.ADC/1529216.

Horo, S. and Topno, S. 2015. Study and analysis of nutritional value of some wild and semiwild edible plants consumed by 'Ho' tribes of W. Singbhum district, Jharkhand, *India. Int. J. Herb. Med., 3*: 25–32.

Horwath, M. and Benin, V. 2011. Theoretical investigation of a reported antibiotic from the "Miracle tree" *Moringa oleifera. Comput. Theor. Chem., 965*: 196–201. https://doi.org/10.1016/j.comptc.2011.01.045.

Ibrahim, M., Hussain, I., Imran, M., Hussain, N., Hussain, A. and Mahboob, T. 2013. Corniculatin A, a new flavonoid glucoside from *Oxalia corniculata. Rev. Bras., 23*: 630–634. https://doi.org/10.1590/S0102-695X2013005000059.

Iwashina, T., Konishi, T., Takayama, A., Fukuda, M. and Ootani, S. 1999. Isolation and identification of the flavonoids in the leaves of taro. *Ann. Tsukuba Bot. Gard., 18*: 71–74.

Jiang, M.Y., Lu, H., Pu, X.Y., Lie, Y.H., Tian, K., Xiong,Y., Wang, W. and Huang, X.Z. 2020. Laxative metabolites from the leaves of *Moringa oleifera. J. Agric. Food Chem., 68*: 7850–7860. https://doi.org/10.1021/acs.jafc.0c01564.

Jiao, Z.Z., Yue, S., Sun, H.X., Jin, T.Y., Wang, H.N., Zhu, R.X. and Xiang, L. 2015. Indoline amide glucosides from Portulaca oleracea: Isolation, structure, and DPPH radical scavenging activity. *J. Nat. Prod., 78*: 2588–2597. https://doi.org/10.1021/acs.jnatprod.5b00524.

Jimoh, T.O. 2018. Enzymes inhibitory and radical scavenging potentials of two selected tropical vegetables (*Moringa oleifera* and *Telfaria accidentalis*) leaves relevant to Type-2 diabetes mellitus. *Rev. Bras. Farmacogn., 28*: 73–79. https://doi.org/10.1016/j.bjp.2017.0.003.

Jin, T.Y., Li, S.Q., Jin, C.R., Shan, H., Wang, R.M., Zhou, M.X., Li, A.L., Li, L.Y., Hu, S.Y., Shen, T. and Xiang, L. 2018. Catecholic isoquinolines from *Portulaca oleracea* and their anti-inflammatory and β2-adrenergic receptor agonist activity. *J. Nat. Prod., 81*: 768–777. https://doi.org/10.1021/acs.jnatprod.7b00762.

Jung, I.L., Lee, J.H. and Kang, S.C. 2015. A potential oral anticancer drug candidate, *Moringa oleifera* leaf extract, induced the apoptosis of human hepatocellular carcinoma cells. *Oncol. Lett., 10*: 1597–1604. https://doi.org/10.3892/ol.2015.3482.

Kashiwada, Y., Ahmed, F.A., Kurimoto, S., Kim, S.Y., Shibata, H., Fujoka, T. and Takaishi, Y. 2012. New α-glucosides of caffeoyl quinic acid from the leaves of *Moringa oleifera* Lam. *J. Nat. Med., 66*: 217–221. https://doi.org/10.1007/s11418-011-0563-5.

Khan, F., Pandey, P., Jha, N.K., Jafri, A. and Khan, I. 2020. Antiproliferative effect of *Moringa oleifera* leaf extract by down-regulating of Notch singling in DU145 prostate cancer cells. *Gene Rep., 19*: 100619. https://doi.org/10.1016/j.genrep.2020.100619.

Khan, M.R., Marium, A., Shabbir, M., Saeed, N. and Bokhari, J. 2012. Antioxidant and hepatoprotective effects of *Oxalis corniculata* against carbon tetrachloride (CCl_4) induced injuries in rat. *Afr. J. Pharm. Pharmcol., 6*: 2255–2267. https://doi.org/10.5897/AJPP11.370.

Khan, W., Praveen, R., Chester, K., Praveen, S. and Ahmad, S. 2017. Hypoglycaemic potential of aqueous extract of *Moringa oleifera* leaf and *in vitro* GC-MS metabolomics. *Front. Pharmacol., 8*: 577. https://doi.org/10.3389/fphar.2017.00577.

Khude, V.S. 2023. Qualitative phytochemical analysis of ten different leafy wild vegetables. *Sci. Arch., 4*: 99–101. https://doi.org/10.47587/SA.2023.4205.

Kiran, K.S., Kameshwar, V.H., Kumar Munakudu-Nagaraju, K., Nagalambika, P., Kavitha, R., Karthick, A., Guru Kumar, D., Swamy, S.N., Krishna, K.L. and Kumar, J.R. 2023. Diosmin: a *Doboia russelii* venom PLA2s inhibitor-purified, and characterized from *Oxalis corniculata* L. medicinal plant. *J. Ethanopharmacol.*, https://doi.org/10.1016/J.JEP.2023.116977.

Koukounaras, A., Bantis, F., Karotolos, N., Melissas, C. and Vezyroglou, A. 2020. Influence of pre-harvest factors on postharvest quality of fresh-cut and baby leafy vegetables. *Agronomy, 10*: 172. https://doi.org/10.3390/agronomy10020172.

Lee, A.S., Kim, J.S., Lee, Y.J., Kang, D.G. and Lee, H.S. 2012. Anti-TNF-α activity of *Portulaca oleracea* in vascular endothelial cells. *Int. J. Mol. Sci., 13*: 5628–5644. https://doi.org/10.3390/ijms13055628.

Lee, S., Kim, K.H., Park, C., Lee, J.S., Kim, Y.H. 2014. *Portulaca oleracea* extracts protect human keratinocytes and fibroblasts from UV-induced apoptosis. *Exp. Dermatol., 1*: 13–17. https://doi.org/10.1111/exd.12396.

Lee, J.L., Oh, J.H., Kong, C.S. and Seo, Y. 2019. Evaluation of anti-adipogenic active homoisoflavonoids from *Portulaca oleracea. Z. Naturforsch*, *74*: 265–273. https://doi.org/10.1515/znc-2018-0114.

Leong, A.C., Kinjo, Y., Tako, M., Iwasaki, H., Oku, H. and Tamaki, H. 2010. Flavonoid glycosides in the shoot system of Okinawa Taumu (*Colocasia esculenta* S.) *Food Chem.*, *119*: 630–635. https://doi.org/10.1016/j.foodchem.2009.07.004.

Li, H.M., Hwang, S.H., Kang, B.G., Hong, J.S. and Lim, S.S. 2014. Inhibitory effects of *Colocasia* (L.) Schott constituents on aldose reductase. *Molecules*, *19*: 13212–13224. https://doi.org/10.3390/molecules190913212.

Li, Y.P., Yao, L.H., Wu, G.J., Pi, X.F., Gong, Y.C., Ye, R.S. and Wang, C.X. 2014. Antioxidant activities of novel small-molecule polysaccharide fractions purified from *Portulaca oleracea* L. *Food Sci. Biotechnol.*, *23*: 2045–2052. https://doi.org/10.1007/s10068-014-0278-y.

Lim, T.K. 2012. *Moringa oleifera*. In: *Edible Medicinal and Non-Medicinal Plants*. Springer, The Netherlands. pp. 435–485. https://doi.org/10.1007/978-94-007-2534-8_61.

Liu R, Liu J, Huang Q, Liu S, Jiang Y. 2022. *Morginga oleifera:* a systematic review of its botany, traditional uses, phytochemistry, pharmacology and toxicity. *J. Pharm. Pharmacol.*, 74: 296-320. https://doi.org/10.1093/jpp/rgab131.

Liu, X., Wu,H., Tao, X., Ying, X. and Stien, D. 2021. Two amide glycosides from *Portulaca oleracea* L. and its bioactivities. *Nat. Prod. Res.*, *35*: 2655–2659. https://doi.org/10.1080/14786419.2019.1660333.

Liu, D.Y., Shen, T. and Xiang, L. 2011. Two antioxidant alkaloids from *Portulaca oleracea* L., Helv. *Chim. Acta.*, *94*: 497–501. https://doi.org/10.1002/hlca.201000250.

Longvah, T., Ananthan, R., Bhaskarachary, K. and Venkaiah, K. 2017. *Indian Food Composition Tables*. National Institute of Nutrition, Hyderabad, India.

Ma, Y., Li, X., Zhang, W., Ying, X. and Stien, D. 2019. A trace alkaloid, oleraisoindole A, from *Portulaca oleracea* L. and its anticholinesterase effect. *Nat. Prod. Res.*, *35*: 350–353. https://doi.org/10.1080/14786419.2019.1627356.

Mahaman, Y.A.R., Huang, F., Wu,M., Wang, Y., Wei, Z., Bao, J., Salissou, M.T.M., Ke, D., Want Q. and Liu, R. 2018. *Moringa oleifera* alleviates homocysteine-induced Alzheimer's disease-like pathology and cognitive impairments. *J. Alzheimers Dis.*, *63*: 1141–1159. https://doi.org/10.3233/JAD-180091.

Manandhar, S., Luitel, S. and Dahal, R.K.. 2019. *In vitro* antimicrobial activity of some medicinal plants against human pathogenic bacteria. *J. Trop. Med.*, *2019*: 1895340 https://doi.org/10.1155/2019/1895340.

Manguro, L.O. and Lemmen, P. 2007. Phenolics of *Moringa oleifera* leaves. *Nat. Prod. Res.*, *21*: 56–68. https://doi.org/10.1080/14786410601035811.

Masiala, A., Vingadassalon, A. and Aurore, G.. 2024. Polyphenols in edible plant leaves: An overview of their occurrence and health properties. *Food Funct.*, *15*: 6847–6882. https://doi.org/10.1039/D4FO00509K.

Mei-Xian, G., Yan-Shuang, W., Gui-Rong, S. and Lei, S.. 2014. Anti-inflammatory and analgesic effect of *Oxalis corniculata* L. on acute peritonitis in mice. *J. Dali Univ.*, *13*: 6.

Miekus, N., Marszalek, K., Podlacha, M., Iqbal, A., Puchalski, C. and Swiergierl, A.H. 2020. Health benefits of plant-derived sulfur compounds, glucosinolates, and organosulfur compounds. *Molecules*, *25*: 3804. https://doi.org/10.3390/molecules25173804.

Mizokami, H., Tomita-Yokotani, K. andYoshitama, K. 2008. Flavonoids in the leaves of *Oxalis corniculata* and sequestration of the flavonoids in the wing scale of the pale grass blue butterfly, *Pseudozizeeria maha*. *J. Plant Res.*, *121*: 133–136. https://doi.org/10.1007/s10265-007-0132-x.

Mosha, T.C. and Gaga, H.E.. 1999. Nutritive value and effect of blanching on the trypsin and chymotrypsin inhibitor activities of selected leafy vegetables. *Plant Foods Hum. Nutr.*, *38*: 513–517. https://doi.org/10.1023/a:1008157508445.

Moyo, B., Masika, P.J., Hugo, A. and Muchenje, V. 2011. Nutritional characterization of moringa (*Moringa oleifera* Lam.) leaves. *Afr. J. Biotechnol.*, *10*: 12912–12933. https://doi.org/10.5897/AJB10.1599.

Mukherjee, S., Pal, S., Chakraborty, R., Koley, H. and Dhar, P. 2018. Biochemical assessment of extract from *Oxalis corniculata* L.: Its role in food preservation, antimicrobial, and antioxidative paradigms using *in situ* and *in vitro* models. *Indian J. Expt. Biol.*, *56*: 230–243.

Mungofa, N., Sibanyoni, J.J., Mashau, M.E. and Beswa, D. 2022. Prospective role of indigenous leafy vegetables as functional food ingredients. *Molecules*, *27*: 7995. https://doi.org/10.3390/molecules27227995.

Musa, K.Y., Ahmed, A., Ibrahim, G., Ojonugwa, O.E., Bisalla, M., Musa, H. and Danmalam, U.H. 2007. Toxicity studies on the methanolic extract of *Portulaca oleracea* L. (Fam. Portulacaceae). *J. Biol. Sci.*, *7*: 1293–1295. https://doi.org/10.3923/jbs.2007.

Natesh, H.N., Abbey, L. and Asiedu, S.K. 2017. An overview of nutritional and anti-nutritional factors in green leafy vegetables. *Horticult. Int. J.*, *1*: 58–65. https://doi.org/10.15406/hij.2017.01.00011.

Neela, S. and Fanta, S.W. (2019) Review on nutritional composition of orange-flushed sweet potato and its role in management of vitamin A deficiency. *Food Sci. Nutr.*, *7*: 1920–1945. https://doi.org/10.1002/fsn3.1063.

Nemzer, B., Al-Taher, F. and Abashiru, N. 2020. Phytochemical composition and nutritional value of different plant parts in two cultivated and wild purslane (*Portulaca oleracea* L.) genotypes. *Food Chem.*, *320*: 126621. https://doi.org/10.1016/j.foodchem.2020.

Niko, K.I., Mpolokeng, T.G., Mokgau, K., Asong, J.A., Omotyo, A.O. and Aremu, A.O. 2024. Ethnobotanical, biological, phytochemical qualities of locally sourced leafy vegetables for food security, good health, and general well-being in South Africa. *S. Afr. J. Bot.*, *172*: 52–68. https://doi.org/10.1016/j.sajb.2024.06.048.

Nouman, W.F., Anwar, F., Gull, T., Newton, A., Rosa, E. and Dominguez-Perles, R. 2016. Profiling of polyphenolics, nutrients, and antioxidant potential of germplasm's leaves from seven cultivars of *Moringa oleifera* Lam. *Ind. Crops Prod.*, *83*: 166–176. https://doi.org/10.1016/j.indcrop.2015.12.032.

Nayaka, H.B., Londankar, R.L., Umesh, M.K. and Tukappa, A. 2014. Antibacterial attributes of apigenin, isolated from *Portulaca oleracea* L. *Int. J. Bacteriol.*, *2014*: 175851. https://doi.org/10.1155/2014/175851.

Obeng, E., Kpodo, F.M., Tettey, C.O., Essuman, E.K. and Adzinyo, O.A. 2020. Antioxidant, total phenol, and proximate constituents of four tropical leafy vegetables. *Sci. Afr.*, *7*: e00227. https://doi.org/10.1016/j.sciaf.2019.e00227.

Oguntibeju, O.O., Aboua, G.Y. and Omodanisi, E.I. 2020. Effects of *Moringa oleifera* on oxidative stress, apoptotic, and inflammatory biomarkers in streptozotocin-induced diabetic animal model. *S. Afr. J. Bot.*, *129*: 354–365. https://doi.org/10.1016/j.sajb.2019.08.039.

Palaniswamy, U.R., McAvoy, R.J. and Bible, B.B. 2001. Stage of harvest and polyunsaturated essential fatty acid concentration in purslane (*Portulaca oleracea* L.) leaves. *J. Agric. Food Chem.*, *49*: 3490–3493. https://doi.org/10.1021/jf0102113.

Park, J.E., Seo, Y. and Han, J.S. 2019a. HM-chromanone isolated from *Portulaca oleracea* L. protects INS-1 pancreatic β cells against glucotoxicity-induced apoptosis. *Nutrients*, *11*: 404. https://doi.org/10.3390/nu11020404.

Park, J.E., Park, J.Y., Seo, Y. and Han, J.S. 2019b. A new chromanone isolated from *Portulaca oleracea* L. increases glucose uptake by stimulating GLUT4 translocation to the plasma membrane in 3T3-L1 adipocytes. *Int. J. Biol. Marcromol.*, *123*: 26–34. https://doi.org/10.1016/j.ijbiomac.2018.10.206.

Pandey, P.B., Pradhan, S.P. and Adhikari, K. 2020. LC-ESI-QTOF-MS for the profiling of the metabolites and *in vitro* enzymes inhibition activity of *Bryophyllum pinnatum* and *Oxalis corniculata* collected from Ramechhap District of Nepal. *Chem. Biodivers.*, *17*: e2000155. https://doi.org/10.1002/CBDV.202000155.

Putriani, N., Perdana, J., Meiliana and Nugrahedi, P.Y. 2022. Effect of thermal processing on key phytochemical compounds in green leafy vegetables: A review. *Food Rev. Int.*, *38*: 783–811. https://doi.org/10.1080/87559129.2020.1745826.

Radha, Kuamr, M., Puri, S., Pundir, A., Bangar, S.P., Changan, S., Choudhary, P., Parameswari, E., Alhariri, A., Samota, M.K., Damle, R.D., Singh, S., Berwal, M.K., Dhumal, S., Bhoite, A.G., Senapathy, M., Sharma, A., Bhusan B. and Mekhemar, M. 2021. Evaluation of nutritional, phytochemical, and mineral composition of selected medicinal plants for therapeutic uses from cold desert of western Himalaya. *Plants*, *10*: 1429. https://doi.org/10.3390/plants10071429.

Rehaman, A., Rehman, A. and Ahmad, I. 2015. Antibacterial, antifungal, and insecticidal potential of *Oxalis corniculata* and its isolated compounds. *Int. J. Anal. Chem.*, *2015*: 842468. https://doi.org/10.1155/2015/842468.

Rocchetti, G., Pagnossa, J.P., Blasi, F., Cossignani, L., Piccoli, R.H., Zengin, G., Montesano, D., Cocconcelli P.S. and Lucini, L. 2020. Phenolic profiling and *in vitro* bioactivity of *Moringa oleifera* leaves as affected by different extraction solvents. *Food Res. Int.*, *127*: 108712. https://doi.org/10.1016/j.foodres.2019.108712.

Sa, A.G.A., Moreno, Y.M.F. and Carciofi, B.A.M. 2020. Food Processing for the improvement of plant protein digestibility. *Crit. Rev. Food Sci. Nutr.*, *60*: 3367–3386. https://doi.org/10.1080/10408398.2019.1688249.

Sahakitpichan, P., Mahidol, C., Disadee, W., Ruchirawat, S, and Kanchanapoom, T. 2011. Unusual glycosides of pyrrole alkaloid and 4′-hydroxyphenylethanamide from leaves of *Moringa oleifera*. *Phytochemistry*, *72*: 791–795. https://doi.org/10.1016/j.phytochem.2011.02.021.

Sakat, S.S., Tupe, P. and Juvekar, A. 2012. Gastroprotective effect of *Oxalis corniculata* (whole plant) on experimentally induced gastric ulceration in Wistar rats. *Indian J. Pharm. Sci.*, *74*: 48–53. https://doi.org/10.4103/0250-474X.102543.

Saleem, A., Saleem, M. and Akhtar, M.F. 2020. Antioxidant, anti-inflammatory, antiarthritic potential of *Moringa oleifera* Lam.: An ethnomedicinal plant of Moringaceae family. *S. Afr. J. Bot.*, *129*: 134–137. https://doi.org/10.1016/j.sajb.2019.11.023.

Singh, S., Singh, D.R., Salim, K.M., Srivastava, A., Singh L.B. and Srivastava, R.C. 2011. Estimation of proximate composition, micronutrients, and phytochemical compounds in traditional vegetables from Andaman and Nicobar Islands. *Int. J. Food Sci. Nutr.*, *62*: 765–773. https://doi.org/10.3109/09637486.2011.585961.

Sivakumar, D., Chen, L. and Sultanbawa, Y. 2018. A comprehensive review on beneficial dietary phytochemical in common traditional southern African leafy vegetables. *Food Sci. Nutr.*, *6*: 714–727. https://doi.org/10.1002/fsn3.643.

Sjamsudin, E., Muhatty, A., Raiwan, L. and Priosoeryanto, B.P. (2021). The efficacy of taro leaf extract on wound healing contaminated with *Staphylococcus aures*. *Padjadjaran J. Dentistry*, *33*: 199–209. https://doi.org/10.24198/pjd.vol33no3.21325.

Song, M., Ying, Z., Ying, X., Jia, L. and Yang, G. 2022. Three novel alkaloids from *Portulaca oleracea* L. and their anti-inflammatory bioactivities. *Fitoterapia*, *156*: 105087. https://doi.org/10.1016/j.fitote.2021.105087.

Sultana B. and Anwar F. 2008. Flavonoids (kaempeferol, quercetin, myricetin) contents of selected fruits, vegetables and medicinal plants. *Food Chem.*, 108: https://doi.org/10.1016/j.foodchem.2007.11.053.

Swamy, G.M., Ramesh, G., Prasad, R.D. and Meriga, B. 2022. Astragalin, (3-O-glucoside of kaempferol), isolated from *Moringa oleifera* leaves modulates leptin adiponectin secretion and inhibits adipogenesis in 3T3-L1 adipocytes. *Arch. Physiol. Biochem.*, *128*: 938944. https://doi.org/10.1080/13813455.2020.1740742.

Teixeria, E.M.B., Carvalho, M.R.B., Neves, V.A., Silva, M.A. and Artes-Pereira, L. 2014. Chemical characteristics and fractionation of protein from *Moringa oleifera* Lam. leaves. *Food Chem., 147*: 51–54. https://doi.org/10.1016/j.foodchem.2013.09.135.

Tewari, A. and Bhakuni, B.S. 2006. Thiocarbamates from *Moringa oleifera. Nat. Prod. Commun., 1*: 721–795. https://doi.org/10.1177/1934578X0600100905.

Tian, J., Ying, Z., Lan, X., Li, Y., Leng, A. and Ying, X. 2022. Two new metabolites from *Portulaca oleracea* and their anti-inflammatory activities. *Phytochem. Lett., 48*: 114–119. https://doi.org/10.1016/j.phytol.2022.02.007.

Tian, J.L., Liang, X., Gao, P.Y., Li, D.Q., Sun, Q., Li, L.Z. and Song, S.J. 2014. Two new alkaloids from *Portulaca oleracea* and their cytotoxic activities. *J. Asian Nat. Prod. Res., 16*: 259–264. https://doi.org/10.1080/10286020.2013.866948.

Tiaz, L., Zeiger, E., Moller, I.M. and Murphy, A. 2015. *Plant Physiology and Development* (6th Edition). Sinauer Associates, Sunderland, USA.

Tleubayeva, M.I., Datkhayev, U.M., Alimzhanova, M., Ishmuratova, M.Y., Korotetskaya, N.V., Abdullabekova, R.M., Flisyuk, E.V. and Gemejiyeva, N.G. 2021. Component composition and antimicrobial activity of CO_2 extract of *Portulaca oleracea*, growing in the territory of Kazakhstan. *Sci. World J., 2021*, Article ID 5434525. https://doi.org/10.1155/2021/5434525.

Tshahalala, T., Ndhlala, A.R., Ncube, B., Addelgadir, H.A. and Van Staden, J. 2020. Potential substitution of the root with the leaf in the use of *Moringa oleifera* for antimicrobial, antidiabetic, and antioxidant properties. *S. Afr. J. Bot., 129*: 106–112. https://doi.org/10.1016/j.sajb.2019.01.029.

Vasant, O.K., Vijay, B.G., Virbhadrappa, S.R., Dilip, N.T., Ramahari, M.V. and Laxamanrao, B.S. 2012. Antihypertensive and diuretic effects of the aqueous extract of *Colocasia esculenta* Linn. Leaves in experimental paradigms. *Iran. J. Phama. Res., 11*: 621. https://doi.org/10.22037/ijpr.2012.1092.

Vongsak, B., Sithisarn, P. and Girtsanapan, W. 2014. Simultaneous HPLC quantitative analysis of active compounds of leaves of *Moringa oleifera* Lam. *J. Chromatogr. Sci., 52*: 641–645. https://doi.org/10.1093/chromsci/bmt093.

Wang, C., Guo, S., Tian, J., Liu, P., Song, M., Zhang, W. and Ying, X. 2022. Two new lignans with their biological activities in *Portulaca oleracea* L. *Phytochem. Lett., 50*: 95–99. https://doi.org/10.1016/j.phytol.2022.06.003.

Wang, Y., Gao, Y., Ding, H., Liu, S., Han, X., Gui, J. and Liu, D. 2017. Subcritical ethanol extraction of flavonoids from *Moringa oleifera* leaf and evaluation of antioxidant activity. *Food Chem., 218*: 152–158. https://doi.org/10.1016/j.foodchem.2016.09.058.

World Health Organization (WHO). 2000. *Nutrition for Health and Development: A Global Agenda for Combining Malnutrition*. Document WHO/NHD/00.6. WHO, Geneva.

Xiu, F., Li, X., Zhang, W., He, F., Ying, X. and Stein, D. 2019. A new alkaloid from *Portulaca oleracea* L. and its antiacetylcholinestrase activity. *Nat. Prod. Res., 33*: 2583–2590. https://doi.org/10.1080/10.1080/14786419.2018.1460833.

Xu., W., Ying, Z., Tao, X., Ying, X. and Yang, G. 2021. Two new amide alkaloids from *Portulaca oleracea* L. and their anticholinesterase activities. *Nat. Prod. Res., 35*: 3794–3800. https://doi.org/10.1080/14786419.2020.1739040.

Yadav, S.K. and Sehgal, S. 2003. Effect of domestic processing and cooking on selected anti- nutrient contents of some green leafy vegetables. *Plant Food Hum. Nutr., 58*: 1–11. https://iris.who.int/handle/10665/66509.

Yan, J., Sun, L.R., Zhou, Z.Y., Chen, Y.C., Zhang, W.M., Dai, H.F. and Tan, J.W. 2012. Homoisoflavonoids from the medicinal plant *Portulaca oleracea. Phytochemistry, 80*: 37–41. https://doi.org/10.1016/j.phytochem.2012.05.014.

Yang, X., Zhang, W., Ying, X. and Stein, D. 2018. New flavonoids from *Portulaca oleracea* L. and their activities. *Fitoterapia, 127*: 257–262. https://doi.org/10.1016/j.fitote.2018.02.032.

Yang, Z., Liu, C., Xiang, L. and Zheng, Y. 2009. Phenolic alkaloids as a new class of antioxidants in *Portulaca oleracea*. *Phytother. Res*., *23*: 1032–1035. https://doi.org/10.102/ptr.2742.

Yang, R.Y., Lin, S. and Kuo, G. 2008. Content and distribution of flavonoids among 91 edible plant species. *Asia Pac. J. Clin. Nutr*., *17*: 275–279. https://doi.org/10.6133/apjcn.2008.17.s1.66.

Yue, S., Jiao,, Z.Z., Sun, H.X., Jin, T.Y. and Xiang, L. 2015. A new tricyclic alkaloid from *Portulaca oleracea* L. *Helv. Chim Acta*., *98*: 961–966. https://doi.org/10.1002/hlca.201400374.

Zhao, C., Zhang, C., He, F., Zhang, W., Leng, A. and Ying, X. 2019. Two new alkaloids from *Portulaca oleracea* L. and their bioactivities. *Fitoterapia*, *136*: 104166. https://doi.org/10.1016/j.fitote.2019.05.005.

Zhao, C., Ying, Z., Tao, X., Jiang, M., Ying, X. and Yang, G. 2018. A new lactam alkaloid from *Portulaca oleracea* L. and its cytotoxicity. *Nat. Prod. Res*., *32*: 1548–1553. https://doi.org/10.1080/14786419.2017.1385022.

Zhang, M., Hettiarachchy, N.S., Horax, R., Kannan, A., Praisoody, A., Muhudan, A. and Mallangi, C.R. 2011. Phytochemicals, antioxidant, and antimicrobial activity *Hibiscus sabdariffa*, *Centella asiatica*, *Moringa oleifera,* and *Murrya koenigii* leaves. *J. Med. Plants Res., 5*: 6673–6679. https://doi.org/10.5897/jmpr11.621.

2 Bioactive Compounds and Biological Activities of *Brassica oleracea* var. *italica*

Iris Gudiño[1,2], Rocío Casquete[1,2*], Alberto Martín[1,2], Yuanfeng Wu[3] and María José Benito[1,2]

Introduction

Broccoli (*Brassica oleracea* var. *italica*) is a leafy vegetable whose consumption has become widespread worldwide. However, only a small fraction of the plant, the sprout, is normally eaten, while stems and leaves represent 60–70% of the total biomass and are generated as by products during cultivation. These portions are typically discarded or diverted to composting and animal feed. Broccoli leaves possess significant nutritional and functional potential owing to their high levels of bioactive compounds. They are rich in phenolic compounds, particularly flavonoids, characteristic Brassica glucosinolates and their derivatives (isothiocyanates), as well as vitamins, proteins, minerals, and dietary fiber (Borja-Martínez et al., 2020; Ferreira et al., 2020; Rivas et al., 2022). This complex matrix of bioactives endows the leaves with antioxidant, antimicrobial, and anticancer activities, among other health-promoting properties (Abdelhalim et al., 2023; Gudiño et al., 2022; Le et al., 2020). Consequently, broccoli leaves feature in the daily diets of many regions, appearing in soups, stews, and salads, underscoring their suitability for direct consumption or as concentrated extracts.

To effectively harness these compounds, it is essential to select the most appropriate extraction technique (such as solvent extraction, ultrasound-assisted extraction, or supercritical fluid extraction) and to optimize biomass pretreatment, while also selecting broccoli varieties with higher bioactive compound content and enhanced bioavailability (Manríquez-Zúñiga et al., 2023; Gudiño et al., 2024). These measures contribute to maximizing process yield and preserving the functional integrity of the resulting extracts. Furthermore, broccoli leaf-derived

[1] Nutrición y Bromatología, Escuela de Ingenierías Agrarias, Universidad de Extremadura, Avd. Adolfo Suárez s/n, 06007 Badajoz, Spain.
[2] Instituto Universitario de Investigación en Recursos Agrarios (INURA), Universidad de Extremadura, Avd. de la Investigación, 06006 Badajoz, Spain.
[3] School of Biological and Chemical Engineering, Zhejiang University of Science and Technology, Hangzhou 310023, China.
* Corresponding author: e-mail: rociocp@unex.es

DOI: 10.1201/9781003687832-2

ingredients exhibit significant potential across various sectors: as natural additives in the food industry, as nutraceutical components with health-promoting properties, and as biostimulants in agricultural applications. This comprehensive valorization strategy illustrates how full plant utilization can drive sustainability and foster innovation.

Growing interest in broccoli leaves is driven by the need to promote more sustainable agricultural resource use and to reduce waste. Accordingly, this work provides an exhaustive review of the chemical profile, functional properties, and potential applications of broccoli leaves, as well as the challenges and prospects for their integration into the food, nutraceutical, and other industries. Valorizing these plant residues not only holds potential to enhance human health, but also to encourage more responsible and sustainable farming practices.

Functional Compounds in Broccoli Leaves

Broccoli leaves are abundant sources of various bioactive compounds that contribute to their functional properties, including antioxidant, anti-inflammatory, and antimicrobial activities. The extraction and characterization of these compounds are vital to harnessing their full potential for application in food, pharmaceutical, and cosmetic industries. In Table 1, relevant information on the functional compounds present in broccoli leaves, as well as the associated extraction and analysis methods, is presented.

Glucosinolates and Sulforaphane

Glucosinolates are characteristic compounds of the Brassica family. Broccoli leaves exhibit a diverse composition of glucosinolates, including aliphatic, indolic, and aromatic types. Among the aliphatic glucosinolates, glucoraphanin is the most abundant, with contents ranging from 2.77 μmol/g to over 1100 μg/g in cultivars such as "Green Magic" and 'Parthenon', respectively (Ares et al., 2015; Liu et al., 2018; Gudiño et al., 2022). Other aliphatic glucosinolates identified in the leaves include glucoiberin, with concentrations up to 173 μg/g (Ares et al., 2015; Liu et al., 2018), while glucoerucin was present at low levels near the limit of quantification in cultivars such as Parthenon, Nubia, and Naxos (Ares et al., 2015). However, gluconapin was not detected in any of the leaf samples from these cultivars, reflecting a strong cultivar-dependent influence on glucosinolate accumulation. Additional aliphatic glucosinolates, such as 6-(methylsulfonyl)hexylglucosinolate and 3-butenyl glucosinolate, have also been detected in broccoli leaves (Pei et al., 2019).

Among the aromatic glucosinolates, gluconasturtiin was detected at concentrations ranging from 51–119 μg/g, with higher levels observed in 'Parthenon' leaves. Several indolic glucosinolates have also been identified in broccoli leaves, with glucobrassicin being particularly prominent, reaching up to 106 μg/g in the 'Nubia' variety (Ares et al., 2015) and 0.24 μmol/g in "Green

Magic" (Liu et al., 2018). Methoxyglucobrassicin, 4-methoxyglucobrassicin, and neoglucobrassicin have also been confirmed in various cultivars (Ares et al., 2015; Liu et al., 2018; Gudiño et al., 2022). The distribution of these compounds showed clear variability among cultivars.

Upon tissue damage by internal or external agents, the enzyme myrosinase catalyzes the hydrolysis of glucosinolates, generating bioactive compounds such as isothiocyanates. One of the most prominent isothiocyanates in broccoli is sulforaphane, derived from the hydrolysis of glucoraphanin, which is highly abundant in leaves. Another isothiocyanate identified in fresh broccoli leaves is benzyl isothiocyanate (BITC), a degradation product of glucotropaeolin, which has been reported to account for up to 16.44% of the extract in specific studies (Abdelhalim et al., 2023).

In summary, broccoli leaves are notable for their high glucosinolate content, particularly glucoraphanin and glucobrassicin, underscoring their potential value as an agricultural byproduct with functional and nutraceutical applications.

Phenolic Compounds and Flavonoids (Quercetin, Kaempferol)

Broccoli leaves represent a valuable source of phenolic compounds, exhibiting a wide diversity and high concentrations of these bioactive constituents. Reported total phenolic content (TPC) values vary significantly, ranging from 112.95– 2435 mg GAE/100 g DW, depending on cultivar and sample conditions (Gudiño et al., 2022; Manríquez-Zúñiga et al., 2023). Such variability has been attributed to genotypic differences and pre-processing conditions of the leaves, including drying methods and extractive protocols (Liu et al., 2018; Le et al., 2020; Rivas et al., 2022; Zhao et al., 2024). In samples comprising both leaves and stems, high TPC values have also been documented, such as 15086.7 μg/g DW in freeze-dried material (Ferreira et al., 2020), 21.32 mg/kg in fresh samples (Martínez et al., 2023), and up to 74.31 mg GAE/g in oven-dried mixtures (Borja-Martínez et al., 2020). These findings emphasize the importance of sample pre treatment in enhancing phenolic recovery.

Exclusively foliar samples have shown considerable flavonoid levels, with total flavonoid contents ranging from 8–10 mg CE/g DW. Specific flavonoids, such as myricetin (2.77 mg/g extract) and quercetin (0.97 mg/g extract), have been quantified in leaf extracts (Le et al., 2020), while Zhao et al. (2024) reported flavonoid concentrations around 3.4 mg/g in freeze-dried broccoli leaves. Among these, kaempferol is one of the most relevant flavonoids, predominantly present as glycosylated derivatives. Ferreira et al. (2020) identified abundant forms such as kaempferol-3-O-diglucoside-7-O-glucoside (K-3-dG-7-G) and kaempferol-3-O-malonyl-diglucoside-7-O-glucoside (K-3-MC-dG-7-G). Similarly, Gudiño et al. (2022) reported a broad profile of kaempferol derivatives, including km-diglucoside, km-3-diglucoside-7-diglucoside, and km-3-feruloyl-sophorotrioside, as well as quercetin conjugates such as qn-3-diglucoside-7-glucoside and qn-3-sinapoyl-sophorotrioside.

In addition to flavonoids, coumarins have also been detected in relevant amounts in broccoli leaf tissues (Le et al., 2020; Gudiño et al., 2022). Moreover, broccoli leaves are notably rich in phenolic acids, with contents ranging from 3.6 to 5.7 mg/g DW (Berndtsson et al., 2020). Identified compounds include derivatives of caffeic and ferulic acids, such as chlorogenic acid and various caffeoylquinic acids, with concentrations influenced by cultivar and processing conditions (Berndtsson et al., 2020; Le et al., 2020; Martínez et al., 2023). In 'Beneforte' cultivar leaves, complex glycosides like K-3-O-caffeoyl-sophoroside-7-O-diglucoside and trisinapoyl-diglucoside have also been characterized (Berndtsson et al., 2020). Altogether, these findings highlight the rich phenolic profile of broccoli leaves, underscoring the predominance of quercetin and kaempferol derivatives, as well as phenolic acids, supporting their potential application in functional and nutraceutical products.

Pigments, Vitamins, and Minerals

Broccoli leaves exhibit a rich nutritional profile characterized by a diverse composition of pigments, vitamins, and essential minerals. Pigment analysis has revealed notably high levels of total chlorophyll, reaching up to 76.42 mg/100 g DW in leaf samples (Liu et al., 2018; Gudiño et al., 2022), and even higher values (up to 5569.4 μg/g DW) when combined with other plant by-products such as stalks (Borja-Martínez et al., 2020; Ferreira et al., 2020). Both chlorophyll a and b are present in significant amounts, with chlorophyll a reaching up to 4477.9 μg/g DW (Liu et al., 2018). In addition, chlorophyll degradation products and intermediates such as pheophytins, pheophorbide a, and pyropheophytins have also been detected, offering potential applications as functional ingredients or natural colorants (Ferreira et al., 2020). Carotenoids are also abundant, particularly β-carotene, which was reported at concentrations as high as 253.4 mg/g DW in certain leaf/stem mixtures. Other relevant carotenoids include lutein, violaxanthin, and neoxanthin, all known for their antioxidant activity and importance in eye health (Ferreira et al., 2020).

Regarding vitamin content, broccoli leaves are especially rich in vitamin C, with levels ranging from 1.08 to 2.74 mg/g DW depending on the extraction method (Le et al., 2020; Zhao et al., 2024). They also contain vitamin E (total tocopherols), with reported levels around 155 μg/g DW, including α-tocopherol, and vitamin K (phylloquinone) at 24.3 μg/g DW, highlighting the presence of both fat- and water-soluble vitamins in this matrix (Liu et al., 2018; Borja-Martínez et al., 2020).

Mineral analysis indicates that broccoli leaves are particularly rich in calcium (28.99 mg/g DW) and iron (40.50 mg/g DW). Other essential minerals identified include magnesium, zinc, copper, phosphorus, and potassium, which further support the high nutritional value of broccoli leaves (Liu et al., 2018). These findings confirm the potential of broccoli leaves as a valuable source of micronutrients for functional food and nutraceutical applications.

Proteins, Amino Acids, and Lipids

Broccoli leaves exhibit a high protein content, particularly when processed by freeze-drying, with values reaching approximately 22.75 g per 100 g dry weight (DW), compared to around 6.13% in fresh samples (Sedlar et al., 2021; Zhao et al., 2024). When combined with other by-products such as stems, the total protein content can reach up to 19.7% DW, depending on the processing conditions (Shi et al., 2019). Regarding protein fractions, albumin constitutes the predominant class (31.66%), followed by globulin (15.89%) in fresh leaves, indicating a high proportion of water-soluble and storage proteins typical of plant matrices (Sedlar et al., 2021).

Broccoli leaves also contain a broad spectrum of amino acids, including lysine, L-histidine, D-serine, L-asparagine, L-leucine, and L-tryptophan, which contribute significantly to their nutritional value and support essential metabolic and cellular functions (Manríquez-Zúñiga et al., 2023).

Lipids represent a minor component in broccoli leaves, with total fatty acid content reported at approximately 3.34 g/100 g DW in the ‘Parthenon’ variety (Rivas et al., 2022). The predominant fatty acids identified include palmitic acid (68.38 UAA) and linolenic acid (40.39 UAA) (Gudiño et al., 2022). Lipid content and composition are influenced by the processing method; for instance, air-drying at 60 °C results in less fatty acid degradation compared to higher temperatures or freeze-drying (Sedlar et al., 2021; Shi et al., 2019; Rivas et al., 2022).

Carbohydrates (Fiber and Sugar) and Hydrocarbons

Broccoli leaves contain a significant amount of carbohydrates, primarily composed of dietary fiber, simple sugars, and minor amounts of starch. The total carbohydrate content in the leaves ranges from 26.0–32.6% dry weight (DW), depending on the cultivar and processing conditions (Berndtsson et al., 2020). When leaves are combined with other broccoli by-products, such as stems, the total fiber content can increase to between 36.5– 49.6% DW, reflecting the influence of sample processing (Shi et al., 2019). The cultivar also plays a crucial role in fiber composition. Leaves of the ‘Parthenon’ and ‘Beneforte’ cultivars have been reported to contain high levels of dietary fiber, predominantly in the insoluble fraction, particularly after air-drying (Berndtsson et al., 2020; Rivas et al., 2022). These fractions are rich in cellulose, hemicellulose, pectins, and gums.

In terms of sugar content, broccoli leaves exhibit substantial levels of both soluble and structural carbohydrates. Soluble sugar levels can reach up to 25.39 g/100 g DW in the ‘Parthenon’ variety, and approximately 18 mg/g of total sugars have been reported in freeze-dried samples (Rivas et al., 2022; Zhao et al., 2024). In fresh leaves, sugar content may reach up to 40.31% DW in protein-enriched fractions (Sedlar et al., 2021). In addition, structural sugars such as uronic acids (up to 657.6 mg/g in fiber concentrates), as well as monosaccharides like glucose, rhamnose, and galactose, are found in both soluble and insoluble fiber fractions (Berndtsson et al., 2020; Rivas et al., 2022).

Regarding hydrocarbons, fresh broccoli leaves are rich in long-chain aliphatic compounds. The most abundant component is 3-methyloctacosane, accounting for 58.76% of the total hydrocarbon fraction. Other relevant compounds include 3,3,17,17-tetraethylnonadecane (33.71%) and 15-methyltriacontane (15.09%) (Abdelhalim et al., 2023).

Extraction Techniques for Bioactive Compounds

Solvent Extraction (Methanol, Ethanol, Water)

Conventional solvent extraction is a widely used technique for recovering bioactive compounds from plant matrices due to its simplicity, low cost, and ease of application. In broccoli leaves, solvents such as water, ethanol, and methanol have been applied under controlled heating or stirring to efficiently extract a broad range of compounds, mainly phenolics and organic acids, as well as glucosinolates, pigments, proteins, and vitamins.

Phenolic compounds, including flavonoids and individual phenolic acids (gallic, caffeic, chlorogenic, and ferulic acids), are commonly extracted using 70–80% ethanol or methanol at 40–75 °C for 20–60 minutes (Berndtsson et al., 2020; Le et al., 2020; Manríquez-Zúñiga et al., 2023; Martínez et al., 2023). Buffers such as MeOH: Tris-HCl (pH 7.5) can improve extract stability (Borja-Martínez et al., 2020). These conditions yield extracts rich in bioactive flavonols (quercetin, myricetin) and coumarins such as esculetin (Le et al., 2020; Zhao et al., 2024).

Glucosinolates (glucobrassicin, glucoiberin, glucoerucin) are efficiently extracted using water or hydroalcoholic mixtures at elevated temperatures (60–95 °C), which enables the inactivation of the myrosinase enzyme. Both methanol (Liu et al., 2018; Ferreira et al., 2020) and ethanol (Pei et al., 2019) have been used, as well as dichloromethane maceration (Abdelhalim et al., 2023). Pressurized liquid extraction (PLE) with 70% ethanol has also been employed to recover bioactive breakdown products such as sulforaphane (Ares et al., 2015).

Pigments including chlorophylls and carotenoids (β-carotene, lutein) have been extracted using ethanol or, more efficiently, methanol (Ferreira et al., 2020; Le et al., 2020). Acetone and its mixtures with other solvents have also proven effective (Liu et al., 2018; Gudiño et al., 2022). These solvents disrupt chloroplast membranes, facilitating the solubilization of both lipophilic and hydrophilic pigments, while water alone exhibits limited extraction efficiency.

Water-soluble vitamins such as vitamin C are typically extracted with hot water or acidic solutions (oxalic, metaphosphoric, or sulfuric acid), whereas lipophilic vitamins (E and K) are recovered using isopropanol-hexane mixtures under inert atmosphere or alcoholic buffers (Liu et al., 2018; Zhao et al., 2024).

Other compounds of interest include amino acids, which are extracted with ethanol and analyzed via GC-MS (Manríquez-Zúñiga et al., 2023); carbohydrates, extracted with water or ethanol-HCl depending on sugar type (Zhao et al., 2024); and minerals, quantified by ICP-OES following acid digestion (Liu et al., 2018).

Volatile hydrocarbons have been recovered using hexane or dichloromethane (Abdelhalim et al., 2023).

Overall, hydroalcoholic mixtures (70–80% ethanol or methanol) are the most effective and versatile solvents for extracting a wide spectrum of bioactive compounds, while water remains particularly suitable for selectively isolating polar molecules.

Ultrasound-Assisted and Enzyme-Assisted Extraction

Ultrasound-assisted extraction (UAE) and enzyme-assisted extraction (EAE) are two promising techniques for the recovery of bioactive compounds from broccoli leaves, as they offer advantages such as lower solvent consumption, shorter extraction times, and better yields compared to conventional methods.

In the case of UAE, ultrasound enhances cell disruption, facilitating solvent penetration and the release of intracellular compounds. This method has been effectively used to extract phenolic compounds, in particular flavonoids and phenolic acids (Berndtsson et al., 2020; Martínez et al., 2023; Rivas et al., 2022), as well as for the extraction of glucosinolates such as glucobrassicin (Gudiño et al., 2022).

In addition, enzyme-assisted methods have been employed for the extraction and quantification of dietary fiber in broccoli leaves using gravimetric protocols. Enzymes such as α-amylase, protease, and amyloglucosidase are used to remove starch and protein residues, enabling the accurate separation and measurement of insoluble and soluble fiber fractions (Manríquez-Zúñiga et al., 2023; Rivas et al., 2022). Enzymatic-gravimetric extraction methods have also been used to obtain proteins (mainly albumin and globulin) from fresh broccoli leaves (Sedlar et al., 2021). This approach is especially valuable in the characterization of structural polysaccharides in vegetable by-products.

These assisted extraction techniques offer promising alternatives to improve the yield and selectivity of functional compounds, while operating under milder conditions, avoiding compound degradation.

Supercritical Fluid Extraction (SFE)

Supercritical fluid extraction (SFE), particularly using CO_2 as a solvent, is a clean and efficient technique for the selective recovery of non-polar bioactive compounds. In the case of broccoli leaves, SFE has been primarily applied to extract phenolic compounds. This method is particularly effective when combined with co-solvents such as ethanol, yielding between 19.5– 29.2 mg GAE/g depending on the cultivar (Arnáiz et al., 2016; Borja-Martínez et al., 2020). Similarly, SFE using methanol has demonstrated high efficiency in the extraction of amino acids from broccoli leaves of different cultivars (Arnáiz et al., 2012).

Pigments such as chlorophylls and carotenoids, especially lutein and β-carotene, can also be extracted using SFE, although this application is less commonly reported in broccoli leaves. Due to their lipophilic nature, carotenoids are well-suited for supercritical CO_2 extraction, particularly when ethanol is used as a co-solvent to enhance solubility and extraction efficiency (Borja-Martínez et al., 2020). SFE has also been employed in the treatment of dietary fiber derived from broccoli leaves, improving its composition and functional properties (Rivas et al., 2024b).

Although the yields obtained by SFE are generally lower than those achieved through conventional or ultrasound-assisted methods, this technique offers notable advantages, including solvent-free extracts, reduced oxidation, and improved environmental sustainability.

Stability and Compatibility of Extracted Compounds

The stability and compatibility of bioactive compounds extracted from broccoli by-products are influenced by several factors, including the extraction method, processing conditions, and storage environment (Krupa-Kozak et al., 2019; Rivas et al., 2024b). Advanced extraction techniques, such as ultrasound-assisted extraction, SFE, or EAE, can significantly affect the stability of these compounds during and after their recovery (Berndtsson et al., 2020; Devi et al., 2022; Khalil et al., 2022). Ensuring appropriate stabilization strategies, such as encapsulation or controlled storage conditions (low temperature, absence of light and oxygen), is essential to preserve their biological activity, chemical integrity, and functional performance in food and nutraceutical applications (Pei et al., 2019; Zhou et al., 2021).

Among post-extraction preservation strategies, freeze-drying is commonly employed to maintain the stability and bioactivity of phenolic compounds and isothiocyanates such as sulforaphane (Ares et al., 2015; Pei et al., 2019). This method removes water at low temperatures, thereby minimizing thermal degradation and oxidative losses. In addition, microwave hydrodiffusion and gravitational separation techniques have been explored for the dehydration of broccoli waste and the recovery of water-soluble bioactives (Devi et al., 2022; Khalil et al., 2022). These techniques offer rapid extraction and drying with reduced thermal stress, contributing to the preservation of labile compounds.

Combination of Techniques

The combination of various extraction techniques applied to broccoli leaves enables the maximization of bioactive-compound yields and the reduction of processing times, while preserving compound integrity and adhering to sustainability criteria. In several studies, hydroalcoholic mixtures (70–80% ethanol or methanol) under moderate agitation and heating (40–75°C, 20–60 min) have proven highly effective for phenolics and glucosinolates (Berndtsson et al., 2020; Le et al., 2020). The incorporation of ultrasound further enhances cell disruption and shortens the extraction time (Gudiño et al., 2022; Martínez et al., 2023; Rivas et al., 2022). A combined approach employing an initial enzymatic predigestion step (α-amylase, protease, and cellulase) facilitates the release of dietary fiber and protein fractions under milder conditions than conventional extraction (Manríquez-Zúñiga et al., 2023; Sedlar et al., 2021). Finally, supercritical CO_2 extraction with ethanol as a co-solvent selectively recovers lipophilic pigments and tocopherols with minimal solvent residues and without producing organic waste (Ares et al., 2015; Borja-Martínez et al., 2020). The deployment of such combined strategies not only optimizes yields and shortens total processing time but also reduces solvent consumption and waste generation.

All data related to functional compounds present in broccoli leaves, as well as extraction and analysis methods, are listed in Table 1.

Beneficial Properties of Broccoli Leaves

Antioxidant Activities

Broccoli leaves exhibit notable antioxidant and anti-inflammatory activities, primarily attributed to their rich content of bioactive compounds such as phenolics, flavonoids, glucosinolates and their derivatives (glucoraphanin and sulforaphane), as well as pigments including chlorophyll and carotenoids (Borja-Martínez et al., 2020; Gudiño et al., 2022). The antioxidant capacity of broccoli leaves has been demonstrated through various assays, including DPPH, ABTS, and ORAC (Le et al., 2020; Shi et al., 2019). This activity is strongly influenced by the cultivar studied (Borja-Martínez et al., 2020; Gudiño et al., 2022). Additionally, the choice of extraction solvent significantly affects the antioxidant potential, with methanol being particularly effective, especially when combined with ultrasound-assisted extraction (Martínez et al., 2023; Ramesh, 2023; Zhao et al., 2024). The application of supercritical fluid extraction (SFE) has also been explored to enhance the antioxidant properties of fiber extracted from broccoli leaves (Rivas et al., 2024b).

Anti-Inflammatory Effects

Regarding anti-inflammatory properties, several studies have shown that broccoli leaf extracts exhibit significant activity, which has been primarily associated with the presence of phenolic compounds such as quercetin and kaempferol, as well as the isothiocyanate sulforaphane. Ethanolic extracts of broccoli leaves have been shown to inhibit the activity of pro-inflammatory cytokines *in vitro* and suppress key signaling pathways such as NF-κB, leading to reduced expression of inflammation-related genes (Park et al., 2019). These effects have been confirmed both *in vitro* and *in vivo*, where leaf extracts alleviated inflammation and tissue damage in rats (Khalil et al., 2022). In another study, Ramesh et al. (2023) demonstrated that aqueous and organic extracts of broccoli leaves significantly inhibited proteinase activity and stabilized cell membranes, outperforming even aspirin.

Table 1 Extraction methods used to depend on the type of compound in broccoli leaves, as well as the method of analysis applied.

Extraction Method	*Compound*	*Extraction Conditions*	*Analysis Method*	*References*
Solvent Extraction	Phenolic compounds	MeOH: Tris-HCl (50 mM, pH 7.5) (1:1; v/v), Ice bath in darkness (20 min, 4ºC)	UV-Vis	Borja-Martínez et al. (2020)
		MeOH (70%) - EtOH (70%) - Hot water	UV-Vis and HPLC	Le et al. (2020)
		MeOH (70%), 30 min; 70ºC	UHPLC-DAD-ESI/MS	Ferreira et al. (2020)
		MeOH (70%); 75ºC	UV-Vis	Liu et al. (2018)
		MeOH (80%), HCl (1%); 2h	UV-Vis	Zhao et al. (2024)
		Water (1:3 W/W); 100ºC; 60 min	HPLC-DAD	Martínez et al. (2023)
		DM: EtOH (30%); 50ºC; 36,92min	UV-Vis	Manríquez-Zúñiga et al. (2023)
	Organic acids	DM: EtOH (30%); 50ºC; 36,92min	GC-EIMS	Manríquez-Zúñiga et al. (2023)
	Glucosinolates and their breakdown products	MeOH (70%), 30 min; 70ºC	UHPLC-DAD-ESI/MS	Ferreira et al. (2020)
		MeOH (70%); 10 min; 95ºC (i.s. glucosinalbin)	UHPLC-PAD	Liu et al. (2018)
		PLE: EtOH (70%); 5min; 60ºC; 1500psi	UHPLC-qTOF-MS	Ares et al. (2015)
		EtOH (70%); 70ºC; 30 min	UHPLC-qTOF-MS	Pei et al. (20190
		Maceration in dichloromethane (DCM) 7 days	GC-MS	Abdelhalim et al. (2023)

Contd...

Table 1 *Contd...*

Extraction Method	*Compound*	*Extraction Conditions*	*Analysis Method*	*References*
	Pigment (Chlorophyll and carotenoids)	MeOH: Tris-HCl (50 mM, pH 7.5) (1:1; v/v), Ice bath in darkness (20 min, 4ºC)	UV-Vis	Borja-Martínez et al. (2020)
	Chlorophyll	Acetone (4ºC, 15 min)	UV-Vis	Gudiño et al., (2022)
	(Chlorophyll and carotenoids)	0.1% butylated hydroxytoluene - EtOH 100% (24h)	UHPLC-DAD-ESI/MS	Ferreira et al., (2020)
	(Chlorophyll)	Acetone/MeOH (2:1 v/v, 0.5% BHT)	UHPLC	Liu et al. (2018)
	(Carotenoids)	acetone/MeOH (2:1 v/v, 0.5% BHT); hexane (i.s. -apo-8'-carotenal)	UHPLC	Liu et al. (2018)
	Vitamin C	MeOH (70%) - EtOH (70%) - Hot water	UV-Vis	Le et al. (2020)
		oxalic acid/EDTA; metaphosphoric acid-acetic; sulfuric acid (80%, 10 min)	UV-Vis	Zhao et al. (2024)
	Vitamin E	isopropanol/hexane (3:2 v/v), extracted with nitrogen	UHPLC	Liu et al. (2018)
		MeOH: Tris-HCl (50 mM, pH 7.5) (1:1; v/v), Ice bath in darkness (20 min, 4ºC)	GC-MS	Borja-Martínez et al. (2020)
	Vitamin K	isopropanol/hexane (3:2 v/v), extracted with nitrogen	UHPLC	Liu et al. (2018)
	Minerals (Mg, Na, Ca, Mn)	ashed 550ºC; HNO_3 (1N)	ICP-OES	Liu et al. (2018)
	Amino acid	EtOH (30%); 50ºC; 36,92 min	GC-EIMS	Manríquez-Zúñiga et al. (2023)

Contd...

Table 1 *Contd...*

Extraction Method	*Compound*	*Extraction Conditions*	*Analysis Method*	*References*
	Carbohydrates (Sugar)	EtOH-HCl	UV-Vis	Zhao et al. (2024)
		distilled water	HPLC-DAD	Rivas et al. (2022)
	Hydrocarbons	Hexane	GC-MS	Abdelhalim et al. (2023)
		Maceration in DCM 7 days	GC-MS	Abdelhalim et al. (2023)
Ultrasound-Assisted and Enzyme-Assisted Extraction	Phenolic compounds	UAE: EtOH (80% v/v); 45°C; 1h	UV-Vis	Gudiño et al. (2022)
		UAE: EtOH (80% v/v); 45°C; 1h	UV-Vis	Rivas et al. (2022)
		UAE: Water (1:3 w/w); 75°C; 60 min	UV-Vis	Martínez et al. (2023)
		UAE: MeOH (60%); 35°C, 1h; alkaline hydrolysis	HPLC-DAD_ESI_MS	Berndtsson et al. (2020)
		EE: water (1:3 w/w), cellulose enzyme (0,01%)	UV-Vis	Martínez et al. (2023)
	Organic acids (Gluconic acid)	UAE: EtOH (80% v/v); 45°C; 1h	HPLC-ESI-MS	Gudiño et al. (2022)
	Glucosinolates (Glucoraphanin)	UAE: EtOH (80% v/v); 45°C; 1h	HPLC-ESI-MS	Gudiño et al. (2022)
	Protein (Albumin, Globulin)	Water- NaCl (5%)- NaOH- EtOH (70%)	Enzymatic-Gravimetric method	Sedlar et al. (2021)

Contd...

Table 1 *Contd...*

Extraction Method	*Compound*	*Extraction Conditions*	*Analysis Method*	*References*
	Lipids (Palmitic acid, Linolenic acid)	UAE: EtOH (80% v/v); 45ºC; 1h	HPLC-ESI-MS	Gudiño et al. (2022)
	Dietary fibre	AIR method (EtOH)	Enzymatic-gravimetric method	Rivas et al. (2022)
Supercritical Fluid Extraction	Phenolic compounds	SFE: CO_2, 40ºC, 443 bar, 7% eToh, 31 g/min of flow rate (68 min)	UV-Vis	Borja-Martínez et al. (2020)
		SFE: CO_2, 35 °C, 150 bars, 20 % EtOH, 10 min dynamic extraction time	UV-Vis	Arnáiz et al. (2016)
	Pigments (Chlorophyll and carotenoids)	SFE: CO_2, 40ºC, 443 bar, 7% EtOH, 31 g/min of flow rate (68 min)	UV-Vis	Borja-Martínez et al. (2020)
	Vitamin (α-Tocopherol)	SFE: CO_2, 40ºC, 443 bar, 7% EtOH, 31 g/min of flow rate (68 min)	UV-Vis	Borja-Martínez et al. (2020)
	Amino acid	SFE: CO_2, 70ºC, 250bar, 35% MeOH, 2mL/min flow rate, 5min and 30min (static and dynamic extraction times)	GC-MS	Arnáiz et al. (2012)
	Dietary fibre	SFE: CO2,40 °C, 191 bar, 1 hour	HPLC-DAD	Rivas et al. (2024a)

Notes: i.s.: internal standard; UAE: ultrasonic-assisted extraction; SFE: supercritical fluid extraction; EE: enzymatic extraction; DCM: dichloromethane; BHT: butylated hydroxytoluene; EDTA: ethylenediaminetetraacetic acid; ICP-OES: Inductively Coupled Plasma Optical Emission Spectrometry.

The observed anti-inflammatory effects are closely linked to the antioxidant properties of these compounds, as both flavonoids and isothiocyanates contribute to the neutralization of oxidative stress, a known trigger and amplifier of inflammatory processes. This relationship suggests a synergistic mechanism of action between the antioxidant and anti-inflammatory properties of broccoli leaf bioactives.

Anticancer Properties

Previous studies have demonstrated that broccoli leaves exhibit cytotoxic and antiproliferative effects against various cancer cell lines. These effects have been primarily attributed to the presence of phenolic compounds (such as flavonoids and tannins), vitamin C, and glucosinolates, like glucoraphanin, and its breakdown product, sulforaphane. Sulforaphane is a potent inducer of Phase II detoxification enzymes, which contributes to the elimination of carcinogenic compounds. Methanolic extracts from broccoli leaves have shown significant growth inhibition in liver (HepG2), breast (MCF-7), and colon (HCT116) cancer cell lines at a concentration of 100 µg/mL, achieving up to 60% inhibition (Gaafar et al., 2020). Similarly, ethanolic extracts of broccoli leaves have been reported to induce cell cycle arrest, promote apoptosis, and suppress tumor development in several cancer types, including lung, colon, and prostate cancers (Le et al., 2020).

Antimicrobial Properties

Extracts from broccoli (*Brassica oleracea* var. *italica*) leaves have demonstrated broad-spectrum activity against pathogenic bacteria and fungi. This antimicrobial effect is attributed to their high content of bioactive phenolic compounds, which varies markedly depending on the solvent and extraction technique employed. Among these, chlorogenic acid is particularly effective at inhibiting *Listeria* spp. and *Salmonella* spp. in enzyme- and ultrasound-assisted aqueous extractions, owing to its high abundance in the lyophilized extract (Martínez et al., 2023). Dominant flavonoids, notably quercetin and kaempferol, disrupt bacterial membrane permeability and contribute to the inhibition of *Bacillus subtilis, Bacillus cereus*, *Staphylococcus aureus, Salmonella typhimurium*, and *Listeria innocua* in ethanolic extracts, especially under ultrasound assistance (Gudiño et al., 2022; Le et al., 2020).

Furthermore, isothiocyanates, especially benzyl isothiocyanate derived from glucosinolates, exhibit synergistic biocidal activity against *S. aureus* and *Escherichia coli* when extracted with apolar solvents such as hexane or dichloromethane (Abdelhalim et al., 2023). Although aqueous maceration of fresh leaves and stems yields only moderate activity against *S. aureus* (Ramesh et al., 2023), this can be substantially enhanced by ultrasound-assisted extraction in combination with a deep eutectic solvent (UAE + ChCl–PG). This approach optimizes recovery of phenylpropanoids and isothiocyanates, outperforming conventional methods in inhibiting *S. aureus*, *E. coli*, and *Salmonella* spp. (Cao et al., 2023).

Overall, the phytochemical profile of broccoli leaves underlies their remarkable antimicrobial properties, underscoring their potential as natural agents for food preservation and therapeutic applications.

Cardiovascular and Hepatic Health Benefits

Bioactive compounds present in broccoli leaves, particularly flavonoids and isothiocyanates, are primarily responsible for the observed cardiovascular and hepatic health benefits associated with this vegetable. Park et al. (2019) demonstrated that the ethanolic extract of broccoli leaves inhibits TNF-α and IL-6 secretion in cells, suggesting a potential anti-inflammatory mechanism that may contribute both to vascular protection, by reducing endothelial activation and atheroma formation, and to the prevention of inflammation-associated liver injury. Moreover, bioactive peptides isolated from stem and leaf proteins of broccoli potently inhibit angiotensin-converting enzyme (ACE), suggesting a clear antihypertensive effect (Zhou et al., 2021). Likewise, Gudiño et al. (2024) demonstrated that this ACE-inhibitory activity is localized exclusively in broccoli leaf extracts and is attributable to specific phenolic compounds, thereby reinforcing the cardioprotective potential of this plant fraction by contributing to blood pressure reduction and improved endothelial function.

Regarding hepatic health, administration of broccoli leaf extract significantly reduced serum ALT, AST, and ALP levels in animal models of induced liver injury, normalized oxidative stress markers, and mitigated the inflammatory response, confirming its hepatoprotective potential (Khalil et al., 2022).

Impact on Gut Health and Disease Prevention

Broccoli leaves and their extracts exert a significant impact on gut health and the prevention of related diseases, owing to their capacity to modulate the intestinal microbiota and attenuate inflammation. Likewise, the high dietary fiber content of broccoli leaves functions as a prebiotic substrate, enhancing microbial diversity and reinforcing intestinal barrier integrity (Rivas et al., 2024a). Berndtsson et al. (2020) demonstrated that the combination of dietary fiber and phenolic acids present in broccoli leaves acts as a prebiotic substrate and releases antioxidants in a controlled manner along the intestinal tract, favoring the production of short-chain fatty acids and mucosal protection, making them a promising ingredient for the prevention of intestinal diseases.

Synergistic Effects of Bioactives

The phenolic compounds, sulforaphane, peptides, dietary fibers, vitamins, and minerals in broccoli leaves synergistically interact to promote health and prevent chronic diseases. Together, these metabolites confer broad-spectrum antimicrobial activity by disrupting microbial membranes and inhibiting key enzymes (Abdelhalim et al., 2023; Le et al., 2020), while simultaneously scavenging reactive oxygen species and modulating pro-inflammatory pathways (Cao et al., 2023). In the cardiovascular system, flavonoids and isothiocyanates act in concert to reduce LDL oxidation and downregulate endothelial adhesion molecules, thereby limiting vascular inflammation and the development of arterial lesions. In the liver, the combined antioxidant and anti-inflammatory effects of chlorogenic acid, sulforaphane, and related phenylpropanoids attenuate cytokine-mediated damage and preserve tissue integrity (Khalil et al., 2022; Martínez et al., 2023). Moreover, the fiber and peptide fractions function as prebiotics, enhancing populations of beneficial gut bacteria, promoting short-chain fatty acid production, and reinforcing mucosal barrier function (Rivas et al., 2024b). This integrated network of activities underscores the superior efficacy of whole-leaf extracts compared to isolated compounds, highlighting the substantial potential of this co-product.

The beneficial properties of broccoli leaves mentioned in the previous sections are summarized in Table 2.

Applications in Food, Cosmetic, and Pharmaceutical Industries

Figure 1 (Valorization of Broccoli Leaves within the Circular Bioeconomy) illustrates the multifunctional potential of bioactive compounds derived from broccoli leaves, which contribute to the extension of food shelf-life, enhancement of nutritional value, improvement of skin protection in cosmetic formulations, and development of pharmaceutical and nutraceutical products.

Use as a Functional Ingredient in Foods

Broccoli leaves are increasingly employed as functional ingredients in food products owing to their rich bioactive profile. Multiple studies have demonstrated that their incorporation into various foods enhances nutritional value, antioxidant capacity, and functional properties. Addition of broccoli leaf powder to cakes, noodles, and gluten-free biscuits significantly increases protein, dietary fiber, mineral content, and phenolic compounds without compromising technological or sensory attributes (Devi et al., 2022; Drabińska et al., 2022). Likewise, leaf extracts rich in phenolics and isothiocyanates act as natural preservatives by delaying lipid oxidation and microbial growth in products such as filled fresh pasta, salad dressings, and spreadable creams, as the lipid matrices facilitate phenolic bioavailability (Angiolillo et al., 2019; Kamiloglu et al., 2022; Lucera et al., 2018).

Broccoli leaves have also been incorporated into extruded snacks and biscuits; the addition of freeze-dried leaf powder delivers a high phenolic load, boosting antioxidant capacity while enriching the product with highly bioavailable fiber and protein (Ying et al., 2021). In functional beverages, fermented drinks fortified with broccoli leaf extracts or powders maintain stable probiotic populations and exhibit enhanced antioxidant profiles when processed using high-pressure or ultrasound

Table 2 Beneficial properties of broccoli leaves.

Plant Material (Information)	*Extract Method*	*Analysis Method*	*Compounds Associated*	*Major Results*	*References*
Antioxidant Activity					
Dried leaves and stems (Parthenon and Naxos)	Conventional Extraction: MeOH: Tris-HCl (50 mM, pH 7.5) (1:1; v/v), Ice bath in darkness (20 min, 4°C) Supercritical Fluid Extraction (SFE): CO_2 40°C, 443 bar, 7% eToh, 31 g/min of flow rate (68 min)	*In vitro*: ABTS, TEAC	Phenolic compounds, carotenoids (Lutein and β-carotene), chlorophyll, and tocopherols.	The Naxos cultivar exhibited the greatest antioxidant activity, measuring 338.69 ± 31.95 mg Trolox/g compared to 262.57 ± 40.57 mg Trolox/g in the Parthenon cultivar.	Borja-Martínez et al. (2020)
Dried Leaves (TSX 007, Monaco, BRO 2047, Parthenon, Summer purple)	EtOH (80% v/v) ultrasound bath (1h) 45°C	*In vitro*: DPPH, ABTS	Phenolic compounds (Kaempferol and Quercetin), chlorophyll a.	Broccoli leaves were found to have the highest antioxidant capacity of all the parts studied, with average values in the DPPH test of 567 mg Trolox/100 g extract and the ABTS test of 1.63 mg Trolox/100 g extract.	Gudiño et al. (2022)
Dried Leaves	EtOH (70%)	*In vitro*: DPPH, Reducing power Absorbance, ABTS	Phenolic Acids (Gallic Acid, Ferulic Acid, Caffeic Acid) and Flavonoids (Esculetin, Myricetin, Quercetin) and Vitamin C	The reducing power and ABTS assays demonstrated that leaves possessed superior antioxidant capacity compared to seeds.	Le et al. (2020)

Contd...

Table 2 *Contd...*

Plant Material (Information)	*Extract Method*	*Analysis Method*	*Compounds Associated*	*Major Results*	*References*
Pureed stems and leaves.	MeOH (80%); 50 min; 60°C	*In vitro*: ORAC	Phenolic compounds (Chlorogenic acid, Neochlorogenic acid, Quinic acid, Caffeic acid) and chlorophylls	The juice fractions exhibited a substantially higher ORAC value (441 µmol TE/g DW) compared to powdered broccoli stems and leaves (311 µmol TE/g DW) and powdered pomace (204 µmol TE/g DW).	Shi et al. (2019)
Pomace (stems and leaves)					
Juice (stems and leaves)					
Leaves (freeze-drying)	PBS/EDTA/PVP	*In vitro*: POD, APX, CAT, SOD, O^{2-}, H_2O_2, DPPH, and ABST 2	Flavonoids, Coumarins and Indoles	The superoxide anion (O_2^-) production rate (U/g) and hydrogen peroxide (H_2O_2) content (µmol/g) were substantially lower in stems than in leaves. Meanwhile, leaves showed greater radical scavenging capacity in the ABTS and DPPH assays (%).	Zhao et al. (2024)
Leaves and stems mix	Water (1:3 W7W); 100°C; 60 min EE: water (1:3 w/w), cellulose enzyme (0,01%) UAE: Water (1:3 w/w); 75°C; 60 min	*In vivo*: Electrical impedance (*Saccharomyces* and *Saccharomyces* subjected to oxidative stress (H_2O_2 1 mmol/L))	Phenolic compounds (Chlorogenic acid, Caffeoylquinic acid, and derivatives) and chlorophylls	The Saccharomyces strain has a growth time of 10.00 hours. When EE broccoli extract is added, the growth time is 8.25 hours, while with AE broccoli extract, the growth time is 8.30 hours, compared to UAE broccoli extract, which reduces the time to 9.36 hours.	Martínez et al. (2023)

Contd...

Table 2 *Contd...*

Plant Material (Information)	*Extract Method*	*Analysis Method*	*Compounds Associated*	*Major Results*	*References*
Fresh Leaves	Maceration: hexane, chloroform, water	*In vivo*: DPPH, FRAP	Phenolic compounds (flavonoids and phenolic hydroxy acids), alkanes and aliphatic alcohols (Nonacosane, Tetratriacontane and n-Tetracosanol-1)	At a concentration of 500 µg, the leaf extract exhibited 52 % activity in both the DPPH antioxidant assay and the FRAP reducing power assay.	Ramesh et al. (2023)
Fibre of dried Leaves	Supercritical Fluid Extraction (SFE): CO_2 40°C, 191 bar, 1h	*In vitro*: DPPH, ABTS	Phenolic compounds bound to the fiber matrix (phenolic acids and flavonoids)	The treatment increased the antioxidant activity of broccoli leaf fiber by 28% (DPPH) and 15% (ABTS) compared to the control.	Rivas et al. (2024a)
Anti-inflammatory Activity					
Leaves (fresh)	Maceration: hexane, chloroform, water	*In vitro*: Inhibition of albumin denaturation	Phenolic compounds (flavonoids and phenolic hydroxy acids), chlorophyll, alkanes, and aliphatic alcohols (Nonacosane, Tetratriacontane, and n-Tetracosanol-1)	Leaf extracts inhibited albumin denaturation by 42%, exceeding the 37% inhibition observed with the aspirin standard.	Ramesh et al. (2023)
		In vitro: Antiproteinase action		At 500 µg/mL, the leaf extracts achieved maximum inhibitions of 71%, surpassing the 62% inhibition observed with the aspirin standard.	
		In vitro: Membrane stabilization		At 500 µg/mL, the leaf sample achieved approximately 58% inhibition.	

Contd...

Table 2 *Contd...*

Plant Material (Information)	*Extract Method*	*Analysis Method*	*Compounds Associated*	*Major Results*	*References*
Leaves	MeOH (70%)	*In vivo* (30 male rats):_RT-PCR: TNF-α, IL-1β, NF-kB	Phenolic compounds (phenolic acids): caffeic acid, gallic acid, and methyl gallate	Broccoli leaf extract treatment significantly reduced TNF-α, IL-1β, and NF-κB expression in both renal and hepatic tissues.	Khalil et al. (2022)
Dried Leaves	EtOH (72h)	*In vitro:* ELISA: PGE_2 and cytokines (TNF-α, IL-6)	Phenolic compounds (flavonoids and tannins) and vitamin C	Treatment with the extract at 50 and 100 µg/mL reduced TNF-α secretion by 12% and 22%, respectively, and IL-6 secretion by 28% and 54%, respectively.	Park et al. (2019)
Anti Cancer Activity					
Dried Leaves	EtOH (70%)	*In vitro* Cytotoxic activity: A549 (lung carcinoma cells), Caco-2 (colorectal adenocarcinoma cells), and HepG2 (hepatocellular carcinoma cells)	Phenolic compounds (Gallic acid, Esculetin, Ferulic acid, Myricetin, Quercetin)	Leave extracts showed potent cytotoxic activity against the A549, Caco-2, and HepG2 cancer cell lines, with IC50 values of 0.191, 0.254, and 0.267 mg/ml, respectively.	Le et al. (2020)

Contd...

Table 2 *Contd...*

Plant Material (Information)	*Extract Method*	*Analysis Method*	*Compounds Associated*	*Major Results*	*References*
Broccoli residues	MeOH	*In vitro* Cytotoxic activity: Cancer cell lines liver HepG2, breast MCF-7, and colon HCT116	Phenolic compounds (catechin, rutin, p-coumaric acid, benzoic acid, luteolin)	The methanol extract of broccoli waste showed strong growth-inhibiting effects on HepG2 liver cancer, MCF-7 breast cancer, and HCT-116 colon cancer cell lines.	Gaafar et al. (2020)
	Water				
Anti microbial Activity					
Fresh leaves	Hexane extracts	*In vitro* (agar-well diffusion method): *B. subtilis, S. aureus, E. coli, K. pneumoniae, H. pylori, C. albicans,* and *A. niger*	Isothiocyanates, fatty acid esters, oxygenated hydrocarbons, and long-chain linear hydrocarbons.	Both extracts showed their most potent antibacterial effects against *Staphylococcus aureus* and *Escherichia coli*, although the DCM extract was significantly more potent (MIC of 1.38 and 1.36 μg/mL versus 31.125 and 62.5 μg/mL, respectively). Both extracts showed moderate activity against *Bacillus subtilis*, *Klebsiella pneumoniae*, and *Candida albicans*. Neither extract was effective against *Aspergillus niger*, with low activity indices of 0.57 (hexane) and 0.66 (DCM).	Abdelhalim et al. (2023)
	Maceration in DCM 7 days				

Contd...

Table 2 *Contd...*

Plant Material (Information)	*Extract Method*	*Analysis Method*	*Compounds Associated*	*Major Results*	*References*
Dried Leaves (TSX 007, Monaco, BRO 2047, Parthenon, Summer purple)	EtOH (80% v/v) ultrasound bath (1h) 45°C	*In vitro*: *B. cereus, S. aureus,* and *L. innocua*	Fatty acids (palmitic acid) and oxidized fatty acids (FA 18:2+3O)	Broccoli leaf extracts significantly inhibited the growth of *Bacillus cereus* (57.5%), *Staphylococcus aureus* (77.8%), and *Listeria innocua* (69.5%)	Gudiño et al. (2022)
Dried Leaves	EtOH (70%)	*In vitro* (well-agar diffusion and MIC): *S. aureus,* Bacillus *subtilis, E. coli,* and *Salmonella typhimurium*	Phenolic compounds (Quercetin, Esculetin, Gallic acid, Ferulic acid), glucosinolates, and organic acids	The leaves showed considerable antibacterial efficacy against Bacillus subtilis and *Salmonella typhimurium* (0.78 mg/ml), along with weak activity against *Staphylococcus aureus* and *Escherichia coli* (3.13 mg/ml).	Le et al. (2020)
Broccoli residues	Water (1:3 W7W); 100°C; 60 min EE: Water (1:3 w/w), cellulose enzyme (0,01%) UAE: Water (1:3 w/w); 75°C; 60 min	*In vivo* (electrical impedance): *Salmonella Enterica, Listeria*	Phenolic compounds (Chlorogenic acid)	The addition of broccoli extract obtained by EE completely inhibits the growth of the *Listeria strain.*	Martínez et al. (2023)

Contd...

Table 2 *Contd...*

Plant Material (Information)	*Extract Method*	*Analysis Method*	*Compounds Associated*	*Major Results*	*References*
Fresh Leaves	Maceration: Water	*In vitro* (agar well diffusion and MIC): *Staphylococcus aureus*	Alkanes, aliphatic alcohols (Nonacosane, Tetratriacontane, and n-Tetracosanol-1), and Polyphenols (gallic acid).	The leaf extract exhibited an MIC of 0.5 mg/mL and an MBC of 0.5 mg/mL.	Ramesh et al. (2023)
Dried Leaves (50 °C, 20 h)	UAE + ChCl-PG (36.35 mL/g, 49.5 °C, 31.4 min, 383 W)	*In vitro* (well-agar diffusion and MIC): *S. aureus*, *E. coli, Salmonella* spp.	Total phenols (neochlorogenic, ferulic, erucic, quinic, chlorogenic, caffeic acid)	The extract exhibited the strongest antimicrobial activity against *S. aureus, E. coli*, and *Salmonella*, with MIC values of 0.125, 0.25, and 0.5 mg/mL, respectively.	Cao et al. (2023)
Cardiovascular and Hepatic Health Benefits					
Dried Leaves	EtOH (72h)	*In vitro* ELISA: PGE_2 and cytokines (TNF-α, IL-6)	Phenolic compounds (flavonoids and tannins) and vitamin C	At concentrations of 50 and 100 µg/mL, the extract dose-dependently inhibited TNF-α (by 12 % and 22 %) and IL-6 (by 28 % and 54 %) and reduced PGE_2 release.	Park et al. (2019)
Fresh Leaves	MeOH (70%)	*In vivo* (30 male rats) RT-PCR: TNF-α, IL-1β, NF-kB	Phenolic compounds (phenolic acids): caffeic acid, gallic acid, and methyl gallate	Treatment with broccoli leaf extract resulted in a significant reduction of TNF-α, IL-1β, and NF-kB expression in renal and hepatic tissues.	Khalil et al. (2022)

Contd...

Table 2 *Contd...*

Plant Material (Information)	*Extract Method*	*Analysis Method*	*Compounds Associated*	*Major Results*	*References*
Peptides of stem and leaf proteins	Enzymatic Hydrolysis (trypsin)	*In vitro*: ACE and DPP-IV Inhibition Assay	Peptides: LPGVLPVA and YLYSPAY	Two peptides were identified: LPGVLPVA, with IC_{50} values of 0.776 μM for ACE inhibition and 392 μM for DPP-IV inhibition, and YLYSPAY, with IC_{50} values of 8.52 μM for ACE and 181 μM for DPP-IV.	Zhou et al. (2021)
Dried Leaves (TSX 007, Monaco, BRO 2047, Parthenon, Summer Purple)	EtOH (80% v/v) ultrasound bath (1h) 45°C	*In vitro*: ACE-inhibitory activity assay	Phenolic compounds (Kaempferol and Quercetin)	Only the extracts from broccoli leaves exhibited antihypertensive activity, with the Parthenon variety displaying the greatest effect.	Gudiño et al. (2022)

Contd...

Table 2 *Contd...*

Plant Material (Information)	*Extract Method*	*Analysis Method*	*Compounds Associated*	*Major Results*	*References*
Impact on Gut Health and Disease Prevention					
Fibre of dried leaves	AIR method	*In vitro*: Human Simulated Digestion	Fiber and phenolic compounds, and oligosaccharides associated with them	The dietary fiber from broccoli leaves retains and protects its antioxidant phenolic compounds throughout digestion, exhibits antiproliferative activity against colon cancer cells, and releases fermentable oligosaccharides that stimulate Bifidobacterium growth and short-chain fatty acid production.	Rivas et al. (2024b)

***Notes*:** TEAC: ascorbic acid equivale; ORAC: oxygen radical absorbance capacity; FRAP: ferric reducing antioxidant potential; DW: dry weight; TE: trolox equivalents; POD: peroxidase; APX: ascorbate peroxidase; CAT: catalase; SOD: superoxide dismutase; EE: enzymatic extraction; AE: aqueous extraction; UAE: ultrasonic-assisted extraction; DCM: dichloromethane; MIC: minimum inhibitory concentration; RT-PCR: Real-Time Reverse Transcription–Polymerase Chain Reactions; ACE: angiotensin I-converting enzyme; TNF-α: tumor necrosis factor alpha: IL-1β: interleukin-1β; NFkB: nuclear factor kappa B; DPP-VI: dipeptidyl peptidase-IV; AIR: alcohol-insoluble residues.

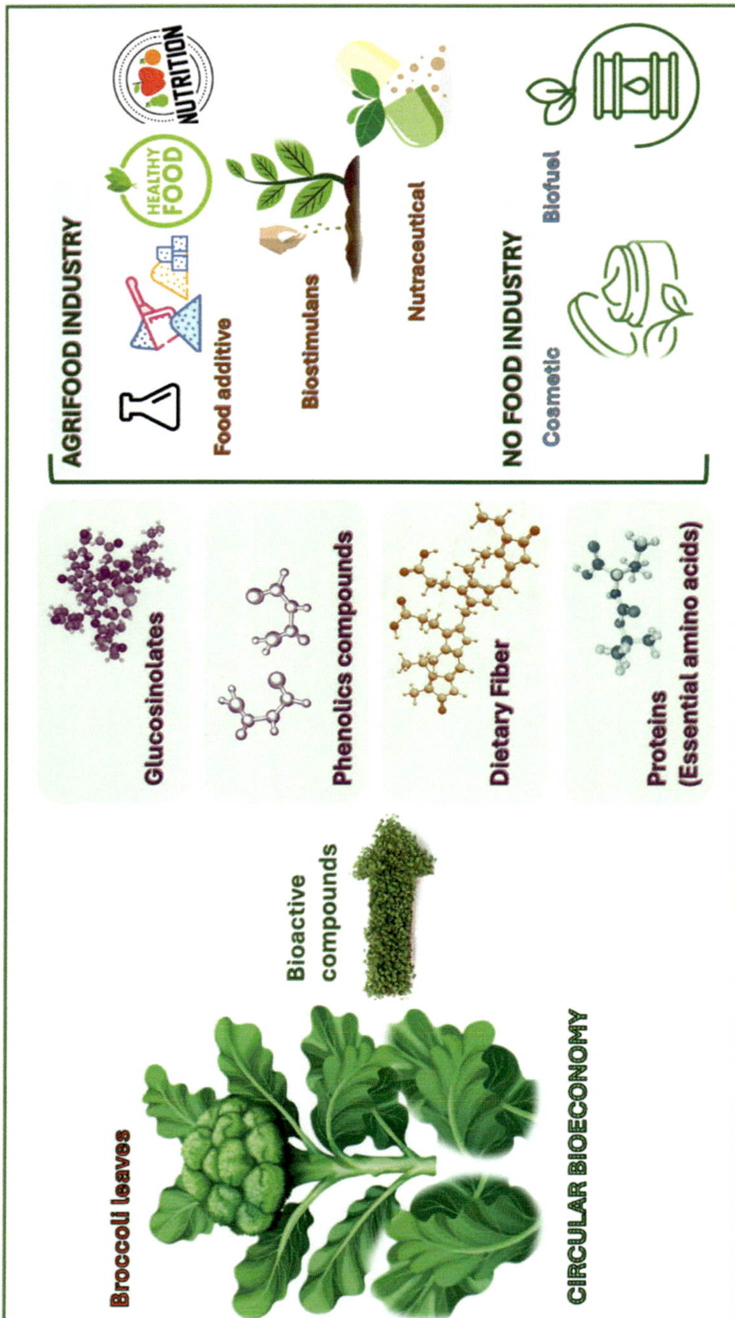

Fig. 1 Outline of the main compounds present in broccoli leaves and their applications in the food industry and beyond.

technologies (Pérez et al., 2024; Salas-Millán et al., 2024). Microencapsulation of leaf extracts yields powders with high flavonoid content that withstand thermal and industrial processing, facilitating their incorporation into matrices such as fish burgers and other processed foods, thereby enriching the product and extending shelf-life (Borja-Martínez et al., 2020).

Incorporation in Cosmetics (Antioxidant and Anti-aging Products)

In the cosmetic industry, the antioxidant and anti-inflammatory properties of broccoli bioactives make them well-suited for inclusion in skincare formulations aimed at reducing skin aging and inflammation. These bioactive compounds, particularly sulforaphane, hold promise for the development of anti-aging and protective skin products.

Broccoli leaf extracts are rich in quercetin, kaempferol, and chlorogenic acid, which reduce oxidative stress in keratinocytes exposed to UV radiation and enhance skin barrier function (Borja-Martínez et al., 2020). In a separate study, Borja-Martínez et al. (2024) also report the incorporation of leaf extracts into creams and gels, highlighting their ability to stabilize cellular membranes and modulate inflammatory mediators, thereby contributing to reduced free radical formation in the skin.

Therapeutic Potential and Nutraceutical Development

Broccoli leaves possess considerable therapeutic potential owing to their rich content of bioactive compounds, which exhibit pronounced anticancer, anti-inflammatory, antioxidant, and antimicrobial activities. Investigations into their ability to inhibit tumor cell proliferation, induce apoptosis, and modulate pathways involved in oxidative stress and inflammation have positioned broccoli leaves as a promising candidate for cancer prevention and therapy (Le et al., 2020).

Furthermore, these same leaf extracts can be leveraged in the development of nutraceuticals, formulated as dietary supplements, functional foods, and beverages aimed at promoting health and mitigating disease risk. Such nutraceuticals have demonstrated efficacy in animal models of metabolic disorders, including diabetes and obesity (Ranaweera et al., 2022), and show potential for the management of cardiovascular and inflammatory diseases. In particular, broccoli leafless, derived formulations have been investigated for their antihypertensive effects (Park et al., 2019; Salas-Millán et al., 2024).

Sustainability and Waste Reduction Perspectives

The valorization of by-products from the broccoli industry aims to meet sustainability and waste-reduction targets within the agricultural and food sectors. Since the portion destined for consumption represents only about 15% of the total broccoli plant, a substantial volume of by-products (stems, leaves, and non-edible plant material) is generated and typically discarded, exacerbating environmental waste.

Upcycling broccoli leaves prevents large volumes of biomass from entering compost or landfill streams, thereby promoting a circular-economy model and more efficient resource use. Recent research emphasizes converting these leaves into value-added products enriched in bioactive compounds such as glucosinolates, phenolics, and dietary fibers, that can replace synthetic additives and serve as functional ingredients in food, cosmetic, and pharmaceutical applications. The application of more selective and efficient extraction techniques preserves the structural integrity of these bioactives and concurrently reduces the environmental footprint associated with waste disposal (Gudiño et al., 2024). Broccoli-leaf extracts have also been applied as biostimulants, enhancing germination and seedling vigor of pak choi (Albaladejo-Marico et al., 2025) and thereby extending their use to sustainable agricultural practices. Overall, this strategy optimizes resource efficiency, mitigates pollution, and generates economic benefits by creating novel market opportunities for broccoli by-products.

Challenges and Considerations

Despite the promising potential of using broccoli leaves to obtain bioactive compounds and sustainable applications, several challenges remain.

Optimization of Extraction Processes

The optimization of extraction methods for bioactive compounds from broccoli leaves is crucial to maximize yield and preserve compound bioactivity, as well as to ensure industrial feasibility and process sustainability. Many of the phytochemicals present in the leaves are highly sensitive to both pre processing and extraction conditions, and may degrade if parameters such as solvent type, temperature, and duration are not precisely controlled (Ares et al., 2015; Le et al., 2020). Techniques such as UAE and SFE can enhance cell disruption and maintain the structural integrity of the bioactives; however, the efficiency of these methods depends on specific variables, such as pressure, ultrasound frequency, and co-solvent selection, that must be tailored to the particular plant matrix (Borja-Martínez et al., 2020; Martínez et al., 2023). Moreover, the widespread use of organic solvents like ethanol and methanol poses environmental and safety concerns, driving the search for greener or solvent-free alternatives.

The multitude of parameters to be managed throughout the extraction process also complicates scale-up: laboratory-optimal conditions do not always translate directly to industrial settings, where equipment constraints, higher energy demands, and elevated costs arise. Therefore, it is essential to develop extraction protocols that deliver high efficiency while meeting economic and environmental criteria, thereby guaranteeing the commercial viability of broccoli by-products.

Stability and Preservation of Bioactive Compounds

The preservation of bioactive compounds in broccoli by products poses significant challenges for their effective use in food and health applications. Many of these compounds are vulnerable to heat, light, pH fluctuations, and oxygen, leading to degradation and loss of bioactivity during processing, storage, and incorporation into products. Thermal treatments such as cooking or dehydration degrade sulforaphane and other glucosinolate derivatives (Costa-Pérez et al., 2024); oxidative reactions diminish the antioxidant capacity of phenolic compounds (Mohamed et al., 2023); and hydrolysis or polymerization under unfavorable pH conditions can reduce compound bioavailability (Zhang et al., 2022). Moreover, light exposure accelerates oxidative degradation, compromising extract functionality (Mohammadi et al., 2020).

Techniques such as encapsulation, stabilized formulations, and protective packaging have been employed to enhance stability and enable controlled release of bioactives (Marinelli et al., 2020; Saavedra-Leos et al., 2021). However, these methods increase both cost and process complexity, as their efficacy varies with the specific compound and final matrix. Additionally, during storage it is essential to optimize and control parameters such as temperature, humidity, and light exposure to preserve bioactive potency (Sánchez-Bravo et al., 2023).

Ultimately, ensuring the stability of these compounds requires advanced conservation strategies capable of maintaining bioactivity throughout processing, storage, and application in functional foods.

Scalability and Commercial Viability

The extraction of bioactive compounds from broccoli leaves must be both scalable and industrially viable, taking into account several critical factors. First, the selected laboratory-scale extraction technique (ultrasound-assisted, supercritical fluid, solvent-based) must be adapted to industrial facilities. This scale-up requires investment in specialized equipment, process optimization, and assurance of consistent quality and yield. Moreover, process profitability is influenced by raw material availability, energy and processing costs (solvent recovery, waste management), and regulatory compliance. The variability in bioactive content due to cultivar, harvest time, and postharvest handling introduces additional challenges to final product consistency (Gudiño et al., 2024).

On the other hand, integrating the resulting extracts into existing production lines or developing new product formats necessitates R&D to ensure stability, sensory quality, and regulatory approval, as well as to assess consumer acceptance and market demand. Finally, employing eco-friendly extraction methods and minimizing waste not only enhances sustainability and reduces costs but also strengthens public perception.

Regulatory and Safety Aspects

Extracts intended for inclusion in foods or dietary supplements must comply with regulatory standards set by agencies such as the U.S. Food and Drug Administration (FDA) and the European Food Safety Authority (EFSA), which require rigorous toxicological evaluations to ensure the absence of contaminants, heavy metals, microbial pathogens, and residual solvents.

The presence of certain bioactive compounds necessitates the establishment of acceptable daily intake levels and safety margins, as excessive doses may be toxic (EFSA, 2023). Moreover, variability in bioactive content arising from cultivar, harvest time, and extraction methods complicates batch-to-batch uniformity; therefore, standardization of product quality and the application of reliable analytical markers are essential.

Finally, the integration of Good Manufacturing Practices (GMP) and quality assurance systems, adherence to packaging and labeling regulations, and substantiation of health claims with robust scientific evidence are imperative to safeguard consumers and facilitate market access.

Conclusions and Future Directions

Broccoli cultivation by products, particularly leaves, possess substantial nutritional and bioactive potential due to their high content of phenolic compounds (kaempferol and quercetin), glucosinolates and their isothiocyanate derivatives, as well as vitamins, minerals, proteins, and dietary fiber. Owing to this compositional profile, broccoli leaves may be incorporated into the diet or formulated into functional foods, nutraceuticals, or natural preservatives. Valorization of these foliar by-products, ordinarily discarded post-harvest, not only generates additional economic value but also contributes to sustainability by reducing agricultural waste.

In the future, optimizing extraction and downstream processing techniques (such as lyophilization or encapsulation) will be necessary to maximize the advantages of broccoli leaf extracts by improving the stability and retention of bioactive components. Likewise, to guarantee batch-to-batch consistency, product quality, and safety requirements for regulatory approval and consumer acceptance, standardization of extraction protocols, and analytical methodologies will be essential.

Future research should investigate the therapeutic and preventive health effects of broccoli leaf extracts, including their anti-inflammatory, antioxidant, and antimicrobial activities, as well as their bioavailability and *in vivo* efficacy. Additionally, studies on the interactions of these bioactives within various food matrices and their stability under processing conditions will inform the development of novel functional food products.

Overall, the promising bioactive profile of broccoli leaves warrants continued investigation to overcome technical and regulatory hurdles, with the ultimate goal of transforming this underutilized biomass into a sustainable source of health-promoting ingredients and functional products.

References

Abdelhalim, M.A., Al-Sayed, E., Shams, K.A. and Eldahshan, O.A. 2023. GC/MS analysis and antimicrobial activities of different extracts of Egyptian sprouting broccoli leaves (*Brassica oleracea* L. var. *italica*) family *Brassicaceae*. *Arch. Pharm. Sci. Ain Shams Univ.*, *7*: 31–40.

Albaladejo-Marico, L., Gómez-Molina, M., García-Ibañez, P., Carvajal, M. and Yepes-Molina, L. 2025. Valorization of broccoli by-products: Seasonal variations in bioactive compounds and their biostimulant effects on pak choi germination. *PLoS One, 20*: e0323848.

Angiolillo, L., Spinelli, S., Conte, A. and Del Nobile, M.A. 2019. Extract from broccoli by-products to increase fresh-filled pasta shelf life. *Foods, 8*: 621.

Ares, A.M., Bernal, J., Nozal, M.J., Turner, C. and Plaza, M. 2015. Fast determination of intact glucosinolates in broccoli leaf by pressurized liquid extraction and ultra-high-performance liquid chromatography coupled to quadrupole time-of-flight mass spectrometry. *Food Res. Int., 76*: 498–505.

Arnáiz, E., Bernal, J., Martín, M.T., Diego, J.C., Bernal, J.L. and Recio, L. 2016. Optimisation of the supercritical fluid extraction of antioxidants from broccoli leaves. *Food Anal., Methods, 9*: 2174–2181.

Arnáiz, E., Bernal, J., Martín, M.T., Nozal, M.J., Bernal, J.L. and Toribio, L. 2012. Supercritical fluid extraction of free amino acids from broccoli leaves. *J. Chromatogr. A, 1250*: 49–53.

Berndtsson, E., Andersson, R., Johansson, E. and Olsson, M.E. 2020. Side streams of broccoli leaves: A climate smart and healthy food ingredient. *Int. J. Environ. Res. Public Health, 17*: 2406.

Borja-Martínez, M., Lozano-Sánchez, J., Borrás-Linares, I., Pedreño, M.A. and Sabater-Jara, A.B. 2020. Revalorization of broccoli by products for cosmetic uses using supercritical fluid extraction. *Antioxidants, 9*: 1195.

Borja-Martínez, M., Pedreño, M.A. and Sabater-Jara, A.B. 2024. Broccoli Byproduct Extracts Attenuate the Expression of UVB-Induced Proinflammatory Cytokines in HaCaT Keratinocytes. *Antioxidants, 13*(12): 1479.

Cao, Y., Song, Z., Dong, C., Ni, W., Xin, K., Yu, Q. and Han, L. 2023. Green ultrasound-assisted natural deep eutectic solvent extraction of phenolic compounds from waste broccoli leaves: Optimization, identification, biological activity, and structural characterization. *LWT, 190*: 115407.

Costa-Pérez, A., Sánchez-Bravo, P., Medina, S., Domínguez-Perles, R. and García-Viguera, C. 2024. Bioaccessible organosulfur compounds in broccoli stalks modulate the inflammatory mediators involved in inflammatory bowel disease. *Int. J. Mol. Sci., 25*: 800.

Devi, M., Soekopitojo, S., Putri, A., Pratikto, H., Wibawa, A.P. and Hamid, M.A. 2022. The addition of dried broccoli (*Brassica oleracea* L.) leaves can improve the nutrition of noodles. *IOP Conf. Ser. Earth Environ. Sci., 1116*: 012024.

Drabińska, N. 2022. The evaluation of amino acid profiles in gluten-free mini sponge cakes fortified with broccoli by-product. *Separations, 9*: 81.

EFSA. 2023. EFSA (European Food Safety Authority) homepage. Accessed [2023]. https://www.efsa.europa.eu/en.

Ferreira, S.S., Monteiro, F., Passos, C.P., Silva, A.M.S., Wessel, D.F., Coimbra, M.A. and Cardoso, S.M. 2020. Blanching impact on pigments, glucosinolates, and phenolics of dehydrated broccoli by-products. *Food Res. Int., 132*: 109055.

Gaafar, A.A., Salama, Z.A., El-Baz and F.K. 2020. A comparative study on the active constituents, antioxidant capacity, and anti-cancer activity of cruciferous vegetable residues. *Baghdad Sci. J., 17*: 743–753.

Gudiño, I., Casquete, R., Martín, A., Wu, Y. and Benito, M.J. 2024. Comprehensive analysis of bioactive compounds, functional properties, and applications of broccoli by-products. *Foods, 13*: 3918.

Gudiño, I., Martín, A., Casquete, R., Prieto, M.H., Ayuso, M.C. and Córdoba, M.G. 2022. Evaluation of broccoli (*Brassica oleracea* var. *italica*) crop by products as sources of bioactive compounds. *Sci. Hortic., 304*: 111284.

Kamiloglu, S., Ozdal, T., Tomas, M. and Capanoglu, E. 2022. Oil matrix modulates the bioaccessibility of polyphenols: A study of salad dressing formulation with industrial broccoli by products and lemon juice. J. *Sci. Food Agric., 102*: 5368–5377.

Khalil, H.E., Abdelwahab, M.F., Emeka, P.M., Badger-Emeka, L.I., Abdel Hafez, S.M.N., AlYahya, K. A., Ahmed, A.S.F., Anter, A.F., Abdel-Wahab, N.M., Matsunami, K. and Ibrahim Selim, A.H. 2022. Chemical composition and valorization of broccoli leaf by products (*Brassica oleracea* L. var. *italica*) to ameliorate reno-hepatic toxicity induced by gentamicin in rats. *Appl. Sci., 12*: 6903.

Krupa-Kozak, U., Drabińska, N., Rosell, C.M., Fadda, C., Anders, A., Jeliński, T. and Ostaszyk, A. 2019. Broccoli leaf powder as an attractive by-product ingredient: Effect on batter behaviour, technological properties, and sensory quality of gluten-free mini sponge cake. *Int. J. Food Sci. Technol., 54*: 1121–1129.

Le, T.N., Sakulsataporn, N., Chiu, C.H. and Hsieh, P.C. 2020. Polyphenolic profile and varied bioactivities of processed Taiwanese grown broccoli: A comparative study of edible and non-edible parts. *Pharmaceuticals, 13*: 82.

Liu, M., Zhang, L., Ser, S.L., Cumming, J.R. and Ku, K.M. 2018. Comparative phytonutrient analysis of broccoli by products: The potentials for broccoli by product utilization. *Molecules, 23*: 900.

Lucera, A., Costa, C., Marinelli, V., Saccotelli, M.A., Del Nobile, M.A. and Conte, A. 2018. Fruit and vegetable by products to fortify spreadable cheese. *Antioxidants, 7*: 61.

Manríquez-Zúñiga, A.N., de la Torre, A.R., Valdés-Santiago, L., Hernández-Bustos, D.A., Cuéllar-Sojo, S., Hernández-Rayas, A., Perez-Vega, S. and Molina-Guerrero, C.E. 2023. Broccoli leaves (*Brassica oleracea* var. *italica*) as a source of bioactive compounds and chemical building blocks: Optimal extraction using dynamic maceration and life cycle assessment. *Sustainability, 15*: 16616.

Marinelli, V., Spinelli, S., Angiolillo, L., Del Nobile, M.A. and Conte, A.. 2020. Emerging techniques applied to by products for food fortification. *J. Food Sci. Technol., 57*: 905–914.

Martínez, D.Q., Lorca, F., Fernández, M.D. and García, P.P. 2023. Broccoli by-product extract as a functional ingredient: Food application. *Open Explor., 1*: 235–243.

Mohammadi, X., Deng, Y., Matinfar, G., Singh, A., Mandal, R. and Pratap-Singh, A. 2020. Impact of three different dehydration methods on nutritional values and sensory quality of dried broccoli, oranges, and carrots. *Foods, 9*: 1464.

Mohammed, A. and Mohammed, H.A. 2023. Beneficial role of broccoli and its active ingredient, sulforaphane, in the treatment of diabetes. *Phytomed. Plus, 3(2)*: 100431.

Park, S.J., An, I., Noh, G.P., Yoo, B.H. and Lee, J.R.. 2019. Inhibitory effect of broccoli leaf extract on PGE 2 production by NF-κB inhibition. *Korea J. Herbology, 34*(6): 117–124.

Pei, P., Tianyan, Fangming, D. and Zhixing, Q. 2019. Identification of glucosinolate in different varieties of broccoli leaves based on HPLC-Q-TOF-MS technology. *Food Mach., 35*: 42–47,119.

Pérez, P., Hashemi, S., Cano-Lamadrid, M., Martínez-Zamora, L., Gómez, P.A. and Artés-Hernández, F. 2024. An innovative beverage made from revalorized broccoli and carrot by products. *Acta Hortic., 1387*: 187–194.

Ramesh, A., Nandi, A., Devanathan, S., Jayaraman, L. and Sivaji, S. 2023. Anti-inflammatory, anti oxidative, and anti microbial activities of the phytochemicals isolated from various parts of broccoli wastes. *J. Adv. Biotechnol. Exp. Ther., 6*: 67– 83.

Ranaweera, S.S., Natraj, P., Rajan, P., Dayarathne, L.A., Mihindukulasooriya, S.P., Dinh, D.T.T., Jee, Y. and Han, C.H.. 2022. Anti-obesity effect of sulforaphane in broccoli leaf extract on 3T3-L1 adipocytes and ob/ob mice. *J. Nutr. Biochem., 100*: 108885.

Rivas, M. Á., Benito, M.J., Martín, A., Córdoba, M. de G., Ruíz-Moyano, S. and Casquete, R. 2022. Improve the functional properties of dietary fiber isolated from broccoli by-products by using different technologies. *Innov. Food Sci. Emerg. Technol., 80*: 103075.

Rivas, M.Á., Benito, M.J., Martín, A., de Guía Córdoba, M., Gizaw, Y. and Casquete, R. 2024a. Development of supercritical technology to obtain improved functional dietary fiber for the valorization of broccoli by-product. *J. Sci. Food Agric., 105*: 2203–2214.

Rivas, M.Á., Ruiz-Moyano, S., Vázquez-Hernández, M., Benito, M.J., Casquete, R., Córdoba, M. de G. and Martín, A.. 2024b. Impact of simulated human gastrointestinal digestion on the functional properties of dietary fibers obtained from broccoli leaves, grape stems, pomegranate, and tomato peels. *Foods, 13*: 2011.

Saavedra-Leos, M.Z., Leyva-Porras, C., Toxqui-Terán, A. and Espinosa-Solis, V. 2021. Physicochemical properties and antioxidant activity of spray-dry broccoli (*Brassica oleracea* var. *italica*) stalk and floret juice powders. *Molecules, 26*: 1973.

Salas-Millán, J.Á., Conesa-Bueno, A. and Aguayo, E. 2024. A novel antidiabetic lactofermented beverage from agro-industrial waste (broccoli leaves): Process optimisation, phytochemical characterisation, and shelf-life through thermal treatment and high hydrostatic pressure. *Food Biosci., 59*: 103999.

Sánchez-Bravo, P., Abellán, Á., Zapata, P.J., García-Viguera, C., Domínguez-Perles, R. and Giménez, M.J. 2023. Broccoli products supplemented beers provide a sustainable source of dietary sulforaphane. *Food Biosci., 51*: 102259.

Sedlar, T., Čakarević, J., Tomić, J. and Popović, L. 2021. Vegetable by-products as new sources of functional proteins. *Plant Foods Hum. Nutr., 76*: 31–36.

Shi, M., Hlaing, M.M., Ying, D.Y., Ye, J.H., Sanguansri, L. and Augustin, M.A. 2019. New food ingredients from broccoli by-products: Physical, chemical, and technological properties. *Int. J. Food Sci. Technol., 54*: 1423–1432.

Ying, D., Sanguansri, L., Cheng, L. and Augustin, M.A. 2021. Nutrient-dense shelf-stable vegetable powders and extruded snacks made from carrots and broccoli. *Foods, 10*: 2298.

Zhang, Y., Jiang, C., Huang, S., Sun, J., Song, X., Nishanbaev, S.Z., Benito, M.J. and Wu, Y.. 2022. Effects of polyphenols and glucosinolates in broccoli extract on human gut microorganisms based on simulation *in vitro*. *ACS Omega, 7*: 45096–45101.

Zhao, Y., Zhang, Y., Yang, H., Xu, Z., Li, Z., Zhang, Z., Zhang, W. and Deng, J. 2024. A comparative metabolomics analysis of phytochemicals and antioxidant activity between broccoli floret and byproducts (leaves and stalks). *Food Chem., 443*: 138517.

Zhou, T., Liu, Z., Pei, J., Pan, D., Gao, X., Dang, Y. and Zhao, Y.. 2021. Novel broccoli-derived peptides hydrolyzed by trypsin with dual-angiotensin I-converting enzymes and dipeptidyl peptidase-IV-inhibitory activities. *J. Agric. Food Chem., 69*: 10885–10892.

3 Nutritive Value, Bioactive Compounds, and Biological Activities of *Basella alba* L.

Suraj P. Devkar[1], Priyanka V. Patil[1], Sachin B. Patil[2], Makarand M. Aitawade[3], Damaji S. Bhise[4], Sagar A. Deshmukh[1*]

Introduction

Leafy vegetables are an irreplaceable part of the human diet for their low calories and rich dietary fiber, vitamins, minerals, and phytochemicals. Consumption of leafy vegetables leads to a reduced risk of chronic and cardiovascular diseases, diabetes, and certain cancers. Rich in micronutrients relative to caloric content, leafy vegetables provide a robust nutritional foundation, earning them the label of "protective foods" in global dietary frameworks (Dias, 2012; Kibr, 2022).

Basella L. is among the four accepted genera of the family Basellaceae, having twining or climbing herbaceous vines commonly known as vine or Malabar spinach or Indian spinach. *Basella* is a type genus of the family consisting of five species distributed to Madagascar, southeastern Africa and Asia, New Guinea, and the Indian subcontinent. *B. alba* L. (Syn- *Basella rubra* Roxb.) found in various native ranges of tropical Asian countries like Bangladesh, Borneo, Cambodia, India, Jawa, Malaya's, Sri Lanka, Jawa and Sumatra, Thailand, Vietnam. while it is also introduced in various other countries of Africa including Ethiopia, Ghana, Kenya, Nigeria, Tanzania, Uganda, Zambia. Asia includes the Andaman Islands, Southeast China, Japan, the Pacific Islands, the Americas, and some parts of Europe. (POWO, 2025). Nutritionally, *B. alba* is an important leafy vegetable due to its phytochemical and medicinal properties, like androgenic, immunomodulatory, antioxidants, antimicrobial, antidiabetic, anti-inflammatory, and anti-ulcer activity. The present book chapter, therefore, aims to provide a comprehensive overview of the taxonomy, distribution, nutritional composition, phytochemical profile, and pharmacological activities of *B. alba*, thereby highlighting its importance as both a food and medicinal plant.

[1] Department of Botany, The New College, Kolhapur 416 012 Maharashtra, India.
[2] Department of Botany, Anandibai Raorane Arts, Commerce and Science College, Vaibhavwadi, Sindhudurg 416 810 Maharashtra, India.
[3] Department of Botany, Prof. Dr. N. D. Patil Mahavidyalaya, Malkapur-Perid 415 101 Maharashtra, India.
[4] Department of Botany, Shri Shivaji Mahavidyalaya, Barshi 413 411 Maharashtra, India.
* Corresponding author: sageraea@newcollege.ac.in

DOI: 10.1201/9781003687832-3

Taxonomic Backgrounds

Basella alba L. (Basellaceae) is a twining vine to procumbent or erect herb up to 10 m long; stem slender, fleshy, succulent, glabrous, green or purplish. Leaves simple, alternate, fleshy, mucilaginous; petiole 5–9 cm long; lamina cordate or sometimes ovate to broadly elliptic, 3–15 x 2.5–12 cm, base cordate to acuminate or obtuse, apex usually acute or somewhat acuminate or obtuse, dark green or purplish. Inflorescence is an unbranched (or with few branches) spike, 1–20 cm long, initially with flowers close to each other, later gradually spreading. Flowers bracteate; bract triangular, acute, bracteolate; bracteole broad, elliptic, blunt, bisexual, sessile, unscented, cleistogamous or sometimes chasmogamous, pentamerous, white, pink, or red colored. Sepals ovate to elliptic, erect, connate at base. Petals ovate to elliptic, erect, connate at base, thick and juicy. Stamens 5, opposite petals; anthers pale yellow. Ovary globose with three styles and linear stigmas. Fruit is a subglobose pseudo-berry, completely and tightly enclosed by fleshy, persistent perianth, 6 mm in diameter, purplish black, juicy, 1-seeded. Seed rough with black and brown in color, globose (Cooke, 1902; Yadav and Sardesai, 2002). The genus *Basella* was named by Linnaeus, who derived it from Malayalam and linked it to "*Hortus Malabaricus*" in his synonymy for *B. rubra* (Cooke, 1902). The species is extensively cultivated and naturalized in tropical and subtropical regions worldwide (Eriksson, 2007), with chromosome counts reported as 2n = 44 (Bolaji et al., 2023) and 2n = 48 (Shaik et al., 2024). Taxonomically, *B. alba* has been debated for centuries due to its morphological variability, particularly in stem and leaf coloration. Linnaeus initially described *B. rubra* and *B. alba* as separate species based on stem color and leaf characters (*Sp. Pl.*, 1753), but Roxburgh (1832) and later workers treated *B. alba*, *B. rubra,* and *B. lucida* in synonymy, recognizing *B. alba* as the correct name (Sidwell, 1999). Recently, Deshmukh and Gaikwad (2020) studied Morphological, anatomical, palynological, and seed surface studies on both forms of *Basella*. Their study found that both forms, i.e., *B. alba* and *B. rubra* belong to the same species (Figure 1).

Uses

B. alba, commonly known as Malabar spinach, has been traditionally utilized in various systems of medicine, especially Ayurveda and traditional Chinese medicine, for its broad spectrum of health benefits. The plant leaves and stems are used as remedies for ailments, such as diabetes, dysentery, skin diseases, eye infection, anemia (Siriwatanametanon et al., 2010; Sushila et al., 2010). Its mucilaginous texture makes it a natural laxative, commonly utilized for gastrointestinal issues, and its extracts are valued for their anti-inflammatory, wound healing, and demulcent properties. In addition to these medicinal uses, *B. alba* has historically been incorporated as a leafy green in culinary traditions across Asia and Africa, and its red variety was even used as a dye in traditional Chinese cosmetics and sealing wax. Modern scientific studies continue to validate and expand upon these traditional uses. Pharmacological research reveals *B. alba* possesses potent antioxidant, antidiabetic, anticancer, antibacterial, and hepatoprotective activities, largely attributed to its rich array of phytochemicals, vitamins, and minerals (Deshmukh and Gaikwad, 2014). The plant is increasingly explored as a functional food supplement for its health-promoting properties, such as improving fertility by stimulating hormone production, supporting wound healing, and being used as a natural food colorant and thickening agent (Reddy et al., 2023). Detailed ethnobotanical applications of *B. alba* are summarized in Table 1.

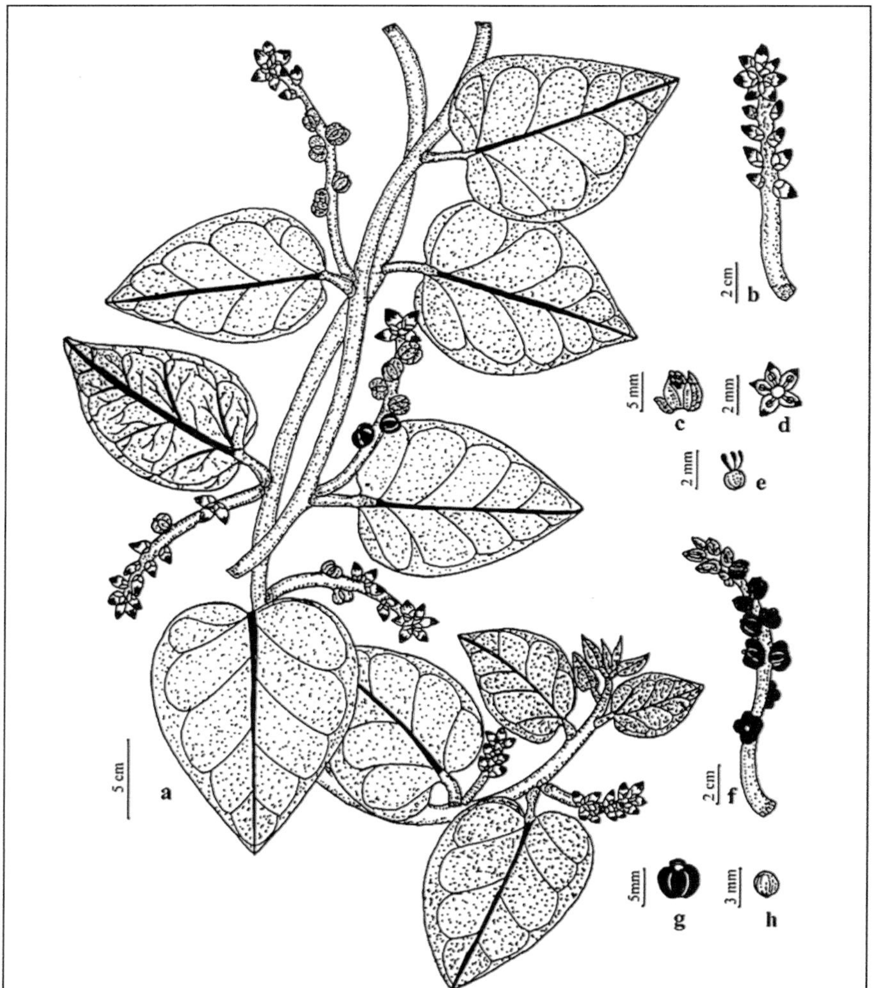

Fig. 1 Flowering twig of *Basella alba* L. a. Flowering twig b. Inflorescence c. Flower d. Androecium e. Gynoecium f. Fruiting twig g. Fruit h. Seed.

Table 1 Ethnobotany of *Basella alba* L.

Country	*Region/Province*	*Local Name*	*Plant part*	*Preparation Method*	*Traditional Use*	*Reference*
India	Assam	NA	Leaves	Juice	Treatment of urticaria	Anand et al. (2022)
	Kolli Hills Namakkal, TN	Pasalai keerai	Leaves	Juice	Drop for eye infection	Murugesan et al. (2011)
	Theni, TN	Pasalai keerai	Leaves	Juice	Externally applied to eye infection	Ignacimuthu et al. (2008); Siriwatanametanon et al. (2010)
	Dharwad, KA	Basale	Leaves	Chewed leaf kept in the mouth for a few minutes	Relief from aphthae (mouth ulcers)	Hebbar et al. (2004)
	Dantewada, CH	Poi	Leaves	NA	Treatment of constipation and gonorrhoea	Sahu et al. (2014)
	North Tripura	Puisaak	Leaves	NA	Externally applied to relief traumatic pain	Choudhury et al. (2014)
	Central Western Ghats, KA	NA	Leaves	NA	Used for the treatment of boils	Bhat et al. (2014)
	Dantewada, CH	Poi	Fruit	Juice	Treat conjunctivitis	Sahu et al. (2014)
	Dantewada, CH	Poi	Root	Decoction	Treat intestinal disorders	Sahu et al. (2014)
Cameroon	Ndop Central Sub-division	Lobekwebefon	Whole plant	Maceration prepared and taken orally	Treatment of infertility, threatened abortion, and spurious labor (false labor)	Focho et al. (2009)

Contd...

Table 1 *Contd...*

Country	*Region/Province*	*Local Name*	*Plant part*	*Preparation Method*	*Traditional Use*	*Reference*
	NA	NA	Leaves	Juice prepared from *B. alba* and *Hibiscus macranthus*	Improves male virility and cures male sexual asthenia	Siriwatanametanon et al. (2010)
	NA	NA	Leaves	Juice	Two spoons of leaf juice are taken to induce abortion, but it may cause vulval lacerations	Siriwatanametanon et al. (2010)
	West Cameroon	NA	NA	NA	Treat sexual asthenia and infertility in men	Nantia et al. (2011)
China	Mien, Nan Province	Dia Chun	Leaves/ Stem	Cooked with chicken	Orally taken as a lactagogue and as a postpartum tonic	Panyaphu et al. (2011)
	NA	NA	Leaves/ Stem	Decoction	Good laxative for pregnant women and children	Sushila et al. (2010)
	NA	NA	Leaves/ Aerial part	Decoction	Treatment of constipation, diuretic, toxicide, gonorrhoea, and balanitis	Sushila et al. (2010)
	NA	NA	Leaves/Young stalks	Mucilaginous liquid	Remedy for habitual headaches	Sushila et al. (2010)
	NA	NA	Flowers	NA	Antidote to poison, diuretic, and febrifuge	Sushila et al. (2010)
	NA	NA	Roots	Paste	Applied to swellings and is also used as a rubefacient	Sushila et al. (2010)
	NA	NA	Roots	NA	Treatment of diarrhea	Sushila et al. (2010)

Contd...

Table 1 *Contd...*

Country	*Region/Province*	*Local Name*	*Plant part*	*Preparation Method*	*Traditional Use*	*Reference*
Kenya	Embu Eastern Province	Murerema	Leaves	Boiled with water	Treatment of diarrhea	Kareru et al. (2007)
	Elgeyo-Marakwet County	Kiraita	Leaves	Boiled	Synergistic, abdominal upsets, joint pains, lumbago, anemia, blood cleanser	Kigen et al. (2017)
	The Nandi community in South Nandi district	Nderemiat	Roots/ Leaves	The plant part is boiled with water and taken internally	Aid in expelling the placenta after childbirth, relieve abdominal discomfort promotes lactation in nursing mothers	Jeruto et al. (2008)
Nepal	NA	NA	Whole plant	NA	Treatment of aphthae	Sushila et al. (2010)
	NA	NA	Leaf	Juice	Treat catarrh	Sushila et al. (2010)
	NA	NA	Leaf	Juice	Applied externally to treat boils	Sushila et al. (2010)
	NA	NA	Sap	NA	Apply to acne-affected areas to reduce irritation	Sushila et al. (2010)
Nigeria	Iwo, Ibadan of South-Western region	Amunututu	Leaf/ Root	Juice	Fresh leaves are washed, blended, and 250 ml is taken 3–4 times daily as a blood tonic	Afolayan et al. (2020)

Contd...

Table 1 *Contd...*

Country	*Region/Province*	*Local Name*	*Plant part*	*Preparation Method*	*Traditional Use*	*Reference*
Pakistan	NA	Saag	Shoot/Root	Decoction	Used for intestinal disorders	Adnan et al. (2015)
Rodrigues Island	NA	Brede gandole	Leaves	Decoction	Drink 1–2 cups to treat urinary tract infection	Samoisy & Mahomoodally (2016)
Tanzania	Kilimanjaro	Inyiri	Leaf	Juice	Drank as treatment for constipation and other ailments	Chacha & Laswai (2020)
Thailand	NA	NA	Aerial parts	NA	Eaten to relieve appendicitis symptoms, reduce smallpox fever, and act as a laxative	Siriwatanametanon et al. (2010)
	NA	NA	Leaves	NA	Crushed and applied to wounds, itching, and abscesses	Siriwatanametanon et al. (2010)
	Farmland, North-Eastern part	NA	Leaves	NA	Crushed with black gram and applied as a paste to boils	Siriwatanametanon et al. (2010)
	Farmland, North-Eastern part	NA	Leaves crushed, mixed with cheese	NA	Treats infected skin/ urticaria or skin burns	Siriwatanametanon et al. (2010)

Contd...

Table 1 *Contd...*

Country	*Region/Province*	*Local Name*	*Plant part*	*Preparation Method*	*Traditional Use*	*Reference*
	Farmland, North-Eastern part	NA	Leaves ground with sour buttermilk and salt	NA	Regular consumption of *Basella rubra* is believed in Ayurveda to alleviate 'Arbuda ', a condition described as a type of tumor.	Siriwatanametanon et al. (2010)
	Farmland, North-Eastern part	NA	Leaves masticated in the mouth	Mastication	Kept in the mouth for some time to relieve aphthae (mouth ulcers)	Siriwatanametanon et al. (2010)
	NA	NA	Flower	Juice	Used in the treatment of smallpox, fevers, and skin inflammations	Siriwatanametanon et al. (2010)
Uganda	Mpigi District	Enderema	Aerial parts/ Leaves	NA	Treatment of AIDS/ HIV	Nyamukuru et al. (2017)
	Northern sector of Kibale National Park	NA	Flower	Decoction	For measles, 500 mL(adults) or 250 mL (children) is taken orally three times daily for 3 days and also applied externally on the skin	Namukobe et al. (2011)
NA	NA	Upodikaa	Leaves	NA	Leaves used as a poultice for sores, urticaria, and gonorrhoea	Maurya & Seth (2014)

Notes: NA- Not Available, TN- Tamil Nadu, KA- Karnataka, CH- Chhattisgarh.

Nutritional Compositions

Basella alba L. (Malabar spinach) is a common green vegetable that has been valued for its high nutritional content (Dahanayaka et al., 2025). Lutein, zeaxanthin, gallic acid, caffeic acid, ferulic acid, β-carotene, antioxidants, phenols, dietary fiber, vitamins, and minerals include the majority of the bioactive compounds present in *B. alba*. The plant species have high levels of protein, vitamin A, vitamin C, calcium, iron, magnesium, potassium, sodium, zinc, copper, and selenium, and they also contain essential amino acids and flavones (Singh et al., 2018; Kumar et al., 2022). Table 2 below shows the proximate composition of *B. alba*. The proximate composition of *B. alba* indicates that it is a high-moisture, low-fat leafy vegetable with appreciable nutritional value. Fresh leaves contain 86.5–98.0 g/100 g moisture, confirming their succulence and suitability as a fresh green vegetable. The protein content ranges from 1.3–8.9 g/100 g, making it a moderate source of plant-based protein compared to other common leafy vegetables. Fat content is very low (0.3–0.5 g/100 g), while ash (0.7–3.3 g/100 g) reflects a good level of mineral elements. Carbohydrates are present in the range of 3.5–7.8 g/100 g and provide some energy value.

The plant is also notable for its crude fiber content (7.5–34.6 g/100 g), suggesting an important role in digestive health and cholesterol regulation. The energy contribution (19–64 kcal/100 g) is relatively low, consistent with its status as a low-calorie vegetable. Overall, the proximate composition demonstrates that *B. alba* is a nutrient-dense, fiber -rich, and mineral-rich leafy vegetable suitable for incorporation into balanced diets and useful in addressing nutritional deficiencies.

Table 2 Proximate composition of *Basella alba* L.

Components (g/100 g FW)			*References*
1	Moisture	86.5–98.0	Abdulrasheed-Adeleke et al. (2021); Singh et al. (2024); Adenegan-Alakinde and Oja (2025)
2	Protein	1.3–8.9	Abdulrasheed-Adeleke et al. (2021); Nur et al. (2023); Singh et al. (2024)
3	Fat	0.3–0.5	Abdulrasheed-Adeleke et al. (2021); Singh et al. (2024)
4	Ash	0.7–3.3	Abdulrasheed-Adeleke et al. (2021); Nur et al. (2023)
5	Carbohydrate	3.5–7.8	Abdulrasheed-Adeleke et al. (2021); Nur et al. (2023)
6	Crude fibre	7.5–34.6	Abdulrasheed-Adeleke et al. (2021); Nur et al. (2023); Adegoke and Ojo (2017)
7	Energy	19–64 kcal	Okouango et al. (2019); Singh et al. (2024)

Contd...

Table 2 *Contd...*

Minerals (mg/100 g FW)			
1	Potassium (K)	389–510	Nur et al. (2023); Singh et al. (2024)
2	Sodium (Na)	24–48	Nur et al. (2023); Singh et al. (2024)
3	Calcium (Ca)	71–109	Nur et al. (2023); Singh et al. (2024)
4	Magnesium Mg)	49–65	Nur et al. (2023); Singh et al. (2024)
5	Phosphorus (P)	13–25	Nur et al. (2023); Singh et al. (2024)
6	Iron (Fe)	1.2–29.1	Nur et al. (2023); Singh et al. (2024)
7	Zinc (Zn)	0.43–0.50	Nur et al. (2023); Singh et al. (2024)
8	Copper (Cu)	0.09	Nur et al. (2023)
Vitamins (mg/100 g FW)			
1	β-carotene	0.02–0.27	Abdulrasheed-Adeleke et al. (2021); Nur et al. (2023)
2	Vitamin C	12.8–95	Abdulrasheed-Adeleke et al. (2021); Nur et al. (2023)
Lipids and Fatty Acids			
1	Crude fat	4.4% of DW (~0.4 g/100 g FW)	Adesina and Adefemi (2017)
	Polyunsaturated fatty acids	72.5% of total fat	Adesina and Adefemi (2017)
3	Saturated fatty acids	26.0% of total fat	Adesina and Adefemi (2017)
4	Monounsaturated fatty acids	1.19% of total fat	Adesina and Adefemi (2017)
5	Sitosterol	196 mg/100 g	Adesina and Adefemi (2017)
6	Phospholipids	1679 mg/100 g	Adesina and Adefemi (2017)

Bioactive Compounds

B. alba L. contains a diverse range of phytochemicals with important nutritional and pharmacological activities, including antioxidant, anti-inflammatory, anti-proliferative, and anticancer effects. These are generally acylated betacyanin, amino acids, anthracene, betacyanin, carotenoids, flavonoids, phenolics, vitamins (Deshmukh and Gaikwad, 2014).

B. alba is a rich source of diverse bioactive compounds, distributed across various chemical classes with broad pharmacological implications. The most notable are acylated betacyanins, such as 3″-hydroxy-butyryl-gomphrenin and several cinnamoyl-derivatives, primarily isolated from the leaves. Alongside these, classic betacyanins (like gomphrenins I–III, isobetanin, betacyanin) occur in both leaf and fruit, bringing vibrant pigmentation and antioxidant activity (Koziol et al., 2024). The plant also harbors significant quantities of key amino acids like isoleucine, leucine, lysine, threonine, tryptophan, and arginine across the whole plant and essential fatty acids, specifically linoleic acid in the seed, underpinning its value as a nutritional and functional food (Deshmukh and Gaikwad, 2014; Singh et al., 2018).

Flavonoids, including kaempferol, acacetin, bioflavonoid rutin, and glycosides, such as betanin, betavulgaroside I, and multiple *Basella* saponins (A–D), are concentrated in leaves and aerial parts, providing antioxidant, anti-inflammatory, and antimicrobial properties (Deshmukh and Gaikwad, 2014; Singh et al., 2018). Leaves are also rich in carotenoids (β-carotene), vitamins C, E, and K, thiamine, niacin, and phenolic acids like ferulic, syringic, vanillin, and uronic acids, further boosting the plant's bioactivity profile (Deshmukh and Gaikwad, 2014; Singh et al., 2018). Additionally, triterpenoids, including basellin, isomalabarin, gandolin series, phytosterols like stigmasterol glucoside, β-sitosterol, lupeol, and anthraquinones are widely distributed through different organs, contributing to a broad range of health-promoting effects (Koziol et al., 2024). The detailed phytochemical isolated from the various plant parts of *B. alba* is summarized in Table 3 and Figure 2.

Table 3 Phytochemicals from Basella alba L.

Class	*Compound Name*	*Plant Parts*	*References*
Acylated betacyanin			
1	3″-Hydroxy-butyryl-gomphrenin	Leaf	Koziol et al. (2024)
2	6′-O-(3,4,5-Trimethoxy-E-cinnamoyl)-gomphrenin	Leaf	Koziol et al. (2024)
3	6′-O-(3,4,5-Trimethoxy-E-cinnamoyl)-isogomphrenin	Leaf	Koziol et al. (2024)
4	6′-O-(3,4-Dimethoxy-E-cinnamoyl)-gomphrenin	Leaf	Koziol et al. (2024)
5	6′-O-(3,4-Dimethoxy-E-cinnamoyl)-isogomphrenin	Leaf	Koziol et al. (2024)
Amino acids			
6	Isoleucine	Whole plant	Deshmukh & Gaikwad (2014); Singh et al. (2018)
7	Leucine	Whole plant	Deshmukh & Gaikwad (2014); Singh et al. (2018)
8	Lysine	Whole plant	Deshmukh & Gaikwad (2014); Singh et al. (2018)
9	Threonine	Whole plant	Deshmukh & Gaikwad (2014); Singh et al. (2018)
10	Tryptophan	Whole plant	Deshmukh & Gaikwad (2014); Singh et al. (2018)

Contd...

Table 3 *Contd...*

Class	*Compound Name*	*Plant Parts*	*References*
11	Arginine	Whole plant	Khare (2007);Deshmukh & Gaikwad (2014)
Anthracene			
12	Anthraquinone	Leaf and stem	Deshmukh & Gaikwad (2014); Singh et al. (2018)
Betacyanins			
13	6′-O-malonyl-gomphrenin	Leaf	Koziol et al. (2024)
14	Gomphrenin I	Fruit	Deshmukh & Gaikwad (2014); Singh et al. (2018)
15	Gomphrenin II	Fruit	Deshmukh & Gaikwad (2014); Singh et al. (2018)
16	Gomphrenin III	Fruit	Deshmukh & Gaikwad (2014); Singh et al. (2018)
17	Isobetanin	Leaf	Koziol et al. (2024)
18	Isobougainvillein-v	Leaf	Koziol et al. (2024)
19	Isogomphrenin	Leaf/ Fruit	Koziol et al. (2024)
Carotenoids			
20	β-Carotene	Leaf	Deshmukh & Gaikwad (2014); Singh et al. (2018)
Fatty acid			
21	Linoleic acid	Seed	Deshmukh & Gaikwad (2014); Singh et al. (2018)
Flavonoids			
22	4,7-Dihydroxy kaempferol	Leaf	Kumar et al. (2013); Deshmukh & Gaikwad (2014)
23	4’-Methoxyisovitexin	Leaf	Kumar et al. (2013); Deshmukh & Gaikwad (2014)
24	Acacetin	Leaf	Deshmukh & Gaikwad (2014); Singh et al. (2018)
25	Kaempferol	Leaf	Deshmukh & Gaikwad (2014); Singh et al. (2018)
Glycosides			
26	Betanin	Leaf	Koziol et al. (2024)
27	Betavulgaroside I	Whole plant	Deshmukh & Gaikwad (2014); Singh et al. (2018)
28	Bioflavonoid (Rutin)	Leaf	Deshmukh & Gaikwad (2014); Singh et al. (2018)

Contd...

Table 3 *Contd...*

Class	*Compound Name*	*Plant Parts*	*References*
29	Basella saponin A	Fresh aerial parts	Deshmukh & Gaikwad, 2014; Singh et al. 2018
30	Basella saponin B	Fresh aerial parts	Deshmukh & Gaikwad (2014); Singh et al. 2018
31	Basella saponin C	Fresh aerial parts	Deshmukh & Gaikwad (2014); Singh et al. (2018)
32	Basella saponin D	Fresh aerial parts	Deshmukh & Gaikwad (2014); Singh et al. (2018)
Phenolics			
33	Ferulic acid	Leaf	Deshmukh & Gaikwad (2014); Singh et al. (2018)
34	Syringic acid	Leaf	Deshmukh & Gaikwad (2014); Singh et al. (2018)
35	Uronic acid	Leaf mucilage	Deshmukh & Gaikwad (2014); Singh et al. (2018)
36	Vanillin	Leaf	Deshmukh and Gaikwad (2014); Singh et al. (2018)
Phytosterols			
37	Stigmasterol glucoside	Leaf	Deshmukh & Gaikwad (2014); Singh et al. (2018)
38	β-Sitosterol	Leaf/ Whole plant	Paul & Singha (2010); Sushila et al. (2010)
39	Lupeol	Whole plant	Sushila et al. (2010)
Pigments			
40	Gomphrenin	Leaf/ Fruit	Koziol et al. (2024)
41	Betacyanin	Fruit	Deshmukh & Gaikwad (2014); Singh et al. (2018)
42	Bougainvillein-v	Leaf	Koziol et al. (2024)
Triterpenoids			
43	Basellin	Leaf/ Fruit	Koziol et al. (2024)
44	Isobasellin	Leaf/ Fruit	Koziol et al. (2024)
45	Z-isomer of Basellin	Leaf/ Fruit	Koziol et al. (2024)
46	Z-isomer of Gandolin	Leaf/ Fruit	Koziol et al. (2024)
47	Gandolin	Leaf/ Fruit	Koziol et al. (2024)
48	Globosin	Leaf/ Fruit	Koziol et al. (2024)
49	Isogandolin	Leaf/ Fruit	Koziol et al. (2024)
50	Isoglobosin	Leaf/ Fruit	Koziol et al. (2024)

Contd...

Table 3 *Contd...*

Class	*Compound Name*	*Plant Parts*	*References*
51	Isomalabarin	Leaf/ Fruit	Koziol et al. (2024)
52	Malabarin	Leaf/ Fruit	Koziol et al. (2024)
53	Momordin II A	Leaf	Deshmukh & Gaikwad (2014); Singh et al. (2018)
54	Momordin II C	Leaf	Deshmukh & Gaikwad (2014); Singh et al. (2018)
Vitamins			
55	Niacin	Whole plant	Deshmukh & Gaikwad (2014); Singh et al. (2018)
56	Thiamine	Whole plant	Deshmukh & Gaikwad (2014); Singh et al. (2018)
57	Vitamin C	Aerial parts	Deshmukh & Gaikwad (2014); Singh et al. (2018)
58	Vitamin E	Leaf	Deshmukh & Gaikwad (2014); Singh et al. (2018)
59	Vitamin K	Leaf	Deshmukh & Gaikwad (2014); Singh et al. (2018)

Biological Activities

Androgenic/Andropogenic Activities

Andropogenic activity was evaluated from the aqueous extract of *B. alba* leaves in male albino Wistar rats. Positive results were observed after seven days of treatment with a dose of 0.108 g/kg body weight, showing enlargement of the seminal vesicles and testes, along with an increase in testosterone levels and prostatic enzyme activity; similarly methanolic extract (10 μg/mL) also showing a similar result where the extract directly increased testosterone (0.63 nanogram/10^6 cells) and Estradiol (88.67 picograms/10^6 cells) in Leydig cells shows both androgenic and estrogenic potential (Moundipa et al., 1999; Nantia et al., 2011). Further, methanolic extract from leaves showing increased serum testosterone and natural fertility in normal and flutamide, a nonsteroidal antiandrogen drug, exposed male Wistar rats, indicating testosterone boosting effect or androgenic activity (Nantia et al., 2011).

Immuno-modulatory Activities

B. alba was evaluated for its *in vitro* immuno-modulatory activity through splenocyte proliferation assays using spleen cells from female BALB/c mice (8–10 weeks old). The plant showed mild to moderate immuno-stimulatory effects, which correlated with its TFC 133.1 ± 26.2 mg QE/100 g FM and TPC 269.0 ± 3.1 mg GAE/100 g FM (Lin and Tang, 2007).

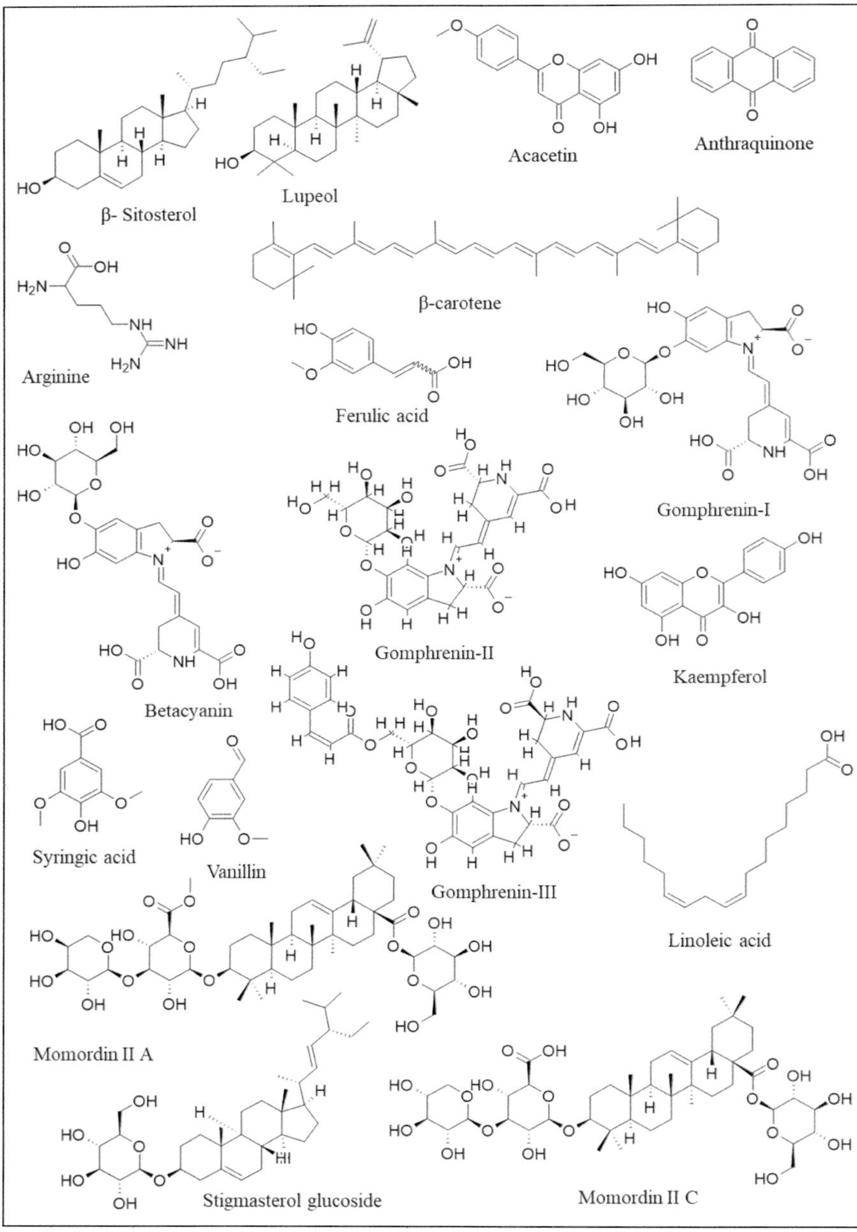

Fig. 2 Chemical structure of some phytochemicals from *Basella alba* L.

Antioxidant Activities

The antioxidant potential of *B. alba* and *B. rubra* was evaluated using multiple assays. In the DPPH free radical scavenging assay, with Trolox (a vitamin E analogue) as the reference standard, the ethyl acetate (EtAc) extract of *B. alba* exhibited moderate antioxidant activity, with an IC_{50} value of 5.32 mg/ml. In contrast, the EtAc extract of *B. rubra* showed weaker activity, with an IC_{50} of 34.58 mg/ml. In the lipid peroxidation inhibition assay, using Trolox and quercetin as positive controls, both species demonstrated low inhibitory activity, with IC_{50} values mostly exceeding 50 mg/ml. The only exception was the EtAc extract of *B. alba*, which showed slightly better activity (IC_{50} = 56.65 mg/ml), though still considered relatively weak. The total phenolic content (TPC), determined using the Folin-Ciocalteu reagent, was highest in the EtAc extract of *B. alba* (7.25 ± 0.76 mg/g), followed by the EtAc extract of *B. rubra* (6.17 ± 0.58 mg/g). A moderate correlation between higher phenolic content and stronger DPPH scavenging activity was observed, particularly in the ethyl acetate extracts. These findings indicate that both *B. alba* and *B. rubra* possess modest antioxidant potential, with the EtAc extract of *B. alba* showing the most promising bioactivity. The results support the role of phenolic compounds in the antioxidant mechanisms of these species and emphasize that the antioxidant efficacy is extract-dependent (Siriwatanametanon et al., 2010). The phenolic content in *B. alba* is considered a major contributor to its antioxidant potential. The aqueous leaf extract contained 6.57 ± 0.6 mg GAE/g of total phenolics, whereas the stem extract showed 2.61 ± 0.1 mg GAE/g. In the DPPH free radical scavenging assay, the leaf extract exhibited an IC_{50} value of 22.85 ± 1.9 μg GAE/mL, while the stem extract demonstrated higher activity with an IC_{50} of 12.61 ± 1.4 μg GAE/mL. In the reducing power assay, the leaf extract recorded an activity of 27.94 ± 2.2 units/mg GAE, and the stem extract showed 40.63 ± 3.8 units/mg GAE. Overall, *B. alba* exhibited moderate to poor antioxidant activity, with DPPH IC_{50} values ranging from 8.2 to 22.8 μg GAE/mL and reducing power values between 27–40 units/mg GAE (Belagihalli and Dharmesh, 2012).

Antimicrobial Activities

The antibacterial activity of *B. alba* has been extensively studied against various microbial strains, particularly Gram-positive and Gram-negative bacteria. Sushila et al. (2010) investigated the antibacterial potential of the whole plant methanolic extract of *B. alba* using a two-fold serial dilution method. The extract was tested against Gram-positive bacteria, including *Staphylococcus aureus*, *Bacillus subtilis,* and *Micrococcus luteus*, as well as Gram-negative strains, such as *Escherichia coli* and *Pseudomonas aeruginosa*. Ciprofloxacin (5 μg/ml) was used as the reference standard. Promising results were observed at higher concentrations of the extract (100 mg/ml), which exhibited significant inhibition zones with MIC 21.02 ± 0.1 mm for *B. subtilis* and *P. aeruginosa,* MIC 22.2 ± 0.1; while 16.3 ± 0.058 mm for *S. aureus*, indicating strong antibacterial activity, suggesting a strain-specific or dose-dependent response. These findings indicate that *B. alba* possesses broad-spectrum antibacterial activity with particularly strong effects against Gram-positive bacteria, such as *B. subtilis* and *S. aureus*, and notable activity against Gram-negative *P. aeruginosa*.

Structural polysaccharides (BRPs) isolated from the aqueous extract of *Basella rubra* aerial parts have demonstrated significant antiviral activities against Herpes Simplex Virus Type 2 (HSV-2), the causative agent of genital herpes, and Influenza virus (IFV) in both *in vitro* and *in vivo* mouse models. Among them, BRP-2 and BRP-4, classified as pectin-type polysaccharides, exhibited the most promising antiviral effects. These polysaccharides primarily act by interfering with the early stages of HSV-2 infection, particularly by blocking virus adsorption and penetration into host cells. *In vivo* studies further confirmed that intravaginal administration of BRP-4 significantly reduced herpetic lesion severity, virus shedding, and improved survival rates in HSV-2-infected mice. Moreover, oral administration of BRP-4 in IFV-infected mice showed moderate efficacy by reducing viral load in the lungs and bronchoalveolar lavage fluid (BALF) while also enhancing the host immune response through increased production of neutralizing antibodies and mucosal IgA levels. These findings suggest that BRP-4 in particular may serve as a potential plant-derived antiviral agent by both directly inhibiting viral entry and modulating mucosal immunity, and represent the first report of antiviral polysaccharides from *Basella rubra* (Dong et al., 2011). *Helicobacter pylori* is a major causal organism in gastric ulcers and gastric cancer. Aerial parts of *B. alba*, including the leaf and stem, have been evaluated for antibacterial activity against *H. pylori*. The extracts showed mild to moderate inhibitory effects, primarily through partial inhibition of the H^+, K^+-ATPase enzyme of *H. pylori*. While in comparative assays, the leaf extract of *B. alba* exhibited 24% inhibition of *H. pylori* growth, while the stem extract showed 19% inhibition, indicating relatively poor activity compared to other tested plants (Belagihalli and Dharmesh, 2012).

Antidiabetic Activities

The aqueous extract of *Basella rubra* (400 mg/kg body weight) significantly reduced blood glucose levels from 252.0 ± 4.16 to 110.1 ± 3.93 mg/dL in streptozotocin (STZ)-induced diabetic albino male rats. It also increased the levels of liver antioxidant enzymes, such as superoxide dismutase (SOD) from 6.08 ± 0.85 to 6.96 ± 0.99 U/mg protein, catalase (CAT) from 62.98 ± 2.96 to 73.90 ± 3.19 U/mg protein, and glutathione peroxidase (GPx) from 8.7 ± 0.25 to 9.1 ± 1.95 U/mg protein. Additionally, only minor changes were observed in non-enzymatic antioxidants, such as vitamin C, vitamin E, and reduced glutathione. The destruction of pancreatic β-cells by STZ leads to hyperglycemia in rats, and these results support the potential antidiabetic activity of the aqueous extract of *Basella rubra* (Nirmala et al., 2011).

Anti-inflammatory Activities

Anti-inflammatory activity was studied using methanolic, petroleum ether, and ethyl acetate extracts of the leaves of *B. alba* and *B. rubra* on human monocyte cells. A moderate inhibitory effect on Interleukin-6 (IL-6) release was observed, with Inhibitory Concentration 50 (IC_{50}) values ranging from 32–46 μg/ml for *B. alba* and 44–45 μg/ml for *B. rubra*. However, these extracts showed no significant inhibition of Prostaglandin E2 (PGE_2), IL-1β, and tumor necrosis factor (TNF-α), as the IC_{50} values exceeded 50 μg/ml, indicating lower potency. Similarly, no significant nuclear factor-Kappa B (NF-κB) inhibition was observed for either species. Overall, *B. alba* and *B. rubra* exhibited mild to moderate anti-inflammatory potential, particularly in suppressing IL-6; their effects on other cytokines and NF-κB were limited at the tested concentrations (Siriwatanametanon et al., 2010). In another study, the *in vitro* anti-inflammatory activity of *B. alba* leaf extracts was assessed using the Human Red Blood Cell (HRBC) membrane stabilization method, with Diclofenac sodium serving as the standard reference drug. Among the ethanolic, methanolic, and aqueous extracts tested, the aqueous extract exhibited the highest membrane-stabilizing effect at increasing concentration 200–400 μg/ml, indicating strong anti-inflammatory potential (58.46 ± 1.61% and 71.89 ± 1.22% protection, respectively) against hypotonicity-induced hemolysis. These values were significantly higher compared to the methanolic extract (57.10 ± 2.18% and 68.38 ± 1.18% at the same concentrations) and relatively close to the standard Diclofenac (70.17 ± 0.22% and 83.54 ± 0.61% protection at 50 and 100 μg/ml). The erythrocyte membrane model used in this study is considered analogous to the lysosomal membrane, suggesting that *B. alba* may exert its anti-inflammatory effects by stabilizing lysosomal membranes and preventing the release of proteolytic enzymes that contribute to tissue inflammation (Kumar et al., 2011). Moreover, Rodda et al. (2012) and Azad et al. (2013) investigated the aqueous leaf extract of *B. alba* for its anti-inflammatory activity in male Wistar albino rats (150–200 g) using two experimental models: carrageenan-induced paw edema and cotton pellet-induced granuloma. In the carrageenan-induced model, the extract at doses of 250 and 500 mg/kg exhibited dose-dependent activity, significantly reducing paw swelling compared to the control group, with a notable percentage inhibition of edema observed. Similarly, in the cotton pellet-induced inflammation model, used to assess chronic inflammation, the extract also demonstrated significant anti-inflammatory activity in a dose-dependent manner, as evidenced by a clear reduction in the weight of dry cotton pellets compared to the control.

Gastroprotective/Anti-ulcer Activities

The gastroprotective activity of *B. alba* leaf extracts was investigated using a pyloric ligation-induced ulcer model in rats. The study compared aqueous and ethanol extracts at two different doses (200 and 400 mg/kg) with famotidine (20 mg/kg) as a standard drug. In the control group, animals exhibited a high volume of gastric juice (9.3 ± 0.12 ml), low pH (2.6 ± 0.12), and elevated levels of free and total acidity (96.2 ± 1.24 and 102.3 ± 1.94 mEq/l, respectively). Treatment with famotidine significantly ($p < 0.01$) reduced gastric juice volume and acidity while increasing pH, confirming its antiulcer efficacy. Notably, both aqueous and ethanolic extracts demonstrated significant gastroprotective effects. Aqueous extract at 400 mg/kg showed greater efficacy than at 200 mg/kg with a significant reduction in gastric volume (4.2 ± 0.12 ml, $p < 0.01$), an increase in pH (6.7 ± 0.08, $p < 0.01$), and a substantial decrease in acidity parameters. Similarly, ethanolic extract at 400 mg/kg also reduced acid output and elevated pH compared to control (Kumar et al., 2012). Furthermore, the aqueous leaf extract of *B. alba* demonstrated moderate antiulcer activity through the inhibition of the H^+, K^+-ATPase enzyme (gastric proton pump), with an IC_{50} (29.08 ± 2.3 μg GAE/mL). The stem extract showed a comparatively lower activity (IC_{50} 45.11 ± 5.0 μg GAE/mL) (Belagihalli and Dharmesh, 2012).

Hepatoprotective Activities

The significant reduction in liver enzyme levels, alkaline phosphatase (ALP: 400.30 ± 0.63 to 308.58 ± 0.58), alanine aminotransferase (ALAT: 57.50 ± 0.44 to 39.70 ± 0.12), and aspartate aminotransferase (ASAT), which showed a slight increase at the lower dose (60 mg/kg) but a significant decrease at the higher dose (57.56 ± 0.36 to 41.08 ± 0.69) suggests that the aqueous leaf extract of *B. alba* supports normal liver function in a dose-dependent manner. These changes indicate the extract's potential hepatoprotective activity and therapeutic value in managing liver disorders in Wistar rats (Bamidele et al., 2010). The high dose of Paracetamol leads to liver damage or hepatotoxicity, Das et al., (2015) studied *B. alba* aqueous leaf extract against Serum bilirubin, SGOT (AST) and SGPT (ALT), Proteins and Alkaline phosphatase (ALP) with histopathology of male albino rats promising results were shown by extract, i.e., *B. alba* aqueous leaf extract (100 mg/kg/day) is strongly hepatoprotective against paracetamol-induced liver damage in rats in comparison with standard silymarin.

Cytotoxicity

The cytotoxic potential of the methanolic extract of *B. alba* whole plant was evaluated using the MTT (tetrazolium) assay against two human cancer cell lines, Jurkat (T-cell leukemia) and A549 (lung carcinoma). The results were compared with Paclitaxel (5 μM), a well-known standard anticancer drug. *B. alba* extract exhibited moderate, dose-dependent cytotoxicity with a more pronounced effect on Jurkat cells than A549 cells. As the concentration of the extract increased from 25 μg/mL to 100 μg/mL, cell viability decreased significantly, particularly in Jurkat cells, suggesting higher sensitivity of this cell line to the extract. In contrast, Paclitaxel demonstrated the strongest cytotoxic effect, reducing cell viability to 34.09% in A549 and 40.72% in Jurkat cells. Although less potent than

Paclitaxel, *B. alba* extract showed promising cytotoxic activity, especially against hematological cancer cells (Sushila et al., 2010).

Conclusion

Basella alba L. is both a nutrient-rich leafy vegetable and a valuable medicinal plant. Its proximate composition highlights high moisture, adequate protein, dietary fiber, essential fatty acids, minerals, and vitamins, confirming its role as a functional food that supports daily nutritional requirements. Beyond its nutritional importance, the plant possesses a wide range of bioactive compounds, including flavonoids, phenolics, betacyanins, triterpenoids, phytosterols, and saponins that are associated with significant pharmacological activities, such as antioxidant, antimicrobial, anti-inflammatory, antidiabetic, androgenic, and immunomodulatory effects. The dual role of *B. alba* as food and medicine strengthens its ethnobotanical and pharmacological relevance. Regular dietary inclusion may aid in disease prevention, health promotion, and nutritional security, particularly in regions where it is traditionally consumed. However, despite promising findings, comprehensive clinical studies and mechanistic investigations remain necessary to validate its therapeutic claims, standardize bioactive compounds, and explore its potential for pharmaceutical or nutraceutical applications. Thus, *B. alba* can be regarded as a multi purpose plant with significant contributions to both nutrition and medicine, warranting greater attention in modern food and health sciences.

References

Abdulrasheed-Adeleke, T., Shaibu, S.O.O., Agboola, A.R., Hamzah, R.U., and Garba, M.H. 2021. Comparative studies on the effect of processing methods on the nutritional value of *Basella alba*, *Talinum triangulare*, *Celosia argentea*, *Amaranthus hybridus*, and *Gnetum africanum* leaves. *GSC Adv. Res. Rev.*, *6*(03): 044–052. http://dx.doi.org/10.30574/gscarr.2021.6.3.0025.

Adegoke, G.O. and Ojo, O.A.. 2017. Phytochemical, antioxidant, and antimicrobial activities in the leaf, stem, and fruit fractions of *Basella alba* and *Basella rubra*. *Science*, *5*(5): 73–79. http://dx.doi.org/10.11648/j.plant.20170505.11.

Adenegan-Alakinde, T.A. and Ojo, F.M. 2025. Proximate, vitamin, and anti-nutritional properties of four forms of the genus *Basella* (Basellaceae) in South-western Nigeria. *Vegetos*, 1–9. http://dx.doi.org/10.1007/s42535-025-01305-y.

Adesina, A.J. and Adefemi, S.O. 2017. Lipid composition of *Basella alba* and *Basella rubra* leaves consumed in South-Western Nigeria: Nutritional implications. *Bangladesh J. Sci. Ind. Res.*, *52*(2): 125–134. http://dx.doi.org/10.3329/bjsir.v52i2.32935.

Adnan, M., Bibi, R., Azizullah, A., Andaleeb, R., Mussarat, S., Tariq, A., Ullah, R., Elsalam, N.M., Khan, A.L. and Begum, S.. 2015. Ethnomedicinal plants used against common digestive problems. *Afr. J. Tradit. Complement. Altern. Med.*, *12*(5): 99–117. http://dx.doi.org/10.21010/ajtcam.v12i5.15.

Afolayan, F.I., Sulaiman, K.A., Okunade, W.T. and Garrido, G. 2020. Ethnobotanical survey of plants used in cancer therapy in Iwo and Ibadan, South-Western Nigeria. *J. Pharm. Pharmacogn. Res.*, *8*(5): 346–367. http://dx.doi.org/10.56499/jppres20.798_8.5.346.

Anand, U., Tudu, C.K., Nandy, S., Sunita, K., Tripathi, V., Loake, G.J., Dey, A. and Prockow, J. 2022. Ethnodermatological use of medicinal plants in India: From ayurvedic formulations to clinical perspectives: A review. *J. Ethnopharmacol.*, *284*: 114744. https://doi.org/10.1016/j.jep.2021.114744.

Azad, A.K., Wan Azizi, W.S., Babar, Z.M., Labu, Z.K. and Zabin, S. 2013. An overview on phytochemical, anti-inflammatory, and anti-bacterial activity of *Basella alba* leaves extract. *Middle-East J. Sci. Res.*, *14*(5): 650–655. http://dx.doi.org/10.5829/idosi.mejsr.2013.14.5.71225.

Bamidele, O., Akinnuga, A.M., Olorunfemi, J.O., Odetola, O.A., Oparaji, C.K. and Ezeigbo, N. 2010. Effects of aqueous extract of *Basella alba* leaves on haematological and biochemical parameters in albino rats. *Afr. J. Biotechnol.*, *9*(41): 6952–6955.

Belagihalli, S.M. and Dharmesh, S.M. 2012. Anti-Helicobacter pylori, proton pump inhibitory, and antioxidant properties of selected dietary/medicinal plants. *Int. J. Phytomed.*, *4*(4): 573.

Bhat, P., Hegde, G.R., Hegde, G. and Mulgund, G.S. 2014. Ethnomedicinal plants to cure skin diseases: An account of the traditional knowledge in the coastal parts of Central Western Ghats, Karnataka, India. *J. Ethnopharmacol.*, *151*(1): 493–502. https://doi.org/10.1016/j.jep.2013.10.062.

Bolaji, A. O., Oladejo, A.S. and Adeniran, O. 2023. Morphological characterization and cytological studies of the green-stemmed and the red-stemmed *Basella alba*. *Not. Sci. Biol.*, *15*(2): 11338. https://doi.org/10.15835/nsb15211338.

Chacha, J.S. and Laswai, H.S. 2020. Traditional practices and consumer habits regarding consumption of under-utilized vegetables in Kilimanjaro and Morogoro regions, Tanzania. *Int. J. Food Sci.*, *2020*(1): 3529434. https://doi.org/10.1155/2020/3529434.

Choudhury, J., Bora, D., Baruah, D., Borah, T. and Bharali, B.K. 2014. Portrayal of folk medicinal practices among the indigenous people of North Tripura District of Tripura, India. *Int. J. Res. Ayurveda Pharm.*, *5*(4); 1–7. https://doi.org/10.7897/2277-4343.05499.

Cooke, T. 1902. *The Flora of the Presidency of Bombay* (Vols. I–III). London: Taylor & Francis. (B.S.I. Reprint, 1958, Calcutta). Retrieved from https://www.biodiversitylibrary.org/.

Dahanayaka, L.W., Mapa, M.M.S.T., Kadigamuwa, C.C. and Udayanga, D. 2025. Impact of geographical variation on nutritional and antioxidant properties of *Basella alba* L. from Sri Lanka. *BMC Plant Biol.*, *25*(1): 118. http://dx.doi.org/10.1186/s12870-025-06143-3.

Das, S., Bandyopadhyay, S., Ramasamy, A. and Mondal, S. 2015. Evaluation of hepatoprotective activity of aqueous extracts of leaves of *Basella alba* in albino rats. *Nat. Prod. Res.*, *29*(11): 1059–1064. http://dx.doi.org/10.1080/14786419.2014.971795.

Deshmukh, S.A. and Gaikwad, D.K. 2014. A review of the taxonomy, ethnobotany, phytochemistry, and pharmacology of *Basella alba* (Basellaceae). *J. Appl. Pharm. Sci.*, *4*(1): 153–165. http://dx.doi.org/10.7324/JAPS.2014.40125.

Deshmukh, S.A. and Gaikwad, D.K. 2020. An attempt to solve the taxonomical conflicts in *Basella alba* L. *Indian J. Tradit. Knowl.*, *19*(3): 669–674.

Dias, J.S. 2012. Nutritional quality and health benefits of vegetables: A review. *Food Nutr. Sci.*, *3*(10): 1354–1374.

Dong, C.X., Hayashi, K., Mizukoshi, Y., Lee, J.B. and Hayashi, T. 2011. Structures of acidic polysaccharides from *Basella rubra* L. and their antiviral effects. *Carbohydr. Polym.*, *84*(3): 1084–1092. http://dx.doi.org/10.1016/j.carbpol.2010.12.073.

Eriksson, R. 2007. A synopsis of Basellaceae. *Kew Bull.*, *62*(2): 297–320.

Focho, D.A., Nkeng, E.A.P., Lucha, C.F., Ndam, W.T. and Afegenui, A. 2009. Ethnobotanical survey of plants used to treat diseases of the reproductive system and preliminary phytochemical screening of some species of Malvaceae in Ndop Central Sub-division, Cameroon. *J. Med. Plants Res.*, *3*(4): 301–314.

Hebbar, S.S., Harsha, V.H., Shripathi, V. and Hegde, G.R. 2004. Ethnomedicine of Dharwad district in Karnataka, India—plants used in oral health care. *J. Ethnopharmacol.*, *94*(2–3): 261–266. http://dx.doi.org/10.1016/j.jep.2004.04.021.

Ignacimuthu, S., Ayyanar, M. and Sankarasivaraman, K. 2008. Ethnobotanical study of medicinal plants used by Paliyar tribals in Theni district of Tamil Nadu, India. *Fitoterapia, 79*(7–8): 562–568. http://dx.doi.org/10.1016/j.fitote.2008.06.003.

Jeruto, P., Lukhoba, C., Ouma, G., Otieno, D. and Mutai, C. 2008. An ethnobotanical study of medicinal plants used by the Nandi people in Kenya. *J. Ethnopharmacol.*, *116*(2): 370–376. https://doi.org/10.1016/j.jep.2007.11.041.

Kareru, P.G., Kenji, G.M., Gachanja, A.N., Keriko, J.M. and Mungai, G. 2007. Traditional medicines among the Embu and Mbeere people of Kenya. *Afr. J. Tradit. Complement. Altern. Med.*, *4*(1): 75–86. http://dx.doi.org/10.4314/ajtcam.v4i1.31193.

Khare, C. 2007. Basella alba Linn. var. rubra Stewart. In: Khare, C. (eds) Indian Medicinal Plants. Springer, New York, NY. https://doi.org/10.1007/978-0-387-70638-2_198.

Kibr, G. (2022). The Health Benefits of Vegetables : Preventive Implications for Chronic Non-Communicable Diseases. *Veg. Crops: Health Benefits Cultiv.*, https://doi.org/10.5772/intechopen.101303.

Kigen, G., Kipkore, W., Wanjohi, B., Haruki, B. and Kemboi, J. 2017. Medicinal plants used by traditional healers in Sangurur, Elgeyo Marakwet County, Kenya. *Pharmacogn. Res.*, *9*(4): 333–347. https://doi.org/10.4103/pr.pr_42_17.

Koziol, L., Knap, M., Sutor-Swieży, K., Gorska, R., Dziedzic, E., Bieniasz, M., Mielczarek, P., Popenda, L., Tyszka-Czochara, M. and Wybraniec, S. 2024. Identification and reactivity of pigments in prominent vegetable leaves of *Basella alba* L. var. '*Rubra*' (Malabar spinach). *Food Chem.*, 445: 138714. https://doi.org/10.1016/j.foodchem.2024.138714.

Kumar, P.C., Oberoi, H.S. and S. Azeez. 2022. *Basella*: An underutilized green leafy vegetable with a potential for functional food development. *Food Rev. Int.*, *38*(1): 456–473. http://dx.doi.org/10.1080/87559129.2021.1874410.

Kumar, S., Prasad, A.K., Iyer, S.V. and Vaidya, S.K. 2013. Systematic pharmacognostical, phytochemical, and pharmacological review on an ethno medicinal plant, *Basella alba* L. *J. Pharmacogn. Phytother.*, *5*(4): 53–58. http://dx.doi.org/10.5897/JPP12.0256.

Kumar, V., Bhat, Z.A., Kumar, D., Bohra, P. and Sheela, S. 2011. *In-vitro* anti-inflammatory activity of leaf extracts of *Basella alba* Linn. var. *alba*. *Int. J. Drug Dev. Res.*, *3*(2): 176–179.

Kumar, V., Bhat, Z.A., Kumar, D., Khan, N.A., Chashoo, I.A. and Ara, I. 2012. Gastroprotective effect of leaf extracts of *Basella alba* var. *alba* against experimental gastric ulcers in rats. *Rev. Bras. Farmacogn.*, *22*: 657–662.

Lin, J.Y. and Tang, C.Y. 2007. Determination of total phenolic and flavonoid contents in selected fruits and vegetables, as well as their stimulatory effects on mouse splenocyte proliferation. *Food Chem.*, *101*(1): 140–147. http://dx.doi.org/10.1016/j.foodchem.2006.01.014.

Maurya, S.K. and Seth, A. 2014. Potential medicinal plants and traditional Ayurvedic approach towards urticaria, an allergic skin disorder. *Int. J. Pharm. Pharm. Sci.*, *6*(5): 172–177.

Moundipa, F.P., Kamtchouing, P., Koueta, N., Tantchou, J., Foyang, N.P.R. and Mbiapo, F.T. 1999. Effects of aqueous extracts of *Hibiscus macranthus* and *Basella alba* in mature rat testis function. *J. Ethnopharmacol.*, *65*(2): 133–139. http://dx.doi.org/10.1016/S0378-8741(98)00207-4.

Murugesan, P., Raja, G., Marx, S.K. and Selvam, B.P. 2011. Ethnobotanical study of medicinal plants used by villagers in Kolli Hills of Namakkal district of Tamil Nadu, India. *Int. J. Pharm. Sci. Rev. Res.*, *10*(1): 170–173.

Namukobe, J., Kasenene, J.M., Kiremire, B.T., Byamukama, R., Kamatenesi-Mugisha, M., Krief, S., Dumontet, V. and Kabasa, J.D. 2011. Traditional plants used for medicinal purposes by local communities around the Northern sector of Kibale National Park, Uganda. *J. Ethnopharmacol.*, *136*(1): 236–245. https://doi.org/10.1016/j.jep.2011.04.044.

Nantia, E A., Travert, C., Manfo, F.P.T., Carreau, S., Monsees, T.K. and Moundipa, P.F. 2011. Effects of the methanol extract of *Basella alba* L (Basellaceae) on steroid production in Leydig cells. *Int. J. Mol. Sci.*, *12*(1); 376–384. http://dx.doi.org/10.3390/ijms12010376.

Nirmala, A., Saroja, S. and Devi, G.G. 2011. Antidiabetic activity of *Basella rubra* and its relationship with the antioxidant property. *Br. Biotechnol. J.*, *1*(1): 1–9. http://dx.doi.org/10.9734/BBJ/2011/223.

Nur, M.A., Khan, M., Satter, M.A., Rahman, M.M. and Amin, M.Z. 2023. Assessment of physicochemical properties, nutrient contents, and colorant stability of the two varieties of Malabar spinach (*Basella alba* L.) fruits. *Biocatal. Agric. Biotechnol.*, *51*: 102746. http://dx.doi.org/10.1016/j.bcab.2023.102746.

Nyamukuru, A., Tabuti, J.R., Lamorde, M., Kato, B., Sekagya, Y. and Aduma, P.R. 2017. Medicinal plants and traditional treatment practices used in the management of HIV/AIDS clients in Mpigi District, Uganda. *J. Herb. Med.*, *7*: 51–58. http://dx.doi.org/10.1016/j.hermed.2016.10.001.

Okouango, I., Vital, M., Michel, E. and Abdoulaye, H. 2019. Evaluation of the consumption and nutritional quality of *Basella alba* L. in Afr. *J. Food Sci.13*(7): 143–151.

Panyaphu, K., Van On, T., Sirisa-Ard, P., Srisa-Nga, P., ChansaKaow, S. and Nathakarnkitkul, S. 2011. Medicinal plants of the Mien (Yao) in Northern Thailand and their potential value in the primary healthcare of postpartum women. *J. Ethnopharmacol.*, *135*(2): 226–237. http://dx.doi.org/10.1016/j.jep.2011.03.050.

Paul, S.B. and Singha, S. 2010. Isolation and identification of physiologically important sterols and sterol glucosides from *Basella rubra* Linn. *Assam Univ. J. Sci. Technol.*, *5*(1): 120–122. http://dx.doi.org/10.29322/IJSRP.12.01.2022.p12114.

Plants of the World Online (POWO). 2025. *Facilitated by the Royal Botanic Gardens, Kew*. Retrieved August 15, 2025, from https://powo.science.kew.org/.

Reddy, G. Jyothi, Reddi, T., Madhavi, Appa Rao, C., Nithya, S. and Rao, P.N. 2023. An updated phytochemical and pharmacological review on Malabar spinach (*Basella alba* or *Basella rubra*). *GSC Biol. Pharm. Sci.*, *24*(2): 161–169. https://doi.org/10.30574/gscbps.2023.24.2.0312.

Rodda, R., Kota, A., Sindhuri, T., Kumar, S.A. and Gnananath, K. 2012. Investigation on anti-inflammatory property of *Basella alba* Linn leaf extract. *Int. J. Pharm. Pharm. Sci.*, *4*(1): 452–454.

Roxburgh, W. 1832. *Flora Indica; or Descriptions of Indian Plants* (Vol. 2), W. Carey & N. Wallich (Eds.). Serampore: Mission Press.

Sahu, P. K., Masih, V., Gupta, S., Sen, D.L. and Tiwari, A. 2014. Ethnomedicinal plants used in the healthcare systems of tribes of Dantewada, Chhattisgarh, India. *Am. J. Plant Sci.*, *5*: 1632–1643. https://doi.org/10.4236/ajps.2014.511177.

Samoisy, A.K. and Mahomoodally, F. 2016. Ethnopharmacological appraisal of culturally important medicinal plants and polyherbal formulas used against communicable diseases in Rodrigues Island. *J. Ethnopharmacol.*, *194*: 803–818. http://dx.doi.org/10.1016/j.jep.2016.10.041.

Shaik, M., Manoj, B.P., Karishma, P., Sai, K. and Rao, T. 2024. *Basella*. In: *Textbook of Underexploited Vegetable Crop* (Chapter 7, pp. 63–70). P.K. Publishers & Distributors.

Sidwell K. 1999. Typification of two Linnaean names in the Basellaceae. *Novon.*, *9*: 562–563.

Singh, A., Dubey, P.K., Chaurasiya, R., Mathur, N., Kumar, G., Bharati, S. and Abhilash, P.C. 2018. Indian spinach: An underutilized perennial leafy vegetable for nutritional security in developing world. *Energy Ecol. Environ.*, *3*(3): 195–205. http://dx.doi.org/10.1007/s40974-018-0091-1.

Singh, A., Mal, D., Swain, A.R. and Mohapatra, A. 2024. Exploring the Nutritional and Medicinal Potential of Indian Spinach (*Basella rubra* L.): A Comprehensive Review. *J. Adv. Biol. Biotechnol.*, *27*(7): 437–446. http://dx.doi.org/10.9734/jabb/2024/v27i71004.

Siriwatanametanon, N., Fiebich, B.L., Efferth, T., Prieto, J.M. and Heinrich, M. 2010. Traditionally used Thai medicinal plants: *In vitro* anti-inflammatory, anticancer, and antioxidant activities. *J. Ethnopharmacol.*, *130*(2): 196–207. https://doi.org/10.1016/j.jep.2010.04.036.

Sushila, R., Deepti, A., Permender, R., Madhavi, T., Dharmender, R. and Rathee, D. 2010. Cytotoxic and antibacterial activity of *Basella alba* whole plant: A relatively unexplored plant. *Pharmacology online, 3*: 651–658.

Yadav, S.R. and Sardesai, M.M. 2002. *Flora of Kolhapur District, Maharashtra*. Shivaji University, Kolhapur.

4 Nutritive Value, Bioactive Compounds, and Biological Activities of Swiss Chard (*Beta vulgaris* subsp. *vulgaris*, Cicla Group and Flavescens Group)

Shrinivas Lamani[1*] and Hosakatte Niranjana Murthy[2-4*]

Introduction

Plants play a potential role in drug development programs as a source of pioneer compounds. People have been using plants for years to treat many different kinds of diseases. They are regarded as a rich supply of raw materials in the form of plant-based constituents, which can be used in modern drugs, and there are currently global attempts to identify such undiscovered resources. Vegetables are one of the sources for the plant-based nutritional and bioactive compounds, including proteins, carbohydrates, dietary fibers, minerals, sugars, organic acids, fatty acids, and lipids. Leafy vegetables are practically a rich source of carotenes, minerals, and dietary fibers. Most of the leafy vegetables are highly perishable products that have to be either consumed raw within a few days after harvest or subjected to preservation methods to extend their shelf life (Aguero et al., 2005).

Consumption of green leafy vegetables has been highly recommended in recent years to attain the nutritional requirements of the body because they contain low energy and high micronutrients. It is recommended that the increased consumption of green vegetables can prevent coronary heart disease by inhibiting the development of atherosclerosis, and this has been experimentally proven by Sener et al. (2002). One such herbaceous leafy vegetable, *Beta vulgaris* subsp. *vulgaris* var. cicla and flavescens is an edible plant of the Amaranthaceae family. This family includes the most important food crops, such as *Spinacia oleracea* (spinach), which is the most consumed leafy vegetable in many countries (Ruales and Nair, 1994). It is commonly known as Swiss chard, a biannual cultivar,

[1] Department of Botany, DVS College of Arts and Science, Shimoga 577201, India.
[2] Department of Botany, Karnatak University, Dharwad 580003, India.
[3] Department of Horticultural Science, Chungbuk National University, Cheongju 28644, Republic of Korea.
[4] Department of Life Sciences, CHRIST University, Bengaluru 560039.
* Corresponding authors: shrinivasgl92@gmail.com[1]; hnmurthy60@gmail.com[2]

DOI: 10.1201/9781003687832-4

originating from Mediterranean regions. Presently, Swiss chard is cultivated in different parts of the world, especially in North-Central America and Europe. Swiss chard is employed as a leafy green vegetable for its availability in all seasons and low cost (Ustundag et al., 2016). The dark green leaves are not only eaten raw as part of a salad, or cooked alone or along with the stems in a similar way as spinach. Moreover, it is also used as a potential source for the medicinal potential due to its rich bioactive constituents (Kim et al., 2004). Its commercial success is due to the ease of cultivation, the adaptability to diverse environmental conditions, and its nutraceutical properties (Gamba et al., 2021). This vegetable contains abundant bioactive natural pigments and secondary metabolites, such as betalains, flavonoids, and other phenolic compounds which have strong antioxidant properties and the capability to counteract free radicals, which is related to multiple health benefits when consumed in the diet (Mzoughi et al., 2020).

Both chard, cicla and the flavescens variety have a long traditional medicinal history as remedies for liver/ kidney complaints, wound healing, immune or blood-related uses, and as antidiabetic/antihypertensive folk remedies in some regions (Almeida et al., 2025). While these benefits are notable, Swiss chard remains relatively underexplored in comparison to other leafy greens. Various findings highlighted the bioactive compounds and their potential health benefits of wild Swiss chard as a source of nutritional and bioactive compounds. The bioactive compositions remain unexplored to date; however, to the best of our knowledge, and based on a comprehensive literature review, no attempt has been documented to assemble material that covers all of the essential aspects. Thus, the current chapter aims to document and critically review current evidence on bioactive components and biological activities.

Taxonomical Classification and Description

The taxonomic classification of *Beta vulgaris* subsp. *vulgaris* var. cicla and flavescens are:

Kingdom	: Plantae
Division	: Tracheophyta
Class	: Magnoliopsida
Order	: Caryphyllales
Family	: Amaranthaceae
Genus	: *Beta*
Species	: *vulgaris*
Sub species	: *vulgaris*

Beet belongs to the family Amaranthaceae, order Caryophyllales, and genus *Beta* (Duke, 1983). The genus is divided into two groups: wild maritime beets and cultivated beets. The wild group is represented solely by the sea beet (*Beta vulgaris* subsp. *maritima*), the ancestral progenitor of all modern beets, valued as a genetic reservoir for disease resistance, stress tolerance, and adaptability. The cultivated group comprises four principal types: the sugar beet (*B. vulgaris* var.

saccharifera), the main source of sucrose in temperate regions; the fodder beet (*B. vulgaris* var. *crassa*), used as livestock feed; the leaf beet (*B. vulgaris* subsp. *cicla*), or Swiss chard, grown for its edible leaves; and the garden beet (*B. vulgaris* var. *rubra*), cultivated as a vegetable and for its betalain pigments (Lewellen et al., 2009 and Lange et al. 1999). Swiss chard is a biennial plant characterized by a strong taproot and large oval leaves with prominent white veins. Generally, it is sown in spring for a summer harvest, and again in late summer to ensure crops for winter and early spring. Harvesting is done by cutting the leaves at the base, allowing the root system to remain in the soil and support the regrowth of foliage. Flowers are arranged in glomerules that form spikes at the apical ends of the stems, and are followed by seed production (Ninfali and Angelino, 2013).

Nutritional Compositions

Proximate Compositions

The nutritional composition of Swiss chard leaves has been investigated by Zeller et al. (1977), Mzoughi et al. (2020), and Almeida et al. (2025), highlighting their richness in essential nutrients and bioactive compounds. Proximate analysis of *Beta vulgaris* subsp. *vulgaris* var. cicla (BVc) leaves, expressed per 100 g fresh weight (fw), revealed high moisture content (93.35 g), along with ash (1.30 g), protein (2.00 g), carbohydrates (2.15 g), total fat (0.09 g), insoluble dietary fiber (2.30 g), and soluble dietary fiber (0.13 g) (Daiss et al., 2008; Mzoughi et al., 2020). Beyond the proximate composition, BVc leaves are notable sources of soluble sugars. Glucose (0.285 g/100 g fw) and sucrose (1.115 g/100 g fw) were reported as the major sugars, whereas fructose (0.046 g), galactose (0.038 g), arabinose (0.003 g), rhamnose (0.018 g), and raffinose (0.061 g) occurred in smaller amounts (Mzoughi et al., 2020). Almeida et al. (2025) further identified trehalose in the leaves of *Beta vulgaris* subsp. *vulgaris* var. flavescens (BVf). Two polyol sugars, inositol (0.182 g) and mannitol (0.082 g), were also detected from the leaves of BVc (Mzoughi et al., 2020), as summarized in Table 1.

Table 1 Nutrient compositions of Swiss chard (*Beta vulgaris* L. *subsp. vulgaris*) leaves.

Sl. No.	*Constituents*	*Variety/ Cultivars*	*Quantity*	*References*
Proximate compositions (g/100 g fw)				
1.	Moisture	Cicla	93.35	Mzoughi et al. (2020)
2.	Crude fat		0.09	
3.	Ash content		1.30	
4.	Protein		2.00	Daiss et al. (2008)
5.	Carbohydrate		2.15	Mzoughi et al. (2020)
6.	Insoluble dietary fiber		2.30	
7.	Soluble dietary fiber		0.13	

Contd..

Table 1 *Contd..*

Sl. No.	*Constituents*	*Variety/ Cultivars*	*Quantity*	*References*
Soluble sugars (g/100 g fw)				
8.	Fructose	Cicla	0.046	
9.	Glucose		0.285	
10.	Galactose		0.038	
11.	Arabinose		0.003	
12.	Rhamnose		0.018	
13.	Sucrose		1.115	
14.	Raffinose		0.061	
15.	Trehalose (g/ 100 g dw)	Flavescens	0.41	Almeida et al. (2025)
Polyols - Sugars (g/100 g fw)				
16.	Inositol	Cicla	0.182	Mzoughi et al. (2020)
17.	Mannitol		0.082	
Organic acids (g/100 g dw)				
18.	Oxalic acid	Flavescens	2.37	Almeida et al. (2025)
19.	Quinic acid		2.54	
20.	Malic acid		1.40	
21.	Citric acid		2.20	
22.	Succinic acid		12.40	
23.	Fumaric acid		Traces	
Fatty acids (%)				
24.	Caproic acid (C6:0)	Cicla	0.37	Mzoughi et al. (2020)
25.	Lauric acid (C12:0)	Cicla	2.17	
		Flavescens	0.10	Almeida et al. (2025)
26.	Tridecanoic acid (C13:0)	Flavescens	0.56	
27.	Myristic acid (C14:0)	Cicla	0.14	Mzoughi et al. (2020)
		Flavescens	0.52	Almeida et al. (2025)
28.	Pentadecanoic acid (C15:0)	Cicla	0.79	Mzoughi et al. (2020)
		Flavescens	0.19	Almeida et al. (2025)
29.	cis-10-pentadecenoic acid (C15:1)	Flavescens	0.19	Almeida et al. (2025)
30.	Palmitic acid (C16:0)	Cicla	22.92	Mzoughi et al. (2020)
		Flavescens	19.60	Almeida et al. (2025)
31.	Palmitoleic acid (C16:1)	Cicla	6.31	Mzoughi et al. (2020)
		Flavescens	1.43	Almeida et al. (2025)

Contd..

Table 1 *Contd..*

Sl. No.	*Constituents*	*Variety/ Cultivars*	*Quantity*	*References*
32.	Margaric acid (C17:0)	Cicla	1.54	Mzoughi et al. (2020)
		Flavescens	0.36	Almeida et al. (2025)
33.	Stearic acid (C18:0)	Cicla	8.17	Mzoughi et al. (2020)
		Flavescens	1.36	Almeida et al. (2025)
34.	Oleic acid (C18:1)	Cicla	19.15	Mzoughi et al. (2020)
		Flavescens	7.67	Almeida et al. (2025)
35.	Arachidic acid (C20:0)	Cicla	3.82	Mzoughi et al. (2020)
36.	Eicosenoic acid (C20:1)	Cicla	0.63	Mzoughi et al. (2020)
		Flavescens	0.45	Almeida et al. (2025)
37.	Behenic acid (C22:0)	Cicla	0.67	Mzoughi et al. (2020)
		Flavescens	0.51	Almeida et al. (2025)
38.	Tricosanoic acid (C23:0)	Cicla	0.81	Mzoughi et al. (2020)
		Flavescens	0.25	Almeida et al. (2025)
39.	Lignoceric acid (C24:0)	Cicla	0.16	Mzoughi et al. (2020)
		Flavescens	0.74	Almeida et al. (2025)
40.	Linoleic acid (C18:2 n6)	Cicla	26.54	Mzoughi et al. (2020)
		Flavescens	17.60	Almeida et al. (2025)
41.	Eicosadienoic acid (C20:2)	Cicla	0.65	Mzoughi et al. (2020)
42.	α-Linolenic acid (C18:3 n3)	Cicla	5.17	
		Flavescens	47.80	Almeida et al. (2025)
43.	γ-Linolenic acid (C18:3 n6)	Flavescens	0.37	
Tocopherols (mg/100 g dw)				
44.	α-Tocopherol	Flavescens	17.23	Almeida et al. (2025)
45.	β-Tocopherol		0.40	
46.	γ-Tocopherol		0.42	
Fatty-related compounds				
47.	3-methyldodecane	Cicla	-	Mzoughi et al. (2020)
48.	(E)-3-octen-1-ol		-	
49.	Methyl decanoate		-	
Lipids				
50.	Phosphatidylcholine (PCH)	Cicla- Glatter Silber	-	Zeller et al. (1977)
51.	Phosphatidylethanolamine (PE)		-	
52.	Phosphatidylglycerol (PG)		-	
53.	Digalactosyldiglyceride (DGG)		-	
54.	Monogalactosyldiglyceride		-	

Organic acid profiling has added further nutritional insight. Almeida et al. (2025) reported oxalic acid (2.37 g/100 g fw), quinic acid (2.54 g), malic acid (1.40 g), citric acid (2.20 g), and succinic acid (12.40 g) in BVf leaves, while fumaric acid was detected only in trace amounts. In addition, Mzoughi et al. (2020) identified fatty-related compounds, including 3-methyldodecane, (E)-3-octen-1-ol, and methyl decanoate, further reflecting the chemical diversity of Swiss chard leaves. Together, these findings demonstrate that Swiss chard is not only a valuable leafy vegetable due to its macronutrient and micronutrient profile, but also a reservoir of functional biomolecules, underscoring its potential applications in human nutrition, functional foods, and health-promoting diets.

Fatty acid and lipid diversity have been extensively reported in the leaves of BVc and BVf. Mzoughi et al. (2020) identified a total of 19 fatty acids in BVc leaves. These included short- to long-chain saturated and unsaturated fatty acids, such as caproic acid (0.37%), lauric acid (2.17%), tridecanoic acid (0.14%), myristic acid (0.14%), pentadecanoic acid (0.79%), palmitic acid (22.92%), palmitoleic acid (6.31%), margaric acid (1.54%), stearic acid (8.17%), oleic acid (19.15%), arachidic acid (3.82%), eicosenoic acid (0.63%), behenic acid (0.67%), tricosanoic acid (0.81%), lignoceric acid (0.16%), linoleic acid (26.54%), eicosadienoic acid (0.65%), and α-linolenic acid (5.17%). In comparison, Almeida et al. (2025) reported that BVf leaves contained higher amounts of several fatty acids, with notable differences in unsaturated fatty acid content. These included lauric acid (0.10%), tridecanoic acid (0.56%), myristic acid (0.52%), pentadecanoic acid (0.19%), cis-10-pentadecenoic acid (0.19%), palmitic acid (19.60%), palmitoleic acid (1.43%), margaric acid (0.36%), stearic acid (1.36%), oleic acid (7.67%), eicosenoic acid (0.45%), behenic acid (0.51%), tricosanoic acid (0.25%), lignoceric acid (0.74%), linoleic acid (17.60%), α-linolenic acid (47.80%), and γ-linolenic acid (0.37%). Particularly noteworthy is the exceptionally high proportion of α-linolenic acid in BVf, indicating its potential value as a dietary source of omega-3 fatty acids.

In addition to fatty acids, Almeida et al. (2025) also documented the presence of tocopherols (mg/100 g dry weight), including α-tocopherol (17.23), β-tocopherol (0.40), and γ-tocopherol (0.42), highlighting the antioxidant potential of BVf leaves. Earlier, Zeller et al. (1977) reported key lipid classes in BVc leaves, including phosphatidylcholine (PCH), phosphatidylethanolamine (PE), phosphatidylglycerol (PG), digalactosyldiglyceride (DGG), and monogalactosyldiglyceride (MGG), which are essential components of plant membranes and play important roles in photosynthesis and metabolic regulation. However, these studies emphasize that Swiss chard is not only a valuable source of essential fatty acids and lipid-soluble antioxidants, but also a promising dietary component for cardiovascular health, inflammation modulation, and overall nutritional well-being.

Mineral Compositions

The mineral nutrients, pigments, amino acids, vitamins, and enzymes identified in Swiss chard revealed the diverse nutritional and biochemical richness of leafy vegetables. As summarized in Table 2, extensive mineral profiling across different cultivars has demonstrated that Swiss chard serves as an excellent source of both macro- and micro-elements essential for human health.

Table 2 Mineral, amino acid, vitamin, and enzyme composition of Swiss chard (*Beta vulgaris* L. subsp. *vulgaris*) leaves.

Sl. No.	*Constituents*	*Variety-Cultivars*	*Plant Parts*	*References*
Mineral nutrients				
1.	Calcium- Ca	Cicla	Leaves and stalks	Pokluda & Kuben (2002)
		Cicla - Lukullus	Leaves	Dzida & Pitura (2008)
		Cicla - 5 cultivars	Leaves	Kolota et al. (2010)
		Cicla - Agila	Leaves	Colonna et al. (2016)
		Cicla - Fordhook Giant	-	Moyo et al. (2018)
		Cicla - Verca F1 hybrid	Tissue (leaves and stems)	Ivanovic et al. (2019)
		Cicla	Leaves	Mzoughi et al. (2020)
2.	Cobal t- Co	Cicla - Verca F1 hybrid	Tissue (leaves and stems)	Ivanovic et al. (2019)
3.	Chromium - Cr	Cicla - Verca F1 hybrid	Tissue (leaves and stems)	Ivanovic et al. (2019)
4.	Copper - Cu	Cicla - Fordhook Giant	-	Moyo et al. (2018)
		Cicla - Verca F1 hybrid	Tissue (leaves and stems)	Ivanovic et al. (2019)
		Cicla	Leaves	Mzoughi et al. (2020)
5.	Iron - Fe	Cicla - Fordhook Giant	-	Moyo et al. (2018)
		Cicla - Verca F1 hybrid	Tissue (leaves and stems)	Ivanovic et al., 2019
		Cicla	Leaves	Mzoughi et al. (2020)
6.	Potassium- K	Cicla	Leaves and stalks	Pokluda & Kuben (2002)
		Cicla - Lukullus	Leaves	Dzida & Pitura (2008)
		Cicla	Leaves	Kolota et al. (2010)
		Cicla - Agila	Leaves	Colonna et al. (2016)
		Cicla - Fordhook Giant	-	Moyo et al. (2018)

Contd..

Table 2 *Contd..*

Sl. No.	*Constituents*	*Variety-Cultivars*	*Plant Parts*	*References*
		Cicla - Verca F1 hybrid	Tissue (leaves and stems)	Ivanovic et al. (2019)
		Cicla	Leaves	Mzoughi et al. (2020)
		Cicla - Magenta sunset	Leaves	Singh et al. (2019)
7.	Magnesium - Mg	Cicla	Leaves and stalks	Pokluda & Kuben (2002)
		Cicla - Lukullus	Leaves	Dzida & Pitura (2008)
		Cicla	Leaves	Kolota et al. (2010)
		Cicla - Agila	Leaves	Colonna et al. (2016)
		Cicla - Fordhook Giant	-	Moyo et al. (2018)
		Cicla - Verca F1 hybrid	Tissue (leaves and stems)	Ivanovic et al. (2019)
		Cicla	Leaves	Mzoughi et al. (2020)
8.	Manganese - Mn	Cicla - Fordhook Giant	-	Moyo et al. (2018)
		Cicla - Verca F1 hybrid	Tissue (leaves and stems)	Ivanovic et al. (2019)
		Cicla	Leaves	Mzoughi et al. (2020)
9.	Sodium - Na	Cicla	Leaves and stalks	Pokluda & Kuben (2002)
		Cicla - Fordhook Giant	-	Moyo et al. (2018)
		Cicla - Verca F1 hybrid	Tissue (leaves and stems)	Ivanovic et al. (2019)
		Cicla	Leaves	Mzoughi et al. (2020)
10.	Phosphorus - P	Cicla - Lukullus	Leaves	Dzida & Pitura (2008)
		Cicla	Leaves	Kolota et al. (2010)
		Cicla	Leaves	Bozokalfa et al. (2011)
		Cicla - Agila	Leaves	Colonna et al. (2016)
		Cicla - Fordhook Giant	-	Moyo et al. (2018)
		Cicla - Verca F1 hybrid	Tissue (leaves and stems)	Ivanovic et al. (2019)
		Cicla - Magenta sunset	Leaves	Singh et al. (2019)

Contd..

Table 2 *Contd..*

Sl. No.	*Constituents*	*Variety-Cultivars*	*Plant Parts*	*References*
11.	Zinc - Zn	Cicla - Fordhook Giant	-	Moyo et al. (2018)
		Cicla - Verca F1 hybrid	Tissue (leaves and stems)	Ivanovic et al. (2019)
		Cicla	Leaves	Mzoughi et al. (2020)
Pigments				
12.	Chlorophyll a	Cicla - Verca F1 hybrid	Leaves	Ivanovic et al. (2019)
		Cicla		Mzoughi et al. (2020)
13.	Chlorophyll b	Cicla - Verca F1 hybrid	Leaves	Ivanovic et al. (2019)
		Cicla		Mzoughi et al. (2020)
Carotenoids				
14.	β-carotene	Flavescens - Berac and Charlotte	Leaves	Reif et al. (2013)
		Cicla - Fordhook Giant	-	Moyo et al. (2018)
		Cicla	Leaves	Mzoughi et al. (2020)
15.	Lutein	Flavescens - Berac and Charlotte	Leaves (Flavescens)	Reif et al. (2013)
16.	Lycopene	Cicla	Leaves	Mzoughi et al. (2020)
Amino acids and Vitamins				
17.	Proline	Cicla	Leaves	Sacan & Yanardag (2010)
18.	Vitamin C (Ascorbic acid)	Cicla - Green and Yellow	Leaves	Gil et al. (1998)
		Cicla - 11 cultivars	Leaves	Pokluda & Kuben (2002)
		Cicla - Lukullus	Leaves	Dzida & Pitura (2008)
		Cicla - Verca F1 hybrid	Tissue (leaves and stems)	Ivanovic et al. (2019)
19.		Cicla - Bressanne	Leaves	Moreira et al. (2003)
		Cicla - Verde da taglio	Leaves	Miceli & Miceli (2014)
		Cicla - Fordhook Giant	-	Moyo et al. (2018)
20.	Vitamin K1 (Phylloquinone)	Cicla	-	Ferland & Sadowski (1992)

Contd..

Table 2 *Contd..*

Sl. No.	*Constituents*	*Variety-Cultivars*	*Plant Parts*	*References*
Enzymes				
21.	Catalase	Cicla	Leaves	Dincler & Aydemir (2001)
22.	Polyphenol oxidase	Cicla	Leaves	Gao et al. (2009)

Microelements like potassium, calcium, sodium, magnesium, and phosphorus are reported at varying concentrations in the leaves of most of the cultivars, including lukullus, agila, fordhook giant, magenta sunset, and the verca F1 hybrid (Pokluda and Kuben, 2002; Dzida and Pitura, 2008; Colonna et al., 2016; Singh et al., 2019; Ivanovic et al., 2019). These elements play crucial roles in bone development, muscle contraction, enzymatic functions, and electrolyte balance (Mzoughi et al., 2020). In addition, micronutrients including copper, iron, zinc, and manganese were found across different cultivars (Moyo et al., 2018; Mzoughi et al., 2020), with iron and zinc being of particular importance due to their roles in hemoglobin synthesis and immune function. Interestingly, rare trace elements like cobalt and chromium were also documented in the Verca F1 hybrid (Ivanovic et al., 2019), indicating that Swiss chard may be a dietary source of lesser-known but physiologically relevant minerals.

Beyond minerals, Swiss chard is equally rich in pigments that contribute both to its coloration and bioactivity. Chlorophyll a and b are reported in Verca F1 hybrid and other cultivars (Ivanovic et al., 2019; Mzoughi et al., 2020), which not only support photosynthesis but also exhibit antioxidant and detoxifying properties in human health. Studies have reported abundant chlorophyll a and b in both the Verca F1 hybrid and other cultivars (Ivanovic et al., 2019; Mzoughi et al., 2020). Carotenoids, including β-carotene, lutein, and lycopene, were also detected, particularly in the flavescens cultivars, such as 'Berac' and 'Charlotte' (Reif et al., 2013). These pigments are well-known for their role in human health, offering protective effects against oxidative stress and age-related diseases.

The presence of amino acids and vitamins additionally enhances the nutritional quality of Swiss chard. Proline, an amino acid often associated with stress tolerance, has been detected in the leaves (Sacan and Yanardag, 2010), suggesting its importance in both plant defense and human nutrition. Among vitamins, vitamin C has been consistently reported across several cultivars, including Green, Yellow, Bressanne, Verde da Taglio, Lukullus, and Fordhook Giant (Gil et al., 1998; Pokluda and Kuben, 2002; Moreira et al., 2003; Dzida and Pitura, 2008; Miceli and Miceli, 2014; Moyo et al., 2018; Ivanovic et al., 2019). Ascorbic acid plays a vital role as an antioxidant, supports immune function, and aids in collagen synthesis. Vitamin K1 (phylloquinone), though less frequently studied, has also been identified (Ferland and Sadowski, 1992), highlighting its relevance to blood coagulation and bone metabolism. In addition to these reports, enzyme studies by Dincler and Aydemir (2001) and Gao et al. (2009) revealed the presence of catalase and polyphenol oxidase, respectively, in BVc, both of which are associated with oxidative stress regulation and phenolic metabolism. The reports highlight the biochemical versatility of Swiss chard and emphasize its potential to support human health when incorporated regularly into the diet. Overall, the diverse array of minerals, pigments, amino acids, vitamins, and enzymes establishes Swiss chard as a functional food with significant nutritional and therapeutic value, suitable for daily diets and promising for nutraceutical and preventive applications.

Specialized Bioactive Compounds

Swiss chard is reported to possess diverse bioactive compounds, detected in the leaves. All the phytochemicals reported were highly potent for the biological activities. Listed below are a few of the identified phytochemicals found in the leaves of different cultivars/varieties of Swiss chard. Betalains are the major compounds observed in this species (Kugler et al., 2004, 2007). The phytochemical analysis of Swiss chard revealed the presence of a variety of bioactive constituents, including flavonoids, phenolic compounds, terpenes, carboxylic acids, alkanes, aldehydes, ketones, and other compounds (Kugler et al., 2004; Pyo et al., 2004; Kugler et al., 2007; Hala et al., 2019; Mzoughi et al., 2020). The identified bioactive constituents are summarized in Table 3, and the structures of some potent compounds are illustrated in Figure 1.

Table 3 Bioactive chemical constituents of Swiss chard leaves (*Beta vulgaris* L. subsp. *vulgaris*).

Sl. No.	*Constituents*	*Variety-Cultivars*	*Plant Part*	*References*
Betalains				
Betaxanthins				
1.	Betaxanthin	Cicla	Petiole	Kugler et al. (2004)
2.	3-methoxytyramine-betaxanthin	Cicla	Petiole	Kugler et al. (2004); Kugler et al. (2007)
3.	Alanine-betaxanthin	Cicla	Petiole	Kugler et al. (2004); Kugler et al. (2007)
4.	Asparagine-betaxanthin (vulgaxanthin III)	Cicla	Petiole	Kugler et al. (2004); Kugler et al. (2007)
5.	Aspartic acid (miraxanthin II)	Cicla	Petiole	Kugler et al. (2007)
6.	Dopa-betaxanthin (dopaxanthin)	Cicla	Petiole	Kugler et al. (2007)
7.	Dopamine-betaxanthin (miraxanthin V)	Cicla	Petiole	Kugler et al. (2004); Kugler et al. (2007)

Contd..

Table 3 *Contd..*

Sl. No.	*Constituents*	*Variety-Cultivars*	*Plant Part*	*References*
8.	Ethanolamine-betaxanthin	Cicla	Petiole	Kugler et al., 2007
9.	γ-aminobutyric acid-betaxanthin	Cicla	Petiole	Kugler et al. (2004); Kugler et al. (2007)
10.	Glutamic acid (vulgaxanthin II)	Cicla	Petiole	Kugler et al. (2004); Kugler et al. (2007)
11.	Glutamine-betaxanthin (vulgaxanthin I)	Cicla	Petiole	Kugler et al. (2004); Kugler et al. (2007)
12.	Glycine-betaxanthin (portulacaxanthin III)	Cicla	Petiole	Kugler et al. (2004); Kugler et al. (2007)
13.	Histamine-betaxanthin	Cicla	Petiole	Kugler et al. (2004); Kugler et al. (2007)
14.	Histidine-betaxanthin (muscaaurin VII)	Cicla	Petiole	Kugler et al. (2004); Kugler et al. (2007)
15.	Isoleucine-betaxanthin	Cicla	Petiole Petiole Leaves	Kugler et al. (2004); Kugler et al. (2007); Franzoni et al. (2024)
16.	Leucine-betaxanthin (vulgaxanthin IV)	Cicla	Petiole Petiole Leaves	Kugler et al. (2004); Kugler et al. (2007); Franzoni et al. (2024)
17.	Lysine-betaxanthin	Cicla	Petiole	Kugler et al. (2007)
18.	Methionine-betaxanthin	Cicla	Petiole	Kugler et al. (2007)
19.	Phenylalanine-betaxanthin	Cicla	Petiole	Kugler et al. (2004); Kugler et al. (2007)
20.	Proline-betaxanthin (indicaxanthin)	Cicla	Petiole	Kugler et al. (2004); Kugler et al. (2007)
21.	Serine-betaxanthin	Cicla	Petiole	Kugler et al. (2004); Kugler et al. (2007)
22.	Threonine-betaxanthin	Cicla	Petiole	Kugler et al. (2007)
23.	Tryptophan-betaxanthin	Cicla	Petiole	Kugler et al. (2004); Kugler et al. (2007)
24.	Tyramine-betaxanthin (miraxanthin III)	Cicla	Petiole	Kugler et al. (2004); Kugler et al. (2007)
25.	Tyrosine-betaxanthin (portulacaxanthin II)	Cicla	Petiole Petiole Leaves	Kugler et al. (2004); Kugler et al. (2007); Franzoni et al. (2024)
26.	Valine-betaxanthin	Cicla	Petiole Petiole Leaves	Kugler et al. (2004); Kugler et al. (2007); Franzoni et al. (2024)

Contd..

Table 3 *Contd..*

Sl. No.	*Constituents*	*Variety-Cultivars*	*Plant Part*	*References*
Betacyanins				
27.	Betacyanin	Cicla	Leaves	Ali et al. (2009)
28.	Betanidin	Cicla	Petiole Leaves	Kugler et al. (2004); Franzoni et al. (2024)
29.	Betanidin-monoferuloyl-5-O-β-diglucoside	Cicla	Petiole	Kugler et al. (2004)
30.	Betanin	Cicla	Petiole Leaves	Kugler et al. (2004); Franzoni et al. (2024)
		Flavescens	Leaves	Almeida et al. (2025)
31.	Isobetanidin	Cicla	Petiole Leaves	Kugler et al. (2004); Franzoni et al. (2024)
32.	Isobetanidin-monoferuloyl-5-O-β-diglucoside	Cicla	Petiole Leaves	Kugler et al. (2004)
33.	Isobetanin	Cicla	Petiole Leaves	Kugler et al. (2004); Franzoni et al. (2024)
		Flavescens	Leaves	Almeida et al. (2025)
34.	Isolampranthin II	Cicla	Petiole	Kugler et al. (2004)
35.	Lampranthin II	Cicla	Petiole	
36.	Phyllocactin	Cicla	Petiole	
37.	17-Decarboxy-neobetanin	Cicla	Leaves	Franzoni et al. (2024)
		Flavescens	Leaves	Almeida et al. (2025)
38.	Neobetanin	Cicla	Leaves	Franzoni et al. (2024)
		Flavescens	Leaves	Almeida et al. (2025)
39.	Miraxanthin-V	Flavescens	Leaves	
40.	Decarboxy-betanin	Flavescens	Leaves	
41.	Decarboxy-isobetanin	Flavescens	Leaves	
42.	Feruloyl-hexosyl betanin	Flavescens	Leaves	
Flavonoids and Derivatives				
43.	2”-Oxylopyranosylvitexin	Flavescens	Leaves	Hala et al. (2019)
44.	6,8-di-C-β-D-glucopyranosylapigenin (vecinin-II)			
45.	Acacetin 8-C-β-D-glucopyranoside			
46.	Acacetin 8-C-α-L-rhamnoside			
47.	(+) Catechin	Cicla - Large white ribbed	Leaves	Pyo et al. (2004)

Contd..

Table 3 *Contd..*

Sl. No.	*Constituents*	*Variety-Cultivars*	*Plant Part*	*References*
48.	Isorhamnetin 3-gentiobioside	Cicla - Green and Yellow	Leaves	Gil et al. (1998)
		Cicla	Leaves	Ninfali et al. (2007)
49.	Isorhamnetin 3-vicianoside	Cicla - Green and Yellow	Leaves	Gil et al. (1998)
50.	Kaempferol	Cicla - Large white ribbed	Leaves	Pyo et al. (2004)
51.	Kaempferol 3-gentiobioside	Cicla - Green and Yellow	Leaves	Gil et al. (1998)
52.	Myricetin	Cicla - Large white ribbed	Leaves	Pyo et al. (2004)
53.	Quercetin	Cicla - Large white ribbed	Leaves	Pyo et al. (2004)
54.	Rutin	Cicla	Leaves	Ninfali et al. (2007)
55.	Vitexin (Apigenin 8-C-β-D-glucopyranoside)	Flavescens	Leaves	Hala et al. (2019)
56.	Vitexin 2”-O-β-D-glucopyranoside	Cicla	Leaves	Kim et al. (2004)
57.	Vitexin 2”-O-rhamnoside	Cicla	Leaves	Ninfali et al. (2007)
58.	Vitexin 7-O-β-D-glucopyranoside	Cicla	Leaves	Kim et al. (2004)
59.	2”-xylosylvitexin	Cicla- Green and Yellow	Leaves	Gil et al. (1998)
60.	2”-O-Xylosylvitexin	Cicla	Leaves	Ninfali et al. (2007)
61.	(+)-Dehydrovomifoliol	Cicla	Leaves	Kim et al. (2004)
62.	3-Hydroxy-5α, 6α-epoxy-β-ionone	Cicla	Leaves	
63.	6”-Malonyl-2”-xylosyl vitexin	Cicla	Leaves	Gil et al. (1998)
64.	2”’,2”’-di-O-α-Rhamnopyranosyl-vicenin II	Flavescens	Leaves	Hashem et al. (2016)
65.	di-C-Glycosyl flavone	Flavescens	Leaves	
66.	Herbacetin 3-O-β-xylopyranosyl-(1”’→2”)-O-β-glucopyranoside	Flavescens	Leaves	
67.	Vitexin pentoside	Flavescens	Leaves	Almeida et al. (2025)
68.	Vitexin deoxyhexoside	Flavescens	Leaves	
69.	Isorhamnetin dihexoside	Flavescens	Leaves	

Contd..

Table 3 *Contd..*

Sl. No.	*Constituents*	*Variety-Cultivars*	*Plant Part*	*References*
Phenolic Compounds				
70.	Caffeic acid	Cicla - Large white ribbed and CXS 2550	Leaves and stems	Pyo et al. (2004)
		Cicla - Fordhook Giant	-	Moyo et al. (2018)
		Cicla	Leaves	Franzoni et al. (2024)
71.	Chlorogenic acid	Cicla - Large white ribbed and CXS 2550	Leaves and stems	Pyo et al. (2004)
72.	Ferulic acid	Cicla - Large white ribbed and CXS 2550	Leaves and stems	
		Cicla - Fordhook Giant	-	Moyo et al. (2018)
		Cicla	Leaves	Franzoni et al. (2024)
73.	Gallic acid	Cicla - Large white ribbed and CXS 2550	Leaves and stems	Pyo et al. (2004)
		Cicla	Leaves	Franzoni et al. (2024)
74.	Myricitrin acid	Cicla	Leaves	Mzoughi et al. (2020)
75.	p-Coumaric acid	Cicla –Large white ribbed and CXS 2550	Leaves and stems	Pyo et al. (2004)
		Cicla - Fordhook Giant	-	Moyo et al. (2018)
		Cicla	Leaves	Mzoughi et al. (2020)
76.	p-Hydroxybenzoic acid	Cicla - Large white ribbed and CXS 2550	Leaves and stems	Pyo et al. (2004)
		Cicla - Fordhook Giant	-	Moyo et al. (2018)
77.	Protocatechuic acid	Cicla - Large white ribbed and CXS 2550	Leaves and stems	Pyo et al. (2004)
		Cicla - Fordhook Giant	-	Moyo et al. (2018)
78.	Rosmarinic acid	Cicla	Leaves	Mzoughi et al. (2020)

Contd..

Table 3 *Contd..*

Sl. No.	*Constituents*	*Variety-Cultivars*	*Plant Part*	*References*
79.	Salicylic acid	Cicla - Fordhook Giant	-	Moyo et al. (2018)
80.	Sinapic acid	Cicla - Fordhook Giant	-	
		Cicla	Leaves	Franzoni et al. (2024)
81.	Syringic acid	Cicla - Large white ribbed and CXS 2550	Leaves and stems	Pyo et al. (2004)
82.	Vanillic acid	Cicla - Large white ribbed and CXS 2550	Leaves and stems	Pyo et al. (2004)
Terpenes and Derivatives				
83.	3-hydroxy-5α,6α-epoxy-β-ionone	Cicla	Leaves	Kim et al. (2004)
84.	(+)-Dehydrovomifoliol	Cicla	Leaves	Kim et al. (2004)
85.	α-Pinene	Cicla	Leaves	Mzoughi et al. (2020)
86.	α-Terpineol	Cicla	Leaves	
87.	α-terpinyl acetate	Cicla	Leaves	
88.	β-caryophyllene	Cicla	Leaves	
89.	β-cyclocitral	Cicla	Leaves	
90.	β-Pinene	Cicla	Leaves	
91.	Carvone	Cicla	Leaves	
92.	Citronellol	Cicla	Leaves	
93.	Cis-dihydrocarvone	Cicla	Leaves	
94.	(E)-β-ionone	Cicla	Leaves	
95.	Geraniol	Cicla	Leaves	
96.	Limonene	Cicla	Leaves	
97.	Linalool	Cicla	Leaves	
98.	Verbenone	Cicla	Leaves	
99.	Menthol	Cicla	Leaves	
Aldehydes				
100.	Benzaldehyde	Cicla	Leaves	Mzoughi et al. (2020)
101.	Decanal	Cicla	Leaves	
102.	Nonanal	Cicla	Leaves	

Contd..

Table 3 *Contd..*

Sl. No.	*Constituents*	*Variety-Cultivars*	*Plant Part*	*References*
Alkanes				
103.	*n*-undecane	Cicla	Leaves	Mzoughi et al. (2020)
104.	*n*-dodecane	Cicla	Leaves	
105.	*n*-tridecane	Cicla	Leaves	
106.	*n*-tetradecane	Cicla	Leaves	
107.	*n*-pentadecane	Cicla	Leaves	
Heterocyclic Compounds				
108.	Cis-linalool oxide (furanoid)	Cicla	Leaves	Mzoughi et al. (2020)
Ketones				
109.	6-methyl-5-hepten-2-one	Cicla	Leaves	Mzoughi et al. (2020)
110.	Acetophenone	Cicla	Leaves	
111.	Camphor	Cicla	Leaves	
Carboxylic Acids				
112.	Exo-fenchyl acetate	Cicla	Leaves	Mzoughi et al. (2020)
113.	Isovaleric acid	Cicla	Leaves	
114.	Methyl octanoate	Cicla	Leaves	
115.	N-transferuloyl 3-O-methyldopamine	Cicla	Leaves	
116.	N-trans-feruloyl tyramine	Cicla	Leaves	
117.	Octanoic acid	Cicla	Leaves	
Other Compounds				
118.	(E)-anethole	Cicla	Leaves	Mzoughi et al. (2020)
119.	Δ8,9-dehydro-4-hydroxythymol dimethylether	Cicla	Leaves	

Betalains

The order Caryophyllales is well known for the presence of betalains, water-soluble pigments that impart the characteristic yellow-orange (betaxanthins) and red-violet (betacyanins) colors to plants. In addition to their widespread use as natural food colorants, betalains are valued for their antioxidant and antimicrobial activities. These properties not only enhance the nutritional quality of betalain-rich foods but also establish them as promising candidates for functional foods, nutraceuticals, and dietary supplements (Kugler et al., 2007; Mzoughi et al., 2020). In total, 42 betalains were detected in Swiss chard leaves (Table 3).

Betaxanthins

Betalains are metabolites detected in leaves, with a total of 26 betaxanthins reported (Table 3). Betaxanthins are yellow-to-orange, water-soluble pigments derived from betalamic acid conjugated with various amino acids or amines, and they contribute to the coloration of plants, such as yellow beets and amaranths. Beyond their role as pigments, betaxanthins exhibit strong antioxidant activity and are widely recognized for their potential health-promoting effects, including antimicrobial and anticancer properties.

Phytochemical studies on the leaves and petioles of Swiss chard revealed a diverse range of betaxanthins. Swiss chard, notably the Cicla group, is reported as a major source of betalains, with betaxanthins constituting a major class of these water-soluble pigments. Comprehensive phytochemical analysis of the petioles and leaves has uncovered a wide diversity of amino acid-derived betaxanthins, reflecting the rich biochemical nature of this leafy vegetable. Studies by Kugler et al. (2004, 2007) and more recently Franzoni et al. (2024) identified over 26 different betaxanthin derivatives, many of which are formed through the condensation of betalamic acid with free amino acids or amines. Among these, common amino acid conjugates, such as betaxanthin, glutamine-betaxanthin (vulgaxanthin I), glutamic acid-betaxanthin (vulgaxanthin II), and leucine-betaxanthin (vulgaxanthin IV) were consistently reported, while rarer forms like histidine-betaxanthin (muscaaurin VII), tyrosine-betaxanthin (portulacaxanthin II), and proline-betaxanthin (indicaxanthin) add to the chemical diversity. In addition, conjugates with biogenic amines, including dopamine-betaxanthin (miraxanthin V), tyramine-betaxanthin (miraxanthin III), and histamine-betaxanthin, highlight the wide metabolic interactions within the plant. More recently, Franzoni et al. (2004) confirmed the presence of isoleucine-, valine-, and tyrosine-derived betaxanthins in both petioles and leaves, suggesting tissue-specific distribution. Overall, this variety of betaxanthins gives Swiss chard its distinctive color while also boosting its antioxidant activity and possible health benefits.

Betacyanins

The other major group of betalains found in Swiss chard is betacyanins, a class of water-soluble, red-violet pigments responsible for the characteristic coloration of beets and other Caryophyllales plants. In addition to their role as natural food colorants, betacyanins exhibit strong antioxidant activity. They differ from yellow betaxanthins in their nitrogen-containing structure and their formation through condensation with cyclo-3,4-dihydroxyphenyl-L-alanine (cyclo-DOPA). However, betacyanins are chemically unstable and prone to degradation under high temperatures and pH fluctuations, which limits their application.

Studies on the leaves and petioles of Swiss chard by Kugler et al. (2004), Ali et al. (2009), Franzoni et al. (2024), and Almeida et al. (2025) identified a diverse spectrum of betacyanins. These include: betacyanin, betanidin, betanin, betanidin-monoferuloyl-5-O-β-diglucoside, isobetanidin, isobetanin, isobetanidin-monoferuloyl-5-O-β-diglucoside, isolampranthin II, lampranthin II, phyllocactin, neobetanin, 17-decarboxy-neobetanin, decarboxy-betanin, decarboxy-isobetanin, miraxanthin V, and feruloyl-hexosyl betanin (Table 3).

Flavonoids and Derivatives

Polyphenols represent a broad class of secondary metabolites that include both flavonoid and non-flavonoid compounds. Among them, flavonoids form the largest subgroup and naturally occur in plants as aglycones, glycosides, methylated derivatives, and esters. This group comprises flavanols, flavanones, flavones, and isoflavones, which are well recognized for their antioxidant, antimicrobial, and anti-inflammatory properties. In addition to contributing to plant coloration and signaling, polyphenols are extensively explored for their therapeutic applications in medicine, food, and agriculture (Lamani & Murthy, 2025b). In total, 22 reported flavonoid compounds identified in the leaves of BVc and BVf, along with references, are presented in Table 3.

The leaves of BVc, green and yellow Swiss chard, are rich sources of flavonoids with diverse structural variations. Phytochemical analyses have identified compounds, such as isorhamnetin 3-gentiobioside, isorhamnetin 3-vicianoside, kaempferol 3-gentiobioside, 2''-xylosylvitexin, and 6''-malonyl-2''-xylosylvitexin (Gil et al., 1998; Ninfali et al., 2007). In white-ribbed BVc, Pyo et al. (2004) reported the occurrence of (+)-catechin, kaempferol, myricetin, and quercetin, while Kim et al. (2004) further documented (+)-dehydrovomifoliol, 3-hydroxy-5α, 6α-epoxy-β-ionone, vitexin 7-O-β-D-glucopyranoside and vitexin 2''-O-β-D-glucopyranoside. Additional flavonoids, such as rutin, vitexin 2''-O-rhamnoside, and 2''-O-xylosylvitexin were also reported in BVc leaves (Ninfali et al., 2007). Similarly, BVf leaves have been shown to contain a distinct flavonoid profile, including 2''-oxylopyranosylvitexin, 6,8-di-C-β-D-glucopyranosylapigenin (vicenin-II), acacetin 8-C-β-D-glucopyranoside, acacetin 8-C-α-L-rhamnoside, and apigenin 8-C-β-D-glucopyranoside (vitexin) (Hala et al., 2019). A study on BVf leaves by Hashem et al. (2016) reported three new flavonoids, including 2''',2'''-di-O-α-rhamnopyranosyl-vicenin II, di-C-Glycosyl flavone, and herbacetin 3-O-β-xylopyranosyl-(1'''→2'')-O-β-glucopyranoside. More recently, Almeida et al. (2025) expanded this profile by reporting three additional flavonoids in BVf leaves: vitexin pentoside, vitexin deoxyhexoside, and isorhamnetin dihexoside. Beyond their phytochemical diversity, flavonoids in Swiss chard contribute significantly to its medicinal and therapeutic potential. The flavonoid derivatives exert antioxidant, anti-inflammatory, cardioprotective, anticancer, neuroprotective, and anti diabetic activities, hepatoprotective benefits (Almeida et al., 2025). Collectively, the flavonoid richness of BVc and BVf leaves enhances their nutritional value and positions Swiss chard as a promising source of bioactive compounds for functional foods, nutraceuticals, and preventive medicine.

Phenolic Compounds

Non-flavonoids primarily comprise phenolic acids, derived from benzoic and cinnamic acids. These compounds occur in free or conjugated forms and are well known for their antioxidant, anti-inflammatory, antimicrobial, cardioprotective, and anticancer activities, making them valuable dietary phytochemicals with applications in functional foods and nutraceuticals (Lamani & Murthy, 2025a).

The leaves of Swiss chard (BVc) have been extensively investigated for their phenolic acid composition, with a total of 13 phenolic acids reported. Pyo et al. (2004) identified caffeic acid, chlorogenic acid, ferulic acid, gallic acid, p-coumaric acid, p-hydroxybenzoic acid, protocatechuic acid, syringic acid, and vanillic acid in the leaves of the large white ribbed and CXS 2550 varieties. Similarly, Moyo et al. (2018) reported the presence of caffeic acid, ferulic acid, p-coumaric acid, p-hydroxybenzoic acid, protocatechuic acid, salicylic acid, and sinapic acid in the Fordhook Giant variety. More recently, Mzoughi et al. (2020) expanded the phenolic acid profile by reporting two additional compounds, namely myricitrin and rosmarinic acid. Franzoni et al. (2024) further confirmed the occurrence of caffeic acid, ferulic acid, gallic acid, and sinapic acid in BVc leaves, thus highlighting the chemical diversity of phenolic metabolites in Swiss chard leaves, which are associated with a wide range of biological activities. These compounds are potent antioxidants that scavenge free radicals, cardioprotective, neuroprotective effects, anti-inflammatory and anticancer properties, and also possess antimicrobial, hepatoprotective, and anti diabetic effects (Mzoughi et al., 2020). Collectively, the phenolic acid profile of BVc leaves not only enhances their nutritional and sensory qualities but also positions Swiss chard as a valuable source of bioactive phytochemicals with promising applications in functional foods, nutraceuticals, and preventive medicine.

Terpenes and Derivatives

The leaves of Swiss chard (BVc) are reported to contain a diverse array of terpenes and their derivatives, which contribute to the plant's aroma, flavor, and bioactive properties. In total, 16 terpenes and their derivatives were reported by Kim et al. (2004) and Mzoughi et al. (2020) from BVc leaves. Kim et al. (2004) reported oxygenated terpene derivatives include 3-hydroxy-5α, 6α-epoxy-β-ionone and (+)-dehydrovomifoliol. Studies of Mzoughi et al. (2020) revealed the presence of some key monoterpenes, including α-pinene, β-pinene, limonene, α-terpineol, α-terpinyl acetate, carvone, citronellol, linalool, geraniol, verbenone, and menthol, while notable sesquiterpenes include β-caryophyllene. Additionally, oxygenated derivatives, such as β-cyclocitral and cis-dihydrocarvone were also detected, along with (E)-β-ionone. These compounds are known to exhibit a range of biological activities, including antioxidant, antimicrobial, neuroprotective, anti-inflammatory, and cytoprotective effects (Mzoughi et al., 2020). Collectively, the terpene profile of BVc leaves not only contributes to their organoleptic qualities but also underscores their potential as a source of bioactive phytochemicals with applications in functional foods, nutraceuticals, and therapeutic formulations.

Carboxylic Acids

Carboxylic acids are essential biomolecules, like fatty acids and amino acids. In the food industry, they act as flavorings and preservatives, while in pharmaceuticals, they are active ingredients in anti-inflammatory drugs like aspirin and are used as drug delivery agents (Gamba et al., 2021). Mzoughi et al. (2020) reported the presence of six carboxylic acid derivatives in the leaves of BVc. These included exo-fenchyl acetate, isovaleric acid, methyl octanoate, N-transferuloyl 3-O-methyldopamine, N-trans-feruloyl tyramine, and octanoic acid. Such compounds are of particular interest due to their bioactive roles: antioxidant, anti-inflammatory, and antimicrobial.

Alkanes

Mzoughi et al. (2020) reported five alkanes in the leaves of BVc, they are n-undecane, n-dodecane, n-tridecane, n-tetradecane, and n-pentadecane. As key components of cuticular waxes, alkanes reduce water loss, enhance drought tolerance, and protect against pathogens. They are also known for antimicrobial, antifungal, and insect-repellent activities.

Other Compounds

Mzoughi et al. (2020) identified additional bioactive compounds in the leaves of BVc, including three aldehydes (benzaldehyde, decanal, nonanal), three ketones (6-methyl-5-hepten-2-one, acetophenone, camphor), a heterocyclic compound (cis-linalool oxide (furanoid)), and other metabolites, such as (E)-anethole and Δ8,9-dehydro-4-hydroxythymol dimethyl ether. These compounds serve as important sources of antioxidants, antimicrobial, and anti-inflammatory agents.

Bioactivities of Compounds Detected

The extracts and phytochemical constituents of Swiss chard showed various biological activities, which include antioxidant activity, cytotoxic/ anticancer activity, antidiabetic activity, and antimicrobial activity. Some of the existing articles on the major pharmacological potential are discussed in this chapter. The biological activities are summarized in Table 4, while the structures of some potent bioactive constituents are illustrated in Figure 1.

Table 4 Biological activities of Swiss chard (*Beta vulgaris* subsp. *vulgaris*) leaf extracts.

Variety/ Cultivars	*Extracts Used*	*Bioactivity*	*References*
Antioxidant activity			
Cicla- Red leaf variety	Methanol extract	DPPH, 87% inhibition at 500 mg/ml	Pyo et al. (2004)
Cicla	Distilled water extract	DPPH, EC_{50}= 23.85 µg/ml; ABTS, EC_{50}= 1093.14 µg/ml; HRSA, EC_{50}= 157.94 µg/ml; DMPD, EC_{50}= 45.21 µg/ml	Sacan & Yanardag (2010)
Cicla	Ethanol extract	DPPH, EC_{50}= 0.75 mg/ml; ABTS, EC_{50}= 1.22 mg/ml; Reducing power, EC_{50}= 0.21 mg/ml; β-carotene bleaching inhibition, EC_{50}= 0.10 mg/ml; TBARS (Thiobarbituric acid reactive substances) inhibition, EC_{50} 0.08 mg/ml	Mzoughi et al. (2020)
Cicla	Aqueous extract	DPPH, IC_{50}= 4.44 mg/ml	Ivanovic et al. (2019)
Cicla	Aqueous extract	DPPH, 4.7 mg TE (Trolox equivalent) 100 g/ fw; FRAP, 29.6 mg TE 100 g/ fw	Libutti et al. (2023)
Flavescens	Leaves	TBARS, EC_{50}= 0.88 mg/ml	Almeida et al. (2025)
Antidiabetic activity			
Cicla	Aqueous extract	In diabetic rabbits, 2, 4, and 8 g/ kg extract had a significant decrease in blood glucose levels by 12.5, 10.6, and 15.1%, respectively, ($P < 0.1$)	Yanardag & Colak (2011)
Cicla	Distilled water extract	Hypoglycemic agents by diabetic patients in Turkey. Extract on pancreatic B cells in streptozotocin-diabetic rats	Bolkent et al. (2000)
Cicla	Distilled water extract	Hypoglycemic agents used by diabetic patients in Turkey. Diabetes-induced free radical-mediated injury in rat aorta and heart tissues	Sener et al. (2002)
Cicla	Aqueous extract	In STZ diabetic rats, the extract significantly decreased the level of LPO in the diabetic rats.	Ozsoy-Sacan et al., 2004
Cicla	Ethanol extract	α-Glucosidase inhibitory activity, IC_{50}= 0.13 mg/ml; α-Amylase inhibitory activity, IC_{50}= 1.03 mg/ml	Mzoughi et al. (2020)

Contd..

Table 4 *Contd..*

Variety/ Cultivars	*Extracts Used*	*Bioactivity*	*References*
Anticancer activity			
Cicla	Hydroalcoholic extract	Antimitotic activity on MCF-7 human breast cancer cells and for toxicity to human lymphocytes and macrophages. A 5, 10, 15, and 20 μg/ml of P2 fraction or combinations of its components were tested for 72 h	Ninfali et al. (2007)
Hepatoprotective activity			
Cicla	Methanol extract	Effective against the release of glutamic pyruvic transaminase (GPT) from the primary cultures of rat hepatocytes injured by CCI4	Kim et al. (2004)
Flavescens	70% Methanol extract	The extract shows hepatoprotective effects in rats intoxicated by administration of acetaminophen. The extract at 200 mg/kg attenuated the increased level of ALT, AST, and ALP by 37.44, 42.07, and 48.01 %, respectively, as compared with the acetaminophen-treated rats.	Hashem et al. (2016)
Neurotropic activity (Acetylcholinesterase inhibitor activity)			
Cicla	Distilled water extract	A high AChE inhibition (20.98%) was seen in 6 μg/ml with EC_{50} value 14.43 μg/ml	Sacan & Yanardag (2010)
Antimicrobial activity			
Antibacterial activity Flavescens	Hydroethanolic extracts	Determined by a rapid colorimetric method by microdilution and using the p-iodonitrotetrazolium chloride (INT) dye. Extracts used against three gram-positive and five gram-negative bacteria. Each one exhibited antibacterial capacity against the strains tested, except *Enterobacter cloacae* and *Pseudomonas aeruginosa*.	Almeida et al. (2025)
Antifungal activity Flavescens		Extracts used against two fungi, *Aspergillus fumigatus* and *A. brasiliensis*, each one which exhibited antifungal capacity against the strains tested.	

Antioxidant Activity

Plants have long been used in traditional medicine to treat various diseases, and many are rich sources of phytochemicals with potent antioxidant properties. Antioxidants mitigate the harmful effects of free radicals on cells, thereby protecting against oxidative stress. The antioxidant potential of Swiss chard (BVc) has been extensively studied. Pyo et al. (2004) evaluated the red leaf variety using DPPH and thiocyanate assays at concentrations of 500 and 1000 mg/ml, respectively. The methanolic extract of red leaves exhibited exceptionally high activity, with 87.0% inhibition at 500 mg/ml, significantly surpassing BHT (65.3%) and α-tocopherol (80.4%) at the same concentration, likely due to its rich flavonoid and phenolic content (Table 1, Figure 1).

Sacan and Yanardag (2010) further demonstrated the strong antioxidant potential of Swiss chard leaves using various assays. The aqueous extract showed DPPH activity with an EC_{50} of 23.85 μg/ml (compared to 47.61 and 80.83 μg/ml for BHA and BHT), ABTS activity with an EC_{50} of 1093.14 μg/ml (113.70 μg/ml against standard rutin), HRSA activity with an EC_{50} of 157.94 μg/ml (lower than trolox, 46.76, and ascorbic acid, 116.04), and DMPD activity with an EC_{50} of 45.21 μg/ml (9.87 and 10.38 μg/ml for standard cysteine and trolox). These results suggest that phenolic acids are major contributors to antioxidant activity. Overall, the water extract of Swiss chard, as typically consumed in Turkey, represents a readily accessible natural source of antioxidants with potential health-promoting benefits.

The antioxidant potential of the ethanol extract from Swiss chard (BVc) leaves was evaluated using five complementary *in vitro* assays (Mzoughi et al., 2020). These included free radical scavenging activity against DPPH and ABTS, ferric reducing power, β-carotene bleaching inhibition, and TBARS inhibition (linoleic acid peroxidation). The ethanol extract exhibited remarkable antioxidant activities, with EC_{50} values of 0.75 mg/ml (DPPH), 1.22 mg/ml (ABTS), 0.21 mg/ml (reducing power), 0.10 mg/ml (β-carotene bleaching), and 0.08 mg/ml (TBARS). Such potent activities are attributed to the high phenolic content of the extract. Additionally, the antioxidant capacity was confirmed by the DNA nicking assay. DNA from hydroxyl radical-induced damage, preserving its native forms. These findings highlight Swiss chard as a rich source of natural antioxidants with significant potential for protecting biomolecules against oxidative stress. Ivanovic et al. (2019) evaluated the antioxidant activity of the aqueous extract of BVc leaves grown under different fertilization and irrigation treatments. The extract of control I100F0 demonstrated notable free radical scavenging capacity in the DPPH assay, with an IC_{50} value of 4.44 mg/ml, indicating its potential as a natural antioxidant source. Libutti et al. (2023) investigated the antioxidant properties of the aqueous extract of BVc. The extract showed a DPPH radical scavenging activity of 4.7 mg Trolox equivalents (TE) per 100 g fw, and a FRAP of 29.6 mg TE per 100 g fw, highlighting its contribution as a dietary source of natural antioxidants. Almeida et al. (2025) evaluated the antioxidant potential of BVc leaves using the TBARS

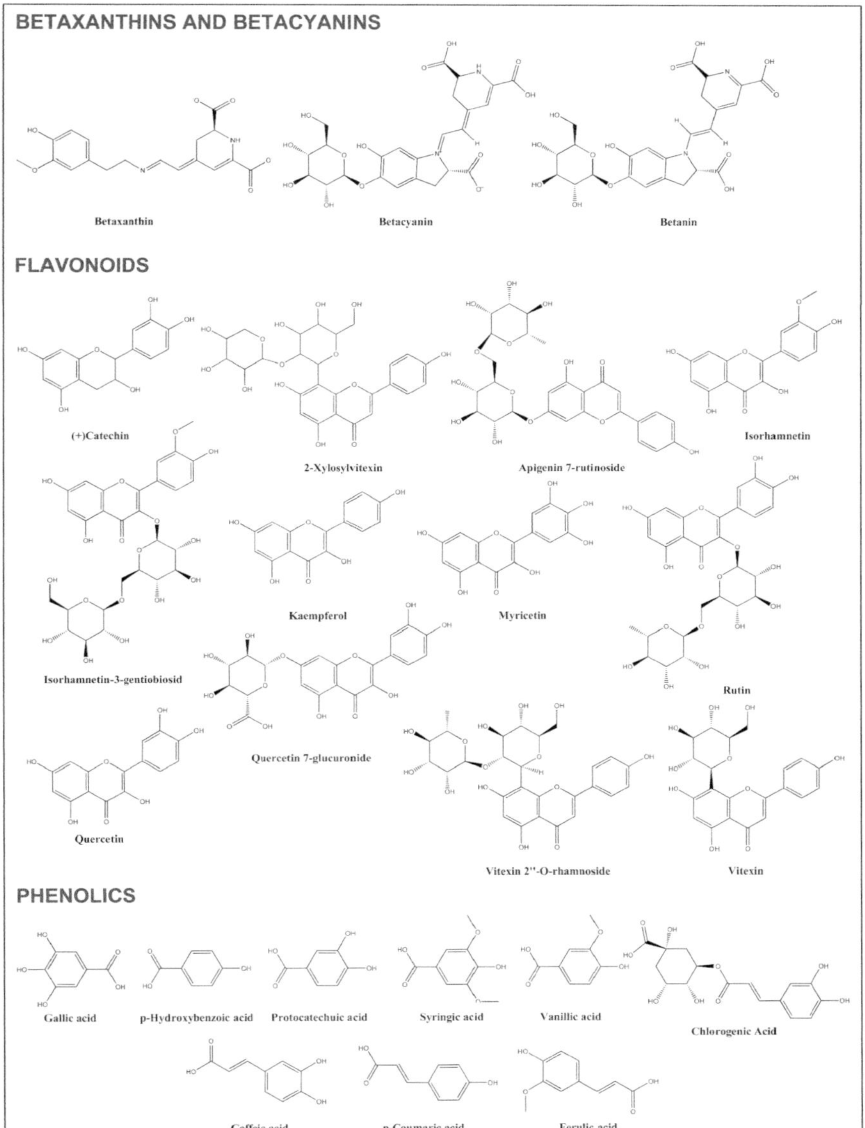

Fig. 1 Chemical structures of some potent bioactive compounds of Swiss chard (*Beta vulgaris* subsp. *vulgaris*): Betalains - Betaxanthin, and Betacyanin. Flavonoids - (+)Catechin, 2-Xylosylvitexin, Apigenin-7-rutinoside, Isorhamnetin, Isorhamnetin-3-gentiobiosid, Kaempferol, Myricetin, Rutin, Quercetin, Quercetin-7-glucuronide, Vitexin 2"-O-rhamnoside, and Vitexin, Phenolics - Gallic acid, p-Hydroxybenzoic acid, Protocatechuic acid, Syringic acid, Vanillic acid, Chlorogenic acid, Caffeic acid, p-Coumaric acid, and Ferulic acid.

assay. The extract demonstrated notable lipid peroxidation inhibitory activity, with an EC_{50} value of 0.88 mg/ml, suggesting its effectiveness in protecting against oxidative stress-induced cellular damage.

Antidiabetic Activity

Swiss chard (BVc) is widely recognized for its anti diabetic properties and has been traditionally used in folk medicine to manage hyperglycemia in diabetic patients (Yanardag and Colak, 1998). Despite its longstanding use, only a limited number of scientific studies have systematically investigated its potential as a therapeutic agent for diabetes. Diabetes mellitus promotes excessive free radical production, leading to a higher risk of atherosclerosis and cardiovascular diseases. *Diabetes mellitus* is a major global health concern, affecting 2.5–3% of the population worldwide and exceeding 7% in some countries. It is a metabolic disorder characterized by chronic hyperglycemia, which leads to damage of vital organs, such as the heart, kidneys, liver, eyes, and blood vessels through impaired glycemic control and inflammation (Ozsoy-Sacan et al., 2004).

Antidiabetic activity of aqueous extracts of BVc leaves was carried out by Yanardag and Colak (1998) in normal and alloxan-induced diabetic rabbits. Initially rabbits were administered 150 mm/kg of alloxan dissolved in distilled water intravenously. The effect of oral administration of varying doses of aqueous and methanolic chard extracts on blood glucose levels was evaluated. In normal rabbits, oral administration of aqueous chard extract at a dose of 2 g/kg resulted in a significant reduction ($P < 0.01$) in blood glucose levels 4 h post-treatment. In diabetic rabbits, administration of 2, 4, and 8 g/kg extract produced reductions in blood glucose levels of 12.5%, 10.6%, and 15.1%, respectively, ($P < 0.1$). The observed hypoglycemic activity of chard leaves is likely attributable to their bioactive constituents, particularly flavonoids, saponins, and polysaccharides. Bolkent et al. (2000) reported antidiabetic activity of distilled water extracts of BVc leaves by using Swiss albino rats of age 6–6.5 months and 150–200 g weight. They were divided into four groups: control, control + chard extract, diabetic, and diabetic + chard extract. The extract was administered orally (2 g/kg/day) for 28 days. The rats were further subjected to histological and biochemical analysis. The histological analysis of diabetic rats showed reduced β-cell numbers, diminished secretory granules, and structural damage within the islets of Langerhans. Treatment with chard extract, however, increased β-cell counts and granule content, alongside hypertrophic Golgi apparatus, indicating enhanced β-cell activity. Although the extract had no effect on normal animals, it significantly lowered blood glucose and improved body weight in streptozotocin-induced diabetic animals, with the greatest reduction in glucose observed on the 42nd day. These results indicate that chard extract may act by stimulating insulin secretion in diabetic rats.

Another report regarding the antidiabetic property of Swiss chard suggested that the aqueous extracts of leaves are a potent drug against diabetic animals (Sener et al., 2002). Chard (BVc) has long been used in Turkey as a traditional hypoglycemic remedy. They investigated the effect of chard extract on diabetes-induced oxidative damage in rat aorta and heart tissues. Female Swiss albino rats are 6–6.5 months old, weighing 150–170 g and are randomly divided into four groups: group I, control (non-diabetic); group II, control + chard extract; group III, diabetic; and group IV, diabetic + chard extract. Diabetes was induced by injection of streptozotocin (65 mg/kg, dissolved in citrate buffer, pH 4.5). Rats with plasma glucose levels of 150–350 mg/100 ml within 48 h were classified as diabetic. Starting from day 14, groups II and IV received chard extract (2 g/kg/day) by gavage for 28 days. The results observed that in diabetic rats treated with chard (D + chard group), aortic lipid peroxidation was significantly reduced compared with untreated diabetics ($P < 0.001$), reaching levels comparable to controls. Similarly, chard treatment markedly lowered lipid peroxidation in heart tissue ($P < 0.01$). It is observed that the chard extract reduced lipid peroxidation and restored glutathione levels in the heart and aorta of STZ-diabetic rats, indicating protection against diabetes-induced oxidative damage through its hypoglycemic and antioxidant effects. Chard (BVc), a leafy vegetable traditionally used in Turkey for managing diabetes, was evaluated for its hepatoprotective potential in an experimental model by Ozsoy-Sacan et al. (2004). In this study, diabetes was induced in female Swiss albino rats (6–6.5 months old, 150–200 g) by a single intraperitoneal injection of streptozotocin (65 mg/kg body weight). The rats were then divided into four groups. Diabetes induction led to marked hepatic degeneration, hyperglycemia, elevated serum transaminases, alkaline phosphatase, total lipids, sialic acid, uric acid, increased lipid peroxidation and non-enzymatic glycosylation, along with decreased body weight and glutathione levels. Daily oral administration of chard extract (2 g/kg) for 28 days significantly attenuated these alterations, lowering blood glucose, oxidative stress markers, and biochemical abnormalities, while restoring glutathione levels and improving liver morphology. These findings suggest that chard exerts hepatoprotective effects against diabetes-induced oxidative and metabolic damage, likely through its combined hypoglycemic and antioxidant activities. In line with earlier findings, Mzoughi et al. (2020) demonstrated that the ethanolic extract of BVc leaves exhibits strong inhibitory effects on α-glucosidase (IC_{50} = 0.13 mg/ml) and α-amylase (IC_{50} = 1.03 mg/ml). Since these enzymes play a key role in carbohydrate digestion and glucose absorption, their inhibition represents an important therapeutic strategy for managing postprandial hyperglycemia. Plant-based enzyme inhibitors have attracted considerable attention as safer alternatives to synthetic drugs, and B. vulgaris is increasingly recognized as a promising antidiabetic herb (Ozsoy-Sacan et al., 2004).

Anticancer/Cytotoxic Activity

Considering its anti-cancer and cytotoxic qualities, Swiss chard has impressed investigators for the past ten years. These activities are emphasized in several articles that include crude extracts, as listed in Table 4. Some potent bioactive compounds structures shown in Figure 1. Ninfali et al. (2007) evaluated the anticancer activity of a hydroalcoholic leaf extract of BVc on MCF-7 human breast cancer cells and its safety on immune cells. The flavonoid-rich P2 fraction (5–20 μg/ml, 72 h) significantly inhibited MCF-7 proliferation (IC_{50}= 9 μg/ml) without inducing apoptosis, showing no toxicity to lymphocytes and only mild toxicity to macrophages. Vitexin-2''-O-rhamnoside, the major component (> 66%), was a potent antioxidant and inhibitor of DNA synthesis, while its demethylated form, 2''-xylosylvitexin, showed weaker activity and acted as an activator. Minor constituents, such as rutin, quercetin-7-glucuronide, and apigenin-7-rutinoside enhanced the overall antimitotic effect. Hence this study revealed the synergistic interactions among P2 components contribute to its selective anticancer activity, supporting the studied species as a promising source of natural anticancer agents.

Hepatoprotective Activity

Kim et al. (2004) isolated four flavonoid compounds from BVc leaves: (+)-dehydrovomifoliol, 3-hydroxy-5α, 6α-epoxy-β-ionone, vitexin 7-O-β-D-glucopyranoside, and vitexin 2''-O-β-D-glucopyranoside. These compounds were evaluated for their hepatoprotective potential using primary cultured rat hepatocytes exposed to carbon tetrachloride (CCl_4)-induced toxicity, a well-established model of chemical liver injury. Among the isolated compounds, vitexin 7-O-β-D-glucopyranoside and vitexin 2''-O-β-D glucopyranoside showed the strongest protective effects, reducing the release of glutamic pyruvic transaminase (GPT) by 65.8% and 56.1%, respectively, compared to untreated controls. The ability of these flavonoids to inhibit GPT release indicates their role in stabilizing hepatocyte membranes and preventing enzyme leakage, thereby mitigating liver cell damage. Remarkably, their protective effects were comparable to silibinin (69.8%), a well-known reference hepatoprotective agent. These findings highlight the therapeutic potential of chard flavonoids, particularly vitexin glycosides, as natural hepatoprotective agents against chemically induced liver injury.

Three novel flavonoids, including 2''',2'''-di-O-α-rhamnopyranosyl-vicenin II, a di-C-glycosyl flavone, and herbacetin 3-O-β-xylopyranosyl-(1'''→2'')-O-β-glucopyranoside, were reported by Hashem et al. (2016) from BVf leaves. The study also investigated the hepatoprotective and hypolipidemic effects of a 70% methanolic leaf extract. The leaf extract demonstrated hepatoprotective effects in acetaminophen-intoxicated rats and hypolipidemic activity in high-fat diet- induced hypercholesterolemic rats, as evidenced by improved liver function enzymes (aspartate (AST) and alanine aminotransferases (ALT) and alkaline phosphatase (ALP), lipid profile, and supportive histopathological findings. At a dose of 200 mg/kg, the extract significantly reduced serum ALT, AST, and ALP levels by 37.44%, 42.07%, and 48.01%, respectively, compared to acetaminophen-treated controls. In high-fat diet–induced hypercholesterolemic rats, the extract significantly modulated lipid metabolism, with elevations in total cholesterol (41.71%), HDL-C (47.19%), LDL-C (127.19%), and triglycerides (79.71%)

compared with normal controls. In high-fat diet–induced hypercholesterolemic rats, the extract significantly modulated lipid metabolism, with elevations in total cholesterol (41.71%), HDL-C (47.19%), LDL-C (127.19%), and triglycerides (79.71%) compared with normal controls. These findings highlight the modulatory role of BVf on lipid homeostasis, suggesting its potential as a natural agent in managing hyperlipidemia and associated cardiovascular risks.

Neurotropic Activity

Plants have traditionally been used to support cognitive function and mitigate symptoms of neurodegenerative disorders. Inhibition of acetylcholinesterase (AChE), the enzyme that degrades acetylcholine (ACh), is a key therapeutic strategy for conditions, such as Alzheimer's disease, ataxia, and Parkinson's disease (Mukherjee et al., 2007). Sacan and Yanardag (2010) reported that the distilled water extract of BVc exhibited significant AChE inhibitory activity. In their study, AChE inhibition increased in a dose-dependent manner, with 20.98% inhibition observed at 6 μg/ml and an EC_{50} value of 14.43 μg/ml. The anti-acetylcholinesterase activity of chard is attributed to its bioactive constituents, including terpenes, oils, flavonoids, alkaloids, and other phenolic compounds, which act potentially to inhibit the enzyme. Although AChE inhibitory activity has been documented in this species, its potential efficacy against other neurological disorders has yet to be investigated.

Antimicrobial Activity

Almeida et al. (2025) evaluated the antibacterial potential of hydroethanolic extracts from BVf leaves using a rapid colorimetric microdilution method with p-iodonitrotetrazolium chloride (INT) as a growth indicator. The extracts were tested against a panel of eight bacterial strains, comprising three Gram-positive (*Staphylococcus aureus, Bacillus cereus* and *Listeria monocytogenes*) and five Gram-negative species (*Escherichia coli, Salmonella enterocolitis, Pseudomonas aeruginosa, Yersinia enterocolitica* and *Enterobacter cloacae*). The results demonstrated that the extracts exhibited notable antibacterial activity against most of the tested strains, with the exceptions of *Enterobacter cloacae* and *Pseudomonas aeruginosa*, which were resistant to the extract. These findings suggest that BVf leaves contain bioactive compounds including phenolic acids, fatty acids, and tocopherols, which are capable of inhibiting the growth of several pathogenic bacteria, highlighting their potential as natural antibacterial agents. In addition, the hydroethanolic extracts were assessed for antifungal activity against *Aspergillus fumigatus* and *Aspergillus brasiliensis*. The extracts displayed significant inhibitory effects against both fungal strains, indicating a broad-spectrum antifungal potential. These results support the traditional use of Swiss chard in managing microbial infections and underscore the value of BVf leaf extracts as a promising source of natural antimicrobial compounds.

Conclusion

This chapter provides a comprehensive overview of the bioactive compounds and biological activities of Swiss chard, a leafy vegetable of Amaranthaceae, highlighting its pharmacological potential and therapeutic benefits, and emphasizing its role as a functional food in nutrition, preventive medicine, and drug development. Swiss chard BVc and BVf are nutritionally and pharmacologically valuable leafy vegetables, rich in essential macronutrients, micronutrients, fatty acids, polyols, organic acids, and tocopherols. Its leaves are a reservoir of diverse bioactive compounds, including 42 betalains (betaxanthins and betacyanins), and 77 other bioactive compounds including flavonoids, phenolic acids, terpenes, carboxylic acids, and other phytochemicals, which collectively contribute to its wide spectrum of biological activities. These include antioxidant, antidiabetic, anticancer/cytotoxic, hepatoprotective, neuroprotective, and antimicrobial activities, as demonstrated in various *in vitro* and *in vivo* studies. The leaves exhibit strong free radical scavenging, enzyme inhibitory, and lipid-modulatory effects, alongside protective actions against chemically induced liver and oxidative damage. Swiss chard's phytochemicals, particularly flavonoids, betalains, and phenolic acids, underlie its medicinal potential, supporting its use in functional foods, nutraceuticals, and preventive medicine. Overall, these findings point to Swiss chard as a rich natural source of nutrients and bioactive compounds that may help in managing cardiovascular health, metabolic disorders, neurodegenerative diseases, infections, and conditions linked to oxidative stress.

Acknowledgments

Hosakatte Niranjana Murthy is thankful for the "Brain Pool" fellowship (Grant No. RS-2025-25425249) of the National Research Foundation, Republic of Korea.

References

Aguero, M.V., Pereda, J., Roura, S.I., Moreira, M.R. and del Valle, C.E. 2005. Sensory and biochemical changes in Swiss chard (*Beta vulgaris*) during blanching. *LWT - Food Sci. Technol., 38*(7): 772–778. https://doi.org/10.1016/j.lwt.2004.07.018.

Ali, B., Khandaker, L. and Oba, S. 2009. Comparative study on functional components, antioxidant activity, and color parameters of selected colored leafy vegetables as affected by photoperiods. *J. Food Agric Environ., 7*(3): 392–398. https://doi.org/10.1234/4.2009.2605.

Almeida, D., Petropoulos, S.A., da Silveira, T.F.F., Pires, T.C.S.P., Ferreira, I.C.F.R., Fernandes, A. and Barros, L. 2025. Exploring the Biochemical Profile of *Beta vulgaris* L.: A Comparative Study of Beetroots and Swiss chard. *Plants, 14*(4): 591. https://doi.org/10.3390/plants14040591.

Bolkent, S., Yanardag, R., Tabakoglu-Oguz, A. and Ozsoy-Sacan, O. 2000. Effects of chard (*Beta vulgaris* L. var. cicla) extract on pancreatic B cells in streptozotocin-diabetic rats: A morphological and biochemical study. *J. Ethnopharmacol., 73*(1–2): 251–259. https://doi.org/10.1016/s0378-8741(00)00328-7.

Bozokalfa, M.K., Yagmur, B., Asciogul, T.K. and Esiyok, D. 2011. Diversity in nutritional composition of Swiss chard (*Beta vulgaris* subsp. L. var. cicla) accessions revealed by multivariate analysis. *Plant Genet. Res., 9*(4): 557–566. https://doi.org/10.1017/S1479262111000876.

Colonna, E., Rouphael, Y., Barbieri, G. and De Pascale, S. 2016. Nutritional quality of ten leafy vegetables harvested at two light intensities. *Food Chem.*, *199*: 702–710. https://doi.org/10.1016/j.foodchem.2015.12.068.

Daiss, N., Lobo, M.G., Socorro, A.R., Bruckner, U., Heller, J. and Gonzalez, M. 2008. The effect of three organic pre harvest treatments on Swiss chard (*Beta vulgaris* L. var. cycla L.) quality. *Eur. Food Res. Technol.*, *226*: 345–353. https://doi.org/10.1007/s00217-006-0543-2.

Dincler, A. and Aydemir, T. 2001. Purification and Characterization of Catalase from Chard (*Beta vulgaris* var. cicla). *J. Enzyme Inhib.*, *16*(2): 165–175. https://doi.org/10.1080/14756360109162366.

Duke, J.A. 1983. *The Handbook of Energy Crops*. http://www.hort.purdue.edu/newcrop/duke_energy/dukeindex.html.

Dzida, K. and Pitura, K. 2008. The influence of varied nitrogen fertilization on yield and chemical composition of Swiss chard (*Beta vulgaris* L. var. cicla L.). *Acta Sci. Pol., Hortorum Cultus*, *73*: 15–24.

Ferland, G. and Sadowski, J.A. 1992. Vitamin K1 (phylloquinone) content of green vegetables: Effects of plant maturation and geographical growth location. *J. Agric. Food Chem.*, *40* (10): 1874–1877. https://doi.org/10.1021/jf00022a029.

Franzoni, G., Guffanti, D., Ferrante, A., Cejudo-Bastante, M.J., Rodriguez-Pulido, F.J. Gordillo, B. and Cocetta, G. 2024. Evaluation of Swiss chard (*Beta vulgaris* L. ssp. cicla) physiological and qualitative responses to water deficit and salicylic acid treatment. *J. Agr. Food Res.*, *18*: 101524. https://doi.org/10.1016/j.jafr.2024.101524.

Gamba, M., Raguindin, P.F., Asllanaj, E., Merlo, F., Glisic, M., Minder, B., Bussler, W., Metzger, B., Kern, H. and Muka, T. 2021. Bioactive compounds and nutritional composition of Swiss chard (*Beta vulgaris* L. var. Cicla and Flavescens): A systematic review. *Crit. Rev. Food Sci. Nutr.*, *61*: 3465–3480. https://doi.org/10.1080/10408398.2020.1799326.

Gao, Z., Han, X.-H. and Xiao, X.-G. 2009. Purification and characterization of polyphenol oxidase from red Swiss chard (*Beta vulgaris* subspecies cicla) leaves. *Food Chem.*, *117* (2): 342–348. https://doi.org/10.1016/j.foodchem.2009.04.013.

Gil, M.I., Ferreres, F. and Tomas-Barberan, F.A. 1998. Effect of modified atmosphere packaging on the flavonoids and vitamin C content of minimally processed Swiss chard (*Beta vulgaris* subspecies cycla). *Journal of Agricultural and Food Chemistry*, *46*(5): 2007–2012. https://doi.org/10.1021/jf970924e.

Hala, S., Mohammed, M., Abdel-Aziz, M., Marwa, S.A., Amal, M.S., Mona, A.M. and Mosad, A.G. 2019. Antibacterial and Potential Antidiabetic Activities of Flavone C-glycosides Isolated from *Beta vulgaris* Subspecies cicla L. var. Flavescens (Amaranthaceae) Cultivated in Egypt. *Curr. Pharm. Biotechnol.*, *20*(7): 595–604. http://dx.doi.org/10.2174/1389201020666190613161212.

Hashem, A.N., Soliman, M.S., Hamed, M.A., Swilam, N.F., Lindequist, U. and Nawwar, M.A. 2016. *Beta vulgaris* subspecies cicla var. flavescens (Swiss chard): Flavonoids, hepatoprotective and hypolipidemic activities. *Die Pharmazie*, *71*(4): 227–232. http://dx.doi.org/10.1691/ph.2016.5821

Ivanovic, L., Milasevic, I., Topalovic, A., Durovic, D., Mugosa, B., Knezevic, M. and Vrvic, M. 2019. Nutritional and phytochemical content of Swiss chard from Montenegro, under different fertilization and irrigation treatments. *Br. Food J.*, *121*(2): 411–425. https://doi.org/10.1108/Bfj-03-2018-0142.

Kim, I., Chin, Y.-W., Lim, S.W., Kim, Y.C. and Kim, J. 2004. Norisoprenoids and hepatoprotective flavone glycosides from the aerial parts of *Beta vulgaris* var. cicla. *Arch. Pharm. Res.*, *27*(6): 600–603. https://doi.org/10.1007/BF02980156.

Kolota, E., Sowinska, K. and Czerniak, K. 2010. Yield and nutritional value of Swiss chard grown for summer and autumn harvest. *J. Agric. Sci.*, *2*(4): 120. https://doi.org/10.5539/jas.v2n4p120.

Kugler, F., Stintzing, F. and Carle, R. 2004. Identification of betalains from petioles of differently colored Swiss chard (*Beta vulgaris* L. ssp. cicla [L.] Alef. Cv. bright lights) by high-performance liquid chromatography-electrospray ionization mass spectrometry. *J. Agric. Food Chem.*, *52*(10): 2975–2981. https://doi.org/10.1021/jf035491w.

Kugler, F., Graneis, S., Stintzing, F.C. and Carle, R. 2007. Studies on betaxanthin profiles of vegetables and fruits from the chenopodiaceae and cactaceae. *Z. Naturforsch., C: J. Biosci.*, *62*(5–6): 311–318. https://doi.org/10.1515/znc-2007-5-601.

Lamani, S. and H.N. Murthy. 2025a. Genetic diversity of Wood apple (*Limonia acidissima* L.) and Sustainable utilization. (pp. 78–94). In: Murthy H.N. [Eds.] . *Genetic Diversity of Fruits and Nuts*. Taylor and Francis Group. CRC Press. http://dx.doi.org/10.1201/9781003492207-5.

Lamani, S. and Murthy, H.N. 2025b. Phytochemical and Biological activities of Ceriops tagal (Perr.). C. B. Rob. (pp. 1–27). In: Murthy H.N. [Eds.], *Bioactive Compounds in Mangroves and Their Associates, Reference Series in Phytochemistry*. Springer Nature Switzerland AG. http://dx.doi.org/10.1007/978-3-031-63920-3_10-1.

Lange, W., Brandeburg, W.A. and De Bock, T.S.M. 1999. Taxonomy and cultonomy of beet (*Beta vulgaris* L.). *Bot. J. Linn. Soc.*, *130*: 81–96. https://doi.org/10.1111/j.1095-8339.1999.tb00785.x.

Lewellen, R., Panella, R. and R. Harveson. 2009. Introduction: Botany of the beet plant. In: Harveson, R., Hanson, L.E. and Hein, G.O. (Eds.), *Compendium of the Beet Diseases and Insects*. St. Paul, MN: APS Press.

Libutti, A., Russo, D., Lela, L., Ponticelli, M., Milella, L. and R. Rivelli. 2023. Enhancement of Yield, Phytochemical Content, and Biological Activity of a Leafy Vegetable (*Beta vulgaris* L. var. cycla) by Using Organic Amendments as an Alternative to Chemical Fertilizer. *Plants*, *12*(3): 569. https://doi.org/10.3390/plants12030569.

Miceli, A. and C. Miceli. 2014. Effect of nitrogen fertilization on the quality of Swiss chard at harvest and during storage as minimally processed produce. *J. Food Qual.*, *37*(2): 125–234. https://doi.org/10.1111/jfq.12073.

Moreira, M.dR., Roura, S.I. and del Valle, C.E. 2003. Quality of Swiss chard produced by conventional and organic methods. *LWT- Food Science and Technology*, *36*(1): 135–141. https://doi.org/10.1016/S0023-6438(02)00207-4.

Moyo, M., Amoo, S.O., Aremu, A.O., Gruz, J., Subrtova, M., Jarosova, M., Tarkowski, P. and K. Dolezal. 2018. Determination of mineral constituents, phytochemicals, and antioxidant qualities of *Cleome gynandra*, compared to *Brassica oleracea* and *Beta vulgaris*. *Front. Chem.*, *5*: 128. https://doi.org/10.3389/fchem.2017.00128.

Mukherjee, P.K., Kumar, V., Mal, M. and Houghton, P.J. 2007. Acetylcholinesterase inhibitor from plants. *Phytomedicine*, *14*, 289–300. https://doi.org/10.1016/j.phymed.2007.02.002.

Mzoughi, Z., Chahdoura, H., Chakroun, Y., Camara, M., Fernandez-Ruiz, V., Morales, P., Mosbah, H., Flamini, G., Snoussi, M. and Majdoub, H. 2020. Wild edible Swiss chard leaves (*Beta vulgaris* L. var. cicla): Nutritional, phytochemical composition, and biological activities. *Food Res. Int.*, *119*: 612–621. https://doi.org/10.1016/j.foodres.2018.10.039.

Ninfali, P. and D. Angelino. 2013. Nutritional and functional potential of *Beta vulgaris* cicla and rubra. *Fitoterapia*, *89*:188–199. https://doi.org/10.1016/j.fitote.2013.06.004.

Ninfali, P., Bacchiocca, M., Antonelli, A., Biagiotti, E., Di Gioacchino, A.M., Piccoli, G., Stocchi, V. and Brandi, G. 2007. Characterization and biological activity of the main flavonoids from Swiss chard (*Beta vulgaris* subspecies cycla). *Phytomedicine*, *14*(2–3): 216–221. https://doi.org/10.1016/j.phymed.2006.03.006.

Ozsoy-Sacan, O., Karabulut-Bulan, O., Bolkent, S., Yanardag, R. and Ozgey, Y. 2004. Effects of chard (*Beta vulgaris* L. var. cicla) on the liver of the diabetic rats: A morphological and biochemical study. *Biosci. Biotechnol. Biochem.*, *68*(8): 1640–1648. https://doi.org/10.1271/bbb.68.1640.

Pokluda, R. and Kuben, J. 2002. Comparison of selected Swiss chard (*Beta vulgaris* ssp. cicla L.) varieties. *Hortic. Sci.*, *29*(3): 114–118. https://doi.org/10.17221/4473-HORTSCI.

Pyo, Y.-H., Lee, T.-C., Logendra, L. and Rosen, R.T. 2004. Antioxidant activity and phenolic compounds of Swiss chard (*Beta vulgaris* subspecies cycla) extracts. *Food Chem.*, *85* (1): 19–26. https://doi.org/10.1016/S0308-8146(03)00294-2.

Reif, C., Arrigoni, E., Scharer, H., Nystrom, L. and Hurrell, R.F. 2013. Carotenoid database of commonly eaten Swiss vegetables and their estimated contribution to carotenoid intake. *J. Food Compos. Anal.*, *29*(1): 64–72. https://doi.org/10.1016/j.jfca.2012.10.005.

Ruales, J. and Nair, B.M. 1994. Properties of starch and dietary fiber in raw and processed quinoa (*Chenopodium quinoa*, Willd) seeds. *Plant Foods Hum. Nutr.*, *45*: 223–246. https://doi.org/10.1007/BF01094092.

Sacan, O. and Yanardag, R. 2010. Antioxidant and antiacetylcholinesterase activities of chard (*Beta vulgaris* L. var. cicla). *Food Chem. Toxicol.*, *48*(5): 1275–1280. https://doi.org/10.1016/j.fct.2010.02.022.

Sener, G., Sacan, O., Yanardag, R. and Ayanoglu-Dulger, G. 2002. Effects of chard (*Beta vulgaris* L. var. cicla) extract on oxidative injury in the aorta and heart of streptozotocin-diabetic rats. *J. Med. Food.*, *5*(1): 37–42. https://doi.org/10.1089/109662002753723205.

Singh, H., Dunn, B. and Payton, M. 2019. Hydroponic pH modifiers affect plant growth and nutrient content in leafy greens. *J. Hort. Res.*, *27*(1): 31–36. https://doi.org/10.2478/johr-2019-0004.

Ustundag, U.V., Tunali, S., Alev, B., Ipekci, H., Emekli-Alturfan, E., Akbay, T.T. and Yarat, A. 2016. Effects of Chard (*Beta vulgaris* L. var. Cicla) on Cardiac Damage in Valproic Acid-Induced Toxicity. *J. Food Biochem.*, *40*(2): 132–139. https://doi.org/10.1111/jfbc.12202.

Yanardag, R. and Colak, H. 2011. Effect of Chard (*Beta vulgaris* L. var. cicla) on Blood Glucose Levels in Normal and Alloxan-induced Diabetic Rabbits. *Pharm. Pharmacol. Commun.*, *4*: 309–311. https://doi.org/10.1111/j.2042-7158.1998.tb00702.x.

Zeller, W., Rudolph, K. and Hoppe, H.H. 1977. Effect of the *Pseudomonas phaseolicola*-toxin on the composition of lipids in leaves of Swiss chard (*Beta vulgaris* L.) II. Changes in the fatty acid spectrum of phospholipids and glycolipids. *J. Phytopathol.*, *89*(4): 296–305. https://doi.org/10.1111/j.1439-0434.1977.tb02870.x.

5 Bioactive Compounds in Sweet Potato Leaves (*Ipomoea batatas* (L.) Lam.) and Their Biological Activities

Tshilongo L.[1] and Sivakumar D.[2,3*]

Introduction

In various regions across the globe, sweet potato leaves (SPLs) are widely recognised as leafy vegetables and are regarded as a potential solution to poverty eradication and chronic micronutrient malnutrition. SPLs are a valuable crop to cope with fluctuations in the demand and supply of leafy vegetables (Dinu et al., 2021). Encouraging its consumption will mitigate the public health concern of chronic disease, which severely affects low-income groups in developing countries (Johnson and Pace, 2010). The SP is the second most produced starchy crop after potatoes, with the world's highest total production of 93,519,977.74 tonnes (FAOSTAT, 2023). Today, 500 decades after SP was domesticated in Latin America and distributed across the globe through human migration (Drapal et al., 2019), 33.3% of the total production is produced in African countries, mainly Malawi, Tanzania, Nigeria, Angola, and Rwanda, after Asia, where the world's leading country, China is located (FAOSTAT, 2023). As a result of typhoons damaging rice fields, the Japanese used SP as a survival crop (International Potato Center: SP FACTS AND FIGURES, Sweetpotato Facts and Figures - International Potato Center).

Because of its low production inputs and labour requirements, SP makes a good staple food (International Potato Center). The plants are tolerant of high temperatures, salt, drought, and low-fertility soil (Hossain et al., 2022). Since SPLs are rich in vitamins, minerals, and antioxidants, they make a significant food source, contributing to a balanced diet and health benefits like anticancer, antioxidant, antidiabetic, anti-inflammatory, antimicrobial, and antimutagenic. Due to these unique properties, the SPLs are gaining popularity in pharmaceuticals, cosmetics, nutraceuticals, and processed foods. SPLs are rich in bioactive

[1] Department of Crop Science, Tshwane University of Technology, Pretoria, South Africa.
[2] South African Research in Phytochemical Food Network, Tshwane University of Technology, Pretoria, South Africa.
[3] Centre for Nutrition & Food Sciences, Queensland Alliance for Agriculture and Food Innovation, The University of Queensland, QLD 4108, Queensland, Australia.
* Corresponding author: d.sivakumar@uq.edu.au; sivakumarD@tut.ac.za

DOI: 10.1201/9781003687832-5

compounds, especially phenolics and carotenoids, both of which are powerful health-promoting compounds. The bioactive compounds in sweet potato leaves make them a dual-purpose crop. Leaves differ morphologically from immature to mature colours, sizes, and pigmentation (Rahajeng et al., 2023). A difference can be seen in leaf lobe type, central leaf lobe shape, vine tip pubescence (Naidoo et al., 2020), leaf outline, leaf lobe number (Laurie et al., 2013), and leaf arrangement on the stems (Jiang et al., 2022). Leaf edges, leaf blades, and veins of young edible leaves are green, purple, or green with purple colouration (Tshilongo et al., 2024).

Promoting SPLs as a valuable food source can significantly contribute to achieving several Sustainable Development Goals (SDGs), such as reducing hidden hunger, improving nutrition, ensuring food security, and promoting sustainable agriculture (Joko et al., 2025). Therefore, commercialising leaves in the formal market requires strategic intervention. This would benefit more smallholders. In general, sweet potato leaves could significantly contribute to health outcomes and economic opportunities. To promote the consumption of SPLs, this chapter discusses phytonutrients and minerals present in SPLs, their biological activities, toxicity, and changes over time due to harvesting maturity and preservation methods.

Taxonomic Background

Sweet potato is scientifically known as *Ipomoea batatas* (L.) Lam., a dicotyledonous plant within the morning glory or bindweed family, belonging to the Convolvulaceae, as presented in Figure 1 (Jiang et al., 2022). It falls under the *Ipomoea* genus, which comprises 600–709 species (Hall and Phatak, 1993). Only 13 wild species are related to sweet potatoes (CIP, 2006). Later discoveries indicate 14 wild species in the genus as close relatives of *Ipomoea batatas* (Hall and Phatak, 1993). *Ipomoea trilida* relates closely to sweet potato compared to other species (Srisuwan et al., 2006). Sweet potato, together with these 14 species, is taxonomically grouped within *Batatas* (Jiang et al., 2022). There is still controversy regarding the sweet potato's place of origin. Up to date, North Western and South America (Colombia, Ecuador, and Peru) and parts of Central America (such as Guatemala) are believed to be the primary Centre of diversity of sweet potatoes, because of the existing wide range of native sweet potatoes, weeds, and wild *Ipomoea* (Aguoru, 2015). There are about 6,500 varieties of sweet potatoes, from wild accessions, farmer varieties, to the breeding lines (CIP, 2006). Genome studies suggest that sweet potatoes are a cross between diploid and tetraploid progenitors, followed by a whole genome duplication approximately 0.5–0.8 million years ago. Phylogenetic studies indicate that sweet potatoes are autopolyploids that evolved from *Ipomoea trifida* based on the nuclear DNA and chloroplast genome comparisons between sweet potatoes and their wild relatives (Muñoz-Rodríguez et al., 2018). However, according to (Gao et al., 2020), sweet potatoes are neither autohexaploids nor allohexaploids, but rather segmental allohexaploids.

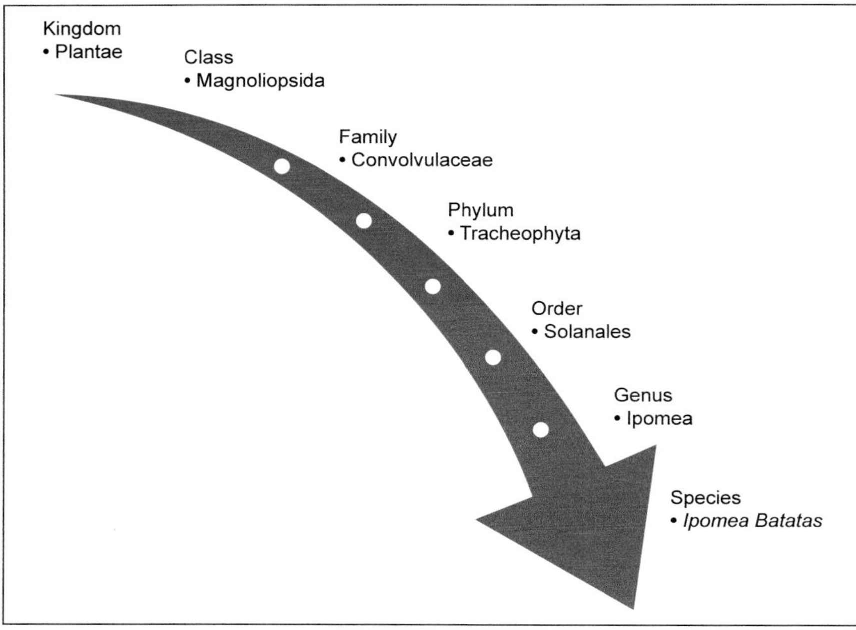

Fig. 1 The taxonomy classification of sweet potato.

Traditional Uses and Modern Uses

The use of sweet potato leaves depends on consumers' preferences or cultural knowledge. SPLs have a texture similar to spinach as a leafy vegetable with a slightly bitter taste (Baylis, 2023). In Asian countries like Fiji, the Philippines, Vietnam, and China, SPLs are normally stir-fried or sautéed and added to the soups, stews/curries, or cooked with other vegetables for extra texture and flavour (Baylis, 2023). Apart from their culinary uses, SPLs serve as a digestive aid and are used in home remedies for fever, asthma, bug bites, burns, upset stomach, diarrhoea, boils, and acne (Ooi and Loke, 2012). Due to their polyphenol, flavonoid, and antioxidant content, they are also applied as an ointment for mouth and throat irritations (Patel and Patel, 2020). Today, SPLs are used for anti diabetic, anticancer, anti-inflammatory, cholesterol-lowering, and immune-boosting properties (Nguyen et al., 2021).

Fibre in the leaves absorbs water, reduces pouches in the colon, and prevents cancer (Enyiukwu et al., 2018). Additionally, they are utilised in agriculture as livestock feed and green manure to enhance soil fertility (Odongo et al., 2015; Swami, 2024). In today's modernized healthy food, SPLs are used in green smoothies, salads, and juices (Anonymous, 2024), processed as dietary supplements, powders, or functional ingredients in noodles, flakes, snacks, and herbal teas (Odongo et al., 2015). As natural colourants, leaves can also be used as food additives (Vishnu et al., 2019).

Also, the SPL powders are used to fortify foods, enhancing nutritional content. Adding SPLs to food products increases their nutritional value, such as dietary fibre and phenolic compounds, without significantly affecting taste or texture. For example, incorporating 2–10% SPL powder into wheat flour steam bread improves its nutritional profile by increasing dietary fibre and phenolic acids, particularly caffeoylquinic acids, while maintaining consumer acceptability (Hu et al., 2022). A beverage made from ultrafine SPL powder, combined with ingredients like xanthan gum, calcium lactate, ascorbic acid, maltodextrin, xylitol, and the essence of apple, was found to be highly acceptable to consumers (Luo et al., 2020). In addition, SPLs can replace starch in starch noodles, improving cooking properties, texture, and sensory attributes (Li et al., 2023). They are also used as a substitute for water leaves in soups, enhancing appearance, taste, flavour, and texture (Nkongho et al., 2014; Omoba et al., 2020). Purple SPLs, rich in anthocyanins, serve as natural colourants, offering a safer alternative to synthetic food colourants and preservatives (Mbaeyi-Nwaoha and Emejulu, 2013; Vishnu et al., 2019).

Nutritional Composition

SPLs exhibit diverse proximate compositions depending on the variety. Generally, they are rich in moisture, crude protein, crude fibre, and ash, with varying levels of crude fat and carbohydrates, low in calories, and fat-free (Baylis, 2023). Table 1 shows the proximate composition in the leaves of 40 different SP cultivars. For example, Xushu No. 053601, Shingshu No. 19 (spring), Shi No. 5, Pushu No. 17, and Shushu No.14 exhibited the highest moisture (88.92 g 100 g^{-1} FW), carbohydrates (61.36 mg 100 g^{-1} DW), crude protein (31.08 mg 100 g^{-1} DW), crude fibre (14.36 mg 100 g^{-1} DW), and Ash (14.66 mg 100 g^{-1} DW), respectively (Sun et al., 2014b). SPLs are rich in mineral elements like calcium (Ca), potassium (K), phosphorus (P), magnesium (Mg), sodium (Na), iron (Fe), manganese (Mn), zinc (Zn), and copper (Cu) (Table 1). The presence of Ca makes SPLs an alternative source of Ca in households that cannot afford milk and kapenta (Chirwa-Moonga et al., 2020). Ca was found to be the most abundant mineral element in SPLs by Tilistyák et al. (2022). In contrast, Ayimbire et al. (2021) reported potassium (K) in the highest concentration among all other minerals in consistent with findings in Table 1. K was accumulated in high quantities (479.3–4280.6 mg 100g^{-1} DW) in SPLs of 40 cultivars in the outcomes of the study by Sun et al. (2014b). The Jishu cultivar contains high K in comparison with the leaves grown in Ethiopia (3608.854 mg 100 g^{-1} DW) (Awol, 2014). K is responsible for maintaining the fluid and electrolyte balance in body cells, especially during pregnancy (Ranteallo et al., 2023), whereas a shortage leads to hypokalema. On the other hand, Fe and Zn are the highest micro minerals accumulated in SPLs (Tshilongo et al., 2024). The leaves of BARI SP-04 cultivar have higher Fe (32.82 mg 100 g^{-1} DW) but lower Zn (2.1 mg 100 g^{-1} DW) (Kuddus et al., 2020). The leaves of SP grown in

Ethiopia are comparably higher in Fe (73.881 mg 100 g^{-1} DW) and Zn (5.647 mg 100 g^{-1}) (Awol, 2014). Fe contributes to the formation of haemoglobin cells in the blood (Enyiukwu et al., 2018), and Zn is needed in the human body for the development and functioning of cells. It also acts as an antioxidant and can stabilise cell membranes (Shankar and Prasad, 1998).

SPLs have various vitamins such as ascorbic acid (Vitamin C), Vitamin A, Vitamin K, and Vitamin B, including B6, Thiamine, Niacin, and Riboflavin (Anonymous, 2018). Vitamin B had the highest concentration, ranging from 856 to 948 mg 100 g^{-1} DW followed by vitamin A (273–400 mg g^{-1} DW) and vitamin B5 (320–660 DW) (Ishida et al., 2000). Thus, SPLs can address the Vitamin A deficiency problem, particularly in developing countries.

According to Table 1, SPLs contain essential and non-essential amino acids (AA). These AA were identified in the leaves of Bonita, Beauregard, Burgundy, and Bellevue (Tilistyák et al., 2022) Glutamic acid was present in large quantities (3.22 g 100 g^{-1} DW), followed by aspartic acid (2.75 g 100 g^{-1} DW) and leucine (2.23 g 100 g^{-1}DW) compared to other AA in the leaves of Simon No.1 cultivar (Luo et al., 2013). Glutamic acid is a precursor of gamma-aminobutyric acid (GABA) (Suzuki and Iwata, 2021). GABA is an anti-insomnia, anticancer, anti-depression, anti diabetes, anti-hypertensive, and anti-inflammatory, and provides neuroprotection (Dutta et al., 2013; Hou et al., 2024). It plays a role in preventing neutral diseases like Type 1 diabetes, cancer, immunological disorders, and asthma (Diez-Gutiérrez et al., 2020). Amino acid derivatives such as L-DOPA, S-Methyl-L-methionine, Creatinine, N(alpha)-gamma-L-Glutamylhistamine, and tranexamic acid were found to relate to the leaf shape (Tan et al., 2024).

Bioactive Compounds

Polyphenols

Polyphenols are known for their antioxidant and anti-inflammatory properties and accumulate in response to various biotic and abiotic stresses. Phenolic compounds are a diverse group of plant metabolites with a common chemical structure: an aromatic ring with one or more hydroxyl (-OH) groups. SPLs and roots of sweet potatoes are rich in polyphenols, including flavonoids and chlorogenic phenolic acids (Salam et al., 2023). SPLs contain phenolic compounds, which are antioxidants. There is a difference in levels among varieties of sweet potatoes depending on the growth stages of the leaves (Phahlane et al., 2021).

Phenolic Acids

Sweet potato leaves are a valuable source of these beneficial compounds for the nutrition of both humans and animals (Ojong et al., 2008). The phenolic acids present in SPLs include the chlorogenic acids like 3-CQA, 4-CQA, and 5-CQA (mono-CQAs), 3,4-diCQA, 3,5-diCQA, and 45-CWA (Di-CQAs), and 3,4,5-CQA

Table 1 Nutritional composition in sweet potato leaves.

Proximate Principles mg 100 g^{-1} Dry weight (DW)		*Minerals mg 100 g^{-1} DW*		*Vitamins mg 100 g^{-1}*		*Amino Acids g 100 g^{-1} DW*			
Moisture fresh weight (FW)	84.09–88.92	Ca	229.7–958.1	A	273–400	Essential AA		Non-essential AA	
Protein	16.9–31.08	K	479.3–4280.6	C	62.7–81	Aspartic acid	2.75	Threonine	1.22
Fiber	$9.15^{-1}4.26$	P	131.1–2639.8	B1 (Thiamine)	53–28	Glutamic acid	3.22	Serine	1.15
Fat	2.08–5.28	Mg	220.2–910.5	B2 (Riboflavin)	248–254	Proline	1.30	Valine	1.43
Carbohydrates	42.03–61.36	Na	8.06–832.32	B3 (Niacin)	$856^{-1}498$	Glycine	1.41	Isoleucine	1.15
Ash	$7.39^{-1}4.66$	Fe	1.92–21.77	B6 (Pyridoxine)	120–329	Alanine	1.53	Leucine	2.23
(Sun et al., 2014b)		Mn	1.71–0.92	Biotin	3–8	Arginine	0.85	Histidine	0.60
		Zn	1.20–3.23	E	1.39–2.81	Phenylalanine	1.46	Lysine	1.90
		Cu	0.67–.86	Pantothenic acid (B5)	320–660	Arginine	0.85	Methionine	0.15
		(Sun et al., 2014b)				Total AA		23.80	
				(Ishida et al., 2000)		T essential AA		10.35	
						T acid AA		5.97	
						T basic AA		3.99	
						(Luo et al., 2020)			

(Tri-CQA). However, the number of phenolic acids in SPLs is up to 32 according to Tan et al. (2024). 3,5-diCQA is often the major phenolic compound in SPLs (Ayeleso et al., 2016). For instance, 3,5-di-CQA occurred as the major phenolic acid, ranging from 1855 in Shanshu 19 to 3338 mg 100 g^{-1} DW in Yizu No. 7 cultivar (Makori et al., 2020). This is lower than 229.10 mg g^{-1} DW found in SP leaf ethyl acetate extract in Zhecaishu 726 China cultivar (Zhang et al., 2019). 3,5-diCQA, along with 1,4-diCQA, 3- CQA were found to be the predominant CQA derivatives in the leaves of six South African SP cultivars, particularly Bophelo OFSP (Phahlane et al., 2021). In the leaves of 14 cultivars, either 4,5 or 3,5 CQA was most dominant, ranging between 691.2–2011.1 mg 100 g^{-1} and 677.7–1091.1 mg 100 g^{-1}, accumulated in quantity in Jeungmi and Gogeonmi cultivars (Li et al., 2017). The difference in the level of 3.5 CQA could be attributed to the cultivar and location. Their accumulation differs with cultivar, maturity stage, and the condition of the surrounding environment (Makori et al., 2020). Cultivars with lobed or heart-shaped leaves are rich in polyphenols; hence, 3-(3,4-dihydroxy-5-methoxy)-2-propenoic acid, 4-hydroxybenzoic acid, and methyl cinnamate phenolic acids showed a relationship with leaf shape (Sharma, 2023; Tan et al., 2024). Therefore, the identification of SPLs rich in phenolic acids can be done through the leaf shape.

Flavonoids

Phytochemical compounds called flavonoids provide biological activity that protects plants from extreme conditions and provides dietary properties to reduce free radical damage (Khalid et al., 2019). SPLs contain different types of flavonoids, including myricitrin, isoquercitrin, astragalin (kaempferol-3-O-glucoside), quercitrin, tiliroside, quercetin, apigenin, kaempferol, and luteolin (Makori et al., 2020; Nguyen et al., 2021). The derivatives of quercetin and luteolin are also found, and these include quercetin-3-O-hexoside, hyperoside (quercetin-3-O-galactoside), rutin (quercetin-3-O-rutinoside), and luteolin-7-O-glucoside (Phahlane et al., 2021; Zhang et al., 2019). Haematoxylin, kaempferol 3-sophorotrioside, myricetin, and luteolin 7-glucoside are a few of the 41 identified flavonoids showing a relation to the SPL colour (Tan et al., 2024). According to Table 2, astragalin was the predominant flavonoid in SPLs, ranging between 491.3–1729.9 mg 100 g^{-1} DW, highest in Yizu No. 7 purple-flesh cultivar followed by either isoquercitrin (126.8–268.3 mg 100 g^{-1}) or quercitrin (22.6–72.6 mg 100 g^{-1}) (Makori et al., 2020). However, quercetin-3-O-hexoside (44.43 mg g^{-1}) was the highest followed by luteolin-7-O-glucoside (21.77 mg g^{-1}), while kaempferol-3-O-glucoside exhibited the lowest concentration (18.46 mg g^{-1}) in the leaves.

Anthocyanin

SPLs are rich in anthocyanin, superior to the storage roots (Sui et al., 2019). Anthocyanins are a subclass of flavonoids with an intense pigmentation for purple, red, and blue colour in fruit and vegetables (Tan et al., 2024), grouped into six types including cyanidin, delphinidin, peonidin, pelargonidin, petunidin and malvidin depending on the chemical structure (Qi et al., 2023). Table 2 shows 15 anthocyanins of the cyanidin and peonidin in green leaves of p40 purple-flesh, Bonita white flesh, and Beauregard orange flesh (Su et al., 2019a). Pelargonidin content ranging between 0.64–0.90 mg.kg $^{-1}$ DW was present in Simon No. 1, Yuzi No. 7, and Pushu 32, but was not detected in Shangshu 19 cultivar (Makori et al., 2020). These compounds are located in the plant cell sap, in the cell layer of palisade parenchyma, epidermis in immature leaves, and vacuoles of the adaxial and abaxial epidermis in leaves (Merzlyak et al., 2008). Newly formed SPLs of certain cultivars are purple and turn green with maturity through the process called juvenile red facing (Deng et al., 2020). For instance, anthocyanin derivatives namely cyanidin-caffeoyl-sophoroside-glucoside (8.13 mg kg^{-1} DW), cyanidin-caffeoylferuloyl-sophoroside-glucoside (nd-41.51 mg.kg $^{-1}$ DW), peonidin-caffeoyl-hydroxybenzoyl-sophoriside-glucoside (nd-45.24 mg.kg^{-1} DW), and peonidin-caffeoyl-feruloyl-sophoroside-glucoside (nd-24.47 mg kg^{-1} DW) detected in leaves of five-purple-fleshed genotypes grown in South Africa were higher in '2019-11-2' with purple tips and present during vegetative and tuber initiation stage (Tshilongo et al., 2024). Therefore, the harvesting growth stage and the quantity of purple leaves influence the composition of anthocyanins (Jyothi et al., 2005; Su et al., 2019b). Su et al. (2019a) reported cyanidin as a major contributor to the total anthocyanin content, cyanidin 3-p-hydroxybenzoyl sophoroside-5-glucoside exhibiting the highest level, ranging from 2.7–24.4 mg g^{-1} PN3GE DW, higher in leaves of p40 purple-flesh cultivar in Beijing, China. This supports the findings by (Islam et al., 2002), cyanidin 3-p-hydroxybenzoyl sophoroside-5-glucoside dominated with a content of 20.24–27.37 mg g^{-1} powder in leaves grown in Japan, higher in Kyushu No.119. Cyanidin 3-glucoside 5-caffeoylglucoside together with delphinidin 3-(6-p-coumaroyl) glucoside were dominant in purple leaves (Tan et al., 2024). Anthocyanins protect plants from UV radiation, promote disease tolerance, and provide protection against herbivores and pathogens (Hatier and Gould, 2009; Lev-Yadun and Gould, 2009). Anthocyanins have antioxidative and antimicrobial properties, improve visual and neurological health, and protect against various noncommunicable diseases (Khoo et al., 2017). Therefore, China recommends a daily allowance of 50 mg day^{-1} (Wallace and Giusti, 2015). Tan et al. (2024) found that cyanidin 3-O-(6-O-malonyl-beta-D-glucoside and cyanidin 3-glucoside 5-caffeoylglucoside are responsible for purple pigmentation in SPL tips.

Tannins

Table 2 shows that tannin content ranged from 2,280–4,460 mg 100 g^{-1} on a dry weight basis in 14 different South Korean cultivars. In descending order, Geonpungmi had the highest tannin (4,460) > Daeyumi (3,800) Jeungmi (3,780) ≥ Yeonhwangmi (3,670) > Andong (3,390) ≥ Sinchunmi (3,370) ≥ Yeonmi (3,360)

Table 2 Bioactive compounds in sweet potato leaves from different cultivars.

Cultivars	*Origin*	*Sample Extraction*	*Analytical Method*	*Bioactive Compounds*	*References*
Green aerial GSC-2 GCS-5 Green-purple aerial Ziyang $GS^{-1}5$-28 $GS^{-1}6^{-1}1$ $GS^{-1}7$-3 $GS^{-1}7$-5 $GS^{-1}7^{-1}0$ $GS^{-1}7$-21 $GS^{-1}7$-22 $GS^{-1}7$-23	China	95% alcohol acidified with 1.5 N HCl (85:15, v/v)	Folin–Ciocalteu assay	Total phenolic content ($48.43^{-1}48.36$ mg 100 g^{-1} Gallic acid equivalent fresh weight (FW)) Total anthocyanin content (3.22–09.60 mg 100 g^{-1} FW)	Jia et al. (2022)
Beauregard, Georgia Jet, Centennial	USA	7% acetone		Total phenolic content (98.153–10.721 µg g^{-1} dry weight (DW)) Total beta-carotene content (218–305 µg g^{-1}DW)	Hapke et al. (2024)
SPK 004 Kakamega Zapallo Jewel 40331 Jewel 566638 Jonathan Tainung Beea local, Mbouda, Santchou	Cameroon	Acetone	Spectrophotometer	Total carotenoid (1414.6–3556.9 µg g^{-1} FW)	Nkongho et al. (2014)

Contd...

Table 2 *Contd...*

Cultivars	*Origin*	*Sample Extraction*	*Analytical Method*	*Bioactive Compounds*	*References*
Bophelo Monate Ndou Blesbok USA's Beauregard Peru's 1999062.1	South Africa	80% methanol	Folin–Ciocalteu Ultra-performance liquid chromatograph (UPLC)	Total phenolic content (2319.10–1322.76 mg kg^{-1})	Phahlane et al. (2021)
				Phenolic acids (mg kg^{-1})	
				CA (26.76–65.07) Neochlorogenic acid (5-CQA) (70.11–80.49) Chlorogenic acid (3-CQA) (33.74–53.76) 3,5-di-CQA (43.06–47.90) 1,3-diCQA (21.56–29.76) 1,4-diCQA (44.92–49.69) 4,5-diCQA (12.62–29.09)	
				Flavonoids (mg Kg^{-1})	
				5-Hydroxy-6-methoxycoumarin 7-glucoside (5.33-9.86) Neoeriocitrin (eriodictyol 7-O neohesperidoside) (9.35–40.26) Quercetin 3-glucosyl-(1–>2)-galactoside (5.46–29.52) Quercetin 3-O-galactoside (22.09–25.04) Quercetin-3-O-rutinoside (rutin) (Nd-32.27)	
				Carotenoids (mg kg^{-1})	
				Lutein (1.29–9.50) Zeaxanthin (0.14–5.02) B-carotene (3.67^{-1}0.270	

Contd...

Table 2 *Contd...*

Cultivars	*Origin*	*Sample Extraction*	*Analytical Method*	*Bioactive Compounds*	*References*
Simon No. 1 Yuzi No. 7 Shangshu 19 Pushu 32	Beijing, China	80% methanol	High-performance liquid chromatography (HPLC)	TPC (6050^{-1}2080 mg 100 g^{-1} DW CAE) TFC (1560–4210 mg 100 g^{-1} QE DW) TAC (790^{-1}010 mg 100 g^{-1} CGE DW)	Makori et al. (2020)
				Phenolic acids (mg 100 g^{-1} DW)	
				CA (11–28) 5-CQA (93–502) 3-CQA (110–349) 4-CQA (290–607) 3,5-di-CQA (1855–3338) 3,4-CQA (202^{-1}316) 4,5-diCQA (813–2231) 3,4,5-CQA (77–310)	
				Flavonoids (mg 100 g^{-1} DW)	
				Myricetrin (36.7–11.8) Isoquercitrin (126.8–268.3) Astragalin (491.3–056.6) Quercitrin (22.6–72.6) Tiliroside (545) Quercetin (29.6–87.7) Apigenin (10.6-57.7) Kaemferol (16.6-60.9)	
				Cyanidin (149.22–763.10) Perlagonidin (0.64–0.90) Peonidin (126.16–619.40)	

Contd...

Table 2 *Contd...*

Cultivars	*Origin*	*Sample Extraction*	*Analytical Method*	*Bioactive Compounds*	*References*
P40 Bonita Beauregard	USA	Acidified water (1 N of formic acid)	HPLC-MS/MS	Cyanidin-3-sophoroside-5-glucoside (3.8–9.4) Peonidin-3-sophoroside-5-glucoside (0.8–4.9) P-hydro benzoylated-(Cyanidin-3-sophoreside-5-glucoside) (2.7–24.4) Caffeoylated-(Cyanidin-3-sophoroside-5-glucoside) (3.3–48) P-hydroxybenzoylated-(Peonidin 3-sophoroside-5-glucoside) (1.0–7.4) Caffeoylated-(Peonidin 3-sopheroside-5-glucoside) (1.0–23.5) Feruloylated (Cyanidin 3-sophoroside-5-glucoside) (1.1–6.6) Cyanidin-3-(6,60-dicaffeoyl-sophoroside)-5-glucoside (11.6–86.1) Cyanidin-3-(6,60-caffeoylphydroxybenzoyl sophoroside)-5-glucoside (9.8–65.8) Cyanidin-3-(6,60-caffeoylferuloylsophoroside)-5-glucoside ($1.9^{-1}1.7$) Peonidin-3-(6,60-dicaffeoylsophoroside)-5-glucoside (2.3–55.4) Peonidin-3-(6,60-caffeoylphydroxybenzoyl-sophoreside)-5-glucoside (1.3–24.5) Peonidin-3-(6,60-caffeoylferuloylsophoroside)-5-glucoside (1.0–7.1) Peonidin 3-caffeoyl-p-coumaryl sophoroside-5-glucoside (1.1–5.1) TAC (38–448 mg kg^{-1}PN3GE DW) TPC (36.8–46.7 mg g^{-1} GAE DW)	Su et al. (2019a)

Contd...

Table 2 *Contd...*

Cultivars	*Origin*	*Sample Extraction*	*Analytical Method*	*Bioactive Compounds*	*References*
Simon no.1	Beijing, China		HPLC	Vitamin C (77.69 mg 100^{-1} DW)	Luo et al. (2020)
Guangcai2 Guangcai5 Ecai1 Ecai10 Zhecai1 Zhecai726 Fu18 Fu22 Fu23 Tainong71 Shulv1 Pushu53 Ningcai	China	0.1M hydrochloric acid 6 ml of 2.5 M sodium acetate and 1 ml of 10% (w/v) taka-diastase solution	HPLC	10.78 ˉ52.92 mg $100g^{-1}$	Hong et al. (2020)
$Local^{-1}$ Local-2 Local-5 Local-8 $Exotic^{-1}$ Exotic-2 Exotic-3 Exotic-4 BARI SP-4	Sylhet Sadar, Bangladesh	3% metaphosphoric acid and acetic acid solution	Titration AOAC	Vitamin C (10.56–8.04 mg 100 g^{-1})	Hossain et al. (2019)

Contd...

Table 2 *Contd...*

Cultivars	*Origin*	*Sample Extraction*	*Analytical Method*	*Bioactive Compounds*	*References*
Geonpungmi Daeyumi Yulmi Jinhongmi Healthymi Manami Sinchunmi Gunmi Yeonhwangmi Yeonmi Andong Singeonmi Gogeonmi Jeungmi	South Korea	80% (v/v) ethanol	UV-Vis Spectrophotometer	TPC (2640.2–4200.9 mg $100g^{-1}$) **Phenolic acids** Caffeic acid (13.7–32.5 mg $100g^{-1}$) Fer ulic acid (29–85.5 mg $100g^{-1}$) Chlorogenic acid (330.5–03.8 mg 100 g-1) 3,5-di-CQA ($677.7^{-1}091.1$) 3,4-CQA (123–395.5 mg $100g^{-1}$) 4,5-diCQA (619.2–2011.1 mg $100g^{-1}$) **Carotenoids** Lutein (19.01–28.85 mg 100 g^{-1}) β-carotene (36.86–49.94 mg 100 g^{-1}) Tannin (2.280–4.460 mg 100 g^{-1} DW)	Li et al. (2017)

≥ Gogeonmi (3,280) ≥ Gunmi 3,240 > Healthymi (3,230) > Singeonmi (3,090) > Manami (2,940) > Jinhongmi (2,560) and Yulmi exhibited the lowest value (2,280) (Li et al., 2017). The presence of tannin in large quantities inhibits the bioavailability of proteins, Zn and Fe and affects the palatability, growth rate, and digestive enzymes (Essack et al., 2017). However, with fermentation, the level of tannin reduces and under different cooking/processing methods (Jubaedah and Hermawan, 2024). The application of tannin in medicine can prevent the risk of coronary heart disease (Anbuselvi and Muthumani, 2014).

Coumarins

Coumarins in SPLs exhibit antioxidant properties and potential anticancer benefits. Tan et al. (2024) identify 10 coumarins, including esculin, esculetin, scopoletin, scopoletin-7-O-glucoside, and isofraxidin. Simple coumarins like 7-hydroxycoumarin, 6-nitro-7-hydroxycoumarin, scopoletin, and esculetin are potential treatments for cancer cells (Emami and Dadashpour, 2015). Additionally, rutarin, a coumarin linked to leaf shape, offers a potential marker for identifying leaves with high coumarin content.

Terpenoids

SPLs contain various terpenoids, which are a class of organic compounds and these terpenoids contribute to the aroma, flavour, and potential health benefits. Terpenoids in SPLs are accumulated in the membrane of guard and side cells of stomata as well as cell wall segments that face the stomatal fissure (Kirakosyan et al., 2023). Shi et al. (2025) and Tan et al. (2024) identified 8 and 13 terpenoids in SPLs, respectively. Some of the terpenoids in SPLs are digiprolactone, α-onocerin, and tussilagonone (Armadany et al., 2024).

Carotenoids

Carotenoids are a large group of pigmented tetraterpenoids naturally occurring in chloroplasts of green plant tissues (Yuan et al., 2015). Depending on the chemical structure, carotenoids are classified as xanthophylls or carotenes. SPLs contain carotenoids such as lutein, zeaxanthin, β-cryptoxanthin, cis-β-Carotene, All-trans β-carotene, 9-cis-β-carotene, and violaxanthin (Drapal et al., 2019; Ooko Abong' et al., 2020). β-carotene is responsible for red-orange colour (Ayeleso et al., 2016), and a precursor of vitamin A, absorbed in the small intestine and distributed through the blood to the liver, where it is converted to vitamin A whenever required in the human/animal body (Maoka, 2020). Table 2 shows the SPLs grown in South Africa and South Korea with β-carotene content ranging from 3.67–10.27 mg kg^{-1} DW (0.367–1.027 mg 100 g^{-1}) and 36.86–52.01 mg 100 g^{-1} DW, as the most accumulated compound in the Blesbok white-flesh cultivar and Healthymi cultivar, respectively (Li et al., 2017; Phahlane et al., 2021). The level of all-trans-β-carotene content in leaves harvested from five purple-fleshed genotypes in South Africa during the vegetative stage (22.2–60.20 mg 100 g^{-1} DW) and tuber initiation stage (17.19–53.52 mg 100 g^{-1}), 8 and 12 weeks after planting, is similar to Li et al. (2017). Whereas the last harvesting stage (9.11–33.16 mg 100 g^{-1}) had minimum values which highlights the role of the harvesting stage on the accumulation of carotenoids (Tshilongo et al., 2024). The highest recommended daily allowance (RDA) of this compound, which is 15 mg/day in older people, 10 g of SPL powder will contribute 18.69–40.13%. It will therefore take a minimum of 24.92 g to meet 100% of the RDA. Therefore, SPLs have the potential to help alleviate the vitamin A deficiency. Lutein was higher in deep, lobed cultivars (YelloSP with yellow flesh and 08–21P USA cultivar with purple and orange flesh). 10 g of these two cultivars will supplement < 52 % and > 52 % exceeding 100 % of the lutein RDA, respectively. Lutein, along with zeaxanthin, is regarded as antioxidants that protect the eyes from sunlight and ultraviolet rays, offering long-term sight by minimising age-related diseases.

Alkaloids

SPLs contain various secondary metabolites, including alkaloids called ipomine A, which has been found in the tuberous root and pyropheophorbide A and methyl-pyropheophorbide A, were isolated in purple SPLs using LC-MS/MS analysis (Armadany et al., 2024). Shi et al. (2025) reported 18 existing alkaloids in SPLs, whereas Tan et al. (2024) identified 19 alkaloids. Alkanoids present in SPLs have anti-alopecia activity in *in silico* studies through molecular simulation, inhibiting androgen receptors (Armadany et al., 2024). Additionally, alkaloids such as strictosidine aglycone showed a relation to the leaf shape Tan et al. (2024).

Biological Activities

SPLs exhibit numerous biological activities, including antioxidant, antidiabetic, anticancer, anti-inflammatory, antimicrobial, hepatoprotective, and immunomodulatory properties, attributed to their high content of beneficial bioactive compounds like polyphenols, flavonoids, carotenoids, and anthocyanins. SPLs are a valuable functional food and a source of natural compounds for various applications because of these activities (Rahman, 2007).

Antioxidant Activities

Antioxidants can prevent or inhibit the oxidation of molecules, a process during which a substance loses a hydrogen atom or an electron (Mehta and Gowder, 2015), and SPLs are regarded as natural antioxidants (Zhang et al., 2019). Purple-fleshed cultivars utilise chlorogenic acid rather than anthocyanins to scavenge DPPH radicals. Yellow and orange sweet potatoes, which are rich in carotenoids with conjugated double bonds, demonstrate strong antioxidant capacity and serve as a source of provitamin A (Santos et al., 2024). Total phenolic content (TPC), total antioxidant capacity (TAC), and total carotenoids (TC) show positive correlations with DPPH, ABTS, and FRAP assays, while TC is specifically correlated with ABTS (Jia et al., 2022). Table 3 illustrates that the antioxidant activity of SPLs is attributed to compounds such as 4,5-di-CQA, 3,4-di-COA, 3,5-di-CQA, and 3,4,5-tri-CQA (Zhang et al., 2019). Notably, chlorogenic acid (CA) exhibited the highest antioxidant activity compared to mono and di-CQA, while 3,4,5-CQA showed the weakest antioxidant activity during *in vitro* digestion (Sun et al. 2018). CGA's polyhydroxyl structure allows for a multifaceted antioxidant mechanism; it directly neutralises free radicals, activates antioxidant signalling pathways, modulates gene expression, and enhances overall antioxidant capacity. Additionally, it regulates the endogenous oxidase system and its associated proteins (Wang et al., 2022). Tshilongo et al. (2024) reported a positive correlation between antioxidant activity and compounds such as cyanidin, peonidin, and rutin, indicating that phenolics are related to multiple health benefits. Furthermore, beta-carotene and vitamin C are also considered antioxidants (Ooko Abong' et al., 2020). Additionally, tannins demonstrated a strong positive correlation ($r = 0.719$) with antioxidant activity (Manasathien and Khanema, 2022) due to their scavenging action against free radicals (Li et al., 2017).

Table 3 Derivatives of phenolics, flavonoids, carotenoids, and vitamin C with antioxidant activity.

Bioactive Compounds		*References*
Phenolic acids	Di- (4,5-di-CQAs, 3,5-di-CQA, 3,4-di-CQAs) and tri- 3,4,5-tri-CQAs Mono-CQAs (3-caffeoylquinic acid, 5-caffeoylquinic acid, 4-caffeoylquinic acid) and di-CQAs	Zhang et al. (2019) Shi et al. (2025)
Flavonoids	Cyanidin-caffeoyl-feruloyl sophoroside-glucoside Peonidin-caffeoyl-hydroxybenzoyl-sophoriside glucoside Peonidin feruloyl-sophoroside-glucoside Peonidin-caffeoyl-feruloyl-sophoroside-glucoside Rutin	Tshilongo et al. (2024)
Carotenoids	β-carotene	Ooko Abong' et al. (2020)
Vitamin	Ascorbic acid (Vitamin C)	Cioloca et al. (2021); Ooko Abong' et al. (2020)
Tannin	Tannin	Manasathien & Khanema, (2022)

Antimicrobial Activities

SPLs possess antioxidants with inhibitory effects against virus, fungus and bacteria pathogens. Methanolic extracts from SPLs were effective against *S. aureus, L. monocytogenes, E. coli, S. mutans, S. dysenteriae,* and *C. albicans* (Sultana et al., 2024). SPL extract exhibited a potent antifungal activity against *Malassezia furfur*, a fungus naturally found on human skin surface (Armadany et al., 2024). SPL extract is a promising antimicrobial agent to treat infections caused by *Bacillus cereus and Salmonella Typhimurium* bacteria, and in food contaminated by these pathogens (Das et al., 2020). The antimicrobial activity in SPLs can be attributed to the gallic acid, 3,4-dihydroxybenzoic acid, and synapic acid phenolic acids and flavonoids such as naringenin, kaempferol, and quercetin (Gilmar Freire da Costa et al., 2021).

Anticancer Activities

SPLs are rich in anthocyanins of the peonidin and cyanidin type (Su et al., 2019a). Cyanidin 3-glucoside and peonidin 3-glucoside were found to inhibit cell growth of breast cancer HS578T cells by arresting the G2/M phase of the cell cycle (Chen et al., 2005). Anthocyanins in SPLs possess higher activity against colon and cervical cancer cells (Vishnu et al., 2019). Sugata et al. (2015) found the SPL extract to inhibit the cancer lines' growth responsible for breast, gastric, and colon adenocarcinoma cancer, in a concentration and dose-dependent manner. SPLs extracts increase the apoptosis in the Caco-2 cell model as a result of degrading the colon cancer cells' proliferation (Bolanos, 2020; Karna et al., 2011). The *in vitro* and *in vivo* effects of SPL extract to inhibit the growth and progression of prostate tumour through the oral administration of 400 mg/kg. Therefore, SPLs are a cheap source of bioactive compounds against the risk of cancer.

Antidiabetic Activities

SPLs contain antidiabetic properties with the ability to alleviate diabetic mellitus by reducing the level of blood glucose (Almoraie, 2019). A dose of 200 mg/kg BW exhibited a better reduction effect on the levels of blood glucose and pancreatic histopathological features compared to other fractions (Solihah et al., 2023). SPLs mitigated the diabetic effect at 5% concentration mixture with the root powder (Refaat et al., 2020). The antidiabetic inhibitory effect can be attributed to the predominant compound, chlorogenic acids. Chlorogenic acids from SPLs had inhibitory effects on aldose reductase, which plays important roles in diabetic complications, wherein 3,4,5-O-caffeoylquinic acid exhibited the maximum inhibitory effect, indicating its medical potential value to treat diabetes (Kurata et al., 2007). CGA can influence glucose metabolism, potentially improve insulin sensitivity, and help manage blood sugar levels, which is beneficial for those with diabetes or at risk of developing it. They are involved in the activity of enzymes in glucose and lipid metabolism, affecting insulin sensitivity and fat storage (Wang et al., 2022).

Anti-inflammatory Activities

Previous findings indicate the anti-inflammatory effect of SPLs. Extract from SPL was found to suppress NF-κB nuclear translocation, IKK-α and IκB-α phosphorylation, and iNOS expression in LPS-stimulated RAW 264.7 cells, hereby inducing an anti-inflammatory effect (Cho et al., 2021). Sun et al. (2022) indicated that polyphenols from purple SPLs have an important role in the intervention of hyperuricaemia and inflammation related to hyperuricaemia or renal damage and can be utilised in the application of functional foods. Sweet potato phenolic extracts have inhibiting on inflammatory Caco-2 monolayers and also decrease pro-inflammatory cytokines like tumour necrosis factor alpha (TNF-α) expression in the aorta (Waskito et al., 2022; Zhang et al., 2022).

Anti-Ulcer/Wound Healing Activities

Crushed SPLs are used medicinally for wound healing by directly applying them to cuts and wounds. On the other hand, *Ipomoea batatas* L. leaf extract was not effective on healing traumatic ulcers in the oral mucosa (Rezeki et al., 2017).

Hepatoprotective Activates

The SPL extract offers hepatoprotective protection against liver damage due to high-fat feeding induction (Mahfudh et al., 2022) and alcohol-induced hepatotoxicity and fatty liver (Kang and Lee, 2021). The liver protection is associated with quercetin flavonoid, which has an antioxidant effect as a free radical scavenger, protecting the liver from oxidative damage by decreasing serum glutamic oxaloacetic transaminase (SGOT) and serum glutamic pyruvate transaminase (SGPT). The protection against alcohol liver damage is linked to the Chlorogenic acid derived from purple SPL extracts (Kang and Lee, 2021). However, the protective effect increases with an increase in dose treatment, achievable with 200 mg/kg BW of SPLE treatment (Mahfudh et al., 2022). Similarly, a dose of 400 mg/kg body weight of SPL crude extract had a hepatoprotective effect (Caviteño et al., 2010). Anthocyanins in purple sweet potatoes exhibit hepatoprotective effects by inhibiting lipid peroxidation and scavenging free radicals. It was proven that in healthy men with borderline hepatitis, consuming purple sweet potato drinks reduced liver enzyme levels, particularly gamma-glutamyl transferase (GGT) (Suda et al., 2008). Animal studies further demonstrate protective effects against carbon tetrachloride (CCl_4)-induced acute liver damage, oxidative stress, and inflammation caused by D-galactose, as well as improved antioxidant status following ischemia-reperfusion liver injury. However, human studies are needed to validate these effects (Santos et al., 2024).

Neuroprotective Activates

The treatment of diabetic groups with SPL extract showed similar brain weight when compared to the diabetic control group, suggesting a protective effect against hyperglycaemia induced neuronal damage. Therefore, SPL extract possesses neuroprotective effects and has the ability to lessen diabetes-induced cognitive dysfunction (Siame et al., 2025).

Larvicidal/ Insecticidal Activities

The concern of high amounts of artificial chemical larvicides is the long-standing residual effect and chronic effects on the environment and non-target organisms, respectively. *Ipomoea batatas* leaves-mediated silver nanoparticles (AgNPs) showed larvicidal activity against mosquito vectors at different concentrations, making it an eco-friendly control effective on dengue, malaria, and filariasis-causing vectors (Pavithra Bharathi et al., 2017).

Toxicity and Safety Profiles and Related Studies

SPLs contain a variety of secondary metabolites that are not nutritive, such as phytic acid, oxalates, tannins, and cyanide, which may pose health risks when consumed in large quantities. They interfere with nutrient absorption, reducing their bioavailability (Auwal et al., 2019). Oxalates bind calcium and magnesium, increasing kidney stone risk, while phytic acid binds phosphorus and tannins to iron (Ooko Abong' et al., 2020). The toxicity of SPLs depends on the quantity consumed and frequency of intake, as these compounds are present in low amounts in raw, dried, or pulverised leaves. Lutein-rich SPL extracts have been deemed safe up to 2,000 mg kg^{-1}day in toxicological evaluations, showing no adverse effects on vital organs (Hisamuddin et al., 2023). However, high concentrations of SPL macerate may lower red blood cell levels and VGM/MCV (mean corpuscular volume), although they are orally non-toxic and do not harm vital or reproductive organs (Ano et al., 2024). At doses above 1,000 mg kg^{-1}, SPL ethanol extracts may cause toxicity to vital organs, particularly the liver and kidneys (Imafidon et al., 2015). SPLs are considered safe in weight-reducing formulas at doses of 10–100 mg kg^{-1}, but higher doses require caution.

Harvesting

The ideal SPLs to harvest are bright green and smooth (Baylis, 2023). Martin (1998) instead used "not fully developed", whereas Islam et al. (2003) used "new and young tender leaves within 10 cm on the growth tips". Concentrated sources of phenolic antioxidants may be young, immature leaves (Padda & Picha, 2007). Figure 2 demonstrates the leaf harvesting for different cooking and processing of products. Harvesting of leaves must include the apical buds to supply more nutritional compounds (Nkongho et al., 2014). Apical buds contain higher TPC and stronger AA than tender leaves (Jia et al., 2022). Stage or period and age during harvesting affect the bioactive compounds and antioxidant activity (Nguyen et al., 2021), as well as the taste, the astringency, and bitterness attributed to polyphenolic compounds (Sasaki et al., 2015). Tshilongo et al. (2024) showed a significant difference between the harvesting stages/period and recommended genotype 08-21p and 20191-2 for the harvesting of young tender leaves for leaf consumption rich in carotenoids and minerals during the vegetative stage, 8 weeks after planting and leaves with high phenolics at the tuber initiation stage, 12 weeks after planting, respectively. The effect of the harvesting stage was reported on proximate composition, phenolic acids, as well as vitamin C (Suárez et al., 2020). Although multiple harvesting can increase the economic yield significantly, the content of CQAs, particularly 4-CQA and 5-CQA is affected (Sasaki et al., 2015).

Cooking

Historically, they were a staple for the Aztecs and Incas, and young leaves are often used in Asian cuisine. SPLs can be eaten raw in salads and smoothies or cooked in soups and curries. Cooking techniques like boiling, steaming, and stir-frying were used during preparation (Joko et al., 2025). Cooking impacts the bioactive compounds and antioxidant activity (AA) of SPLs, with variations depending on the cooking method and leaf variety (Sun et al., 2014a). Traditional cooking involves boiling the leaves in water for 15–20 minutes, followed by washing, seasoning, and serving (Martin, 1998). However, boiling can lead to the loss of certain water-soluble compounds, as they leach into the water during the process. Modern cooking methods, such as steaming, frying, baking, and microwaving, offer better retention or release of phytonutrients compared to traditional boiling. Steaming is particularly effective in retaining or enhancing total phenolic content (TPC) and antioxidant activity (AA) compared to frying, baking, microwaving, and boiling. Boiling causes the leakage of polyphenols into the cooking water, leading to a decrease in bioactive compounds (Jang & Koh, 2019). Steaming also disrupts carotenoid-protein complexes, facilitating the release of carotenoids like β-carotene (Bernhardt and Schlich, 2006). Maximum β-carotene content is achieved by steaming for 15 min, as opposed to 10 minutes (Chirwa-Moonga et al., 2020). Baylis. (2023) recommends blanching SPLs for a few minutes before cooking. However, a reduction of antioxidant content and activity resulted from conventional and microwave blanching due to thermal degradation and leaching (Jang and Koh, 2019). Preservation of phytonutrients depends highly on cooking time. The number of phenolic compounds *found* in conventionally blanched water was higher than in microwave-blanch blanched because of the longer cooking time (Jang and Koh, 2019). Therefore, using a suitable steam-cooking process will be a preferable method to enhance or preserve SPLs bioactive compounds (Hung et al., 2024).

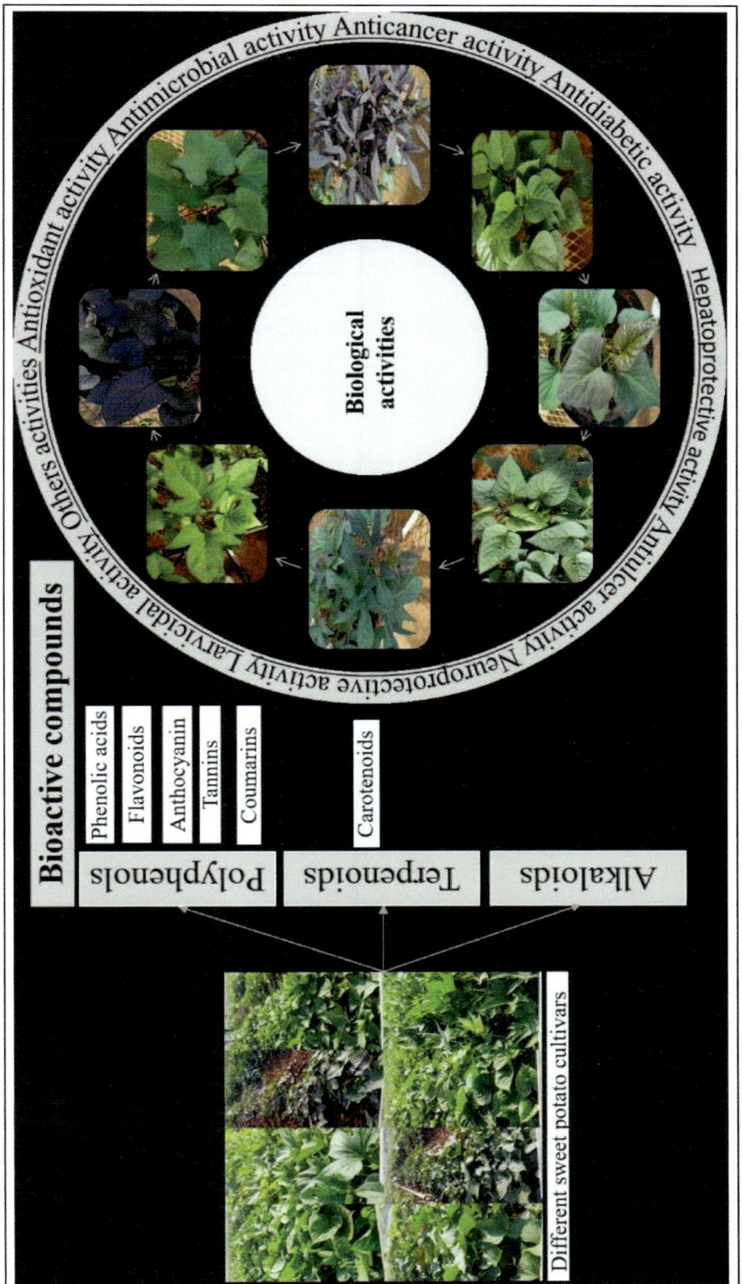

Fig. 2 The harvesting of SPLs for processing of different products.

Post-harvest Preservation

Drying

The common drying techniques include sun drying, an indigenous traditional method from the past and currently at industrial level, freeze-drying, and hot air drying are used. Microwaving drying is an alternative option to freeze-drying due to comparable higher physicochemical properties, drying rate, and lower cost (Sui et al., 2019). During drying, preservation of bioactive compounds and antioxidant activity relies on the drying period, leaf age, and heat level in terms of temperature. Freeze-drying is a preferred method, despite the cost, as it removes water content without compromising the quality and composition of the leaves (Hariyadi, 2013). For instance, freeze-drying showed the highest preservation of the CQA derivatives and antioxidant property (Jeng et al., 2015).

Storage

The SPLs require cooler storage temperatures to preserve their quality. They can be stored without risk for the duration of 14 days at 6°C, but under 12°C, they should not be stored for more than 7 days (Tilistyák et al., 2022). Changes in colour and moisture content indicate a decline in quality. Fresh SPLs can be stored in the refrigerator for a week, either wrapped in paper or in an airtight container (Baylis 2023). For dried SPLs, packaging them in PA with aluminum foil (90% vacuum) maintains the quality for up to 3 months (Van Tai et al., 2023).

Conclusion

Young tender SPLs are rich in bioactive compounds, particularly 3.5-CQA phenolic acid, quercetin flavonoid, cyanidin anthocyanin, lutein carotenoid, and glutamic acid, which is a precursor to gamma-aminobutyric acid (GABA). These compounds are linked with various biological activities that offer significant health benefits, such as antioxidant, anti-inflammatory, antimicrobial, anticancer, and hepatoprotective effects. This makes SPLs a prominent substitute leafy vegetable for consumption. Cultivars characterized by deep lobes or heart shapes are recommended for leaf consumption, as they contain high levels of polyphenols. Leaves with purple pigmentation on the vine tips, such as those from the GS-17-22 cultivar from China and the 2019-11-2 ARC-SA cultivar, are particularly noteworthy due to their high total anthocyanin content. Additionally, cultivars like Blesbok and Healthymi, from South Africa and South Korea, are ideal for their high β-carotene content. SPLs are safe for consumption, showing no signs of cytotoxicity. The concentration of these bioactive compounds varies among cultivars and is significantly influenced by the stage of harvesting and the cooking or processing methods used. Utilising SPLs can enhance food availability through multiple harvests, making them readily accessible throughout the season. This contributes to food security, helps in ending hunger and malnutrition, and promotes sustainable agriculture, aligning with key aspects of Sustainable Development Goal 2 (SDG 2) of zero hunger, as well as SDG 1 (Ending Poverty) and SDG 3 (Good Health and Well-Being). Future studies should focus on isolating and quantifying additional polyphenols, including coumarins, alkaloids, and lignans, and determining the specific compounds responsible for each biological activity.

References

Aguoru, C.U. 2015. Varietal characterisation and taxonomic evaluation of sweet potato (*Ipomoea batatas*) using macro-and micromorphological evidence. *Open Access Library Journal*, *2*(08), 1. http://dx.doi.org/10.4236/oalib.1101757.

Almoraie, N.M. 2019. The role of *Ipomoea batatas* leaves extract on the treatment of diabetes induced by streptozotocin. *Pharmacophore*, *10*(3), 14–20.

Anbuselvi, S. and Muthumani, S. 2014. Phytochemical and antinutritional constituents of sweet potato. *J. Chem. Pharm. Res.*, *6*, 380–383.

Ano, E., Siapo, Y., Onsiyor, E., Achi, D. and Tahiri, Y. 2024. Evaluation of phytochemical screening and acute toxicity of sweet potato leaf macerate (*Ipomoea batatas*) on female rats. January 2024. DOI: 10.13040/IJPSR.0975-8232.IJP.11(8).438-44.

Anonymous. 2018. *Sweet Potato Leaves' (Talbos ng Kamote) Amazing Health Benefits* https://region5.healthresearch.ph/index.php/library-health-news/132-sweet-potato-leaves-talbos-ng-kamote-amazing-health-benefits.

Anonymous. 2024. *Benefits of Sweet Potato Leaves*. https://www.bing.com/ck/a?!&&p=e8fd2b6abf835ac0e4f1d058aa79d994ee0ae3056784e1fc542d175ab192f88dJmltdHM9MTc2MDIyNzIwMA&ptn=3&ver=2&hsh=4&fclid=0d191ba8-76f8-6528-0a57-0ead7705647a&psq= Benefits+of+Sweet+Potato+Leaves&u=a1aHR0cHM6Ly9vbmNxdWVzdGx hYnMuY29tL2Jsb2cvYmVuZWZpdHMtb2Ytc3dlZXQ tcG90YXRvLWxlYXZlcy8.

Armadany, F.I., Sopyan, I. and R. Mustarichie, R. 2024. Antifungal activity and hair growth stimulation of purple sweet potato leaf fraction (*Ipomoea batatas* (L.) Lamk.) and its molecular mechanism through androgen receptor inhibition. *Pharmacia*, *71*: 1–14. https://doi.org/10.3897/pharmacia.71.e119384.

Auwal, M.M., Yelwa, M., Abubakar, I., Umar, J.B., Anchau, G. and Tanimu, B. 2019. The levels of antinutritional factors in *Moringa oleifera* and *Vernomia amygdalina* leaves found in some part of plateau state, Nigeria. *Orient. J. Phys. Sci.*, *4*(2): 1–5. http://dx.doi.org/10.13005/OJPS04.02.06.

Awol, A. 2014. Phytochemical screening, proximate and mineral composition of sweet potato leaves grown in Tepi Provision, South-west of Ethiopia. *Sci. Technol. Arts Res. J.*, *3*(3): 112–115. http://dx.doi.org/10.4314/star.v3i3.1.

Ayeleso, T.B., Ramachela, K. and Mukwevho, E. 2016.. A review of therapeutic potentials of sweet potato: Pharmacological activities and influence of the cultivar. *Trop. J. Pharm. Res.*, *15*(12): 2751–2761. http://dx.doi.org/10.4314/tjpr.v15i12.31.

Ayimbire, A., Asumboya, G., Atinga, C., Akazotiyele, R. and Bodieu, M. 2021. Growth and yield characteristics of five different sweet potato varieties grown in the Guinea Savanna Ecological Zone of Ghana. *J. Exp. Agri. Int.*, *43*(2): 44–55. https://doi.org/10.9734/jeai/2021/v43i330664.

Baylis, M. 2023. *Sweet Potato Leaves: Edible, Flavor, Nutritional and Medicinal Benefits* https://gardenersmag.com/sweet-potato-leaves/.

Bernhardt, S. and Schlich, E. 2006. Impact of different cooking methods on food quality: Retention of lipophilic vitamins in fresh and frozen vegetables. *J. Food Eng.*, *77*(2): 327–333. https://doi.org/10.1016/j.jfoodeng.2005.06.040.

Bolanos, J. 2020. *Antioxidant Properties and Anti-cancer Effects of Polyphenols in Sweet Potato Leaves*. University of Arkansas.

Caviteňo, G.R.M., Manalo, C.M., Eduardo, G. and Muli, I. 2010. Hepatoprotective activity of the crude leaf extract of *Ipomoea batatas* Linn. against acetaminophen and alcohol induced liver damage on Sprague-dawley rats. *The Seth. 4*, 41–49. http://research.lpubatangas.edu.ph/wp-content/uploads/2014/10/STETH-4.5.pdf.

Chen, C.-M., Li, S.-C., Lin, Y.-L., Hsu, C.-Y., Shieh, M.-J. and Liu, J.-F. (2005). Consumption of purple sweet potato leaves modulates human immune response: T-lymphocyte functions, lytic activity of natural killer cell, and antibody production. *World Journal of Gastroenterology: WJG*, *11*(37), 5777–5781.

Chirwa-Moonga, T., Muzungaile, T., Siyumbano, N., Moonga, H.B. and Nyau, V. 2020. Nutrient composition of raw and steamed, green, and purple sweet potato leaf varieties (*Ipomoea batatas*). *J. Med. Active Plants*, *9*(4): 253–261. https://doi.org/10.7275/9ns0-1640.

Cho, H.D., Brownmiller, C., Sorker, H., Islam, S. and Lee, S.O. 2021. Sweetpotato leaves inhibit lipopolysaccharide-induced inflammation in RAW 264.7 macrophages via suppression of NF-kappaB signaling pathway. *Foods*, *10*(9): 2051. https://doi.org/10.3390/foods10092051.

Cioloca, M., Bădărău, C.L., Tican, A. and Popa, M. 2021. Antioxidant potential, total vitamin C, phenolic and flavonoids content of sweet potato leaves. *Rom. Biotechol. Lett.*, *26*, 2441–2447. https://doi.org/10.25083/rbl/26.2/2441.2447.

da Costa, G.F., Girsi, C.V.B., Meireles, Bruno, R.L.D.A., de Sousa, S. and Cordeiro, A.M.T.M. 2021. Extracts of Sweet Potato leaf (*Ipomoea batatas*) and Taioba (*Xanthosoma Sagittifolium*): New Sources of Natural Bioactives for the Food Industry. *Acta Sci. Nutr. Health*, *6*(4): 21–33.

Das, A.K., Islam, M.N., Faruk, M.O., Ashaduzzaman, M. and Dungani, R. 2020. Review on tannins: Extraction processes, applications, and possibilities. *S. Afr. J. Bot.*, *135*, 58–70. https://doi.org/10.1016/j.sajb.2020.08.008.

Deng, J., Wu, D., Shi, J., Balfour, K., Wang, H., Zhu, G., Liu, Y., Wang, J. and Zhu, Z. 2020. Multiple MYB activators and repressors collaboratively regulate the juvenile red fading in leaves of sweetpotato. *Fron. Plant Sci.*, *11*: 941. https://doi.org/10.3389/fpls.2020.00941.

Diez-Gutiérrez, L., San Vicente, L., Barrón, L.J.R., del Carmen Villarán, M. and Chávarri, M. 2020. Gamma-aminobutyric acid and probiotics: Multiple health benefits and their future in the global functional food and nutraceuticals market. *J. Funct. Foods*, *64*: 103669. https://doi.org/10.1016/j.jff.2019.103669.

Dinu, M., Soare, R., Hoza, G., Becherescu, A.D. and Băbeanu, C. (2021). Bioactive compounds content and antioxidant activity in the leaves of some sweet potato cultivars (Ipomoea batatas L.). *Scientific Papers. Series B. Horticulture*, *65*(1): 415–422.

Drapal, M., Rossel, G., Heider, B. and Fraser, P.D. 2019. Metabolic diversity in sweet potato (*Ipomoea batatas* Lam.) leaves and storage roots. *Hortic. Res.*, *6*: 1–9. https://doi.org/10.1038/s41438-018-0075-5.

Dutta, S., Ray, S. and Nagarajan, K. 2013. Glutamic acid as anticancer agent: An overview. *Saudi Pharma. J.*, *21*(4): 337–343. https://doi.org/10.1016/j.jsps.2012.12.007.

Emami, S. and Dadashpour, S. 2015. Current developments of coumarin-based anti-cancer agents in medicinal chemistry. *Eur. J. Med. Chem.*, *102*: 611–630. https://doi.org/10.1016/j.ejmech.2015.08.033.

Enyiukwu, D., Amadioha, A. and Ononuju, C. 2018. Nutritional significance of cowpea leaves for human consumption. *Greener Trends Food Sci. Nutr.*, *1*: 1–10. https://doi.org/10.15580/GTFSN.2018.1.061818085.

Essack, H., Odhav, B. and Mellem, J.J. 2017. Screening of traditional South African leafy vegetables for specific anti-nutritional factors before and after processing. *Food Sci. Technol.*, *37*(3): 462–471. https://doi.org/10.1590/1678-457x.20416.

FAO. 2023. FAOSTAT. Retrieved on 11 June 2025. FAOSTAT.

Gao, M., Soriano, S. F., Cao, Q., Yang, X. and Lu, G. 2020. Hexaploid sweetpotato (*Ipomoea batatas* (L.) Lam.) may not be a true type to either auto-or allopolyploid. *PloS One*, *15*(3): e0229624. https://doi.org/10.1371/journal.pone.0229624.

Hall, M.R. and Phatak, S.C. 1993. Sweet potato: *Ipomoea batatas* (L) Lam. In: *Genetic Improvement of Vegetable Crops* (pp. 693–708). Elsevier.
Hapke, K.L., Azad, M.A.K., Sultana, T. and Islam, S. 2024. Comparison of antioxidant, β-carotene, and phenolic levels between roots and leaves across three popular sweet potato (*Ipomoea batatas* L.) varieties. *Agric. and Food Secur.*, *13*(1): 1–10. https://doi.org/10.1186/s40066-024-00516-x.
Hariyadi, P. 2013. Freeze Drying Technology: For better quality and flavour of dried products. *Foodreview Indonesia*, *8*(2): 52–57.
Hatier, J.-H. B. and Gould, K.S. 2009. Anthocyanin Function in Vegetative Organs. In C. Winefield, K. Davies and K. Gould (Eds.), *Anthocyanins: Biosynthesis, Functions, and Applications* (pp. 1–19). Springer New York. https://doi.org/10.1007/978-0-387-77335-3_1.
Hisamuddin, A. S.-d. B., Naomi, R., Bin Manan, K.A., Bahari, H., Yazid, M.D., Othman, F., Embong, H., Hadizah Jumidil, S., Hussain, M.K. and Zakaria, Z.A. 2023. Phytochemical component and toxicological evaluation of purple sweet potato leaf extract in male Sprague–Dawley rats. *Front. Pharmacol.*, *14*: 1132087. https://doi.org/10.3389/fphar.2023.1132087.
Hong, J., Mu, T., Sun, H., Richel, A. and Blecker, C. 2020. Valorization of the green waste parts from sweet potato (*Impomea batatas* L.): Nutritional, phytochemical composition, and bioactivity evaluation. *Food Sci. Nutr.*, *8*(8): 4086–4097. https://doi.org/10.1002/fsn3.1675.
Hossain, M.A.S., Saiful Islam, A.F.M., Noor, M., Miah, M.N.H. and Khan, M. 2019. Leaf Chlorophylls and Carotenoids Status and Their Correlation with Storage Root Weight of Some Local and Exotic Sweetpotato Genotypes. *Asian J. Res. Bot.*, *2*(4): 1–11.
Hossain, M.M., Rahim, M.A., Moutosi, H.N. and Das, L. (2022). Evaluation of the growth, storage root yield, proximate composition, and mineral content of coloured sweet potato genotypes. *Journal of Agriculture and Food Research*, *8*, 100289. https://doi.org/10.1016/j.jafr.2022.100289.
Hou, D., Tang, J., Feng, Q., Niu, Z., Shen, Q., Wang, L. and Zhou, S. 2024. Gamma-aminobutyric acid (GABA): A comprehensive review of dietary sources, enrichment technologies, processing effects, health benefits, and its applications. *Crit. Rev. Food Sci. Nutr.*, *64*(24): 8852–8874. doi: https://doi.org/10.1080/10408398.2023.2204373.
Hu, Y., Sun, H. and Mu, T. 2022. Effects of sweet potato leaf powder on sensory, texture, nutrition, and digestive characteristics of steamed bread. *J. Food Process. Preserv.*, *46*(7): e16697. https://doi.org/10.1111/jfpp.16697.
Hung, C.-Y., Jiang, Y.-R. and Chen, Y.-C. 2024. Total Phenolic Contents and Antioxidant Activity in Sweet Potato Leaves and Carrots after Steam-Cooking. *J. Taiwan Agric. Res.*, *73*(3): 219–233.
International Potato Center: *Sweetpotato Facts and Figures*. International Potato Center._Retrieved on 11 June 2025.
Imafidon, K.E., Durojaiye, M. and Daniel, B. 2015. Safety assessment of *Ipomoea batatas* (L). Lam leaf extract, a traditional haematinic; in male Wistar rats. *J. Ghana Sci. Assoc.*, *16*(2): 45–51.
Ishida, H., Suzuno, H., Sugiyama, N., Innami, S., Tadokoro, T. and Maekawa, A. 2000. Nutritive evaluation on chemical components of leaves, stalks, and stems of sweet potatoes (*Ipomoea batatas* poir). *Food Chem.*, *68*(3): 359–367. https://doi.org/10.1016/S0308-8146(99)00206-X.
Islam, M.S., Yoshimoto, M. and Yamakawa, O. 2003. Distribution and physiological functions of caffeoylquinic acid derivatives in leaves of sweetpotato genotypes. *J. Food Sci.*, *68*(1): 111–116. https://doi.org/10.1111/j.1365-2621.2003.tb14124.x.
Islam, S., Yoshimoto, M., Ishiguro, K. and Yamakawa, O. 2002. Bioactive compounds in *Ipomoea batatas* leaves. *Acta Hortic.*, *628*: 693–699. https://doi.org/10.17660/ActaHortic.2003.628.88.
Jang, Y. and Koh, E. 2019. Antioxidant content and activity in leaves and petioles of six sweet potato (*Ipomoea batatas* L.) and antioxidant properties of blanched leaves. *Food Sci. Biotechnol.*, *28*: 337–345. https://doi.org/10.1007/s10068-018-0481-3.
Jeng, T.L., Lai, C.C., Liao, T.C., Lin, S.Y. and Sung, J.M. 2015. Effects of drying on caffeoylquinic acid derivative content and antioxidant capacity of sweet potato leaves. *J. Food Drug Anal.*, *23*(4): 701–708. http://dx.doi.org/10.1016/j.jfda.2014.07.002.

Jia, R., Tang, C., Chen, J., Zhang, X. and Wang, Z. 2022. Total phenolics and anthocyanins contents, and antioxidant activity in four different aerial parts of leafy sweet potato (*Ipomoea batatas* L.). *Molecules*, *27*(10): 3117. http://dx.doi.org/10.3390/molecules27103117.

Jiang, T., Ye, S., Liao, W., Wu, M., He, J., Mateus, N. and Oliveira, H. 2022. The botanical profile, phytochemistry, biological activities, and protected-delivery systems for purple sweet potato (*Ipomoea batatas* (L.) Lam.): An up-to-date review. *Food Res. Int.*, *161*: 111811. http://dx.doi.org/DOI: 10.1016/j.foodres.2022.111811.

Johnson, M. and Pace, R.D. (2010). Sweet potato leaves: Properties and synergistic interactions that promote health and prevent disease. *Nutrition Reviews*, *68*(10), 604–615. https://doi.org/10.1111/j.1753-4887.2010.00320.x.

Joko, L.B., Eiasu, B.K. and Ngwenya, S.M. 2025. Evaluation of Drying Methods and Leaf Age on the Nutrient Composition of Orange-fleshed Sweetpotato (*Ipomoea batatas* L.) Leaves. *HortScience*, *60*(4): 514–518. https://doi.org/10.21273/HORTSCI18356-24.

Jubaedah, E. and Hermawan, M.H. 2024. The effect of giving sweet potato leaf-based functional drink to increase HB levels. *Hum. Nutr. Metab.*, *37*: 200280. https://doi.org/10.1016/j.hnm.2024.200280.

Jyothi, A., Moorthy, S. and Eswariamma, C. 2005. Anthocyanins in sweet potato leaves-varietal screening, growth phase studies, and stability in a model system. *Int. J. Food Prop.*, *8*(2): 221–232. https://doi.org/10.1081/JFP-200063051.

Kang, H. and Lee, S.G. 2021. Protective effect of purple sweet potato leaf (*Ipomoea batatas* Linn Convolvulaceae) against alcohol-induced liver damage in mice. *Trop, J. Pharm. Res.*, *20*(2): 301–308.

Karna, P., Gundala, S.R., Gupta, M.V., Shamsi, S.A., Pace, R.D., Yates, C., Narayan, S. and Aneja, R. 2011. Polyphenol-rich sweet potato greens extract inhibits proliferation and induces apoptosis in prostate cancer cells *in vitro* and *in vivo*. *Carcinogenesis*, *32*(12): 1872–1880. https://doi.org/10.1093/carcin/bgr215.

Khalid, M., Bilal, M. and Huang, D.-F. 2019. Role of flavonoids in plant interactions with the environment and against human pathogens: A review. *J. Integr. Agric.*, *18*(1): 211–230. https://doi.org/10.1016/S2095-3119(19)62555-4.

Khoo, H.E., Azlan, A., Tang, S.T. and Lim, S.M. 2017. Anthocyanidins and anthocyanins: Colored pigments as food, pharmaceutical ingredients, and the potential health benefits. *Food Nutr. Res.*, *61*(1): 1361779. https://doi.org/10.1080/16546628.2017.1361779.

Kirakosyan, R., Golivanov, Y.Y., Trukhachev, V., Zaitseva, S., Khaliluev, M., Baranova, E. and Kalashnikova, E. 2023. Colleter formation on the leaves of sweet potato microclones (*Ipomoea batatas* L.) under *in vitro* conditions. *Russ J. Plant Physiol.*, *70*(6): 115. https://doi.org/10.1134/S1021443723601301.

Kuddus, M.A., Miah, M.M.A., Datta, G.C., Sarker, A.K., Alam, M.H. and Khan, M.M.H. 2020. Proximate and micronutrient analysis of orange fleshed sweet potato leaves grown in Sylhet regions of Bangladesh. *Res. Agric. Livest. Fish.*, *7*(1): 43–50. https://doi.org/ 10.3329/ralf.v7i1.46830.

Kurata, R., Adachi, M., Yamakawa, O. and Yoshimoto, M. 2007. Growth suppression of human cancer cells by polyphenolics from sweetpotato (*Ipomoea batatas* L.) leaves. *J. Agric. Food Chem.*, *55*: 185–190. https://doi.org/10.1021/jf0620259.

Laurie, S., Calitz, F., Adebola, P. and Lezar, A. 2013. Characterization and evaluation of South African sweet potato (*Ipomoea batatas* (L.) LAM) land races. *S. Afr. J. Bot.*, *85*: 10–16. https://doi.org/10.1016/j.sajb.2012.11.004.

Lev-Yadun, S. and Gould, K.S. 2009. Role of anthocyanins in plant defence. In: Winefield, C., Davies, K., Gould, K. (Eds.), *Anthocyanins*, Springer, New York, pp. 22–28. https://doi.org/10.1007/978-0-387-77335-3_2.

Li, G., Gao, X., Wang, Y., He, S., Guo, W. and Huang, J. 2023. Effects of superfine grinding sweet potato leaf powders on physicochemical and structure properties of sweet potato starch noodles. *Food Sci. Nutr.*, *11*(10): 6498–6508. https://doi.org/10.1002/fsn3.3593.

Li, M., Jang, G.Y., Lee, S.H., Kim, M.Y., Hwang, S.G., Sin, H.M., Kim, H.S., Lee, J. and Jeong, H.S. 2017.. Comparison of functional components in various sweet potato leaves and stalks. *Food Sci. Biotechnol.*, *26*: 97–103. https://doi.org/10.1007/s10068-017-0013-6.

Luo, C., Wang, X., Gao, G., Wang, L., Li, Y. and Sun, C. 2013. Identification and quantification of free, conjugate, and total phenolic compounds in leaves of 20 sweet potato cultivars by HPLC–DAD and HPLC–ESI–MS/MS. *Food Chem.*, *141*(3): 2697–2706. https://doi.org/10.1016/j.foodchem.2013.05.009.

Luo, D., Mu, T. H., Sun, H. and Chen, J. 2020. Optimization of the formula and processing of a sweet potato leaf powder-based beverage. *Food Sci. Nutr.*, *8*(6): 2680–2691. https://doi.org/10.1002/fsn3.1555.

Mahfudh, N., Mantali, M.F. and Sulistyani, N. 2022. Antioxidant and antihyperlipidemic effect of purple sweet potato leaf extract (*Ipomoea batatas* L.) and red yeast rice combination in hypercholesterol rats. *Indones. J. Pharm.*, 93–99. https://doi.org/10.22146/ijp.2115.

Makori, S. I., Mu, T.-H. and Sun, H.-N. 2020. Total polyphenol content, antioxidant activity, and individual phenolic composition of different edible parts of 4 sweet potato cultivars. *Nat. Prod. Commun.*, *15*(7): 1–12. https://doi.org/10.1177/1934578X20936931.

Manasathien, J. and Khanema, P. 2022. Nutrition, phytochemicals, antioxidants, and cytotoxicity of sweet potato leaf tea. *Trends Sci. 19*(17): 5766, 1–15. https://doi.org/10.48048/tis.2022.5766.

Maoka, T. 2020. Carotenoids as natural functional pigments. *J. Nat. Med.*, *74*(1): 1–16. https://doi.org/10.1007/s11418-019-01364-x.

Martin, F. (1998). Sweet potato. *Echo Technical Note. Echo*, *17391*. http://resilience-exchange.s3.amazonaws.com/attachments/uploads/3206/original/Sweet_Potato.pdf.

Mbaeyi-Nwaoha, I. and Emejulu, V. 2013. Evaluation of phytochemical composition and antimicrobial activity of sweet potato (*Ipomoea batatas*) leaf. *Pak. J. Nutr.*, *12*(6): 575–586

Mehta, S.K. and Gowder, S.J.T. (2015). Members of antioxidant machinery and their functions. *Basic Principles and Clinical Significance of Oxidative Stress*, *11*, 59–85. http://dx.doi.org/10.5772/61884.

Merzlyak, M.N., Chivkunova, O.B., Solovchenko, A.E. and Naqvi, K.R. 2008. Light absorption by anthocyanins in juvenile, stressed, and senescing leaves. *J. Exp. Bot.*, *59*(14): 3903–3911. https://doi.org/10.1093/jxb/ern230.

Muñoz-Rodríguez, P., Carruthers, T., Wood, J.R., Williams, B.R., Weitemier, K., Kronmiller, B., Ellis, D., Anglin, N.L., Longway, L. and Harris, S.A. 2018. Reconciling conflicting phylogenies in the origin of sweet potato and dispersal to Polynesia. *Curr. Biol.*, *28*(8): 1246–1256. e1212. https://doi.org/10.1016/j.cub.2018.03.020.

Naidoo, S.I., Laurie, S.M., Shimelis, H. and Laing, M.D. 2020. Morpho-agronomical characterisation of local and international sweetpotato germplasm from the South African collection. *S. Afr. J. Plant Soil*, *37*(4): 308–320. https://doi.org/10.1080/02571862.2020.1791983.

Nguyen, H.C., Chen, C.-C., Lin, K.-H., Chao, P.-Y., Lin, H.-H. and Huang, M.-Y. 2021. Bioactive compounds, antioxidants, and health benefits of sweet potato leaves. *Molecules*, *26*(7): 1820. https://doi.org/10.3390/molecules26071820.

Nkongho, G.O., Achidi, A.U., Ntonifor, N.N., Numfor, F.A., Dingha, B.N., Jackai, L. and Bonsi, C.K. 2014. Sweet potatoes in Cameroon: Nutritional profile of leaves and their potential new use in local foods. *Afr. J. Agric. Res.*, *9*(18): 1371–1377. https://doi.org/10.5897/AJAR2014.8611.

Odongo, N.O., Abong, G.O., Okoth, M.W. and Karuri, E.G. 2015. Development of high protein and vitamin a flakes from sweet potato roots and leaves. *O.A. Lib. J.*, *2*, 1–10. https://doi.org/10.4236/oalib.1101573.

Ojong, P.B., Njiti, V., Guo, Z., Gao, M., Besong, S. and Barnes, S.L. 2008. Variation of flavonoid content among sweet potato accessions. *J. Am. Soc. Hortic. Sci.*, *133*(6): 819–824. https://doi.org/10.21273/JASHS.133.6.819.

Omoba, O.S., Oyewole, G.O. and Oloniyo, R.O. 2020. Chemical compositions and antioxidant properties of orange fleshed sweet potato leaves and the consumer acceptability in vegetable soup. *Prev. Nutr. Food Sci.*, *25*(3): 293. https://doi.org/10.3746/pnf.2020.25.3.293.

Ooi, C.P. and Loke, S.C. 2012. Sweet potato for Type 2 diabetes mellitus. *Cochrane Database of Syst. Rev.*, (2), 1–39. https://doi.org/10.1002/14651858.CD009128.pub3.

Ooko Abong', G., Muzhingi, T., Wandayi Okoth, M., Ng'ang'a, F., Ochieng', P.E., Mahuga Mbogo, D., Malavi, D., Akhwale, M. and Ghimire, S. 2020. Phytochemicals in leaves and roots of selected Kenyan orange fleshed sweet potato (OFSP) varieties. *Int. J. Food Sci.*, *2020*(1): 3567972. https://doi.org/10.1155/2020/3567972.

Padda, M.S. and Picha, D. 2007. Antioxidant activity and phenolic composition in 'Beauregard' sweet potato are affected by root size and leaf age. *J. Am. Soc. Hort. Sci.*, *132*(4): 447–451. https://doi.org/10.21273/JASHS.132.4.447.

Patel, Y.K. and Patel, K.K. 2020. From Farm Waste to Functional Food: The Hidden Value of Sweet Potato Leaves. *Int. J. Home Sci.*, *11*(08): 435–442. https://doi.org/10.36537/IJAHS/11.7&8/435-442.

Pavithra Bharathi, V., Ragavendran, C., Murugan, N. and Natarajan, D. (2017). *Ipomoea batatas* (Convolvulaceae)-mediated synthesis of silver nanoparticles for controlling mosquito vectors of *Aedes albopictus*, *Anopheles stephensi*, and *Culex quinquefasciatus* (Diptera: Culicidae). *Artifi. Cells Nanomed. Biotechnol.*, *45*(8): 1568–1580.

Phahlane, C.J., Laurie, S.M., Shoko, T., Manhivi, V.E. and Sivakumar, D. 2021. An evaluation of phenolic compounds, carotenoids, and antioxidant properties in leaves of South African cultivars, Peruvian 199062.1 and USA's Beauregard. *Front. Nutr.*, *8*: 773550. https://doi.org/10.3389/fnut.2021.773550.

Qi, Q., Chu, M., Yu, X., Xie, Y., Li, Y., Du, Y., Liu, X., Zhang, Z., Shi, J. and Yan, N. 2023. Anthocyanins and proanthocyanidins: Chemical structures, food sources, bioactivities, and product development. *Food Rev. Int.*, *39*(7): 4581–4609. https://doi.org/10.1080/87559129.2022.2029479.

Rahajeng, W., Restuono, J. and Indriani, F.C. 2023. The Relationship of Sweet Potato Germplasm Based on Morphological Characters. *J. Biodjati.*, *8*(1): 94–105.

Ranteallo, Y., Ahmad, M., Syam, A. and Nilawati, A. 2023. Identification and quantification of minerals and vitamins of purple sweet potato (*Ipomoea batatas*) leaves. In: *IOP Conference Series: Earth Environ. Sci.*, *1230*: 012134. https://doi.org/10.1088/1755-1315/1230/1/012134.

Refaat, W. A., El-Kholie, E. M. and Abo, E.-N. 2020. AF possible effects of sweet potato leaves, roots and their mixture feeding on alloxan-induced diabetic rats. *J. Home Econ.*, *30*: 1–18.

Rezeki, S., Chismirina, S. and Novita, I. 2017. Effect of *Ipomoea batatas* L. leaves extract on percentage healing of traumatic ulcer in oral mucosa (*Rattus novergicus* Strain Wistar). *Biomed. Pharm. J.*, *10*(4): 1943–1947. https://dx.doi.org/10.13005/bpj/1314.

Salam, U., Ullah, S., Tang, Z.-H., Elateeq, A. A., Khan, Y., Khan, J., Khan, A. and Ali, S. 2023. Plant metabolomics: An overview of the role of primary and secondary metabolites against different environmental stress factors. *Life*, *13*(3): 706. https://doi.org/10.3390/life13030706.

Santos, Y.J.S., Malegori, C., Colnago, L.A. and Vanin, F.M. 2024. Application on infrared spectroscopy for the analysis of total phenolic compounds in fruits. *Crit. Rev. Food Sci. Nutr.*, *64*(10): 2906–2916. https://doi.org/10.1080/10408398.2022.2128036.

Sasaki, K., Oki, T., Kai, Y., Nishiba, Y. and Okuno, S. 2015. Effect of repeated harvesting on the content of caffeic acid and seven species of caffeoylquinic acids in sweet potato leaves. *Biosci. Biotechnol. Biochem.*, *79*(8): 1308–1314. https://doi.org/10.1080/09168451.2015.1025032.

Shankar, A.H. and Prasad, A.S. 1998. Zinc and immune function: The biological basis of altered resistance to infection. *Am. J. Clin. Nutr.*, *68*(2): 447S–463S. https://doi.org/10.1093/ajcn/68.2.447S.

Sharma, S.S.R. 2023. An Update on Pharmacological Activity of *Ipomoea batata*. *J. Emerg. Technol. Innov. Res.*, *10*(8): 518–527.

Shi, J., Fang, D., Sui, Y., Xiong, T., Chen, X., Fan, C., Zhou, D., Cai, F. and Mei, X. 2025. Polyphenol content, antioxidant capacity, and composition in different varieties of sweet potato (*Ipomoea *batatas* L.) leaves during growth stages. *Sci. Hortic.*, *342*: 113925. https://doi.org/10.1016/j.scienta.2024.113925.

Siame, L.M., Mwitwa, K., Kaundu, S., Luambia, I.N., Siabwacha, A. and Yusuf, U.A. 2025. Effects of aqueous extract of *Ipomoea batatas* (Sweet Potato) leaves on the frontal cortex of diabetic Wistar rats. *Eur. J. Med. Health Res.*, *3*(3): 21–31.

Solihah, I., Herlina, E. M., Haryanti, H., Amalia, M. and Puspita, R.S. 2023. The hypoglycemic effect of purple sweet potato leaf fractions in diabetic rats. *J. Adv. Pharm. Edu. Res.*, *13*(3): 65.

Srisuwan, S., Sihachakr, D. and Siljak-Yakovlev, S. 2006. The origin and evolution of sweet potato (*Ipomoea batatas* Lam.) and its wild relatives through the cytogenetic approaches. *Plant Sci.*, *171*(3): 424–433. https://doi.org/10.1016/j.plantsci.2006.05.007.

Su, X., Griffin, J., Xu, J., Ouyang, P., Zhao, Z. and Wang, W. 2019a. Identification and quantification of anthocyanins in purple-fleshed sweet potato leaves. *Heliyon*, *5*(6): 1–6. https://doi.org/10.1016/j.heliyon.2019.e01964.

Su, X., Jia, Z., Tao, F., Shen, J., Xu, J., Griffin, J. and Wang, W. 2019b. Characterization of anthocyanins in sweet potato leaves grown in various stages and conditions. *Eur. J. Nutr. Food Saf.*, *10*(4): 253–262. https://doi.org/10.9734/ejnfs/2019/v10i430119.

Suárez, S., Mu, T., Sun, H. and Añón, M.C. 2020. Antioxidant activity, nutritional, and phenolic composition of sweet potato leaves as affected by harvesting period. *Int. J. Food Prop.*, *23*(1): 178–188. https://doi.org/10.1080/10942912.2020.1716796.

Suda, I., Ishikawa, F., Hatakeyama, M., Miyawaki, M., Kudo, T., Hirano, K., Ito, A., Yamakawa, O. and Horiuchi, S. 2008. Intake of purple sweet potato beverage affects on serum hepatic biomarker levels of healthy adult men with borderline hepatitis. *Eur. J. Clin. Nutr.*, *62*(1): 60–67.

Sugata, M., Lin, C.-Y. and Shih, Y.-C. 2015. Anti-inflammatory and anticancer activities of Taiwanese purple-fleshed sweet potatoes (*Ipomoea batatas* L. Lam) extracts. *BioMed Res. Int.*, *2015*(1): 768093. https://doi.org/10.1155/2015/768093.

Sui, W., Mu, T., Sun, H. and Yang, H. 2019. Effects of different drying methods on nutritional composition, physicochemical and functional properties of sweet potato leaves. *J. Food Process. Preserv.*, *43*(3): e13884. https://doi.org/10.1111/jfpp.13884.

Sultana, T., Islam, S., Azad, M.A.K., Akanda, M.J.H., Rahman, A. and Rahman, M.S. 2024. Phytochemical profiling and antimicrobial properties of various sweet potato (*Ipomoea batatas* L.) leaves assessed by RP-HPLC-DAD. *Foods*, *13*(17): 2787. https://doi.org/10.3390/foods13172787.

Sun, H., Mu, B., Song, Z., Ma, Z. and Mu, T. (2018). The *in vitro* antioxidant activity and inhibition of intracellular reactive oxygen species of sweet potato leaf polyphenols. *Oxidative Medicine and Cellular Longevity*, *2018*(1). https://doi.org/10.1155/2018/9017828.

Sun, H., Mu, T., Xi, L. and Song, Z. 2014a. Effects of domestic cooking methods on polyphenols and antioxidant activity of sweet potato leaves. *J. Agric. Food Chem.*, *62*(36): 8982–8989. https://doi.org/10.1021/jf502328d.

Sun, H., Mu, T., Xi, L., Zhang, M. and Chen, J. 2014b. Sweet potato (*Ipomoea batatas* L.) leaves as nutritional and functional foods. *Food Chem.*, *156*: 380–389. https://doi.org/10.1016/j.foodchem.2014.01.079.

Sun, R., Kan, J., Cai, H., Hong, J., Jin, C. and Zhang, M. 2022. *In vitro* and *in vivo* ameliorative effects of polyphenols from purple potato leaves on renal injury and associated inflammation induced by hyperuricemia. *J. Food Biochem.*, *46*(2): e14049. https://doi.org/10.1111/jfbc.14049.

Suzuki, A. and Iwata, J. 2021. Amino acid metabolism and autophagy in skeletal development and homeostasis. *Bone*, *146*: 115881. https://doi.org/10.1016/j.bone.2021.115881.

Swami, K. 2024. Origin, distribution, taxonomy, botanical description, genetics and cytogenetics, genetic diversity, and breeding of lima bean (*Phaseolus lunatus* L.). *Int. J. Rec. Adv. Multidisc. Res.*, *11*(5), 9956–9972.

Tan, W., Guo, X., Wang, Z., Zhang, R., Tang, C., Jiang, B., Jia, R., Deng, Y., Yang, S. and Chen, J. 2024. Metabolic profiles and morphological characteristics of leaf tips among different sweet potato (*Ipomoea batatas* Lam.) varieties. *J. Integr. Agric.*, *23*(2): 494–510.

Tilistyák, J. T., Cziáky, Z., Hegedüs, I. and Tarek, M. 2022. Bioactive constituents and shelf-life of sweet potato (*Ipomoea batatas* L.) leaves. *Rev. Agric. Rur. Devel.*, *11*(1–2): 73–78.

Tshilongo, L., Mianda, S.M., Seke, F., Laurie, S.M. and Sivakumar, D. 2024. Influence of harvesting stages on phytonutrients and antioxidant properties of leaves of five purple-fleshed sweet potato (*Ipomoea batatas*) genotypes. *Foods*, *13*(11): 1640. https://doi.org/10.3390/foods13111640.

Van Tai, N., Tran, V.T.N. and Loan, L.T.K. 2023. Effect of drying methods and storage conditions on quality of purple sweet potato leaves. *Foods*, *26*: 44. https://doi.org/10.3390/Foods2023-15093.

Vishnu, V.R., Renjith, R.S., Mukherjee, A., Anil, S.R., Sreekumar, J. and Jyothi, A.N. 2019. Comparative study on the chemical structure and *in vitro* antiproliferative activity of anthocyanins in purple root tubers and leaves of sweet potato (*Ipomoea batatas*). *J. Agric. Food Chem.*, *67*(9): 2467–2475. https://doi.org/10.1021/acs.jafc.8b05473.

Wallace, T.C. and Giusti, M.M. 2015. Anthocyanins. *Adv. Nutr.*, *6*(5): 620–622.

Wang, L., Pan, X., Jiang, L., Chu, Y., Gao, S., Jiang, X., Zhang, Y., Chen, Y., Luo, S. and Peng, C. 2022. The biological activity mechanism of chlorogenic acid and its applications in the food industry: A review. *Front. Nutr. 9*: 943911. https://doi.org/10.3389/fnut.2022.943911.

Waskito, B.A., Sargowo, D., Kalsum, U. and Tjokroprawiro, A. 2022. The role of *Ipomoea batatas* leaves extract as a potent antioxidant and anti-inflammatory in rats fed high-fat diet. *Res. J. Pharm. Technol.*, 2395–2401. https://doi.org/10.52711/0974-360x.2022.00398.

Yuan, H., Zhang, J., Nageswaran, D. and Li, L. 2015. Carotenoid metabolism and regulation in horticultural crops. *Horticulture Research*, *2*: 15036. https://doi.org/10.1038/hortres.2015.36.

Zhang, C., Liu, D., Wu, L., Zhang, J., Li, X. and Wu, W. 2019. Chemical characterization and antioxidant properties of ethanolic extract and its fractions from sweet potato (*Ipomoea batatas* L.) leaves. *Foods*, *9*(1):15. https://doi.org/10.3390/foods9010015.

Zhang, Y., Bian, S., Hu, J., Liu, G., Peng, S., Chen, H., Jiang, Z., Wang, T., Ye, Q. and Zhu, H. 2022. Natural deep eutectic solvent-based microwave-assisted extraction of total flavonoid compounds from spent sweet potato (*Ipomoea batatas* L.) leaves: Optimization and antioxidant and bacteriostatic activity. *Molecules 27*(18): 5985. https://doi.org/10.3390/molecules27185985.

6 Nutritive Value, Bioactive Compounds, and Biological Activities of Coriander (*Coriandrum sativum* L.) Leaves

Supreethe Holenarasipura Subramanya, Aayush Sinha, Dhanasekaran Sneha and Praveen Nagella*

Introduction

The plant kingdom has been serving humanity since the time of evolution. Throughout recorded history, the plant kingdom has been considered to be the primary wellspring of nourishment, healing, and advancement (Zhang et al., 2021). Apart from plants merely being considered as indispensable providers of micro- and macro nutrients, they are the treasure of several bioactive molecules that humans are using even without their knowledge, dating back to the origin of life. This primarily includes the secondary metabolites, which are of no use to host plants, but are a great source of various bioactive molecules, including polyphenols, alkaloids, flavonoids, terpenoids, and essential oils, which are synthesized in response to ecological pressures (Scandar et al., 2023). Beyond their ecological function, many of these compounds exert profound pharmacological effects in humans. They are an integral part of ethnopharmacology, traditional systems of medicine followed in various parts of the world, and a primary focus of advanced biomedical research (Sambasivaraju and Za, 2016). Indeed, contemporary nutritional science recognizes that a diet abundant in a few selected essential plants not only meets caloric and nutritional requirements but also delivers an impressive repertoire of phytochemicals with antioxidant, anti-inflammatory, antimicrobial, and anticancer properties, which modulate key physiological pathways involved in chronic and degenerative diseases like cardiovascular disorders, neurodegenerative conditions, diabetes, and certain malignancies. Modern analytical methods, encompassing genomics, metabolomics, and high-throughput screening, have accelerated the discovery, profiling, and utilization of plant bioactive compounds, forging links between traditional knowledge and the frontiers of molecular medicine (Cuevas-Cianca et al., 2023; Manickam et al., 2023).

Department of Life Sciences, School of Sciences, Christ University, Bengaluru 560 029, India.
* Corresponding author: praveen.n@christuniversity.in

DOI: 10.1201/9781003687832-6

Among the rich phytodiversity of the plant kingdom, coriander (*Coriandrum sativum* L.), a member of the Apiaceae family, emerges as an exemplar of botanical richness and cultural resonance (Mahleyuddin et al., 2021). The known usage of coriander is unique with the endowment of culinary, medicinal, and economic significance. It has characteristic feathery, aromatic leaves, which are known as cilantro in many regions of the world, and globose, spicy-sweet seeds. Coriander positions itself among all the continents and cuisines (Maroufi et al., 2010). It is a fast-growing annual, native to southern Europe and western Asia, and has been cultivated for at least 3,000 years, as evidenced by its remains in ancient Egyptian tombs. The foliage of the plant is distinguished by its distinct and polarizing odour that is attributed to complex aldehydes and volatile molecules (Ebrahimi et al., 2010). According to etymology, this is a trait that connects its nomenclature to the Greek word for ' bedbug' due to the aroma of its green parts. Botanically, coriander is an erect, glabrous, and strongly branching herb that reaches heights of up to 1.3 meters when fully grown (Tomar et al., 2014). This is also characterized by delicate compound leaves, umbrella-like inflorescences of tiny white or pink flowers, and a deep tap root system (Talat, 2017). These morphological features not only facilitate wide ecological adaptability from temperate zones to tropical highlands but also influence the accumulation and diversity of bioactive constituents (López et al., 2008). The leaves are specially valued for their high content of vitamins A, C, and K, mineral nutrients like potassium, calcium, magnesium, and volatile oils, predominantly linalool, which is a naturally occurring terpene alcohol that contributes to both sensory attributes and health-promoting effects (Parmar et al., 2011). Recent genomic and agroecological studies suggest intraspecific variation in bioactive composition among coriander cultivars, which may influence both nutritional value and medicinal potential (Meena et al., 2024; Basol et al., 2024). Notably, coriander leaves are distinguished by their rich profile of polyphenolic compounds like quercetin, kaempferol, and rutin, as well as phenolic acids such as caffeic and chlorogenic acid. These molecules have been proven to be ROS-scavengers and elicit cytoprotective effects at a certain concentration range (Maroufi et al., 2010). Furthermore, they effectively demonstrate antimicrobial activity against Gram-positive and Gram-negative bacteria as they possess essential oil fractions containing aldehydes and alcohols (Mandal and Mandal, 2015). In the context of metabolic health, coriander leaf extracts have demonstrated regulatory activity on lipid metabolism and glycaemic control in preclinical models, attributed to the synergistic action of flavonoids and terpenes. Investigations into the immunomodulatory properties of coriander have identified selective modulation of pro- and anti-inflammatory cytokines, indicating potential in anti-inflammatory dietary interventions (Nadeem et al., 2013). The pharmacological and nutritional significance of coriander is further augmented by its low-caloric density and dense micronutrient profile, positioning its leaves as a valued component in both traditional medicines and contemporary health-promoting diets (Zaouali et al., 2020). Scientific literature outlines the capacity of coriander leaf supplementation to enhance antioxidant status, reduce inflammatory markers, and inhibit microbial proliferation, especially when administered as part of an integrated dietary regime. Clinical and animal studies increasingly confirm these effects, supporting the rationale for further research into dosage, safety, and formulation (Gazwi et al., 2022).

The current chapter provides a complete overview of the nutritive content of coriander leaves, like macronutrients, micronutrients, and organoleptic constituents, as well as the identification and quantification of bioactive compounds using advanced phytochemical methods. The focus will also be on mechanistic pathways for antioxidative, anti-inflammatory, antimicrobial, and metabolic health effects and on cultivar choice and agronomic management factors that influence phytochemical yield. The intention is to provide a scientifically sound synthesis that adds to the current body of knowledge concerning coriander leaves as a multifunctional food crop with proven health benefits.

Taxonomic Background and Description of *Coriandrum sativum*

Botanically classified *Coriandrum sativum* L. is anchored in the Apiaceae family, often referred to as Umbelliferae, which is a prominent clade of dicotyledonous flowering plants. The Apiaceae, sometimes referred to as the carrot or parsley family and is notable for its cosmopolitan distribution with over 3,600 species grouped into more than 400 genera, encompassing a diverse assembly of aromatic herbs, shrubs, and climbers (Sobhani et al., 2022). The family is morphologically segregated by the presence of compound umbels, hollow stems, and sheathing, divided leaves, characteristics that have facilitated the adaptation of its members to a wide range of ecological niches. Florally, Apiaceae are recognized by actinomorphic, epigynous blossoms, generally pentamerous in their sepal and petal composition, accompanied by a bicarpellate, inferior ovary, an architectural feature of evolutionary consequence within the order Apiales (Wang et al., 2022). The taxonomic tree of *C. sativum* is presented in Table 1.

Table 1 Taxonomic tree of *Coriandrum sativum* L.

Domain	*Eukaryota*	*References*
Kingdom	Plantae	Khan et al. (2014); Sharma & Dhingra (2024)
Division	Magnoliophyta (Angiospermae)	
Class	Magnoliopsida (Dicotyledonae)	
Order	Apiales	
Family	Apiaceae	
Genus	*Coriandrum*	
Species	*Coriandrum sativum* L.	

Origin and Distribution of *Coriandrum sativum*

Coriander is one of the oldest cultivated herbs, with domestication routes diverging towards the Mediterranean and Southwest Asian region, with archaeological evidence dating back 7,500–8,000 years. The evidence, like desiccated mericarps found in Pre-Pottery Neolithic layers in Israel and abundant seeds from Egyptian sites such as Tutankhamen's tomb, suggests planned cultivation by ancient Egyptians, although the plant is not native to Egypt (Wang et al., 2022). Further, ancient Greek Bronze Age layers at Sitagroi in Macedonia demonstrate early widespread cultivation for both its seeds and leaves, used as spices and herbs, respectively. As the cultural exchange accelerated, Greeks and Romans played a major role in spreading throughout eastern and western parts of the world (Spence, 2023). Due to the regional adaptations and cultivar preferences, diverse fruit and leaf morphologies were expected. Gradual understanding of the medicinal properties of coriander made it an integral part of widely used traditional herbal medicine. From the Mediterranean, coriander established itself in China, India, and subsequently penetrated Latin America (Laribi et al., 2015).

At present, coriander leaves are being cultivated extensively across the world, including temperate, subtropical, and tropical regions, making them a vital component of fresh herb markets on multiple continents. Major producers include Asian countries like India, China, Iran, parts of Europe, North America, South America, and a few parts of Africa, as shown in Figure 1 (Rai et al., 2025). The global production of coriander leaves continues to expand due to rising consumer demand driven by increased awareness of their nutritional benefits and health-promoting phytochemicals. Fresh coriander leaves play a crucial role in diverse global cuisines and traditional health systems, contributing to their status as a multifunctional botanical resource. This worldwide distribution and growing market demand reinforce the importance of coriander leaves as a key agricultural commodity with substantial functional food and nutraceutical potential (Hiremath et al., 2025).

Traditional Uses of *Coriandrum sativum*

Coriander leaves have a well-documented traditional use as a medicine. The history runs through various cultures and continents and is supported by epidemiological evidence and recent phytochemical investigations. Traditionally, in many parts of South Asia, the Middle East, North Africa, and Mediterranean societies, coriander leaves have been valued not only as a culinary herb but as an important herbal medicine, as their bioactive compounds are leveraged in prevention and treatment (Nath et al., 2015). The fresh leaves are commonly prepared as infusions or decoctions and traditionally employed for relieving a wide range of digestive disorders, including abdominal pain, flatulence, nausea, dyspepsia, and diarrhoea. The activities attributed are their carminative, antispasmodic, and weak laxative actions (Das et al., 2023b). This is supported by recent studies documenting the

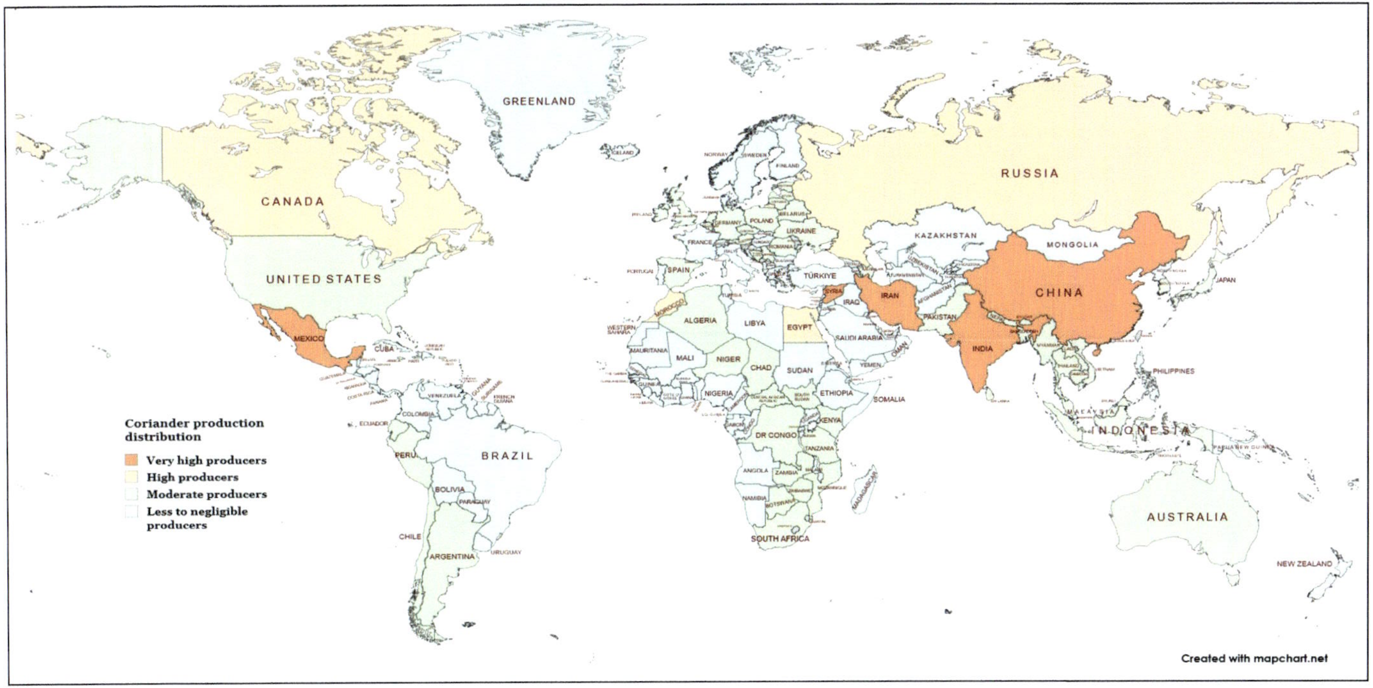

Fig. 1 Geographical distribution of *Coriandrum sativum*.

presence of linalool, borneol, and other monoterpenes of significant activity on gut motility and the enteric nervous system, justifying traditional long practices. Ethnomedical practices in Ayurveda and Unani medicine employ coriander leaves for stimulating appetite, soothing digestive disturbances, and cooling excessive heat in the body, occasionally supplemented with other mild herbs (Sharifi-Rad et al., 2017). In Iran, Turkey, Nepal, and some areas in Africa, coriander leaf infusions are still used for their anxiolytic, diuretic, and detoxifying effects, with applications being extended to the treatment of insomnia, mild hypertension, and urinary tract discomfort. The mild sedative and soothing actions, elicited in animal and clinical experiments, form the basis for the use of leaf preparations as adjuncts in traditional protocols for nervousness and insomnia (Sharma et al., 2025). Coriander leaves are also topically applied in most traditional systems. Fresh-leaved macerations or poultices are employed in the treatment of skin irritations, bites, allergic rashes, minor burns, and mouth ulcers. Such uses are in accordance with evidence for their antimicrobial, wound healing, and anti-inflammatory activity of coriander leaf extracts, due to the synergy between phenolic acids, flavonoids, and essential oils contained in high density in the leaves (Ismail et al., 2025). Leaf-based eyewashes and rinses have also been used traditionally in Persian and North African folk medicine for the reduction of conjunctival and oral inflammation, and evidence exists for their use in post-partum management to promote recovery and regulation of water balance (Sobhani et al., 2022). Coriander leaf juice has been employed by many cultures, e.g., in Central America and the Caribbean, as a supportive therapy for fever, mild infections, and to promote convalescence, taking advantage of the herb's immunomodulatory and nutritive properties (Lee et al., 2025). Selection and adaptation have resulted in the availability of high-leaved coriander with delayed bolting, enhanced aroma, and augmented leaf yield, suggestive of extensive knowledge of continued household and community use. For example, slow-bolting and broad-leaved cultivars are favored in traditional homestead gardens and local markets to provide year-round availability for medicinal and household purposes (Sharangi et al., 2024). The high vitamin and mineral content of the leaves also favors their inclusion in dietary interventions for anaemia, convalescence, and chronic fatigue, especially in resource-poor settings. Furthermore, populations in Europe, the Middle East, and Asia have discovered the use of coriander leaves in the modulation of metabolic health, employing them as weak glycaemic and lipid-lowering agents and as part of dietary interventions to hypertension and cardiovascular health, findings now substantiated by controlled experimental studies (Sarwade et al., 2025).

The differentiating factor of coriander leaf use is its adaptability within traditional systems of medicine, being characterized as a 'balancing' herb: reducing heat, detoxifying toxins, calming the digestive and nervous systems, and promoting gentle detoxification (Bhat et al., 2014). Despite specific preparations, coriander leaf water, fresh juice, or leaf paste is being varied in various locations, the unifying principles converge at its claimed safety, accessibility, and broad spectrum of action. In the various parts of the world, different cultivars are being used and developed to elicit certain prime benefits; these utilities are discussed in Table 2.

Table 2 Traditional uses of various varieties/cultivars of Coriander.

Cultivar of Coriandrum sativum	*Country/Continent*	*Traditional Use*	*References*
Caribe	USA, Latin America	Anti-inflammatory leaf preparations for digestive infusions	Devi et al. (2022)
Jantar	Europe and North America	Used to treat respiratory and digestive issues	Al-Khayri et al. (2023)
Marino	Mediterranean, Europe	Used to treat skin irritation and helps in wound healing	Berkomah et al. (2022)
Santo	Europe, North America	Helps treat oral ulcers and indigestion	Bomme et al. (2007)
Calypso	Europe, Australia	Treats oral inflammations and digestive issues	Alkharpotly et al. (2024)
Slow Bolt	Europe, Australia	Used as a mild sedative and for lipid regulation	Diederichsen et al. (2020)
Thuringer	Germany, Central Europe	Leaf poultices for fever, inflammation	Kuneva et al. (2024)
CIMPO-33	India	Provides digestive aid, leaf essential oils	Singh et al. (2024)
Hisar Anand	India	Leaf extracts are used traditionally for gastrointestinal health	Khan et al. (2017)
Swathi	India	High leaf yield and is used in traditional remedies for digestive health	Pujari et al. (2019)
Green Aroma	India	High leaf biomass cultivar with traditional digestive uses	McAusland et al. (2020)
Vikram	India	Leaf used for digestive and anti-inflammatory applications	Gandepalli & Prasad (2020)
Japanese 57	Canada	Leaf preferred for fresh herb use and traditional digestive remedies	Acharya et al. (2019)
Sugandha	India, Canada	Leaf for culinary and medicinal uses	Joble et al. (2022)

Nutritional Composition of *Coriandrum sativum*

Coriander is not only valued for its culinary purposes but also for its substantial nutritional and medicinal properties. The leaves exhibit a rich nutritional profile that includes essential macronutrients, vital micronutrients, and a broad spectrum of phytochemicals, positioning coriander as a functional food with health benefits widely accepted across cultures (Agarwal et al., 2025). However, the nutritional content of coriander leaves is not uniform and is influenced by varietal genetics, geographic origin, environmental variables, and molecular regulatory mechanisms that govern metabolite biosynthesis. Understanding these multifaceted influences is crucial for optimizing coriander leaf quality in agricultural production, maximizing health benefits, and guiding informed cultivar selection for nutraceutical development (Mauer and El-Sohemy, 2012).

Macro and Micro Nutrients

Coriander leaves are characterized by a low caloric and fat content while delivering appreciable levels of moisture, protein, carbohydrates, and dietary fiber. The moisture content typically ranges from 85% to 92%, influenced by cultivar and environmental factors. Protein content varies between 2– 4% which is an important nutritional feature relative to other herbs, providing essential amino acids vital for human diets, especially in populations consuming predominantly plant-based foods (González-Pérez et al., 2023). The leaves possess approximately 0.3–1.5% lipids, consisting mainly of essential fatty acids and bioactive volatile oils essential for the herb's aroma and biological activity. Carbohydrates, which include soluble sugars and complex dietary fibers, account for 3.5–6% of the leaf content. Fiber assists in digestive regulation and can play a role in blood glucose homeostasis (Bhatti et al., 2019).

Coriander leaves are excellent sources of micronutrients, too. Vitamin A (as β-carotene), C (ascorbic acid), and K. Beta-carotene content can vary widely from 1.2–6.7 mg per 100 g fresh weight, with the highest concentrations found in microgreens and tender young leaves (Abdalla, 2009). Vitamin C content ranges from approximately 20–48 mg per 100 g, providing antioxidant protection and supporting immune function. Vitamin K content ranges from 150 to 250 μg per 100 g, essential for blood clotting and bone metabolism. Essential minerals such as calcium (up to 150 mg/100 g), iron (2– 4 mg/100 g), magnesium, and potassium are also present in significant quantities, supporting enzymatic functions, oxygen transport, and electrolyte balance (Ayran et al., 2017).

Various Cultivars of Coriander and Their Nutritional Profile

The genetic makeup of coriander cultivars substantially influences their nutritional and phytochemical composition. Cultivars bred primarily for leaf production, such as Caribe, Confetti, and Jantar, show a great variation in protein content, vitamin concentration, and antioxidant activity (Chaulagain et al., 2011). Caribe was found to contain higher total phenolics and vitamin C levels, which correlate with its traditional use in digestive health and inflammation modulation (Kalenska et al., 2024). Indian cultivars are unique for their elevated citral and linalool content, enhancing both flavor complexity and therapeutic efficacy, particularly in respiratory and digestive ailments (Agarwal et al., 2025). Indian cultivars like CIMPO-33 and Hisar Anand have been shown to accumulate higher levels of micronutrients and bioactive molecules, aligned with the cultural preference for coriander leaf as both a dietary and medicinal herb (Gantait et al., 2022). These cultivars distinctly vary from European types such as Confetti or Slow Bolt, where longer harvest windows and resistance to bolting maximize fresh leaf availability but may result in varied secondary metabolite content. Genetic analyses reveal that biosynthetic pathways for flavonoids and essential oils exhibit polymorphisms across cultivars, affecting concentrations of key antioxidants like quercetin, kaempferol, and rutin, as well as the major volatile compound linalool. This genetic variability underpins the differences in aroma, taste, and health-promoting efficacy noted among varieties, offering strategic targets for breeding programs focused on nutritional quality and phytochemical robustness (Mandal and Mandal, 2015).

Environmental and Geographic Influences

Coriander leaf composition is modulated by growing conditions, including climate, soil type, nutrient availability, and agricultural practices. Plants grown in arid and Mediterranean climates commonly accumulate increased levels of essential oils and polyphenols, likely as adaptive responses to oxidative stress prompted by high light intensity and temperature. This confers enhanced antimicrobial and antioxidant effects, but it is also accompanied by reduced moisture content and altered textural properties (Jampílek and Kráľová, 2023). Contrastingly, leaves sourced from cooler, temperate regions maintain higher moisture content and vitamin C concentrations, although often with diminished essential oil levels, thus slightly altering the health profile. Soil fertility, especially the availability of nitrogen, magnesium, and potassium, influences coriander leaf protein and phenolic accumulation, with studies showing increased nitrogen correlates positively with flavonoid biosynthesis and antioxidant activity (Kaleem et al., 2025). Water management impacts secondary metabolite synthesis and also, controlled drought stress can boost phenolic and volatile oil production without significantly reducing leaf biomass, suggesting irrigation strategies as tools to optimize phytochemical content. Leaf maturity and harvest timing further affect nutritional composition (Özyazici, 2021). Younger leaves and microgreens are consistently richer in vitamins, antioxidants, and secondary metabolites than fully mature foliage. Thus, harvesting strategies may be designed to align with specific nutritional or therapeutic goals, balancing yield and bioactive concentrations for consumer benefit (Ulusu, 2024).

Bioactive Compounds of *Coriandrum sativum*

The study of bioactive compounds within the leaves of coriander has gained substantial scientific interest due to their multifaceted therapeutic potential and diverse chemical composition. The traditional medicinal systems across various cultures have used coriander leaves for gastrointestinal, inflammatory, and infectious ailments. This initial ethnobotanical knowledge provided a foundation for further research work to derive the functions of all the bioactive molecules (Das et al., 2023a). Early phytochemical studies primarily focused on the essential oil fraction, identifying a complex profile of volatile terpenoids contributing to the characteristic organoleptic properties and preliminary biological activities (Al-Khayri et al., 2023). Subsequent advancements in analytical methodologies like notably chromatographic and spectrometric techniques such as gas chromatography-mass spectrometry (GC-MS), high-performance liquid chromatography (HPLC), and nuclear magnetic resonance (NMR) spectroscopy, facilitated the comprehensive characterization of coriander leaf constituents (Oishi, 2024). These analyses revealed a rich assemblage of secondary metabolites encompassing monoterpenes, sesquiterpenes, flavonoids, phenolic acids, fatty acids, phytosterols, and other specialized metabolites (Eyres et al., 2005).

Contemporary research expanded beyond chemical profiling to elucidate the molecular mechanisms elevating the bioactivities of these compounds. Investigations employing *in vitro* enzymatic assays, cell culture models, and *in vivo* animal studies have demonstrated that coriander-derived phytochemicals modulate key biological pathways, including oxidative stress response via augmentation of endogenous antioxidant enzymes, inflammatory cascades through cytokine and eicosanoid regulation, and cellular signaling pathways involved in apoptosis and metabolic control (Zhang et al., 2015). The identification of structure-activity relationships further refined the understanding of compound-specific therapeutic effects, thus aligning the chemical diversity with observed pharmacological profiles. The translational aspect of these findings has stimulated the development of phytotherapeutics, nutraceutical formulations, and functional foods incorporating standardized coriander leaf extracts (Jin et al., 2025). These products aim to harness the synergistic effects of the phytochemical matrix to mitigate conditions associated with oxidative damage, microbial infections, metabolic dysregulation, and neurodegeneration. Moreover, genomic and metabolomic approaches continue to enhance the comprehension of biosynthetic pathways and varietal differences influencing phytochemical accumulation, which is critical for optimized cultivation and targeted therapeutic utilization (Falade et al., 2025). Overall, the systematic exploration and integration of coriander leaf bioactive molecules and traditional knowledge with rigorous scientific inquiry offers profound implications for botanical drug discovery and evidence-based phytotherapy development. A few notable classes of bioactive molecules found in coriander leaf are given below.

Essential Oils

Essential oils derived from coriander leaves predominantly comprise a complex mixture of volatile terpenoids and monoterpenoids, which contribute not only to the plant's characteristic aroma but also to its diverse therapeutic properties (Cortés-Eslava et al., 2004). The principal monoterpenoid constituent identified in coriander leaf essential oil is linalool (3,7-dimethyl-1,6-octadien-3-ol), which often exceeds 60–70% of the volatile fraction, conferring the herb's prime and characteristic floral scent (Saxena et al., 2022). The bicyclic alcohol structure of linalool exhibits significant antimicrobial, anti-inflammatory, sedative, and anxiolytic bioactivities. Other abundant monoterpenoids include γ-terpinene (1-methyl-4-(1-methylethyl)-1,4-cyclohexadiene), α-pinene (bicyclo [3.1.1] hept-2-ene), terpinolene, camphor (1,7,7-trimethylbicyclo [2.2.1] heptan-2-one), borneol, bornyl acetate, and geranyl acetate. Sesquiterpenoids such as β-caryophyllene also contribute in minor but biologically relevant concentrations (Wei et al., 2019).

Intraspecific and cultivar-dependent variations significantly influence essential oil profiles. In coriander, the caribe cultivar is characterized by higher linalool content, whereas other genotypes express increased proportions of α-pinene or γ-terpinene (Ravi et al., 2007). Environmental factors, including soil composition, temperature, and harvest time, affect the relative abundance of constituents, as observed in comparative analyses between Mediterranean and subtropical ecotypes. Such variability can be seen in both sensory and pharmacological qualities of the essential oils (Katar et al., 2016). The biosynthetic plasticity of coriander essential oils corresponds with ecological functions, including plant defence mechanisms against herbivores, microbes, and attraction of pollinators through volatile signaling (Zhang et al., 2023). Recent molecular studies reveal that the regulation of terpene synthase gene families and methyl jasmonate-induced enhancement of linalool synthase expression, indicating inducible pathways responsive to biotic and abiotic stimuli. Therapeutic applications reveal that coriander essential oil exhibits potent antimicrobial activity against a broad spectrum of pathogens, disrupting microbial membranes and inhibiting biofilm formation (Kianersi et al., 2022). Its anti-inflammatory effects are mediated through the modulation of cytokine production and inhibition of cyclooxygenase enzymes. Linalool and related monoterpenoids have proven to have neuroprotective and anxiolytic potential through interaction with GABAergic and glutamatergic neurotransmission. Recent investigations showed anticancer properties in terms of apoptosis induction and antioxidant activity (Salem et al., 2025). The isomeric and minor volatile components, such as cis and trans linalool oxides, are relatively less explored. It has been understood that these bioactive molecules modulate olfactory perception and bioactivity yet remain under-characterized in coriander (Wei et al., 2024). Figure 2 mentions the structure of major essential oils, and Table 3 gives an inference on the relative abundance of major volatile compounds in coriander essential oils across regions and possible reasons for variability.

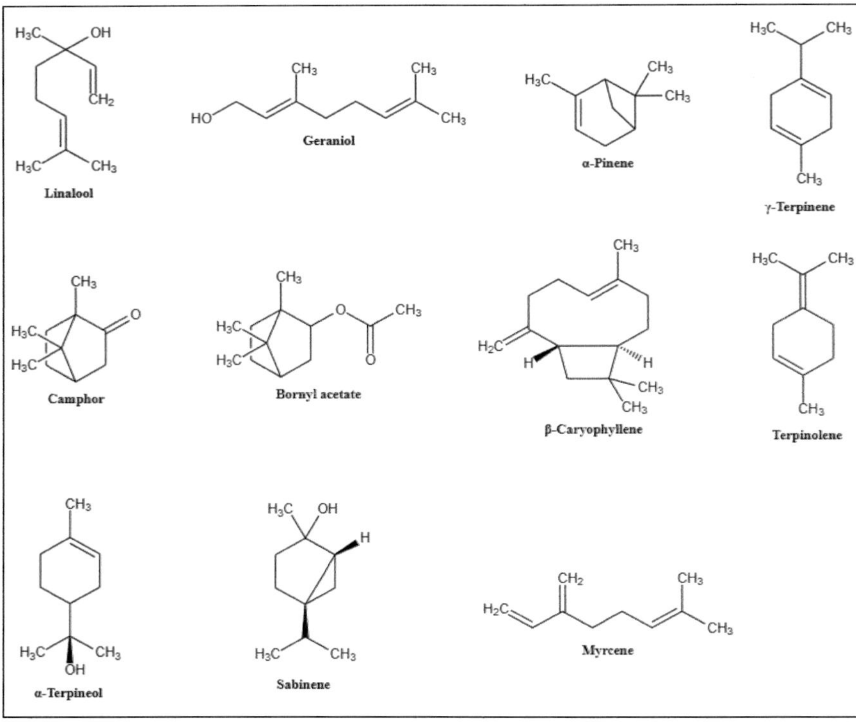

Fig. 2 Chemical structures of major essential oils found in coriander leaves.

Table 3 Relative abundance of major volatile compounds in coriander essential oils across the world.

Volatile Compound	*Quantity Range*	*Study Location and Possible Reason for Variability*	*References*
Linalool	40.9–83.7%	Highest in India and Iran (up to 80%+) due to warm, dry climates that favor biosynthesis, moderate in Europe and Argentina (~60%), lower in Brazil (~40– 50%).	Kumar et al. (2022); Saxena et al. (2022); Wei et al. (2024)
Neryl acetate	2.3–14.2%	Higher in Iran and India (up to 14%), lower or absent in Brazil and Europe; variation linked to geographical location, as Neryl acetate biosynthesis is sensitive to temperature and sunlight intensity.	Santos et al. (2019; Kumar et al. (2022)
γ-Terpinene	0.1–13.6%	Higher in Iran and Europe, moderate in Argentina; lower in tropical Brazil, possibly due to the accumulation of γ-terpinene in cooler climates and chemotype divergence.	Kianersi et al. (2022); Das et al. (2023a)

Contd...

Table 3 *Contd...*

Volatile Compound	*Quantity Range*	*Study Location and Possible Reason for Variability*	*References*
α-Pinene	0.2–11.4%	Predominantly higher in European and Brazilian samples (up to 11%), less in Iran and India (~3-5%). Growth conditions like cooler environments and soil types elevate α-pinene synthesis.	Nejad Ebrahimi et al. (2010); İzgı et al. (2017); McAusland et al. (2020)
Camphor	0–5.5%	Present mainly in European and Argentine coriander, negligible in Iran and Brazil samples; camphor content linked to genetic differences.	Nadjafi et al. (2009; Saxena et al. (2022); Basol et al. (2024)
Geranyl acetate	0.2–14.2%	High in Iran and India, moderate in Argentina, low in Brazil and Europe; linked to genetic factors and post-harvest processing. Sun exposure increases geranyl acetate production.	İzgı et al. (2017); Kianersi et al. (2022)
p-Cymene	0.1–8.1%	Higher in European cultivars, lower in tropical regions; likely linked to chemotype and enzymatic activity in the biosynthetic pathway, influenced by temperature.	Nadjafi et al. (2009); İzgı et al. (2017); Kianersi et al. (2022)
Geraniol	0.6–3.2%	More abundant in Argentina and India; less detected in Brazil and Europe, the eco-climatic zones greatly influence.	El-Nasr et al. (2013; Hadi et al. (2022)
Limonene	0.4–1.3%	Generally low across all regions, slightly higher in Brazilian samples, possibly related to tropical conditions.	Mohamed et al. (2018); Machado et al. (2023); Amer et al. (2024)
2-Decenal	0.16–0.4%	Mostly in Brazilian coriander, trace or absent elsewhere; environment and genetic factors likely responsible for its selective presence.	Priyadarshi et al. (2016); Zaeddi & Sulayman (2018); Kumar et al. (2022)
Linalool oxide (cis & trans)	<0.5%	Minor constituent with slight presence in some Argentine samples; forms by oxidation of linalool during storage or processing.	Kianersi et al. (2022); Banerjee et al. (2024)

Volatile terpenoids and monoterpenes of coriander leaves are commonly extracted using steam or hydrodistillation procedures. However, due to the delicate nature and lower oil content of leaves compared to seeds, microwave-assisted hydro distillation (MAHD) is often preferred as it shortens extraction time, reduces thermal degradation, and also preserves the integrity of heat-sensitive compounds (Shahwar et al., 2012). Supercritical CO_2 extraction is also used to produce high-purity oils by selectively extracting volatiles under mild conditions, though it requires specialized equipment and higher costs. Solvent extraction and ultrasound-assisted techniques are occasionally used to improve yield, but hydro distillation and MAHD remain the most practical approaches for coriander leaf essential oils (Song and Ko, 2022).

Flavonoids

Flavonoids constitute a significant class of bioactive compounds in coriander leaves. They are characterized by their polyphenolic structure, consisting primarily of two aromatic rings which are linked through a three-carbon bridge forming an oxygenated heterocyclic ring. Broadly, they are a 15-carbon skeleton structure represented as C6-C3-C6 that shows structural diversity, including subclasses such as flavonols, flavones, flavanones, and their glycosylated derivatives (Sahib et al., 2013). In coriander leaves, major flavonoids identified include quercetin, kaempferol, luteolin, apigenin, and their respective glycosides. These compounds exhibit various substitution patterns like hydroxylations and glycosylations, which modulate their physicochemical properties and biological activities (Scandar et al., 2023).

Therapeutically, coriander leaf flavonoids are highly used for their broad-spectrum antioxidant, anti-inflammatory, antimicrobial, neuroprotective, and cardioprotective effects. The ROS (reactive oxygen species) scavenging ability of these polyphenols gives protection against oxidative damage of cellular components; they contribute significantly to potential benefits in chronic disease management, including diabetes and neurodegenerative disorders (Mahleyuddin et al., 2021). Experimental studies have revealed flavonoids' ability to modulate signaling pathways such as NF-κB and MAPK, attenuating pro-inflammatory cytokine production. Their antimicrobial efficacy against a spectrum of pathogenic bacteria and fungi further extends their ethnomedicinal use. Pharmacokinetic studies elevate the significance of glycosylation status in bioavailability and metabolism, uncovering the therapeutic importance of specific flavonoid conjugates prevalent in coriander leaves (Lin et al., 2022). Despite this pharmacological potential, commercial or agronomic selection of coriander varieties with enhanced flavonoid profiles is nascent, opening a scope for further research and development (Atrooz et al., 2023).

Extraction of flavonoids from coriander leaves is predominantly done using solvent-based techniques utilizing hydroalcoholic mixtures (ethanol or methanol with water). The optimization of polarity is essential to maximize flavonoid recovery while preserving structural integrity. Maceration, ultrasound-assisted extraction (UAE), and microwave-assisted extraction (MAE) are a few other commonly used methods (Hardiany et al., 2025). UAE and MAE offer enhanced efficiency along with reduced solvent consumption and shorter extraction duration. Post-extraction, flavonoid-rich extracts are subjected to purification typically via chromatographic methods such as HPLC, facilitating qualitative and quantitative analyses. Considerations around solvent choice, extraction parameters like temperature, duration, solvent-to-solid ratio, and pre-treatment procedures like drying, grinding significantly impact yield and composition (Shahwar et al., 2012). Current extraction methods balance yield optimization with maintenance of bioactivity, crucial for downstream applications in nutraceuticals and pharmaceuticals focused on coriander leaf flavonoids (Priyadarshi et al., 2016). A comprehensive table presenting major flavonoids identified in coriander leaves, along with quantitative or qualitative comparisons across various global regions, is discussed in Table 4, along with their structures in Figure 3.

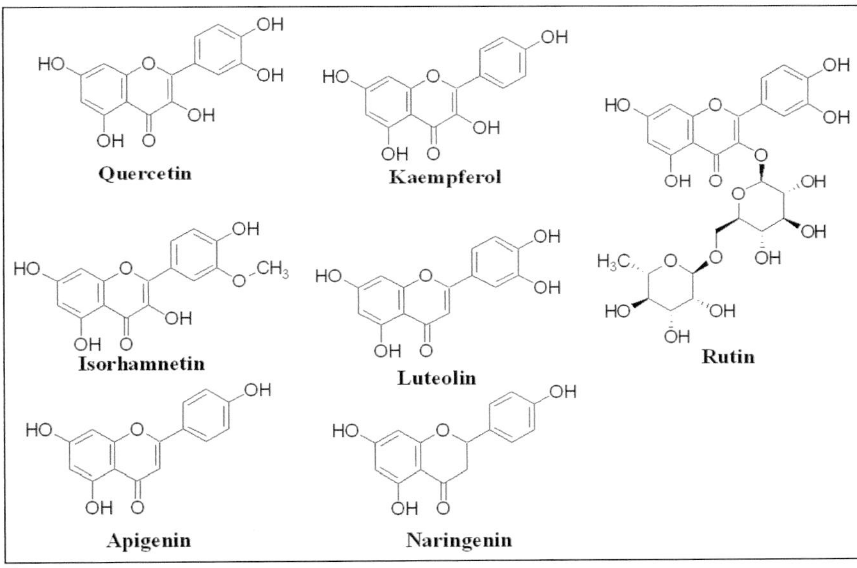

Fig. 3 Chemical structures of major flavonoids found in coriander leaves.

Table 4 Relative abundance of major flavonoids in coriander across the world.

Flavonoid Compound	*Relative Abundance*	*Study Location and Possible Reason for Variability*	*References*
Quercetin	3.5–12.8% (of total flavonoids)	Relatively higher in Indian, Iranian, and European leaf extracts. Temperate and Mediterranean climates favour biosynthesis. Lower in tropical regions like Brazil and Southeast Asia due to environmental modulation.	Khan et al. (2017; Gahory et al. (2022); Tibebe et al. (2024); Elhassaneen et al. (2024)
Kaempferol	2.0–10.4%	Present worldwide. Higher in Iranian and Mediterranean cultivars because of cultivation practices and photoperiodic exposure.	Choudhary et al. (2017); Hadjmohammadi & Sharifi (2020); Elhassaneen et al. (2024)

Contd...

Table 4 *Contd...*

Flavonoid Compound	*Relative Abundance*	*Study Location and Possible Reason for Variability*	*References*
Luteolin	1.5–8.9%	Reported abundantly in European and Middle Eastern coriander leaves. Moderate levels in Indian leaves. Biosynthesis is sensitive to UV light intensity.	Nadjafi et al. (2009); Gahory et al. (2022); Salem et al. (2025)
Apigenin	0.8–6.7%	Detected predominantly in Mediterranean and European leaf extracts. Lower abundance in tropical specimens. Abiotic stresses like drought can elevate the abundance.	Nadeem et al. (2013); Gantait et al. (2022); Nisa et al. (2025)
Iso-quercetin (quercetin-3-O-glucoside)	1.2–7.0%	Higher in Indian and Iranian leaves, contributing to enhanced solubility and pharmacokinetic benefits compared to the aglycone.	Sahib et al. (2013); Sarwade et al. (2025)
Hyperoxide (quercetin-3-O-galactoside)	Trace to 4.5%	Found primarily in Middle Eastern and Mediterranean samples, qualitative reports in India, and rare in Brazil and Southeast Asia.	Wei et al. 2019); Kianersi et al. (2022); Tibebe et al. (2024)
Rutin (quercetin-3-rutinoside)	0.5–3.8%	Variable presence in European and Mediterranean leaves. Very little to negligible in tropical specimens.	Abdul-Jalil (2013); Bhat et al. (2014); Kuri et al. (2015)
Vitexin	Trace–2.0%	Detected in limited studies from Mediterranean cultivars. Very little to negligible in tropical specimens.	Saxena et al. (2015); Elhassaneen et al. (2024)
Isorhamnetin	0.3–1.5%	A minor flavonoid identified in Iranian and European leaves, which is associated with the methylation of quercetin derivatives. Low abundance globally.	Barros et al. (2012); Hardiany et al. (2025); Sawi et al. (2025)

Phenolic Acid and Other Phenolics

Phenolic acids and other phenolic compounds present in coriander leaves contribute a significant class of bioactive secondary metabolites that are primarily characterized by aromatic rings bearing one or more hydroxyl groups (Barros et al., 2012). These compounds are generally categorized into two main types: hydroxybenzoic acids (C6-C1 skeleton) and hydroxycinnamic acids (C6-C3 skeleton). Structurally, hydroxybenzoic acids such as gallic acid and protocatechuic acid consist of a benzene ring substituted with hydroxyl and carboxyl groups. On another hand, hydroxycinnamic acids, including caffeic acid, ferulic acid, p-coumaric acid, and sinapic acid, feature an extended side chain of three carbons terminating in a carboxyl group and often conjugated with sugars or other moieties, giving an enhanced solubility and biological function (Nisa et al., 2025). Additional phenolic derivatives like chlorogenic acid, an ester of caffeic acid and quinic acid, expand the phenolic diversity in coriander leaves and contribute to their valuable bioactivity. Biosynthesis of phenolic acids in coriander leaves primarily proceeds through the shikimate and phenylpropanoid pathways (Tibebe et al., 2024).

From a therapeutic point of view, phenolic acids and related phenolics in coriander leaves exhibit extensive pharmacological activities, mainly through their potent antioxidant properties and interaction with cellular signaling molecules. By scavenging ROS, phenolic acids mitigate oxidative stress implicated in aging, inflammation, neurodegeneration, and carcinogenesis (Nadeem et al., 2013). Caffeic and ferulic acids modulate gene expression through the Nrf2/ARE pathway, enhancing endogenous antioxidant defences (Kaiser et al., 2013). Antimicrobial efficacy against bacteria and fungi also correlates with phenolic content, justifying the traditional use of coriander leaves in infection control. Chlorogenic acid, notably abundant in coriander, has been implicated in glucose metabolism regulation and cardiovascular protection (Xie et al., 2024). Extraction strategies for phenolic acids from coriander leaves commonly employ solvent-based methods using aqueous methanol or ethanol mixtures, which balance polarity to solubilize a broad spectrum of phenolics while preserving bioactivity. Techniques such as UAE and MAE enhance yield and reduce solvent consumption and processing time (Mouhoubi et al., 2023). Subsequent purification and quantification are through chromatographic techniques, including HPLC coupled with photodiode array (PDA) or mass spectrometry (MS) detectors, ensuring accurate profiling of phenolic constituents for nutraceutical or pharmaceutical application development (Mouhoubi et al., 2022). Table 5 presents major phenolic acids and related phenolics identified in coriander leaves with comparative data across various global regions, along with their structures in Figure 4.

Table 5 Relative abundance of major phenolic compounds in coriander across the world.

Phenolic Compound	*Relative Abundance*	*Study Location and Possible Reason for Variability*	*References*
Gallic acid	1.0–7.5%	Prominent in Indian, Iranian, and Mediterranean coriander leaves. Moderately present in European samples, lower in tropical regions such as Brazil and Southeast Asia, likely due to environmental and genotypic factors.	Divya et al. (2012, 2014); Gahory et al. (2022); de Lima et al. (2024)
Caffeic acid	2.5–9.2%	Abundant in leaves from India, Iran, and Europe, whereas intermediate in South American samples. Its biosynthesis is stimulated by UV light exposure and stress conditions.	Nadeem et al. (2013); Demir & Korukluoğlu (2020)

Contd...

Table 5 *Contd...*

Phenolic Compound	*Relative Abundance*	*Study Location and Possible Reason for Variability*	*References*
p-Coumaric acid	0.8–6.1%	Detected broadly in Mediterranean and Middle Eastern leaves with moderate concentrations. Lower in tropical regions, attributable to climatic stress and enzymatic activity differences.	Demir & Korukluoğlu (2020); de Lima et al. (2024)
Ferulic acid	0.7–5.8%	Frequently reported across temperate and subtropical regions, notably in Iranian and European leaves. Reduced levels in tropical cultivars, possibly linked to differential methylation enzyme expression.	Barros et al. (2012); Gantait et al. (2022)
Chlorogenic acid	1.2–8.7%	High in Indian and Mediterranean coriander leaves, important for antioxidant capacity, sporadic reports in tropical regions suggest lower concentrations due to climate and genetic variability.	Clifford (1999); Barros et al. (2012); Demir & Korukluoğlu (2020)
Protocatechuic acid	0.3–3.0%	Present mainly in Middle Eastern and European samples, rare or trace levels in tropical crops, possibly reflecting variations in phenylpropanoid flux.	Gahory et al. (2022); Scandar et al. (2023)
Vanillic acid	Trace–2.2%	Minor phenolic found occasionally in Mediterranean leaves; low to absent elsewhere, production linked to oxidation and degradation of ferulic acid.	Ivanovic et al. (2019); Shoko et al. (2022); Nisa et al. (2025)
Syringic acid	Trace–1.5%	Detected in limited studies from Iranian and Mediterranean leaves, scarcely reported in other regions.	Mouhoubi et al. (2022); Scandar et al. (2023)

Fatty Acids and Phytosterols

Fatty acids present in coriander leaves are predominantly composed of a spectrum of saturated, monounsaturated, and polyunsaturated fatty acids (PUFAs), which are structurally characterized by long aliphatic hydrocarbon chains terminating in a carboxyl group. The leaf fatty acid profile typically includes palmitic acid (C16:0), linoleic acid (C18:2, an omega-6 PUFA), α-linolenic acid (C18:3, an omega-3 PUFA), and oleic acid (C18:1, monounsaturated), shown in Figure 5 (Nguyen, 2015). The relative abundance of ω-3 α-linolenic acid imparts anti-

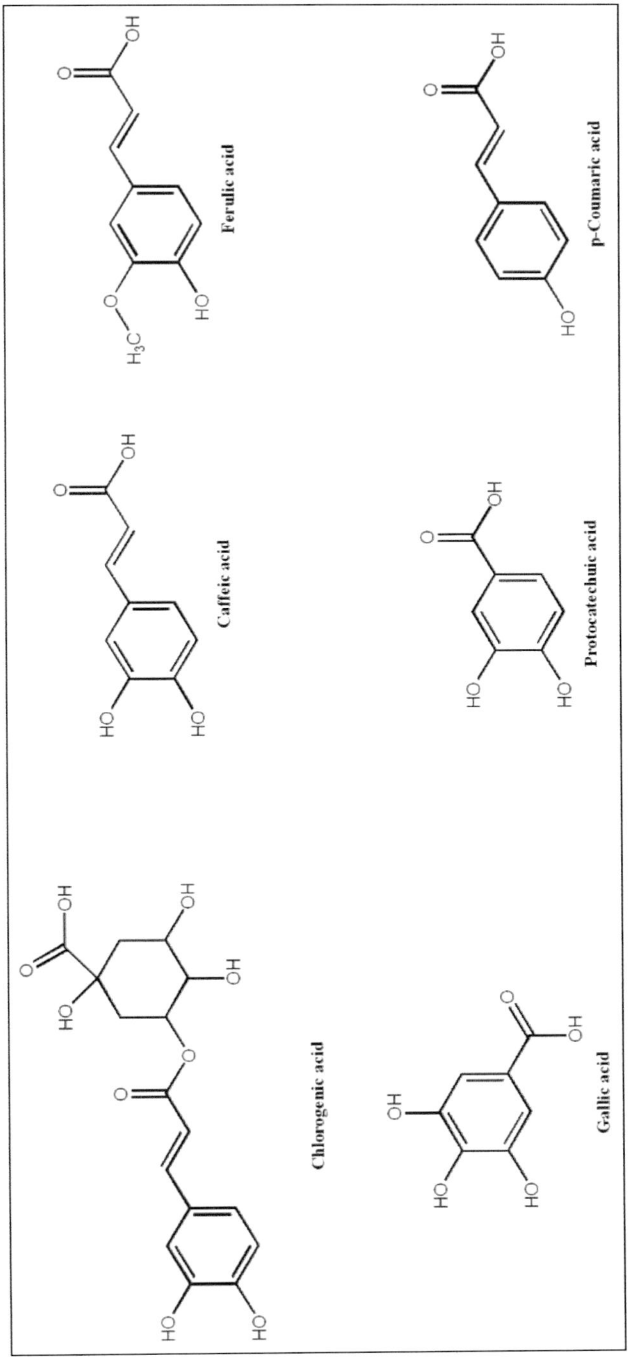

Fig. 4 Chemical structure of major phenolic compounds found in coriander.

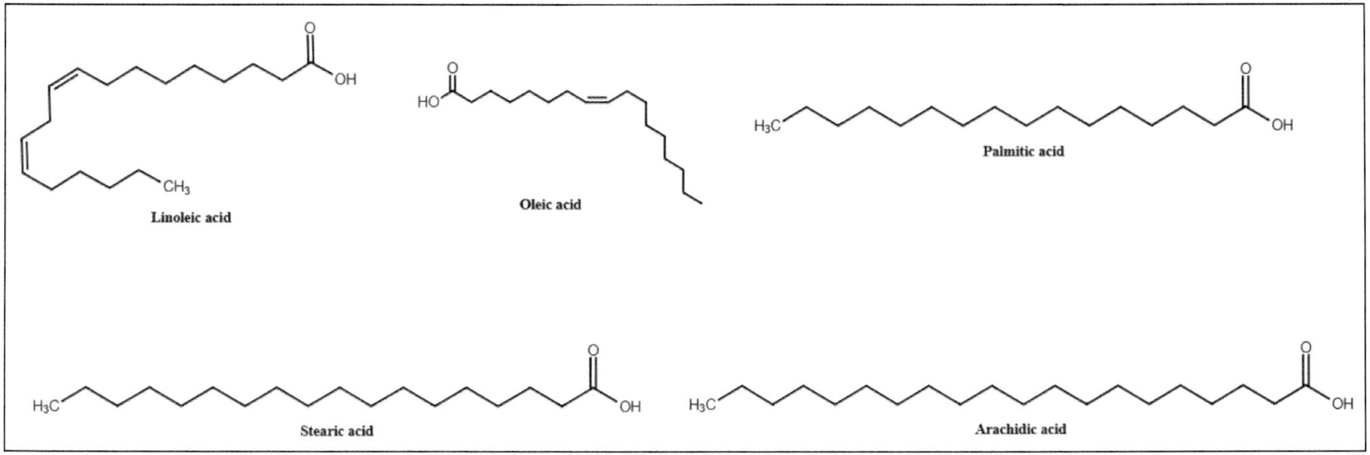

Fig. 5 Chemical structure of major phytosterols found in coriander.

inflammatory properties and nutritional value, while the balanced fatty acid composition yields in physiological functions (Chander, 2010). Extraction of leaf lipids typically employs organic solvent methods such as Soxhlet extraction with hexane or chloroform-methanol mixtures, followed by transesterification for fatty acid methyl ester (FAME) analysis via gas chromatography-mass spectrometry (GC-MS), and thus providing detailed fatty acid profiling essential for assessing nutritional and pharmacological potentials (Amiripour et al., 2021).

Phytosterols identified in coriander leaves belong mainly to the steroidal triterpenoid family. They are structurally analogous to mammalian cholesterol but differ in side-chain composition and saturation degree. Principal phytosterols include β-sitosterol, stigmasterol, and campesterol, characterized by a tetracyclic cyclopenta[a]phenanthrene nucleus with various methyl or ethyl substitutions at C-24 (Yoon et al., 2025). These compounds contribute to membrane structure and signaling in leaves and are implicated in modulating plant responses to environmental stressors. Therapeutically, coriander leaf phytosterols demonstrate cholesterol-lowering effects via inhibition of intestinal cholesterol absorption, anti-inflammatory activities, and potential anticancer properties through modulation of apoptotic pathways (Yoon et al., 2025). Common extraction methods like non-polar solvents are used to isolate total sterols, followed by saponification and purification steps; quantification employs chromatographic techniques such as GC-MS or HPLC with mass spectrometry detection, enabling precise determination of individual sterol constituents important for functional food and pharmaceutical applications (Molina et al., 2020).

Alkaloids, Tannins, and Saponins

Alkaloids identified in coriander leaves are nitrogen-containing heterocyclic compounds, structurally diverse but commonly characterized by one or more nitrogen atoms incorporated within ring systems (Zyad and Ahmed, 2018). These compounds are biosynthesized predominantly via amino acid precursors. Alkaloid biosynthesis involves decarboxylation, methylation, and cyclization steps. In coriander leaves, specific alkaloids such as phenylethylamine-type compounds have been detected in low to moderate concentrations, contributing to the plant's defensive chemistry and pharmacological potential (Nathenial et al., 2019). They exhibit mild antimicrobial, analgesic, and anti-inflammatory properties, although their concentrations are generally insufficient for primary therapeutic effects in crude extracts. After acidified solvent extraction, subsequent purification involves chromatographic techniques such as column chromatography or preparative HPLC, enabling isolation and characterization critical for bioactivity studies (Misharina, 2016).

Tannins in coriander leaves are polyphenolic compounds recognized for their ability to complex and precipitate proteins, grouped into two major categories: condensed tannins (proanthocyanidins) and hydrolysable tannins (gallotannins and ellagitannins) (Rajeshwari et al., 2012a). Structurally, condensed tannins are oligomers or polymers of flavan-3-ols such as catechin, whereas hydrolysable tannins are esters of gallic acid or ellagic acid with a sugar core, typically glucose. Tannins contribute to antioxidant activity and defence mechanisms against herbivory, antimicrobial effects, anti-inflammatory activities, and potential benefits in metabolic syndrome management by modulating oxidative stress. Extraction typically employs aqueous organic solvents under controlled conditions to minimize polymer degradation with subsequent quantification via spectrophotometric assays or advanced chromatographic methods (Hardiany et al., 2025).

Saponins in coriander leaves are amphiphilic glycosides consisting of a hydrophobic aglycone (sapogenin) linked to hydrophilic sugar moieties. Aglycones are commonly triterpenoid skeletons such as oleanane derivatives (Verma et al., 2018). In coriander leaves, triterpenoid saponins like oleanolic acid glycosides have been identified, implicated in the plant's defence and stress adaptation (Uma et al., 2009). Therapeutic properties include cholesterol-lowering effects, immunomodulation, and antimicrobial activities demonstrated *in vitro*. Extraction methods rely on aqueous-alcoholic solvents followed by partitioning into butanol or ethyl acetate to enrich saponins. Purification employs chromatographic separation, with structural elucidation via NMR and MS techniques advancing understanding of bioactive saponin diversity (Nathenial et al., 2019).

Other Bioactive Molecules

Other than these classes of bioactive molecules, coriander leaves also contain notable compounds such as coumarins, organic acids, vitamins, minerals, polysaccharides, and bioactive peptides. Coumarins are benzopyrone derivatives with anticoagulant and anti-inflammatory properties present in minor amounts (Verma et al., 2018). Organic acids like malic, citric, and tartaric acids influence flavour and exhibit mild antimicrobial effects. Vitamins C and K, found in leaves, serve antioxidant and coagulation roles, respectively. Polysaccharides and dietary fibers contribute to gut health and immune modulation, enhancing coriander's nutritional value (Zyad and Ahmed, 2018). Additionally, bioactive peptides with antioxidant and antimicrobial potential have been isolated from leaf extracts, although these are less characterized. These ancillary molecules complement primary phytochemicals. Altogether, they enhance the therapeutic efficacy of coriander leaves.

Biological Activities of *Coriandrum sativum*

Coriander has been traditionally employed to manage a variety of ailments such as digestive issues, metabolic disorders, cardiovascular ailments, infection by pathogens in upper respiratory tract and digestive tract, etc. Recent scientific efforts have elucidated mechanisms and substantiated the biological potential of coriander in treating wide variety of conditions such as antioxidant, antimicrobial, anti-inflammatory, antidiabetic, neuroprotective, anticancer, cardioprotective, hepatoprotective, anxiolytic, anticonvulsant, immunomodulatory, gastrointestinal, and reproductive health effects as shown in Figure 6 (Elhassaneen et al., 2024).

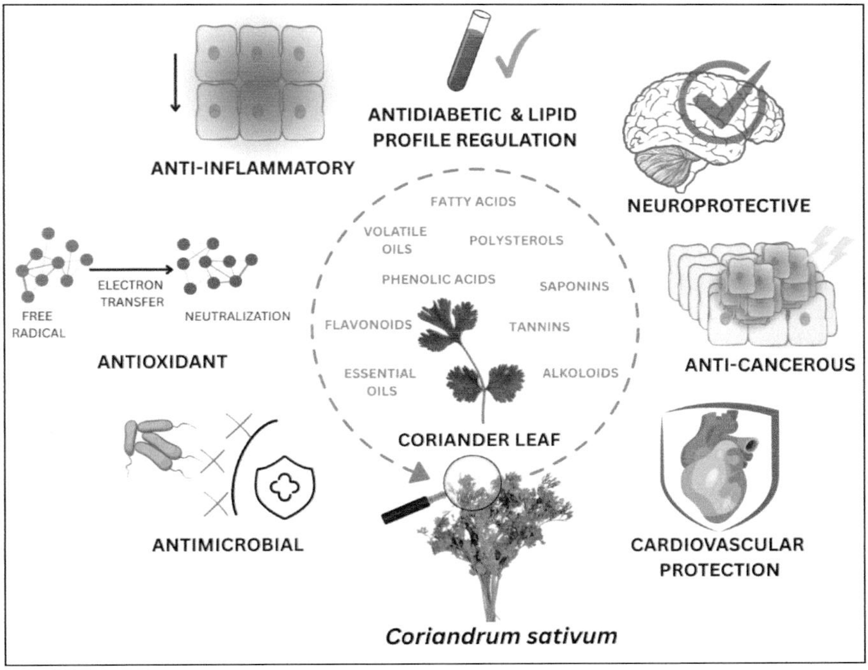

Fig. 6 Phytochemicals and biological activities of *Coriandrum sativum*.

Antioxidant Activity

One of the most intensively studied bioactivities of coriander pertains to its antioxidant potential. Oxidative stress, caused by ROS, is a major factor implicated in the pathogenesis of chronic diseases such as cardiovascular disorders, diabetes, neurodegenerative diseases, and cancer. They are also involved in the spoilage of food products. These ROS could be effectively mitigated by coriander's phytochemicals. Coriander leaves possess significant amounts of flavonoids like quercetin and kaempferol, phenolic acids including chlorogenic and caffeic acids, vitamins (C and E), and terpenoid constituents dominated by linalool. The levels of each phytochemical in coriander vary among different genotypes and can also vary in the same cultivar depending on the growth environment, time of harvesting, etc., but these variations are very minute. Their consumption as food can reduce the oxidative stress in the consumer, and the use of coriander extracts or powder in food items can increase their shelf life (Wei et al., 2019). These compounds collectively scavenge ROS in several *in vitro* models using diphenyl picrylhydrazyl (DPPH), inhibition of 15-lipoxygenase (15-LO), and inhibition of Fe^{2+} induced peroxidation of phospholipids, among others (Wangensteen et al., 2004). *In vivo*, coriander supplementation enhances endogenous antioxidant defences by upregulating enzymes like superoxide dismutase (SOD), catalase, glutathione-

S-transferase (GST) (Rajeshwari et al., 2012b). For instance, animal models like tilapia have shown increased SOD expression, and myeloperoxidase, lysozyme, and antiprotease activities were observed to be significantly higher when the feed was supplemented with coriander oil (Das et al., 2023a).

Antimicrobial Properties

The increasing incidence of drug-resistant pathogens and the toxicity of existing antimicrobial compounds have drawn attention toward the antimicrobial activity of natural products. Coriander's essential oils and leaf extracts exhibit broad-spectrum antimicrobial activity against bacteria, fungi, and certain parasites. The essential oil and extracts from coriander leaves have a high linalool concentration. Linalool has been reported to disrupt bacterial cell membranes, causing permeability loss, leaking of intracellular contents, and inhibition of bacterial growth. Studies consistently report efficacy against *Escherichia coli*, *Staphylococcus aureus*, *Yersinia enterocolitica*, *Salmonella enterica,* and *Vibrio cholerae* and fungal pathogens (Rajeshwari et al., 2012b; Bahrami et al., 2022). Recent research highlights coriander's ability to inhibit biofilm formation by coating the surfaces available to the biofilm-forming micro organisms, which effectively reduces the possibility of cell adhesion, which is the initial stage of biofilm formation. Biofilms are an essential virulence factor for persistent infections; thus, preventing biofilm formation enables existing antibiotics in combating microbial growth, which could be a key approach in the prevention of nosocomial infections (Molina et al., 2020). Other advances include exploration of coriander nanoparticles and nano-emulsions to improve antimicrobial efficacy and stability, promising next-generation natural antimicrobials (Luna et al., 2016). These developments align with global efforts to reduce reliance on synthetic antibiotics and elaborate sustainable, plant-based antimicrobial therapies.

Anti-inflammatory and Analgesic Effects

Inflammation is a crucial mechanism associated with multiple chronic diseases such as rheumatoid arthritis, cardiovascular disease, neurodegeneration, and cancer. Inflammation is the body's response to destroy invading organisms, remove irritants, and prepare for the tissue repair process. Though various steroidal and non-steroidal anti-inflammatory drugs are available, they are often accompanied by an array of side effects. Extracts made from coriander leaves display pronounced anti-inflammatory activity and are less harmful compared to synthetic drugs (Wangensteen et al., 2004). *In vivo* studies confirm coriander's efficacy in reducing paw swelling and edema in animal models of arthritis and acute inflammation. Phytochemicals such as flavonoids, phenolics, and terpenoids suppress nuclear factor kappa B (NF-κB) activation and mitogen-activated protein kinases (MAPK) signaling, which leads to inhibition of pro-inflammatory mediator expression, downregulate cytokines like TNF-α, IL-6, and IL-1β, and this creates an anti-inflammatory response. These findings support the use of coriander-derived formulations for joint pain, backache, and other inflammatory conditions (Agarwal et al., 2025).

Antidiabetic and Hypolipidemic Potential

Metabolic syndromes and diabetes are lifestyle diseases that are escalating global health burdens at a steep rate, with changes in lifestyle like dietary habits, as well as due to genetic factors. Diabetes mellitus is a condition associated with hyperglycaemia, which is a state of increased blood glucose level. There are multiple drugs with different routes of administration and different mechanisms of action to deal with the issue. However, the effectiveness of such drugs varies quite significantly depending on each patient's unique condition. The efficiency of the treatment can be enhanced by the use of conventional diabetes medication along with supplementation of natural antidiabetic products, usually taken as a dietary supplement (Yoon et al., 2025). Coriander is one such natural material that can be consumed along with a regular diet, and it demonstrates promising antidiabetic properties mediated by multiple molecular mechanisms. Bioactive compounds in coriander or coriander extracts show inhibitory action on enzymes like α-glucosidase and α-amylase, which are involved in carbohydrate digestion; this leads to delayed carbohydrate digestion and glucose absorption. Coriander also enhances insulin secretion and sensitivity, protects pancreatic β-cells against oxidative stress, and modulates glucose transporter activity (Kumar et al., 2025). Studies on animal models like mice with diabetes induced using chemical agents have indicated a significant reduction in blood glucose and other lipid parameters such as total cholesterol levels, low density lipopolysaccharide levels, high density lipopolysaccharide levels, etc., when extracts of coriander leaves at dosage of 200– 400 mg/kg of bodyweight was given to such mice for a long period of time. Other studies on similar models showed a significant increase in 2-deoxyglucose transport (1.6-fold), glucose oxidation (1.4-fold), and incorporation of glucose into glycogen and increased secretion of insulin when the aqueous extract of coriander was supplemented in the feed and water of the subjects (Elhassaneen et al., 2024).

Neuroprotective and Anxiolytic Actions

Neurological disorders such as Alzheimer's, Parkinson's, and epilepsy are related to neuroinflammation and oxidative stress. Coriander possesses antioxidant and anti-inflammatory properties, and these help to protect neurons from apoptosis and excitotoxicity. Studies indicate that coriander can improve behavioural problems, as well as reduce cognitive impairment. The flavonoids and terpenes of coriander can also modulate neurotransmitter systems, specifically GABA and serotonin, resulting in anxiolytic, sedative, and anticonvulsant effects (Zein et al., 2014). The bioactive compounds in coriander, including linalool (major), α-pinene, and γ-terpinene, improve chloride conductance without acting as direct agonists, which will minimize the incidence of tolerance. Modulation mostly occurs in limbic areas such as the amygdala and hippocampus. The re-establishment of the serotonergic pathway is positively possible with the help of the linalool and geranyl acetate found in coriander, which helps in the generation of an anti-anxiety effect. Anti-inflammatory neuroprotection can be accomplished through the contextualized modulation of microglial activity to reduce neuroinflammation and maintain the blood-brain barrier (Wangensteen et al., 2004). Glutamate and GABA need to be balanced to reduce excitotoxicity through maintaining normalcy. Linalool interrupts NMDA receptor damage and reinforces GABA transmission. Cholinergic-based support by regulating acetylcholinesterase can protect cognitive skills from toxic damage from β-amyloid. Further, coriander leaf extract can restore the activity of the inhibited mTORC1 pathway when at-risk (lead toxic) neurons are present, enhancing neuronal survival and plasticity, and for this purpose, 250 mg/kg of coriander extract was more efficacious than the 500 mg/kg dose (Rajeshwari et al., 2012b; Eid et al., 2021).

Anticancer Activity

Coriander leaves have components that can prevent cancer by restricting cell growth, inducing the suicide of those cells, and preventing vasculogenesis. The evidence indicates that extracts from coriander lower the detrimental genes caused by Bcl-2 and cyclin D1 while acting on the existential protective genes of p53. Coriander also has identified phenolic components, such as gallic acid, useful for blocking colorectal cancer through the generation of ROS within the mitochondria. Flavonoids and other phenolic acids have been shown to inhibit cancer cell growth (Kostelecka et al., 2024). Linalool induces oxidative stress in cancer cells, which may interfere with cancer cell function and lead to their demise. Coriander also has caffeic, chlorogenic acids, quercetin, and kaempferol that could enhance the cytotoxic effects of chemotherapy, and other components like dodecanal and α-pinene also generate anticancer properties. It acts on the PI3K/Akt/mTOR pathway, central to cancer cell growth, and supplemented by lowering Akt so that PTEN is increased and restores tumour suppression. Acute inhibition of mTOR limits the production of all proteins and stunts growth, while p53 induction can also mediate cell cycle arrest or cell death through TRAIL receptors and caspases (Khwairakpam et al., 2020). Further, decreasing STAT3 activity reduces inflammation and angiogenesis. Coriander leaves additionally prevent the transition of good cells to cancer stem cells, preventing metastasis, and provide antioxidants (Rajeshwari et al., 2012b).

Cardiovascular and Hepatoprotective Effects

Coriander leaves have a number of preventive effects on cardiovascular and hepatic disease owing to their bioactive phytochemicals, including flavonoids and essential oils. Its mechanism of action involves radical scavenging. Coriander heals cardiovascular health by decreasing atherosclerosis, up-regulating blood pressure, and improving endothelial function to prevent the oxidative modification of LDL (low density lipoprotein) and lipid peroxidation to halt atherogenesis. Studies explained decreased lipid peroxidation and inflammation in the heart muscle with coriander leaf extract supplementation and improved nitric oxide (NO) production, which inhibited angiotensin converting enzyme (ACE) to cause vasodilation and lower blood pressure (Choudhary et al., 2017). It has significant hypolipidemic and hepatoprotective effects that reduce liver injury from toxins and oxidative stress by increasing anti-oxidative enzymes and reducing inflammation. It also restores the functions of the liver and many parameters of liver functioning, like ALT, AST, etc. As it includes various phytosterols like stigmasterol and β-sitosterol, it helps reduce total cholesterol by inhibiting absorption from diet and increasing bile-acid synthesis and ultimately lowering cholesterol levels through both total and LDL categories (Zein et al., 2014). In addition, coriander enhances reverse cholesterol transport and lowers the risk of atherosclerosis. Coriander has antihypertensive and antiarrhythmic effects due to flavonoids like apigenin and quercetin that inhibit the ACE and reduce blood pressure. Some essential oils, like linalool, modulate calcium influx to relax vascular smooth muscle, and aqueous extracts of coriander provide diuretic effects which reduce blood volume (Sakurai et al., 2019). Coriander also stimulates the Keap1–Nrf2 pathway, adding an antioxidant defence to vascular and liver cells and protecting them against oxidative damage. When the polyphenols in coriander are encapsulated in nano-carriers, their bioavailability is improved, which supports liver and cardiovascular protection. Recent studies are trying to standardize extracts to increase concentrations of compounds such as linalool and quercetin, which could give a better way to offer a directed dose aimed at reducing cardiovascular risk in nutraceuticals (Truc, 2024).

Toxicity and Safety Profiles

Coriander leaves have a well-established safety profile, characterization indicated by low toxicity in both acute and sub-chronic studies. Experimental evaluations indicate that hydro-methanolic extracts administered orally at doses up to 3000 mg/kg body weight in animal models produce no mortality or significant adverse effects (Sharma et al., 2021). This also includes no detectable alterations in haematological, biochemical, or histopathological parameters of vital organs such as the liver, kidney, and heart. Acute toxicity assessments show no behavioural changes or mortality within 24 hours post-administration, supporting a high median lethal dose (LD_{50}). Sub-chronic exposure at higher doses may occasionally result in decreased food intake and subsequent body weight reduction (Horky et al., 2019). Coriander essential oil, primarily composed of linalool, is not clastogenic or mutagenic and is considered safe at usual dietary levels, though it can cause mild irritation in sensitive individuals. Allergic reactions remain rare but possible, mostly related to contact dermatitis from leaf handling or essential oil exposure (Horky et al., 2019). While no significant genotoxicity has been reported, potential interactions, such as anticoagulant effects, necessitate caution in certain populations. Overall, coriander leaves are safe for dietary and medicinal use within customary consumption limits, but systematic chronic toxicity data for high-dose or pharmaceutical applications remain limited.

Conclusion and Prospects

Coriander leaves are a rich source of diverse bioactive compounds, including essential oils, flavonoids, phenolic acids, fatty acids, phytosterols, alkaloids, tannins, and saponins, each contributing uniquely to the plant's pharmacological and nutritional profile. These phytochemicals exhibit potent antioxidant, antimicrobial, anti-inflammatory, and metabolic regulatory properties, making coriander leaves valuable in traditional medicine and as functional food components. Advances in phytochemical characterization have deepened understanding of how biosynthetic pathways and environmental factors influence the qualitative and quantitative variability of these bioactive compounds across different geographic regions and cultivars. Furthermore, evolving extraction and purification technologies have facilitated more efficient isolation of active compounds while maintaining bioactivity, expanding potential applications within nutraceutical, pharmaceutical, and cosmetic industries. Despite significant progress, limited agronomic and genetic interventions have targeted the enhancement of specific bioactive profiles in coriander leaves, leaving considerable scope for cultivation-based optimization.

Future research prospects for coriander involve integrated approaches combining genomics, metabolomics, and precision agriculture to develop cultivars with elevated levels of desired bioactive compounds tailored for specific therapeutic or industrial purposes. Elucidation of biosynthetic gene networks, regulatory mechanisms, and environmental modulators will be crucial to manipulate metabolic flux toward high-value compounds. Additional long-term safety and efficacy studies of purified extracts and isolated compounds will support clinical translation. Innovations in sustainable extraction methods, including green chemistry and biotechnological advancements, promise improved yield and environmentally friendly processing. Moreover, exploration of synergistic interactions among coriander leaf bioactive molecules could pave the way for novel multi target therapies addressing chronic diseases. Harnessing traditional knowledge alongside modern scientific tools will enable a fuller realization of coriander leaf potential, promoting health benefits, and commercial viability in a broader context.

References

Abdalla, M.Y.A. 2009. Effect of organic, bio- and mineral fertilization on growth, yield, oil productivity, and chemical constituents of coriander plant. *J. Plant Prod.*, *34*: 5195–5208. https://doi.org/10.21608/jpp.2009.118395.

Abdul-Jalil, T.Z. 2013. Screening of rutin from seeds and leaves extracts of dill, coriander, and fennel cultivated in Iraq. *Pharm. Glob.*, *4* :1–6.

Acharya, S., Duwal, A., Nepal, A., et al., 2019. Evaluation of coriander (*Coriandrum sativum* L.) varieties for growth and yield parameters. *Nepalese J. Agri. Sci.*, *18*: 36–46.

Agarwal, U., Kanupriya, Tonk, R.K. and Verma, S. 2025. A comprehensive review of supernatural coriander herb (*Coriandrum sativum*): Phytochemical insights, pharmacological potential and future perspective. *Phytochem. Rev.* https://doi.org/10.1007/s11101-025-10142-5.

Alkharpotly, A.A., Abd-Elkader, D.Y., Salem, M.Z.M. and Hassan, H.S. 2024. Growth, productivity, and phytochemicals of coriander in responses to foliar application of *Acacia saligna* fruit extract as a biostimulant under field conditions. *Sci. Rep.*, *14*: 2921. https://doi.org/10.1038/s41598-024-53378-5.

Al-Khayri, J.M., Banadka, A., Nandhini, M., et al., 2023. Essential oil from *Coriandrum sativum*: A review on its phytochemistry and biological activity. *Molecules*, *28*: 696. https://doi.org/10.3390/molecules28020696.

Amer, H.M., Hussein, M.S., Khalid, K.A. and Mohammad, A.A. 2024. Improving productivity and chemical constituents of *Coriandrum sativum* L. under foliar application of tryptophan. *Vegetos*, *37*: 1989–1996. https://doi.org/10.1007/s42535-024-00875-7.

Amiripour, A., Jahromi, M.G., Soori, M.K. and Torkashvand, A.M. 2021. Changes in essential oil composition and fatty acid profile of coriander (*Coriandrum sativum* L.) leaves under salinity and foliar-applied silicon. *Ind. Crops Prod.*, *168*: 113599. https://doi.org/10.1016/j.indcrop.2021.113599.

Atrooz, O., Al-Nadaf, A., Uysal, H., et al., 2023. Biosynthesis of silver nanoparticles using *Coriandrum sativum* L. extract and evaluation of their antibacterial, anti-inflammatory, and antinociceptive activities. *South Afr. J. Bot.*, *157*: 219–227. https://doi.org/10.1016/j.sajb.2023.04.001

Ayran, İ., Çelık, S.A. and. Kan, A 2017. The elemental characterization of coriander populations (*Coriandrum sativum* L.) cultivated in Turkey. *Ziraat Fakültesi Derg.*, *12*: 110–114.

Bahrami, A., Delshadi, R., Cacciotti, I., et al. 2022. Targeting foodborne pathogens via surface-functionalized nano-antimicrobials. *Adv. Colloid Interface Sci.*, *302*: 102622. https://doi.org/10.1016/j.cis.2022.102622.

Banerjee, S., Sarkar, K. and Sil, P.C. 2024. Coriander leaves and seeds (*Coriandrum sativum*). In: *Medicinal Spice and Condiment Crops*. CRC Press, pp. 225–263.

Barros, L., Dueñas, M., Dias, M.I., et al. 2012. Phenolic profiles of *in vivo* and *in vitro* grown *Coriandrum sativum* L. *Food Chem.*, *132*: 841–848. https://doi.org/10.1016/j.foodchem.2011.11.048.

Basol, A., Yaldiz, G. and Camlica, M. 2024. Genetic diversity, analysis of some agro-morphological and quality traits, and utilization of plant resources of Coriander (*Coriandrum sativum*) supported with cluster and multivariate analyses. *Biology*, *13*: 866. https://doi.org/10.3390/biology13110866.

Berkomah, J., Li, H., Siddiqui, R., et al. 2022. Cilantro and coriander yield as affected by cultivars and row spacings in fall and spring production in Virginia. *HortScience*, *57*(9): 1156– 1158. https://doi.org/10.21273/HORTSCI16759-22.

Bhat, S., Kaushal, P., Kaur, M. and Sharma, H.K. 2014. Coriander (*Coriandrum sativum* L.): Processing, nutritional, and functional aspects. *Afr. J. Plant Sci.*, *8*: 25–33. https://doi.org/10.5897/AJPS2013.1118.

Bhatti, S., Baig, J.A., Kazi, T.G., et al. 2019. Macro and micro mineral composition of Pakistani common spices: A case study. *J. Food Meas. Charact.*, *13*: 2529–2541. https://doi.org/10.1007/s11694-019-00173-w.

Bomme, U., Blum, H. and Rinder, R. 2007. Spring cultivation of coriander (*Coriandrum sativum* L.) for leaf use: Results from several years of variety testing. *J. Med. Spice Plants*, *12*: 24–29.

Chander, A.K. 2010. *Characterisation and Oxidative Stability of Speciality Plant Seed Oils*. PhD, Aston University, UK.

Chaulagain, R., Pant, S.S., Thapa, R.B. and Sharma, M.D. 2011. Performance of coriander cultivars for green leaf production under late sowing condition. *J. Agric. Environ.*, *12*: 67–73. https://doi.org/10.3126/aej.v12i0.7565.

Choudhary, S., Pereira, A., Basu, S. and Verma, A. 2017. Differential antioxidant composition and potential of some commonly used Indian spices. *J. AgriSearch*, *4*: 160–166. https://doi.org/10.21921/jas.v4i03.9015.

Clifford, M.N. 1999. Chlorogenic acids and other cinnamates: Nature, occurrence, and dietary burden. *J. Sci. Food Agric.*, *79*: 362–372.

Cortés-Eslava, J., Gómez-Arroyo, S., Villalobos-Pietrini, R., Espinosa-Aguirre, J.J. 2004. Antimutagenicity of coriander (*Coriandrum sativum*) juice on the mutagenesis produced by plant metabolites of aromatic amines. *Toxicol. Lett.*, *153*: 283–292. https://doi.org/10.1016/j.toxlet.2004.05.011.

Cuevas-Cianca, S.I., Romero-Castillo, C., Gálvez-Romero, J.L., et al. 2023 Antioxidant and anti-inflammatory compounds from edible plants with anti-cancer activity and their potential use as drugs. *Molecules*, *28*: 1488. https://doi.org/10.3390/molecules28031488.

Das, S., Pradhan, C. and Pillai, D. 2023a. Dietary coriander (*Coriandrum sativum* L) oil improves antioxidant and anti-inflammatory activity, innate immune responses, and resistance to *Aeromonas hydrophila* in Nile tilapia (*Oreochromis niloticus*). *Fish Shellfish Immunol.*, *132*: 108486. https://doi.org/10.1016/j.fsi.2022.108486.

Das, S., Pradhan, C., Singh, A.K., et al. 2023b. Dietary coriander (*Coriandrum sativum* L) oil improves growth, nutrient utilization, antioxidant status, tissue histomorphology, and reduces omega-3 fatty acid production in Nile tilapia (*Oreochromis niloticus*). *Anim. Feed Sci. Technol.*, *305*: 115774. https://doi.org/10.1016/j.anifeedsci.2023.115774.

de Lima, D.P., Pinto Júnior, E.S., Menezes, A.V., Souza, D.A., São José, V.P.B., Silva, B.P., Almeida, A.Q. and Carvalho, I.M.M. 2024. Chemical composition, minerals concentration, total phenolic compounds, flavonoids content, and antioxidant capacity in organic and conventional vegetables. *Food Res. Int.*, *175*: 113684. https://doi.org/10.1016/j.foodres.2023.113684.

Demir, S. and Korukluoğlu, M. 2020. A comparative study about antioxidant activity and phenolic composition of cumin (*Cuminum cyminum* L.) and coriander (*Coriandrum sativum* L.). *Indian Journal of Traditional Knowledge*, *19*(2): 383– 393.

Devi, K.B., Ariina, M.S., Singh, T.S., et al. 2022. Performance of coriander (*Coriandrum sativum* L.) on growth and selecting best variety. *Pharma Innov. J.*, *11*: 1612–1613.

Diederichsen, A., Banniza, S., Armstrong-Cho, C. and Sander, T. 2020. Coriandrum sativum L.: Coriander. In: *Medicinal, Aromatic, and Stimulant Plants*, pp. 265–281. Springer International Publishing, Cham.

Divya, P., Puthusseri, B. and Neelwarne, B. 2012. Carotenoid content, its stability during drying and the antioxidant activity of commercial coriander (*Coriandrum sativum* L.) varieties. *Food Res. Int.*, *45*: 342–350. https://doi.org/10.1016/j.foodres.2011.09.021.

Divya, P., Puthusseri, B. and Neelwarne, B. 2014. The effect of plant regulators on the concentration of carotenoids and phenolic compounds in foliage of coriander. *LWT - Food Sci. Technol.*, *56*: 101–110. https://doi.org/10.1016/j.lwt.2013.11.012.

Ebrahimi, S.N., Hadian, J. and Ranjbar, H. 2010. Essential oil compositions of different accessions of *Coriandrum sativum* L. from Iran. *Nat. Prod. Res.*, *24*: 1287–1294. https://doi.org/10.1080/14786410903132316.

Eid, A.M., Issa, L., Al-Kharouf, O., et al. 2021. Development of *Coriandrum sativum* oil nanoemulgel and evaluation of its antimicrobial and anticancer activity. *Bio.Med. Res. Int.*, *2021*: 5247816. https://doi.org/10.1155/2021/5247816.

Elhassaneen, Y., ElBassouny, G., Emam, O. and Aram, E. 2024. Nutrients and nutraceuticals content and *in vitro* biological activities of formulae from plant parts commonly spread in Egyptian markets. *Am. J. Food Nutr.*, *12*: 134–151. https://doi.org/10.12691/ajfn-12-5-2.

El-Nasr, T.H.S.A., Ibrahim, M.M., Aboud, K. and El-Enany, M.A.M. 2013. Assessment of genetic variability for three coriander (*Coriandrum sativum* L.) cultivars grown in Egypt, using morphological characters, essential oil composition, and ISSR markers. *World Appl Sci. J.*, *25*: 839–849. https://doi.org/10.5829/idosi.wasj.2013.25.06.13349.

Eyres, G., Dufour, J.P., Hallifax, G., et al. 2005. Identification of character-impact odorants in coriander and wild coriander leaves using gas chromatography-olfactometry (GCO) and comprehensive two-dimensional gas chromatography–time-of-flight mass spectrometry (GC×GC–TOFMS). *J. Sep. Sci.*, *28*: 1061–1074. https://doi.org/10.1002/jssc.200500012.

Falade, A.O., Akingbasote, M.D., Adewole, K.E., et al. 2025. *Cinnamomum verum* and *Coriandrum sativum* secondary metabolites exhibit neuroprotective potential: Insights from *in vitro*, *ex vivo*, and computational models. *Silico Pharmacol.*, *13*: 120. https://doi.org/10.1007/s40203-025-00412-7.

Gahory, A.M.O., Ayyat, A.M. and Soliman, T.M.A. 2022. Growth, yield and its component of coriander (*Coriandrum sativum* L.) in response to the addition of compost, ascorbic acid, and salicylic acid under Aswan Governorate conditions, *Egypt. J. Plant Prod.*, https://doi.org/10.21608/jpp.2022.177560.1192.

Gandepalli, L. and Prasad, V.M. 2020. Evaluation of coriander (*Coriandrum sativum* L.) varieties in Prayagraj agro-climatic conditions. *Int. J. Curr. Microbiol. Appl. Sci.*, *9*: 1086–1091. https://doi.org/10.20546/ijcmas.2020.912.130.

Gantait, S., Sharangi, A.B., Mahanta, M. and Meena, N.K. 2022. Agri-biotechnology of coriander (*Coriandrum sativum* L.): An inclusive appraisal. *Appl. Microbiol. Biotechnol.*, *106*: 951–969. https://doi.org/10.1007/s00253-022-11787-4.

Gazwi, H.S.S., Mahmoud, M.E. and Toson, E.M.A. 2022. Analysis of the phytochemicals of *Coriandrum sativum* and *Cichorium intybus* aqueous extracts and their biological effects on broiler chickens. *Sci. Rep.*, *12*: 6399. https://doi.org/10.1038/s41598-022-10329-2.

González-Pérez, E., Villalobos-Reyes, S. and C.A. Núñez-Colín. 2023. Nutrient dynamics of macro and micronutrients in coriander. *Acta Agríc. Pecu.*, *9*: 1–10.

Hadjmohammadi, M. and Sharifi, V. 2020. Investigation of optimum extraction conditions for determination of quercetin and kaempferol in coriander (*Coriandrum sativum* L.) by using experimental design and HPLC. *J. Food Drug Anal.*, *17* (4): 293–299. https://doi.org/10.38212/2224-6614.2601.

Hadi, M.R.H.S., Rezakhani, A., Darzi, M.T. 2022. Composition of essential oil in coriander (*Coriandrum sativum* L.) as influenced by manure and amino acid application. *J. Org. Farming Med. Plants*, *1*: 27–35.

Hardiany, N.S., Fadilah, F., de Lima, F.V.I., et al. 2025. Flavonoid content and total antioxidant capacity of coriander (*Coriandrum sativum* L.) seed extract in different level polarity of solvent. *AIP Conf. Proc.*, *3186*: 020063. https://doi.org/10.1063/5.0234691.

Hiremath, R., Hosamath, J.V., Srinivasulu, B. and Hegde, R.V. 2025. Trends, innovations, and sustainable practices in Coriander production: A global perspective. *J. Sci. Res. Rep.*, *31*: 891–907. https://doi.org/10.9734/jsrr/2025/v31i53085.

Horky, P., Skalickova, S., Smerkova, K. and Skladanka, J. 2019. Essential oils as a feed additives: Pharmacokinetics and potential toxicity in monogastric animals. *Animals*, *9*: 352. https://doi.org/10.3390/ani9060352.

Ismail, F., Ismail, W.U., Muhammad, G., et al. 2025. Phytochemistry, pharmacological attributes, and clinical evaluations of *Coriandrum sativum*: A comprehensive review. *Nat. Prod. Commun.*, *20*(1): 1–23. https://doi.org/10.1177/1934578X241312791.

Ivanovic, M., Razboršek M.I., Košir, I.J. and Kolar, M. 2019. Response surface methodology: An optimal design applied for maximum ultrasound-assisted extraction efficiency of phenolic acids from *Coriandrum sativum* L. *J. Appl. Bot. Food Qual.*, 378–387. https://doi.org/10.5073/JABFQ.2019.092.050.

İzgı, M.N., Telci, İ. and Elmastaş, M. 2017. Variation in essential oil composition of coriander (*Coriandrum sativum* L.) varieties cultivated in two different ecologies. *J. Essent. Oil Res.*, *29*: 494–498. https://doi.org/10.1080/10412905.2017.1363090.

Jampílek, J. and Kráľová, K. 2023. Impact of abiotic stresses on production of secondary metabolites in medicinal and aromatic plants. In: Aftab, T. (Ed.), *New Frontiers in Plant-Environment*

Interactions: Innovative Technologies and Developments. Springer Nature Switzerland, Cham, pp. 169–252.

Jin, W., Zhou, H., Zhao, H., et al. 2025. Isolation, *in vitro* antioxidant capacity, hypoglycaemic activity, and immunoactivity evaluation of polysaccharides from *Coriandrum sativum* L. *Antioxidants*, *14*: 149. https://doi.org/10.3390/antiox14020149.

Joble, J., Prasad, V.M. and Bahadur, V. 2022. Evaluation of different varieties of Coriander (*Coriandrum sativum* L.). *Int. J. Plant Soil Sci.*, 1231–1238. https://doi.org/10.9734/ijpss/2022/v34i232537.

Kaiser, A., Kammerer, D.R. and Carle, R. 2013. Impact of blanching on polyphenol stability and antioxidant capacity of innovative coriander (*Coriandrum sativum* L.) pastes. *Food Chem.*, *140*: 332–339. https://doi.org/10.1016/j.foodchem.2013.02.077.

Kaleem, M., Shah, A.A., Usman, S., et al. 2025. Magnesium oxide nanoparticles improved drought resilience in *Coriandrum sativum* L. through lifting antioxidant level, redox balancing, and improving photosynthetic efficiency. *ACS Omega*, *10*: 32813–32828. https://doi.org/10.1021/acsomega.5c00986.

Kalenska, S., Mazurenko, B., Harbar, L., et al. 2024. Seed yield limitations of coriander (*Coriandrum sativum* L.) based on plant structure analysis. *J. Agric. Food Res.*, *18*: 101321. https://doi.org/10.1016/j.jafr.2024.101321.

Katar, D., Kara, N. and Katar, N. 2016. Yields and quality performances of coriander (*Coriandrum sativum* l.) genotypes under different ecological conditions. *Turk J. Field Crops*, *21*: 79–87. https://doi.org/10.17557/tjfc.77478.

Khan, I., Dubey, W. and Gupta, V. 2014. Taxonomical aspects of coriander (*Coriandrum sativum* L.). *Int. J. Curr. Res.*, *06*: 9926–9930.

Khan, I., Verma, P., Rathore, B., et al. 2017. Genetic variability for essential oil, polyphenols, and antioxidant activity of coriander (*Coriandrum sativum* L.) genotypes grown in humid south eastern plain zone V of Rajasthan. *International Journal of Chemical Studies*, *5*: 2292–2297.

Khwairakpam, A.D., Monisha, J., Roy, N.K., et al. 2020. Vietnamese coriander inhibits cell proliferation, survival and migration via suppression of Akt/mTOR pathway in oral squamous cell carcinoma. *J. Basic Clin. Physiol. Pharmacol.*, *31*. https://doi.org/10.1515/jbcpp-2019-0162.

Kianersi, F., Azarm, D.A., Fatemi, F., et al. 2022. Methyl jasmonate induces genes involved in linalool accumulation and increases the content of phenolics in two Iranian coriander (*Coriandrum sativum* L.) ecotypes. *Genes*, *13*: 1717. https://doi.org/10.3390/genes13101717.

Kostelecka, K., Bryliński, Ł., Komar, O., et al. 2024. An overview of the spices used for the prevention and potential treatment of gastric cancer. *Cancers*, *16*: 1611. https://doi.org/10.3390/cancers16081611.

Kumar, S., Ahmad, R., Saeed, S., et al. 2022. Chemical composition of fresh leaves, headspace aroma, and essential oils of four coriander cultivars. *Front Plant Sci.*, *13*. https://doi org/10.3389/fpls.2022.820644.

Kumar, S.Y.R., Lamba, D. and Khanduri, S.N.S. 2025. Potential of *Coriandrum sativum* in management of diabetes and hyperlipidaemia: A comprehensive review of pharmacological evidence and mechanisms. *Int. J. Basic Clin. Pharmacol.*, *14*: 324–329. https://doi.org/10.18203/2319-2003.ijbcp20250496.

Kuneva, V., Manhart, S., Delibaltova, V., et al. 2024. Mathematical approach used for the assessment of foliar application of biostimulant and fertilizers to coriander varieties (*Coriandrum sativum* L.). *Ind. Crops Prod.*, *222*: 119783. https://doi.org/10.1016/j.indcrop.2024.119783

Kuri, B.R., Jat, N.L., Shivran, A.C., et al. 2015. Effect of sowing time, varieties and plant growth regulators on growth, physiological indices, and productivity of coriander (*Coriandrum sativum*). *Indian J. Agron.*, *60*: 464–470. https://doi.org/10.59797/ija.v60i3.4481.

Laribi, B., Kouki, K., M'Hamdi, M. and Bettaieb, T. 2015. Coriander (*Coriandrum sativum* L.) and its bioactive constituents. *Fitoterapia*, *103*: 9–26. https://doi.org/10.1016/j.fitote.2015.03.012.

Lee, E.Y., Li, Y., Wu, X., et al. 2025. The chemical composition of coriander (*Coriandrum sativum* L.) extract and its ability in reducing the risk of SARS-CoV-2 infection and scavenging radicals. *ACS Food Sci. Technol.*, *5*: 2215–2224. https://doi.org/10.1021/acsfoodscitech.5c00088.

Lin, H-H., Lin, K-H., Yang, M-J., et al. 2022. Physiological responses and antioxidant properties of coriander plants (*Coriandrum sativum* L.) under different light intensities of red and blue lights. *Sci. Rep.*, *12*: 21139. https://doi.org/10.1038/s41598-022-25749-3.

López, P.A., Widrlechner, M.P., Simon, P.W., et al. 2008. Assessing phenotypic, biochemical, and molecular diversity in coriander (*Coriandrum sativum* L.) germplasm. *Genet. Resour. Crop Evol.*, *55*: 247–275. https://doi.org/10.1007/s10722-007-9232-7.

Luna, C., Barriga-Castro, E.D., Gómez-Treviño, A., et al. 2016. Microstructural, spectroscopic, and antibacterial properties of silver-based hybrid nanostructures biosynthesized using extracts of coriander leaves and seeds. *Int. J. Nanomedicine*, *11*: 4787–4798. https://doi.org/10.2147/IJN.S105166.

Machado, A.M., Lopes, V., Barata, A.M., et al. 2023. Chemical variability of the essential oils from two Portuguese Apiaceae: *Coriandrum sativum* L. and *Foeniculum vulgare* Mill. *Plants*, *12*: 2749. https://doi.org/10.3390/plants12142749.

Mahleyuddin, N.N., Moshawih, S., Ming, L.C. et al. 2021. *Coriandrum sativum* L.: A review on ethnopharmacology, phytochemistry, and cardiovascular benefits. *Molecules*, *27*: 209. https://doi.org/10.3390/molecules27010209.

Mandal, S. and Mandal, M. 2015. Coriander (*Coriandrum sativum* L.) essential oil: Chemistry and biological activity. *Asian Pac. J. Trop. Biomed.*, *5*: 421–428. https://doi.org/10.1016/j.apjtb.2015.04.001.

Manickam, S., Rajagopalan, V.R., Kambale, R., et al. 2023. Plant metabolomics: Current initiatives and future prospects. *Curr. Issues Mol. Biol.*, *45*: 8894–8906. https://doi.org/10.3390/cimb45110558.

Maroufi, K., Farahani, H.A. and Darvishi, H.H. 2010. Importance of coriander (*Coriandrum sativum* L.) between the medicinal and aromatic plants. *Adv. Environ. Biol.*, *4*: 433–436.

Mauer, L. and El-Sohemy, A. 2012. Prevalence of cilantro (*Coriandrum sativum*) disliking among different ethnocultural groups. *Flavour*, *1*: 8. https://doi.org/10.1186/2044-7248-1-8.

McAusland, L., Lim, M-T., Morris, D.E., et al. 2020. Growth spectrum complexity dictates aromatic intensity in Coriander (*Coriandrum sativum* L.). *Front. Plant Sci.*, *11*: 1– 14. https://doi.org/10.3389/fpls.2020.00462.

Meena, R.S., Praneetha, S., Meena, N.K., et al. 2024. Genetics, genomics, and breeding of Coriander: A Review. In: Kole, C. (Ed.), *Genetics, Genomics, and Breeding of Seed Spices*. Springer Nature, Singapore, pp 45–67.

Misharina, T.A. 2016. Antiradical properties of essential oils and extracts from coriander, cardamom, white, red, and black peppers. *Appl. Biochem. Microbiol.*, *52*: 79–86. https://doi.org/10.1134/S0003683816010087.

Mohamed, M.A., Ibrahim, M.E. and Wahba, H.E. 2018. Flavoring compounds of essential oil isolated from agriculture waste of coriander (*Coriandrum sativum*) plant. *J. Mater. Environ. Sci.*, *9*: 77–82.

Molina, R.D.I., Campos-Silva, R., Macedo, A.J., et al. 2020. Antibiofilm activity of coriander (*Coriander sativum L.*) grown in Argentina against food contaminants and human pathogenic bacteria. *Ind. Crops Prod.*, *151*: 112380. https://doi.org/10.1016/j.indcrop.2020.112380.

Mouhoubi, K., Boulekbache-Makhlouf, L., Madani, K., et al. 2023. Microwave-assisted extraction optimization and conventional extraction of phenolic compounds from coriander leaves: UHPLC characterization and antioxidant activity. *North Afr. J. Food Nutr Res.*, *7*: 69–83.

Mouhoubi, K., Boulekbache-Makhlouf, L., Madani, K., et al. 2022. Phenolic compounds and antioxidant activity are differentially affected by drying processes in celery, coriander, and parsley leaves. *Int. J. Food Sci. Technol.*, *57*: 3467–3476. https://doi.org/10.1111/ijfs.15670.

Nadeem, M., Anjum F.M., Khan, M.I., et al. 2013. Nutritional and medicinal aspects of coriander (*Coriandrum sativum* L.): A review. *Br. Food J.*, *115*: 743–755. https://doi.org/10.1108/00070701311331526.

Nadjafi, F., Damghani, A.M.M. and Ebrahimi, S.N. 2009. Effect of irrigation regimes on yield, yield components, content, and composition of the essential oil of four Iranian land races of coriander (*Coriandrum sativum*). *J. Essent. Oil Bear. Plants*, *12*: 300–309. https://doi.org/10.1080/0972060X.2009.10643724.

Nath, P., Kale, S. and Chauhan, O. 2015. Coriander: A potential medicinal herb. *Indian Food Industry Mag.*, *34* (2): 29–35.

Nathenial, S., Fatima, A., Fatima, R., et al. 2019. Phytochemical study of acetone solvent extract of *Coriandrum sativum*. *Journal of Pharmacognosy and Phytochemistry*, *8*(6): 136–140.

Nguyen, Q.H. 2015. *Study on Bioaccumulation and Integrated Biorefinery of Vegetable Oil and Essential Oil of Coriander* (*Coriandrum sativum* L.). PhD thesis, Institute National Polytechnique de Toulouse - INPT, France.

Nisa, M-U., Umer, M., Hamza, M., et al. 2025. Investigating coriander leaf phenolics with HPLC-UV and their role in modulating nitrogen metabolism. *Food Sci. Nutr.*, *13*(3): e70029. https://doi.org/10.1002/fsn3.70029.

Oishi, O.G. 2024. *GC-MS Employed Identification of Phytoconstituents and Antioxidant Activity Study of Coriander Leaf Extract*. Thesis, Brac University, Dhaka, Bangladesh.

Özyazici, G. 2021. Influence of organic and inorganic fertilizers on coriander (*Coriandrum sativum* L.) agronomic traits, essential oil, and components under semi-arid climate. *Agronomy*, *11*: 1427. https://doi.org/10.3390/agronomy11071427.

Parmar, R.R., Jain, K.R. and Modi, C.K. 2011. Image morphological operation-based quality analysis of coriander seed (*Coriandrum sativum* L). In: *2011 International Conference on Emerging Trends in Networks and Computer Communications* (*ETNCC*), pp. 482–486. IEEE, Udaipur, India.

Priyadarshi, S., Khanum, H., Ravi, R., et al. 2016. Flavour characterisation and free radical scavenging activity of coriander (*Coriandrum sativum* L.) foliage. *J. Food Sci. Technol.*, *53*: 1670–1678. https://doi.org/10.1007/s13197-015-2071-1.

Pujari, R., Sunanda, B.B., Kurubar, A.R., et al. 2019. Collection and evaluation of coriander varieties for growth and seed purpose in UKP command area. *Int. J. Curr. Microbiol. Appl. Sci.*, *8*: 3125–3130. https://doi.org/10.20546/ijcmas.2019.806.372.

Rai, A., Mohanalakshmi, M., Rajashree, V., et al. 2025. Exploring the leafy coriander *Coriandrum sativum* L. (Apiaceae): A comprehensive review on the metabolomics properties *South Afr. J. Bot.*, *181*: 258–271. https://doi.org/10.1016/j.sajb.2025.04.017.

Rajeshwari, C.U., Shobha, R. and Andallu, B. 2012a. Antihaemolytic activity of various fractions of methanolic extract of coriander (*Coriandrum sativum* L.) leaves and seeds: A comparative study. *Pak. J. Biol. Sci.*, *22*: 1–6.

Rajeshwari, C.U., Siri, S. and Andallu, B. 2012b. Antioxidant and antiarthritic potential of coriander (*Coriandrum sativum* L.) leaves. *E-SPEN J.I*, *7*: e223–e228. https://doi.org/10.1016/j.clnme.2012.09.005.

Ravi, R., Prakash, M. and Bhat, K.K. 2007. Aroma characterization of coriander (*Coriandrum sativum* L.) oil samples. *Eur. Food Res. Technol.*, *225*: 367–374. https://doi.org/10.1007/s00217-006-0425-7.

Sahib, N.G., Anwar, F., Gilani, A-H., et al. 2013. Coriander (*Coriandrum sativum* L.): A potential source of high-value components for functional foods and nutraceuticals: A review. *Phytother. Res.*, *27*: 1439–1456. https://doi.org/10.1002/ptr.4897.

Sakurai, J., Izumo, N. and Watanabe, Y. 2019. Effect of *Coriandrum sativum* L. leaf extract on the brain GABA neurons in mice. *J. Nutrition Health Food Sci.*, *7*(2): 1–7. DOI: 10.15226/jnhfs.2018.001154.

Salem, M.A., Khalil, H.M.A., Manaa, E.G., et al. 2025. Antioxidant potential of selected Apiaceae plant extracts: A study focused on the chemical composition and neuroprotective effect of *Coriandrum sativum* L. extract against lead (Pb)-induced neurotoxicity in rats. *Biol. Trace Elem. Res*. https://doi.org/10.1007/s12011-025-04627-9.

Sambasivaraju, D. and Za, F. 2016. Evaluation of antibacterial activity of *Coriandrum sativum* (L.) against gram- positive and gram- negative bacteria. *Int. J. Basic Clin. Pharmacol.*, *5*: 2653–2656. https://doi.org/10.18203/2319-2003.ijbcp20164141.

Santos, M.D.V. dos., Neto, M.F. de C., Melo, A.C.G.R de., et al. 2019. Chemical composition of essential oil of coriander seeds (*Coriandrum sativum*) cultivated in Amazon Savannah, Brazil. *Chemical Engineering Transaction*, *17*: 409– 423. https://doi.org/10.3303/CET1975069.

Sarwade, P.P., Gaisamudre, K.N., Gaikwad, R.S., et al. 2025. A review on *Coriandrum sativum* L.: Its ethnopharmacology, phytochemistry, mechanism, traditional uses, and cardiovascular effects. *Int. J. Pharm. Health Care Res.*, *13*: 280–290.

Sawi, S.E., El-Feky, A.M., El-Sayed, M.I. and El-Sayed, A.F. 2025. Therapeutic potential of isolated flavonoids from Anise and coriander aerial parts in antimicrobial efficacy, molecular docking, ADMET, and dynamic simulations. *Sci. Rep.*, *15*: 26485. https://doi.org/10.1038/s41598-025-10927-w.

Saxena, S.N., Sharma, Y.K., Rathore, S.S., et al. 2015. Effect of cryogenic grinding on volatile oil, oleoresin content, and anti oxidant properties of coriander (*Coriandrum sativum* L.) genotypes. *J. Food Sci. Technol.*, 52: 568–573. https://doi.org/10.1007/s13197-013-1004-0.

Saxena, S.N., Meena, R.S., Vishal, M.K., et al. 2022. Variation in essential oil constituents of coriander (*Coriandrum sativum* L.) germplasm across coriander growing regions in India. *J. Essent. Oil Res.*, *34*: 173–180. https://doi.org/10.1080/10412905.2022.2036644.

Scandar, S., Zadra, C. and Marcotullio, M.C. 2023. Coriander (*Coriandrum sativum*) polyphenols and their nutraceutical value against obesity and metabolic syndrome. *Molecules*, *28*: 4187. https://doi.org/10.3390/molecules28104187.

Shahwar, M.K., El-Ghorab, A.H., Anjum, F.M., et al. 2012. Characterization of coriander (*Coriandrum sativum* L.) seeds and leaves: Volatile and non-volatile extracts. *Int. J. Food Prop.*, *15*: 736–747. https://doi.org/10.1080/10942912.2010.500068.

Sharangi, A., Bal, S., Maji, A., et al. 2024. Coriander (*Coriandrum sativum* L.) diversity analysis: Insights from germplasm resources leading to crop improvement. *Turk J. Agric. For.*, *48*: 876–887. https://doi.org/10.55730/1300-011X.3227.

Sharifi-Rad, J., Sureda, A., Tenore, G.C., et al. 2017. Biological activities of essential oils: From plant chemo ecology to traditional healing systems. *Molecules*, *22*: 70. https://doi.org/10.3390/molecules22010070.

Sharma, N., Tan, M.A. and An, S.S.A. 2021. Mechanistic aspects of Apiaceae family spices in ameliorating Alzheimer's disease. *Antioxidants*, *10*: 1571. https://doi.org/10.3390/antiox10101571.

Sharma, R. and Dhingra, L. 2024. Meta-learning framework for coriander variety classification using CNN and DBN. In: *2024 15th International Conference on Computing Communication and Networking Technologies* (*ICCCNT*), pp. 1–5. IEEE, Bangalore, India.

Sharma, T., Chaudhary, S. and Jalwal, P. 2025. Evaluation of phytoconstituent and pharmacological properties of *Coriandrum sativum* (LINN). *J. Neonatal. Surg.*, *14*: 1083–1093.

Shoko, T., Manhivi, V.E., Mtlhako, M. and Sivakumar, D. 2022. Changes in functional compounds, volatiles, and antioxidant properties of culinary herb coriander leaves (*Coriandrum sativum*) stored under red and blue LED light for different storage times. *Front. Nutr.*, *9*: 856484. https://doi.org/10.3389/fnut.2022.856484.

Singh, P., Singh, D., Wesley, C.J. 2024. Study on genetic variability in coriander (*Coriandrum sativum* L.). *Int. J. Plant Soil Sci.*, *36*: 790–795. https://doi.org/10.9734/ijpss/2024/v36i64685.

Sobhani, Z., Mohtashami, L., Amiri, M.S., et al. 2022. Ethnobotanical and phytochemical aspects of the edible herb *Coriandrum sativum* L. *J. Food Sci.*, *87*: 1386–1422. https://doi.org/10.1111/1750-3841.16085.

Song, E-J. and Ko, M-J. 2022. Extraction of monoterpenes from coriander (*Coriandrum sativum* L.) seeds using subcritical water extraction (SWE) technique. *J. Superscript Fluids*, *188*: 105668. https://doi.org/10.1016/j.supflu.2022.105668.

Spence, C. 2023. Coriander (cilantro): A most divisive herb. *Int. J. Gastron. Food Sci.*, *33*: 100779. https://doi.org/10.1016/j.ijgfs.2023.100779.

Talat, F. 2017. Effects of seed priming on morphological and phonological characteristics of the coriander (*Coriandrum sativum* L.). *Adv. Plants Agric. Res.*, *7*(6): 411–415. https://doi.org/10.15406/apar.2017.07.00275.

Tibebe, D., Belete, A., Kassa, Y., et al. 2024. Evaluation of total phenolic, flavonoid contents, and antioxidant activities of seed extracted from coriander (*Coriandrum sativum* L.) and black cumin (*Nigella sativa*) spices. *Food Anal. Methods*, *17*: 945–955. https://doi.org/10.1007/s12161-024-02622-9.

Tomar, R.S., Kulkarni, G., Parakhia, M., et al. 2014. Genetic diversity analysis in coriander (*Coriandrum sativum*) genotypes through morphological and molecular characterization. *Res. J. Biotechnol.*, *9*: 1–11.

Truc, T.T. 2024. Unleashing nature's epigenetic warriors: Bioactive compounds and the Nrf2/Keap1 system. *Chem. Eng. Trans.*, *113*: 349–354. https://doi.org/10.3303/CET24113059.

Ulusu, F. 2024. Effects of fertilizer type on phenolic compounds, essential oil content, and biological activities of *Coriandrum sativum* L. *Anatol. J. Bot.*, *8*: 106–113. https://doi.org/10.30616/ajb.1455761.

Uma, B., Prabhakar, K., Rajendran, S. and Sarayu, Y. 2009. Antimicrobial activity and phytochemical analysis of *Coriander sativum* against infectious diarrhoea. *Ethnobot. Leafl.*, *13*: 590– 594.

Verma, A., Dhanik, J., Agarwal, D., et al. 2018. Qualitative phytochemical and cluster analysis of genotypic extracts of coriander leaves and seeds from Tarai and Kumaon regions of Uttarakhand, Himalayan state of India. *Int. J. Chem. Stud.*, *6*: 1566–1571.

Wang, X-J., Luo, Q., Li, T., et al. 2022. Origin, evolution, breeding, and omics of Apiaceae: A family of vegetables and medicinal plants. *Hortic. Res.*, *9*: 1– 24. https://doi.org/10.1093/hr/uhac076.

Wangensteen, H., Samuelsen, A.B. and Malterud, K.E. 2004. Antioxidant activity in extracts from coriander. *Food Chem.*, *88*: 293–297. https://doi.org/10.1016/j.foodchem.2004.01.047.

Wei, J-N., Liu, Z-H., Zhao, Y-P., et al. 2019. Phytochemical and bioactive profile of *Coriandrum sativum* L. *Food Chem.*, *286*: 260–267. https://doi.org/10.1016/j.foodchem.2019.01.171.

Wei, S., Wei, L., Xie, B., et al. 2024. Characterization of volatile profile from different coriander (*Coriandrum sativum* L.) varieties via HS-SPME/GC–MS combined with E-nose analysed by chemometrics. *Food Chem.*, *457*: n140128. https://doi.org/10.1016/j.foodchem.2024.140128.

Xie, J., Xiong, S., Li, Y., et al. 2024. Phenolic acids from medicinal and edible homologous plants: A potential anti-inflammatory agent for inflammatory diseases. *Front Immunol.*, *15*: 1– 26. https://doi.org/10.3389/fimmu.2024.1345002.

Yoon, B-H., Truong, V-L. and Jeong, W-S. 2025. Phytosterols: Extraction methods, analytical techniques, and biological activity. *Molecules*, *30*: 2488. https://doi.org/10.3390/molecules30122488.

Zaeddi, E. and Sulayman, H.A. 2018. *Effects of Irrigation Dose, Plant Density, Application of Organic Acids, and Harvest Date on the Quality and Yield of Commercial Aromatic Plants*. Doctoral Thesis, Universidad Miguel Hernández de Elche, Elche, Spain.

Zaouali, W., Mahmoudi, H., Salah, I.B., et al. 2020. Copper-induced changes in growth, photosynthesis, antioxidative system activities, and lipid metabolism of cilantro (*Coriandrum sativum* L.). *Biologia* (*Bratisl*), *75*: 367–380. https://doi.org/10.2478/s11756-020-00419-9.

Zein, N., Elghani, E.A. and Talat, E. 2014. Effect of *Coriandrum sativum* on experimentally induced hepatotoxicity of carbon tetrachloride in rats. *Biochem. Lett.*, *10*: 135–155. https://doi.org/10.21608/blj.2014.63860.

Zhang, C-R., Dissanayake, A.A., Kevseroğlu, K. and Nair, M.G. 2015. Evaluation of coriander spice as a functional food by using *in vitro* bioassays. *Food Chem.*, *167*: 24–29. https://doi.org/10.1016/j.foodchem.2014.06.120.

Zhang, L., Su, Q-F., Wang, L-S., et al. 2023. Linalool: A ubiquitous floral volatile mediating the communication between plants and insects. *J. Syst. Evol.*, *61*: 538–549. https://doi.org/10.1111/jse.12930.

Zhang, W., Li, Z., Wang, L., et al. 2021. Effect of coriander plants on human emotions, brain electrophysiology, and salivary secretion. *Biology*, *10*: 1283. https://doi.org/10.3390/biology10121283.

Zyad, R. and Ahmed, Y. 2018. Detection of some active compounds in the leaves and stems of local coriander plant: *Coriandrum sativum* L. *Tikrit. J. Pure Sci.*, *23*: 6–15. https://doi.org/10.25130/tjps.v23i3.494.

7 Nutritive Value, Bioactive Compounds, and Biological Activities of Lettuce (*Lactuca sativa* L.)

Nqobile Monate Mkolo[1*], Karabo Maselepe Makoana[2] and Clarissa Marcelle Naidoo[3]

Introduction

The cultivated lettuce, scientifically known as *Lactuca sativa* L., belongs to the genus *Lactuca* under the family Asteraceae, which encompasses approximately 100 species distributed worldwide but predominantly in subtropical and temperate regions of North America, North Africa, and Eurasia (Syed et al., 2022). Cultivation of *L. sativa* reportedly occurred in the basin of the Mediterranean from *L. serriola,* a wild relative. In addition, throughout history, it has become remarkably widespread to cultivate leafy vegetables worldwide (Medina-Lozano et al., 2021). Presently, India, China, Spain, and the United States are amongst the major producers of *L. sativa*, underscoring its global agricultural and nutritional significance (Knez et al., 2024; FAOSTAT, 2022).

Nutritionally, lettuce is characterized by its high-water content (90–95%) and low caloric density, with fresh leaves consisting of a caloric density of about 14 kcal/100 g (Noumedem et al., 2017). Despite being often recognized as an insignificant contributor to diet, it offers vital micronutrients including provitamin A carotenoids, vitamin K, folate, and minerals, for instance, iron, potassium, and calcium (Shi et al., 2022). Particularly, darker cultivars, for instance, red and romaine leaf types, comprise greater amounts of minerals and vitamins in comparison with iceberg lettuce (Yang et al., 2022).

In addition to its intrinsic nutritional properties, lettuce comprises diverse bioactive compounds. Foremost are phenolic acids comprising chicoric acid, caffeic acid derivatives, and chlorogenic acid together with flavonoids, for instance, luteolin glycosides and quercetin (Coria-Cayupán et al., 2009; Shi et al., 2022; Yang et al., 2022). Pigmented types accumulate anthocyanins, whereas

[1-3] Department of Biology and Environmental Sciences, Sefako Makgatho Health Sciences University, Pretoria 0204, South Africa.
* Corresponding author: nqobile.mkolo@smu.ac.za

DOI: 10.1201/9781003687832-7

carotenoids, for instance, β-carotene and lutein, predominate in green cultivars. Additionally, lettuce synthesizes sesquiterpene lactones, particularly lactucin, lactucopicrin, and dihydrolactucin, which are linked to the bitterness characteristic and possession of pharmacological properties (Stojanović et al., 2023). The comparative abundance of these compounds differs according to several factors, including cultivars, seasonal factors, and cultivation conditions, underscoring the solid impact of genotype and environment connections on phytochemical composition (de Souza et al., 2022).

The wide spectrum of plant metabolites of lettuce reinforces numerous biological activities that have been proven both *in vitro* and *in vivo*. Lettuce metabolites display antioxidant activities due to the possession of phenolic, anthocyanin, and carotenoid compounds (Shi et al., 2022). Moreover, preclinical studies also validate analgesic and anti-inflammatory properties of seed and leaf extracts of Lettuce (Ismail and Mirza, 2015), in addition to anti-ulcer properties (Syed et al., 2022). Beyond, polyphenol-rich red lettuce consumption has been revealed to enhance glucose metabolism and reduce accumulation of hepatic lipid in obese mice, demonstrating its metabolic health claims (Cheng et al., 2014). Moreover, antimicrobial properties of lettuce have been confirmed to inhibit both fungal and bacterial pathogens, while the results can be influenced by the extraction technique (Edziri et al., 2011). In this chapter, the nutritive value, bioactive compounds, and biological activities of *L. sativa* will be discussed.

Taxonomic Background and Brief History of *Lactuca sativa* L.

Lactuca sativa L., widely known as lettuce, is a leafy vegetable of major economic significance that falls under the family Asteraceae, one of the biggest and greatest diverse families of flowering plants. The name of the genus *Lactuca* is obtained from the Latin word lac, defined as ‘ milk’, whereas the species epithet sativa is interpreted as ‘ cultivated’ in Latin, meaning its domesticated position (Schoch et al., 2020). Lettuce is believed to have originated from the wild species *Lactuca serriola*, which is considered its primary progenitor, even though evidence shows that other wild relatives might also have contributed towards its genetic makeup, signifying a polyphyletic origin (Medina-Lozano et al., 2021; Wei et al., 2021). Historical and archaeological records show that lettuce was domesticated first in the Caucasus region about 6,000 years ago due to its rich seed oil earlier to it spread to ancient Egypt, for both nutritional and medicinal purposes. In addition, it was also linked with religious symbolism (Lindqvist, 1960; Wei et al., 2021). Lettuce was introduced from Egypt to Greece and Rome, where Roman agriculturalists detailed various kinds and presented it in Latin as ‘ *lactuca*’. During the medieval era, lettuce became common in Europe and was introduced to the Americas by Christopher Columbus (Zohary, 1991; De Vries, 1997; Wei et al., 2021). Progressively, selective breeding produces the horticultural groups that are cultivated today, comprising leaf, romaine (cos), stem (celtuce), oilseed lettuce, butterhead, and crisphead (iceberg), individually adapting to different agro-climatic conditions (Mou, 2011; Wei et al., 2021).

Traditional and Modern Practices of *Lactuca sativa* L.

Lactuca sativa (lettuce) has been treasured since ancient times for its cultural, nutritional, and medicinal worth (Figure 1). Traditionally, its cultivation stems from ancient times of Egypt, where it was cultivated as a leafy vegetable, a source of oil-seed, and was symbolically allied with religious and fertility rituals (Wei et al., 2021; Lindqvist, 1960). Roman and Greek societies additionally extended their medicinal and dietary usages; even Roman physicians recommended it for its cooling, digestive, and sedative properties (De Vries, 1997; Zohary, 1991; Wei et al., 2021). In terms of traditional medicine, lettuce parts were used for the treatment of ailments, for instance, digestive disturbances, insomnia, and inflammation, mainly due to its lactucarium (milky latex), which was trusted to possess analgesic and sedative properties (Schoch et al., 2020). In several cultures, comprising Asia and the Middle East, lettuce was likewise appreciated for its properties for reduction of fever, promotion of diuresis, and alleviation of nervousness, making it a therapeutic and food plant (Lebeda et al., 2019).

In the modern era, *L. sativa* has progressed to be an important horticultural crop worldwide, cultivated mainly for fresh consumption (Figure 1). It is represented as the most consumed vegetable salad, valued for its mild flavor, refreshing texture, and tender leaves. Moreover, in this modern era, it is also used for its nutritional value, which contributes to its antioxidant and health-promoting properties (Cao et al., 2024). Furthermore, lettuce plays a pivotal role in the food-health industry, where it can be used as the basis of juices and nutrient-rich foods intended for the management of weight, cardiovascular health, and the inhibition of chronic diseases (Shi et al., 2022; Hung et al., 2024). Besides human nutrition richness, lettuce is also used in contemporary research for studies of crop domestication, plant-microbe interactions, and genetic diversity, owing to its in-depth sequenced genome and adaptation to hydroponic systems (Lebeda et al., 2019). Therefore, the usages of *L. sativa* validate a range from its antique roles in medicine and diet to its contemporary-day applications in scientific research, nutrition, and pharmacology, stressing its timeless economic and cultural worth.

Nutritional Composition of *Lactuca sativa* L.

Lactuca sativa emerges as a novel source of nutraceuticals owing to its medicinal and nutritional value of dietary fiber, and amongst its different forms, leafy type and romaine lettuces offer significantly greater levels of minerals (for instance, iron and folate), vitamins, and bioactive compounds (Dyląg et al., 2023). It is rich in provitamin A, C, and K, and folate lettuce contributes appreciably to micronutrient intake is also present, adding to its dietary value (Kim et al., 2016). Nutritional composition of *L. sativa* per 100 g fresh weight is depicted in Table 1.

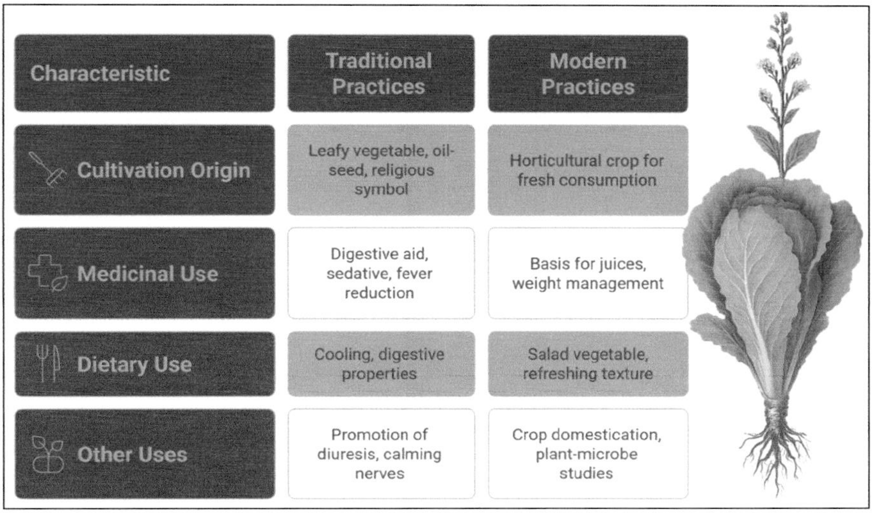

Fig. 1 Traditional and modern practices of *Lactuca sativa* L. (Image created by NM Mkolo).

Table 1 Nutritional composition of *Lactuca sativa* per 100 g fresh weight.

Nutrient	*Value (per 100 g)*	*References*
Proximate composition		
Ash (g)	0.7–1.0	Shi et al. (2022)
Carbohydrates (g)	2.0–3.0	Mou (2009)
Dietary fiber (g)	1.0–1.5	Kim et al. (2016)
Energy (kcal)	15–20	Shi et al. (2022)
Fat (g)	0.2–0.3	Shi et al. (2022)
Moisture (g)	94.0–95.0	Shi et al. (2022)
Protein (g)	1.2–1.5	Mou (2009)
Minerals		
Calcium (mg)	35–40	Mou (2009)
Iron (mg)	1.0–1.2	Shi et al. (2022)
Magnesium (mg)	10–12	Mou (2009)
Phosphorus (mg)	30–35	Kim et al. (2016)
Potassium (mg)	220–250	Kim et al. (2016)
Sodium (mg)	30–33	Kim et al. (2016)
Vitamins		
B-complex vitamins	Thiamine 0.07 mg; Riboflavin 0.08 mg; Niacin 0.3–0.4 mg	Shi et al. (2022)

Contd...

Table 1 *Contd...*

Nutrient	*Value (per 100 g)*	*References*
Folate (µg)	35–40	Mou (2009)
Vitamin A (β-carotene, µg)	4400–4500	Shi et al. (2022)
Vitamin C (mg)	8–10	Kim et al. (2016)
Vitamin E (mg)	0.2–0.3	Shi et al. (2022)
Vitamin K (µg)	120–130	Shi et al. (2022)
Fatty acids		
Linoleic acid (mg)	15–20	Sularz et al. (2020)
Linolenic acid (mg)	40–50	Sularz et al. (2020)
Palmitic acid (mg)	10–12	Sularz et al. (2020)
Amino acids		
Valine (mg/g protein)	40	Mou (2009)
Phenylalanine (mg/g protein)	41	Mou (2009)
Lysine (mg/g protein)	54	Mou (2009)
Leucine (mg/g protein)	66	Mou (2009)

Bioactive Compounds

Polyphenols

Polyphenols are a class of compounds that includes phenolic acids, flavonoids, lignans, stilbenes, tannins, and coumarins. They are a diverse group of naturally occurring bioactive compounds characterized by multiple phenolic rings. In lettuce (*Lactuca sativa*), these compounds serve as secondary metabolites that contribute to plant defense mechanisms while providing significant health benefits when consumed (Shi et al., 2022). Among the various phenolic groups in *Lactuca sativa*, phenolic acids form the most dominating class, particularly hydroxycinnamic acids such as Chicoric acid (dicaffeoyltartaric acid), which represents the most abundant phenolic compound reaching 3.0% dry weight in aerial parts (Michalska et al., 2024) and up to 472.77 mg/kg dried extract (Bouymajane et al., 2024). Chlorogenic acid is another phenolic acid reported at 196.79 mg/kg (Bouymajane et al., 2024) and increases by as much as 320% under saline stress (Ferrantelli et al., 2023). Caffeic acid derivatives such as Caftaric acid were also noteworthy; reported to occur at about 0.4% dry weight (Michalska et al., 2024) and showing increases of 467% under stress conditions (Ferrantelli et al., 2023). Additional phenolic acids have been identified, including 5-caffeoylquinic acid at 171.74 mg/kg and 3,5-di-O-caffeoylquinic acid at 196.79 mg/kg concentrations (Bouymajane et al., 2024).

Latest metabolomic studies have revealed diverse flavonoid profiles in *L. sativa*. A comprehensive analytic study by Šimko (2024) has identified 1905 compounds in *L. sativa*, with flavonoids making up one of the largest groups.

These occur primarily as quercetin derivatives, kaempferol, luteolin, anthocyanins, and apigenin derivatives. Kaempferol, in particular, has been highlighted for its nutritional significance, with certain mutant lines (KfoA and KfoB) selectively bred for high flavonoid content showing elevated kaempferol accumulation under field conditions, alongside quercetin (Cho et al., 2023). Comparative analyses of 22 lettuce cultivars further revealed kaempferol derivatives as the dominant flavonoids, with content varying by cultivar and maturity stage (Assefa et al., 2019). Quantitative assessments using UPLC-PDA-Q-TOF-HDMS and UHPLC have confirmed its prevalence and structural diversity (Assefa et al., 2019; Ribas-Agustí et al., 2011). Other major flavonoids include quercetin 3-O-(6″-O-malonyl) glucoside (45.4–31,121.0 μg/g DW in red-pigmented lettuce (Assefa 2021); luteolin 7-O-glucoside (0.4% dry weight in aerial parts) (Michalska et al., 2024); luteolin 7-O-glucuronide (224.30 mg/kg dried extract) (Bouymajane et al., 2024); apigenin 7-O-glucoside (Michalska et al., 2024); and apigenin 7-O-glucuronide (with reported antibacterial potential) (Bouymajane et al., 2024).

Among the anthocyanins identified in *L. sativa*, cyanidin 3-O-(6″-O-malonyl) glucoside, which is responsible for the pigmentation in red and purple lettuce, has been reported in concentrations ranging from 4.7 to 5,013.6 μg/g DW (Assefa, 2021). In addition to this, it is important to note that flavonoids such as anthocyanins have been demonstrated to have a number of biological effects, such as improving vascular health, reducing bacterial growth, and modifying inflammatory pathways. Lettuce also contains lignans, which are secondary plant metabolites that are a product of two or more phenylpropanoid units that oxidatively couple to generate a wide variety of polyphenolic structures (De Gregorio et al., 2023). Research has identified several lignan compounds in lettuce species, including Syringaresinol and its 4-0-β-glucoside, which has been isolated from *L. sativa* var angustana cv. Gruner stem (Malarz et al., 2020). Dihydrodiconiferyl alcohol was found to have phytotoxic effects on lettuce metabolism, reducing chlorophyll synthesis and affecting carbon and nitrogen metabolism. Lastly, Lariciresinol, which is a phytotoxic lignan that affects lettuce seedling development (Carillo et al., 2010). These compounds also show strong correlations with radical scavenging activities in metabolomic studies (De Gregorio et al., 2023).

Lactuca sativa is among the least examined polyphenolic classes historically and recently. Early studies, such as Chon (2005), only referenced tannins incidentally without structural identification or quantification. Recent studies, including Utami (2023) and Flores et al. (2023), have identified hydrolyzable tannins in specific cultivars, but lacked thorough profiling and reliable concentration data. The most notable contemporary contribution is from Šimko (2024), who reported a remarkably high broad-sense heritability ($H^2 = 0.909$) for tannins in young lettuce plants, indicating strong genetic regulation of their biosynthesis. Nonetheless, no research has verified the presence of condensed tannins, hydrolyzable tannins, or both in lettuce, and no quantitative measurements in standard units have been documented. Similarly, coumarins are also among the least explored polyphenols. The only study providing indirect evidence comes from Zha et al. (2024), who conducted multi-omics analyses of lettuce under UV-A radiation. This research found that UV-A promoted the synthesis of most phenylpropanoids originating from the shikimate pathway, which includes coumarin biosynthesis precursors. However, the study did not specifically identify or quantify individual coumarin compounds (Zha et al., 2024).

Terpenoids

Terpenoids are a class of compounds that includes hemiterpenes, monoterpenes, sesquiterpenes, diterpenes, sesterterpenes, triterpenes, and tetraterpenes. Terpenoids are a large and structurally diverse class of plant secondary metabolites, recognized for their wide range of biological activities and therapeutic potential across medicine, agriculture, and industry (Turatbekova et al., 2024). Research identifies them as secondary metabolites found in plants, distinct from primary metabolites involved in basic cellular functions (Sadino et al., 2024). The most extensively documented terpenoids in *L. sativa* are sesquiterpene lactones, specifically contributing to the characteristic bitter taste of the plant, and representing the predominant class in this species. Key compounds include lactucin, a naturally occurring active sesquiterpene lactone abundantly found in *L. sativa* (Jang et al., 2021), which serves as one of the major contributors to lettuce bitterness and is particularly enriched in specialised cultivars such as *Heukharang* lettuce (Ko et al., 2024). Lactucopicrin, another major sesquiterpene lactone that works synergistically with lactucin and is the second major bitter sesquiterpene lactone found in lettuce, lactucopicrin was reported at concentrations of 199.2 μg/g in green romaine lettuce extracts (Kim et al., 2019); and hanphyllin, a compound isolated from lettuce latex with demonstrated biological activity against insect herbivores (Gu et al., 2024). Additional sesquiterpene lactones, such as 8-α-angeloyloxyleucodin, matricarin, 15-deoxylactucin, and deacetylmatricarin 8-β-glucopyranoside, have also been identified, with some reported as novel to *Lactuca* species (Michalska et al., 2024).

In addition to the sesquiterpene lactones, *L. sativa* contains other terpenoid classes, including monoterpenes and triterpenoids present in the wax layer of leaves that contribute to the plant's surface properties and potentially serve as protective agents against environmental stress (Zhao et al., 2022). Tetraterpenes are another terpenoid class that is also called Carotenoids, specifically β-carotene, which represents the primary tetraterpene compounds found in *L. sativa,* contributing significantly to its nutritional and antioxidant properties (Brychkova et al., 2023). *Lactuca* species also produces terpenoid derivatives known as apocarotenoids, such as loliolide and roseoside, which are involved in stress signaling, defense against herbivores and pathogens, and contribute to the flavor and aroma profile of lettuce (Michalska et al., 2024). Overall, *L. sativa* represents a notable source of structurally diverse terpenoid compounds, with sesquiterpene lactones dominating its chemical profile. According to a study by Šimko (2024), spatio-temporal analysis of *L. sativa* revealed that terpenoids exhibit distinct distribution patterns within the plant, with concentrations declining from the base toward the tip of leaves and varying significantly with plant maturity and environmental conditions. Additionally, the study also revealed that Terpenoid content varies with the developmental stages of the plant and the environmental conditions, which suggests dynamic regulation of biosynthesis.

Alkaloids

Alkaloids are extensive and intricate natural compounds that are distinguished from others by their nitrogenous base (Mohammed et al., 2024). They can be grouped into several alkaloid categories according to their chemical structure, which includes organic amines, Imidazole, Purine, Pyridine, Isoquinoline, Diterpenoid, Pyridine, Indole, and Steroid (Yang et al., 2024). Despite their pharmacological importance, the presence and profile of alkaloids in *L. sativa* are not well characterized in recent literature. This gap persists even though extensive analytical metabolomic studies on *Lactica sativa* L. by Šimko (2024) and Erdoğan et al. (2024) have been conducted, in which alkaloids were neither systematically investigated nor reported as significant components of its phytochemistry. Only one available study provided direct evidence for alkaloids in the genus *Lactuca*, in which Aziz (2024) qualitatively detected alkaloids in a hot aqueous extract of *Lactuca serriola* leaves. However, this study did not specify alkaloid subclasses or provide quantitative data. The lack of systematic research in *L. sativa* suggests that, if alkaloids are present, they are likely to occur in low concentrations and may not constitute a significant component of the lettuce phytochemical profile. Due to the shortage of information on alkaloids in *L. sativa*, further phytochemical studies would be necessary. Such studies may develop specific analytical methods for lettuce alkaloid detection, therefore providing a better understanding of the complete nutritional and phytochemical profile of this widely consumed plant.

Sulfur-containing Compounds

Numerous studies on *L. sativa* have focused on terpenoid and phenolic chemicals. In contrast, a lot less is known about its metabolites that contain sulfur. Among the compounds reported, glutathione (γ-glutamyl-cysteinyl-glycine) is often regarded as the most prominent sulphur antioxidant in lettuce for playing a central role in cellular redox homeostasis, detoxification, and plant responses to abiotic and biotic stress, though there is very limited quantitative data on its concentrations in lettuce tissues (Hill et al., 2025; Abdalla et al., 2022). In line with this, a study by Abdalla et al. (2022) demonstrated that stress conditions and sulfur nutrition can alter both glutathione levels and glutathione S-transferase activity, highlighting the compound's dynamic role in lettuce metabolism. In addition to glutathione, sulfoquinovosyl diacylglycerols (SQDG) are the main thylakoid sulfolipids in lettuce by playing the key roles in membrane structure and photosynthetic efficiency, with oxidized SQDG serving as a biochemical marker of sulfur-related stress (Körber et al., 2023). Using advanced LC-ESI-MS/MS analysis,

Körber et al. (2023) demonstrated that sulfur fertilization markedly increases total SQDG, monogalactosyldiacylglycerol (MGDG), and digalactosyldiacylglycerol (DGDG) contents, while sulfur deprivation enhances oxidized SQDG and elevates antioxidant responses. Likewise, increased amounts of Polyunsaturated SQDG and galactolipid species were linked to sulfur sufficiency, suggesting enhanced membrane fluidity and stress tolerance. Notably, red lettuce exhibited a stronger antioxidant capacity and more pronounced lipid remodeling than green cultivars, underscoring cultivar-dependent responses to sulfur nutrition.

In a study by Zeng et al. (2023), the researchers reported that lettuce either lacks or contains very little of the traditional sulfur compounds such as glucosinolates, thionins, and alliin that are found in Brassicaceae and *Allium* species. Instead, despite their characterization being nonetheless lacking, additional metabolites that include sulfur, such as modified amino acid derivatives and S-methyl cysteine sulfoxide derivatives, have been under-reported (Hill et al., 2025; Abdalla et al., 2022). Altogether, these current findings indicate that glutathione and sulfolipids (notably SQDG) constitute the principal sulfur-containing metabolites characterized in lettuce, with additional sulfur compounds still underexplored. Therefore, future studies should concentrate on thoroughly characterizing these metabolites across cultivars, growing environments, and stressors, as well as evaluating their bioavailability and their use in human nutrition.

Biological Activities

Antioxidant Activities

Lactuca sativa is a leafy vegetable that has gained widespread popularity and is widely consumed across the globe. It is particularly acknowledged for its abundant and diverse composition of bioactive compounds such as polyphenols, terpenoids, and other compounds that collectively play a significant role in enhancing antioxidant capacity (Berríos et al., 2024; Vella et al., 2020). Recently, research has witnessed substantial advancements in both metabolomic profiling methodologies and innovative cultivation techniques, which have substantially contributed to a more comprehensive understanding of the phytochemical diversity and antioxidant potential that is inherent in various cultivars of lettuce (Yang et al., 2018; Choi, 2023). Notably, various studies consistently indicate significant antioxidant activity across their various cultivars, although the extent of variability is affected by the specific cultivar type, the cultivation methodology employed, and the conditions under which processing occurs (Vella et al., 2020; Yang et al., 2018; Utami, 2023). According to study comparisons, red-leaf Lettuce types have higher phenolic compound accumulation and frequently display more robust antioxidant activity than their green-leaf counterparts. For instance, an earlier study by Liu et al. (2007) reported that red cultivars exhibited a quenching capability of 77.9% for DPPH radicals, in contrast to the 65.5% observed in green cultivars, with red lettuce possessing a phenolic content exceeding double that of its green counterpart (52.4 vs. 21.7 mg GAE/g dry mass). At the level of specific cultivars, Galactic (red) demonstrated a phenolic content of 85.7 mg GAE/g DM and an impressive DPPH quenching percentage of 81.8%, while Green Vision (green) displayed a lower phenolic content of 33.7 mg GAE/g DM alongside a quenching percentage of 84.2%. Similarly, a more recent study by Jiangseubchatveera et al. (2023) discovered that the ethyl acetate fraction derived from red coral lettuce exhibited the most pronounced bioactivity, with DPPH IC_{50} = 0.277 ± 0.006 mg/mL and ABTS IC_{50} = 0.300 ± 0.002 mg/mL, a finding that was correlated with its elevated phenolic content (9.747 ± 0.021 mg GAE/g extract). In another study, Negrão et al. (2021) found major variations in antioxidant activity among various farming systems by employing the DPPH radical capture method. Lettuce cultivated within the organic farming system manifested the highest antioxidant activity, quantified at 4075.28 $\mu molTEACg^{-1}$; the hydroponic farming system demonstrated an antioxidant activity of 3074.30 $\mu molTEACg^{-1}$; whereas the conventional farming system exhibited the least antioxidant activity among the three, yielding a measurement of 2136.88 $\mu molTEACg^{-1}$. Moreover, the processing that occurs after harvesting is of significant importance, as research indicates that limited processing and appropriate storage conditions can enhance the bioaccessibility of polyphenols without diminishing their antioxidant efficacy (Lafarga et al., 2020). At the molecular level, extracts derived from *L. sativa* have been demonstrated to initiate the activation of the Nrf2 signaling pathway while concurrently inhibiting NF-κB signaling, leading to the upregulation of antioxidant enzymes, including catalase (CAT), glutathione peroxidase (GPx), and superoxide dismutase (SOD); this biochemical interplay facilitates a reduction in oxidative stress and inflammation (Adesso et al., 2016). Furthermore, the incorporation of glycine nitrogen and selenium biofortification has been documented to augment the biosynthesis of flavonoids and phenolic compounds, subsequently enhancing the antioxidant capacity in both *in vitro* and *in vivo* experimental paradigms (Yang et al., 2017; Buendía-García et al., 2021). Functionally, these antioxidant activities play an important role in the prevention of illnesses. For instance, extracts from lettuce have exhibited hypolipidemic and hepatoprotective properties by diminishing cholesterol, triglycerides, and LDL levels while concurrently elevating HDL levels in animal models (Ayo et al., 2024). Additionally, the use of nitrate-rich lettuce in the human diet has illustrated vascular advantages, including acute reductions in systolic blood pressure alongside enhanced metabolic markers (Qadir et al., 2022). Moreover, a study by Naseem and Ismail (2022) has highlighted an association between lettuce-derived antioxidants with neuroprotective and antidiabetic effects, which proves their therapeutic relevance. These findings establish *L. sativa* as a functional food with robust antioxidant activity. Its efficacy depends not only on intrinsic phytochemical composition but also on agronomic, genetic, and post-harvest factors, which together determine its potential role in mitigating oxidative stress-related disorders such as cardiovascular disease, metabolic syndrome, and neurodegeneration.

Antimicrobial Activities

The antimicrobial properties of *L. sativa* are increasingly acknowledged, showing efficacy against various microbial strains. These qualities come from its wide-ranging phytochemical structure, featuring phenolic compounds, flavonoids, and tannins (Nisa and Tajammul, 2022). Methanolic extracts from lettuce leaves exhibit inhibitory effects on both Gram-positive and Gram-negative bacteria, with inhibition zones of 8.2– 13.8 mm and minimum inhibitory concentrations of 6.25–25 mg/ml (Foudah et al., 2019). Furthermore, extracts from lettuce grown under controlled conditions also displayed significant antibacterial activity against *Bacillus subtilis* and other strains (Zdravković et al., 2012). Notably, lettuce extracts have also exhibited activity against antibiotic-resistant pathogens. A study by Nisa and Tajammul (2022) reported that *L. sativa* displayed substantial inhibitory effects against clarithromycin-resistant Gram-negative bacteria, suggesting its potential as an alternative treatment option in the fight against antimicrobial resistance. In addition to its antibacterial activity, lettuce extracts have also shown antifungal potential. For instance, Zdravković et al. (2012) reported strong inhibitory effects against *Candida albicans* and *Aspergillus niger*, with significant inhibition of *A. niger* at 19.53 μg/ml, pointing to its promise as a natural antifungal agent. However, it is important to understand that the antimicrobial efficacy of *L. sativa* is dependent on environmental conditions and extraction methods. For instance, solvent polarity significantly affects the yield and potency of bioactive compounds, thereby influencing antimicrobial efficacy (Zekkori et al., 2021). Consequently, further research is required to fully elucidate the mechanisms of action and to evaluate the feasibility of developing lettuce-based antimicrobial formulations for pharmaceutical and industrial use.

Anticancer Activities

The anticancer activities of *L. sativa* are primarily related to its variety of phytochemical constituents, including sesquiterpene lactones, polyphenols, flavonoids, and phenolic acids, which have cytotoxic effects by means of mechanisms such as the induction of apoptosis, suppression of cellular proliferation, and modulation of oxidative stress. Notably, Jang et al. (2021) discovered that lactucin, a common sesquiterpene lactone present in romaine lettuce, causes renal carcinoma (Caki-1) cells to undergo apoptosis by generating reactive oxygen species (ROS) and inhibiting anti-apoptotic proteins such as CFLARL and BCL-2. In addition to that, the study further revealed that the treatment with 50 μM of lactucin via a caspase-dependent mechanism has caused up to 72% apoptosis. *In silico* studies further support this activity, with lactucine derivatives demonstrating effective docking interactions with apoptosis-inducing proteins such as CDK-4, suggesting their promise as potential drug leads (Arya et al., 2023). Cultivars that are high in polyphenols, particularly those with red coloring, have significant cytotoxic effects. According to Qin et al. (2018), anthocyanins, flavones, and phenolic acids found in red lettuce extracts successfully stopped the growth of lung and colon cancer cells; the IC_{50} values varied depending on the cell line under investigation, ranging from 45 to 120 μg/mL. A different study by Foudah et al. (2019), the methanolic leaf extracts derived from *L. sativa,* which are abundant in flavonoids, have been shown to impede cellular proliferation and initiate apoptosis across various cancer cell lines. More recently, Erdoğan et al. (2024) confirmed that lettuce extracts exert dose-dependent cytotoxicity against breast (MCF-7) and lung (A549) cancer cells, achieving growth inhibition of over 65% at 200 μg/mL.

For antioxidant activity, a study by Harsha et al. (2013) demonstrated that *L. sativa* leaf extracts exhibit strong radical-scavenging activity, protecting biomolecules such as DNA and proteins from oxidative damage. Supporting this, Ulfah et al. (2019) identified alkaloids, phenols, and flavonoids in both romaine and curly lettuce varieties, with extracts displaying significant antioxidant potential in DPPH assays (IC_{50} values within 20–40 μg/mL). Importantly, biofortification strategies such as iodine enrichment enhanced anticancer activity by promoting ROS-mediated apoptosis in gastrointestinal cancer cells (Harsha et al., 2013). Mechanistic studies provide further evidence of cell-cycle regulation. Gridling et al. (2010) showed that water extracts of *L. sativa* activated checkpoint kinase 2 (Chk2), induced the tumor suppressor p21, and downregulated cyclin D1 in leukemia and breast cancer cells. In addition, some earlier biotechnological approaches have positioned lettuce as a platform for producing anticancer compounds. Such approaches include a study by Kwon et al. (2022), who demonstrated that engineering of germacrene A synthase in *L. sativa* enabled high-yield production of germacrene A, a precursor of β-elemene, and a clinically used anticancer agent. These findings explain and outline the anticancer properties of *L. sativa* via multiple mechanisms, encompassing the induction of apoptosis, modulation of oxidative stress, and regulation of the cell cycle. Nevertheless, the concentrations requisite for substantial *in vitro* effects frequently exceed those attainable through dietary sources, underscoring the necessity for improved extraction, biofortification, and formulation techniques, alongside comprehensive preclinical and clinical validation.

Antidiabetic Activities

Numerous studies have investigated the hypoglycemic effects, insulin sensitization, and antioxidant processes of *L. sativa* antidiabetic biological activities, underscoring their diverse potential in the treatment of diabetes. These multifaceted activities emerge as a consequence of an intricate and varied phytochemical profile, which is characterized by the presence of a wide array of bioactive compounds, including but not limited to flavonoids, carotenoids, polysaccharides, and phenolic acids, all of which operate through complex and synergistic mechanisms that enhance their overall efficacy. In terms of the hypoglycemic effects, the underlying mechanism is fundamentally linked to the inhibition of specific carbohydrate-digesting enzymes, such as α-amylase and α-glucosidase, which play significant roles in carbohydrate metabolism and glucose absorption in the human body (Naseem and Ismail,

2022). For instance, Gopal et al. (2017) reported that polysaccharide extracts of *L. sativa* significantly reduced postprandial blood glucose in diabetic rat models by suppressing α-glucosidase activity, thereby moderating glucose absorption. In comparison, a more recent study conducted by Huang et al. (2024) highlighted the significant inhibitory effect of lactucaxanthin on both α-amylase and α-glucosidase *in vitro*, which is consistent with a decrease in blood glucose levels in experimental animals, thereby confirming its potential as a natural hypoglycemic agent.

In addition to lowering blood sugar, the study also revealed that *L. sativa* can enhance insulin sensitivity through modifications in essential signaling pathways. In cellular models with insulin resistance, Amadi et al. (2021) demonstrated that avicularin from lettuce greatly increased insulin responsiveness by upregulating the PI3K/Akt signaling pathway, which in turn led to better glucose absorption. In agreement, Tian et al. (2025) revealed that polysaccharides isolated from *L. sativa* improved β-cell responsiveness and glucose regulation in diabetic mice, effects attributed to multitarget interactions involving IRS-1 and PPAR-γ. These results imply that the phytochemicals in lettuce both correct damaged insulin signaling and reduce blood glucose levels. In addition, Naseem and Ismail (2022) have also documented antioxidant-related effects in a study, which found that hydroponically cultivated lettuce demonstrated greater radical scavenging ability than soil-grown varieties, linked to increased antioxidant enzyme activity in diabetic rats. In a related study, Marwaha (2022) indicated that carotenoids like lactucaxanthin offered protection against lipid peroxidation and mitigated retinal oxidative stress, benefiting diabetic retinopathy models. These antioxidant properties highlight lettuce's possible contribution to glycemic control and preventing the development of tissue damage associated with diabetes (Marwaha, 2022; Shivam et al., 2023). All of the above studies demonstrate *L. sativa's* therapeutic potential in the treatment of diabetes. Nonetheless, further clinical investigations are necessary to substantiate these results in humans and enhance the application of lettuce-derived compounds in antidiabetic treatment.

Anti-inflammatory Activities

Numerous studies have explored the anti-inflammatory properties of *L. sativa,* with evidence linking its efficacy to the plant's abundant phenolic and flavonoid content. The strength of these effects depends on factors such as genetic modification, cultivation practices, extraction methods, fermentation, and formulation strategies. Genetic modification has been explored as one means of enhancing this activity. For instance, Ismail et al. (2017) demonstrated that transformation of lettuce with the *rol C* gene resulted in a 78.5% increase in anti-inflammatory activity compared to wild-type plants, with the transformed extracts significantly reducing carrageenan-induced paw edema in rats at doses of 200 mg/kg. Expanding on this, the same group later confirmed that lettuce lines engineered with different *rol* genes showed elevated phenolic and flavonoid levels, which translated into improved analgesic and anti-inflammatory outcomes *in vivo*, comparable to indomethacin at 10 mg/kg (Ismail et al., 2017). Extraction methods can also influence the bioactivity profile. Zekkori et al. (2018) reported that hydroalcoholic extracts with a solvent polarity index of 5.8 yielded the highest concentrations of phenolics and flavonoids, producing anti-inflammatory effects that rivaled dexamethasone in reducing paw edema. Similarly, Adesso et al. (2016b) found that lettuce cultivated under mineral fertilization exhibited elevated quercetin content, which strongly correlated with reduced expression of pro-inflammatory markers such as IL-6 and TNF-α *in vitro*, highlighting the role of cultivation practices in modulating therapeutic efficacy. Complementary to these findings, fermentation strategies have been explored as an additional means of improving lettuce's anti-inflammatory potential. Park et al. (2023) showed that fermented lettuce extracts enriched in nitric oxide metabolites significantly attenuated pro-inflammatory cytokine production and suppressed the TGF-β/Smad signaling pathway in both murine models and human fibroblast-like synoviocytes. Along similar lines, Seo and Jeong (2020) demonstrated that green lettuce extracts modulated inflammatory responses through activation of TLR4–MAPK/NF-κB signaling, suggesting a role in immune regulation and the attenuation of chronic inflammation. Beyond processing strategies, comparative studies of cultivars and formulation innovations have further expanded understanding of the plant's therapeutic promise. Zekkori et al. (2018) observed that red lettuce varieties, such as *L. sativa* var. *crispa*, exhibited broader anti-inflammatory effects than green cultivars, which was attributed to higher phytochemical accumulation. In parallel, Gyawali et al. (2020) developed microsphere-loaded gels containing lettuce extracts, which produced dose-dependent inhibition of carrageenan-induced inflammation in animal models, offering a potential route for controlled-release therapeutic applications. When combined, these studies show that lettuce has anti-inflammatory properties through various pathways. However, research is still limited to *in vitro* and *in vivo* tests. Thorough clinical studies are necessary to create standardized formulations for therapeutic use and to validate effectiveness in people.

Anti-Ulcer/Wound Healing Activities

Various research studies (both *in vivo* and in *vitro*) have identified anti-ulcer and wound-healing activities within varying extracts of *L. sativa.* For example, methanolic extracts of lettuce, administered to rats at 200– 400 mg/kg, reduced ethanol-induced gastric ulcers significantly, with as much as 72% inhibition of ulcer formation; most likely attributed to flavonoids and phenolic acids, in combination, which enhance mucosal defense and inhibit oxidative stress (Mandal et al., 2020). In another similar study by Gyawali et al. (2020), aqueous leaf extracts of lettuce stimulated wound-healing in excision models with fast epithelialization and low amounts of scar tissue in as little as 10– 14 days. The process was tied to reminiscence of the stomach mucosa and stimulation of tissue regeneration and is associated with resultant stimulation of collagen production and antioxidant defense (Naseem and Ismail, 2022).

Hepatoprotective Activities

The extracts of *L. sativa* have been identified in a number of studies to contain hepatoprotective benefits through the modulation of oxidative stress, improvement of biochemical markers, and protection of hepatocytes. In an *in vivo* study by Ghanadi et al. (2019), ethanolic extracts *of L. sativa* were administered orally at doses of 250–500 mg/kg to rats with carbon tetrachloride (CCl_4)-induced hepatotoxicity. As a result, the treatment brought the levels of serum alkaline phosphatase (ALP), aspartate aminotransferase (AST), and alanine aminotransferase (ALT) back to baseline. In addition to that, other studies have also revealed that the extracts alternatively enhanced endogenous antioxidant enzymes, including superoxide dismutase (SOD) and catalase, while reducing lipid peroxidation (Cui et al., 2017). These results reveal that in instances of chemical toxicity, lettuce phytochemicals protect hepatocyte membranes and reduce oxidative damage. In addition to chemically induced models, other studies were conducted to look at hepatoprotection in the setting of viral hepatitis. For instance, a study by Cui et al. (2017) revealed that in Hepatitis B virus (HBV)-expressing HepG2.2.15 and HepAD38 cell lines, lettuce extracts inhibited viral antigen secretion, suppressing Hepatitis B surface antigen (HBsAg) release by up to 82.5% and reducing HBV DNA and RNA replication by 35% and 43.9%, respectively, without cytotoxicity at concentrations up to 15 μg/μl. Furthermore, the metabolomic analysis identified luteolin-7-O-glucoside as a principal active compound, achieving similar inhibition of HBsAg secretion (77.4%) and protecting hepatocytes by reducing ROS accumulation and restoring mitochondrial membrane potential. Most importantly, the study further reveals that both the whole lettuce extracts and luteolin-7-O-glucoside demonstrated enhanced antiviral efficacy when combined with interferon-α2b (IFNα-2b) or lamivudine (3TC), highlighting their potential as adjunctive therapies in chronic HBV infection. All these studies point to *L. sativa* as a potentially effective hepatoprotective solution that can prevent liver injury. However, more preclinical and clinical research is necessary to confirm the plant's therapeutic value in the treatment of HBV, even though *in vivo* data clearly support its protective effect against chemical hepatotoxins.

Neuroprotective Activities

Neuroprotective studies are relatively scarce and often lack mechanistic depth regarding signaling pathways and molecular targets. The diversity of *Lactuca* species tested is limited, and literature on long-term safety and efficacy is almost nonexistent; however, the neuroprotective effects of *L. sativa* that have been reported are in relation to both oxidative and excitotoxic stress. Hydroalcoholic lettuce extracts, at doses of 100–200 mg/kg in mice, improved memory and learning in scopolamine-induced amnesia models, with performance comparable to piracetam (Jang et al., 2021). In a study aimed at identifying wound healing and gastroprotective activities of *L. sativa* by Gyawali et al. (2020), the hydroalcoholic extracts mechanistically enhanced acetylcholinesterase inhibition, reduced malondialdehyde accumulation, and increased antioxidant enzyme activities in the brain. In line with this, a more recent study by Malik et al. (2022) has shown that the methanolic lettuce extracts attenuated neuroinflammation by suppressing NF-κB and iNOS signaling, while promoting neuronal survival. Therefore, even though the studies are limited, these findings still indicate that lettuce phytochemicals could offer protective benefits against neurodegenerative conditions such as Alzheimer's disease, but further research is still needed to validate the long-term safety and efficacy in humans.

Larvicidal/ Insecticidal Activities

In addition to the therapeutic and nutritional potential of *L. sativa*, it also exhibits bioactivity against insect pests. In a recent study by Maity et al. (2024), petroleum ether and methanolic lettuce extracts have demonstrated significant larvicidal activity against *Culex quinquefasciatus* larvae, with LC_{50} values ranging from 120 to 180 ppm after 24 hours of exposure. Moreover, aqueous lettuce seed extracts reduced larval survival and disrupted pupation in *Aedes aegypti*. These effects are related to sesquiterpene lactones and phenolic compounds that interfere with larval development and metabolism, positioning lettuce as a potential eco-friendly biopesticide source; however, research is still limited, and mechanistic insights remain unknown, prompting future research.

Other Reported Activities

Lactuca sativa has been reported to possess additional pharmacological and therapeutic benefits beyond its well-established antioxidant, antimicrobial, anticancer, antidiabetic, anti-inflammatory, anti-ulcer, hepatoprotective, neuroprotective, and larvicidal/insecticidal activities. Gyawali et al. (2020) revealed that supplementation with lettuce has been associated with hypolipidemic effects, including reductions in serum cholesterol and triglyceride levels in animal models, while the lettuce extracts have also demonstrated significant anxiolytic activity, with both crude preparations and sustained-release microsphere gels reducing anxiety-like behavior *in vivo*. In addition, a comparative phytochemical study by Mughrbi et al. (2020) revealed notable anticoagulant activity, with lettuce stems exhibiting greater efficacy than leaves, which were instead richer in antioxidant constituents. These diverse findings broaden the pharmacological profile of *L. sativa* and highlight its potential as a versatile medicinal leafy vegetable with many therapeutic uses and nutritional value.

Toxicity and Safety Profiles

Individuals globally have relied on *L. sativa* L. for several applications, including food, pharmaceutical, nutraceutical, cosmetic, agricultural, and industrial and emerging applications (Wani et al., 2020). Due to its wide consumption as a fresh leafy vegetable, this species has often been accepted and recognized as safe (Pour et al., 2018). However, various factors have led to toxicological considerations of this species based on several circumstances, such as anti-nutritional factors, nitrate accumulation, heavy metal and pesticide residues, microbiological risks, allergic reactions, and finally, toxicological studies related to its overall safety profile (Zhang et al., 2013; Noumedem et al., 2017; Tofiq et al., 2024). Despite its well-known uses and accepted origins, the validation of its toxicity and safety profiles is essential to ensure good health and well-being for human health and nutrition (Ivanova et al., 2023). According to literature, the toxicological profile of *L. sativa* L. has been studied extensively using various methods and models, which has resulted in considerable concerns related to environmental contaminants and herbicides. Several studies have indicated that subsequent exposure to these toxins may cause detrimental effects, such as the impairment of physiological processes and ultimately a decline in biomass and oxidative stress (Zhao et al., 2022). A study by Zemanová et al. (2024) recently evaluated the physiological and metabolic processes in *L. sativa* by analyzing the free amino acids and gas exchange parameters in this species. Their results showed that the toxicity of As, Cd, Pb, and Zn significantly affected lettuce growth, causing a reduction in leaf biomass by 40–48% and inhibiting root formation. Overall, the potent toxicity caused oxidative and metabolomic stress, causing leaf sensitivity. Wang et al. (2024) examined the structure-dependent effects of alkyl organophosphate esters in *L. sativa.* Their study focused on the uptake, distribution, and toxicity in hydroponic lettuce, which showed strong uptake and translocation patterns in lettuce, with shorter-chain phosphates being more mobile and less biotransformed, while longer-chain OPEs accumulated in roots and produced hydroxylated metabolites that exhibited stronger toxicity, highlighting the need for environmental risk evaluation of hydroxylated derivatives. In another study, Yu et al. (2022) investigated cadmium toxicity using the PacBio and Illumina sequencing techniques. Their study revealed that exposure to melatonin further exacerbated cadmium stress in lettuce roots, which was confirmed by transcriptome analyses showing distinct gene expression patterns in the various treatments. These studies highlight that heavy metals and chemical contaminants, along with their interactions with plant regulators, can have a substantial impact on lettuce growth, metabolism, and stress responses (Figure 2).

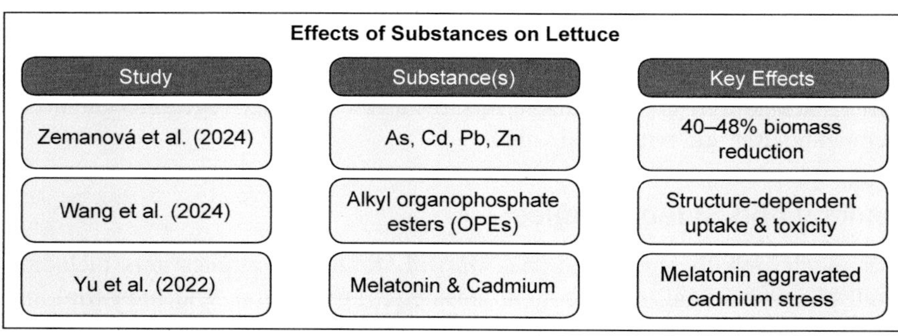

Fig. 2 Responses of *L. sativa* to environmental contaminants. (Image created by CM Naidoo.)

Harvesting, Post-harvesting/ Preservation, Drying, and Processing Methodologies

Maintaining the integrity of *L. sativa* requires careful management from the time of harvesting, post-harvesting, preservation, drying, and processing methodologies (Pandya et al., 2025). The production and quality of nutrients and bioactive substances are greatly impacted by differences in harvesting time, drying strategies, and extraction methodologies. Moreover, these steps are crucial in maintaining quality and extending its shelf-life (Ceccanti, 2020). Harvesting, a crucial step, relies on the lettuce type since crisphead varieties are often harvested upon firm head formation, whereas the loose-leaf type varieties are generally harvested once leaves are mature (Kader, 1992). In terms of post-harvesting, due to the short shelf life, high respiration rates, and water loss of *L. sativa,* strict cooling protocols are often followed, such as vacuum and hydrocooling, to ensure ultimate freshness (Yang et al., 2024). Thereafter, preservation and safety regulations must be followed to ensure compliance and the reduction in microbial load, which is usually conducted by a process of washing with chlorine, peracetic acid, or ozone (Gómez-López, 2012). For extended shelf-life, edible coatings such as chitosan and alginate are applied to preserve phenolics and reduce spoilage (Cofelice et al., 2024).

According to literature, there are various drying methods ranging from hot air drying (AD), hot air-assisted radio frequency drying (ARFD), infrared drying (IRD), and microwave-assisted hot air drying (MAD), all of which yield variable quality based on drying periods (Ogunlade et al., 2023), as observed in Figure 3. A study by Roknul et al. (2014), comparing the four drying methods AD; ARFD; IRD; and MAD for *L. sativa* stem slices, reported that ARFD had the shortest drying time (120 min), and it was found that ARFD produced the best quality dried samples. Lastly, processing often involves the usage of lettuce powders and extracts that are frequently used in food products and fermentation processes due to their antioxidant potential (Mulabagal et al., 2010; Han et al., 2022).

Potential Food, Pharmaceutical, and Value-added Products

According to its widely used culinary uses, *L. sativa* L., commonly referred to as lettuce, is a foundation of various food, medicinal, and value-added goods (Rubatzky et al., 1997). It is a versatile species with extensive applications due to its robust nutritional profile and beneficial qualities. The main features of its applications are described in the sections that follow (Feás et al., 2014). The nutritional and culinary uses of *L. sativa* are exponential due to its frequent consumption in salads, soups, and as a fresh vegetable, where it provides valuable vitamins, minerals, and phytonutrient content, such as flavonoids and phenols, which are essential for good health and well-being (Abbasi et al., 2024). From a pharmacological perspective, lettuce has shown considerable effects, including antimicrobial, antioxidant, and neuroprotective properties (Noumedem et al.,

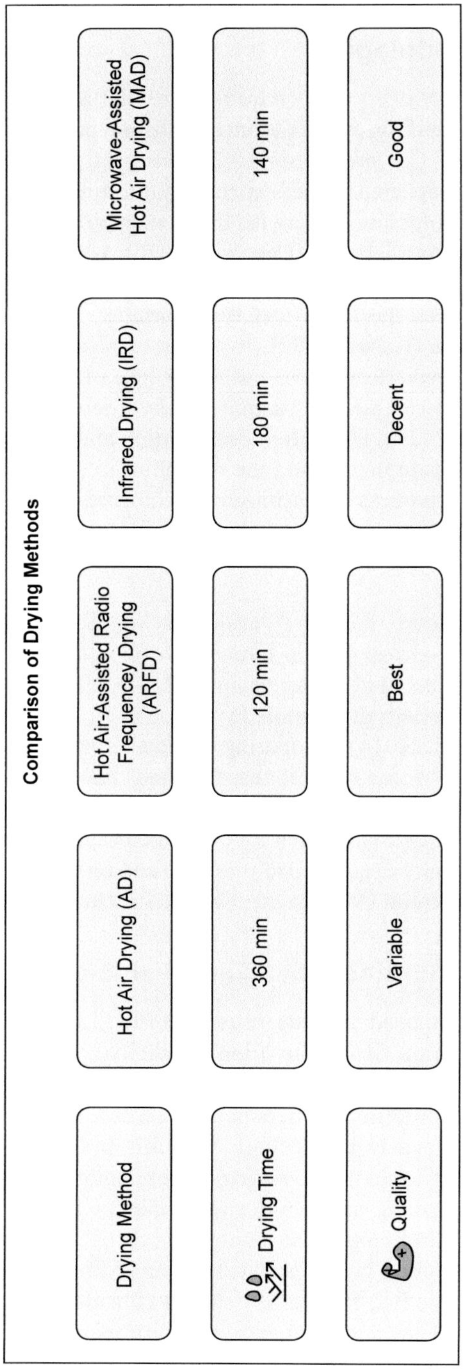

Fig. 3 Drying methods for *L. sativa*, showing trade-offs between time and quality. (Image created by CM Naidoo.)

2017). Furthermore, studies have shown that seeds and leaves contain essential oils with beneficial constituents, such as linoleic and oleic acids, which have possible well-being benefits (Afsharypuor et al., 2018). Finally, with its seeds traditionally used for sedative and analgesic reasons, *L. sativa* L. offers fascinating value-added uses (Akbar, 2020). Customers are able to procure items with a wellbeing emphasis due to contemporary techniques like biofortification. However, challenges, including consumer acceptance of biofortified products and the need for further research on pharmaceutical applications, must be addressed to fully realize its potential in nutrition and health.

Conclusion

Lactuca sativa is considered a dietary staple; however, according to research, this plant has displayed substantial bioactive compounds linked to its pharmacological and therapeutic value. Studies have demonstrated its antioxidant, antimicrobial, and anti-inflammatory potential, which is aligned with its use in traditional and contemporary medicine. Its emerging use in the nutraceutical and pharmaceutical sectors shows potential; however, limitations remain, predominantly in clinical validation, dosage standardization, and toxicological evaluation. Hence, to completely harness the benefits of *L. sativa*, future research should integrate phytochemical analyses with *in vivo* and clinical studies, confirming both efficacy and safety. These efforts will ensure that *L. sativa* remains a valued resource for health advancements and disease management in a sustainable bioeconomy.

Acknowledgments

We sincerely thank the South African Medical Research Council (SAMRC: 2024/2025) through its Division of Research Capacity Development under Research Capacity Development.

References

Abbasi, A., Waheeb, S.A., Dawood, A.I., Zafar, S., Hina, A., Zaman, Q.U., Ali, H.M., Jaremko, M. and Abdelsalam, N.R. 2024. Biofortification of leafy vegetables: Maximizing nutritional potential for Well-being of humans. *In:* Tiwari, S., Singh, B. (Eds.), 2024. *Harnessing Crop Biofortification for Sustainable Agriculture* (pp. 275–294). Springer, Singapore. doi: 10.1007/978-981-97-3438-2_14.

Abdalla, M.A., Famuyide, I.M., Wooding, M. and Mühling, K.H. 2022. Secondary metabolite profile and pharmacological opportunities of lettuce plants following selenium and sulfur enhancement. *Pharmaceutics*, *14*: 2267. doi: 10.3390/pharmaceutics14112267.

Adesso, S., Pepe, G., Sommella, E., Manfra, M., Scopa, A., Sofo, A., Tenore, G.C., Russo, M., Di Gaudio, F., Autore, G., Campiglia, P. and Marzocco, S. 2016. Anti-inflammatory and antioxidant activity of polyphenolic extracts from *Lactuca sativa* (var. Maravilla de Verano) under different farming methods. *J. Sci. Food Agric.*, *96*: 4194–4206. doi: 10.1002/jsfa.7622.

Afsharypuor, S., Ranjbar, M., Mazaheri, M., Shakibaei, F. and Aslani, A. 2018. Essential oil constituents of seeds and fresh leaves of garden lettuce (*Lactuca sativa* L.) grown in Isfahan, Iran. *Res. J. Pharmacogn.*, *5*: 1–5. doi: 10.22127/rjp.2018.64852.

Akbar, S. 2020. *Lactuca sativa* L. (Asteraceae/Compositae). *In: Handbook of 200 Medicinal Plants: A Comprehensive Review of Their Traditional Medical Uses and Scientific Justifications* (pp. 1067–1075). Springer International Publishing, Cham, Switzerland: doi: 10.1007/978-3-030-16807-0_109.

Amadi, J.A., Amadi, P.U., Njoku, U.C. and Onitchi, C.L. 2021. Phytochemicals as insulin sensitizers in Type 2 diabetes. *Iran J. Basic Med. Sci., 24*: 232– 239. doi: 10.22038/ijbms.2020.49770.11368.

Arya, M., Tiwari, A., Singh, D.B. and Taj, G. 2023. Computational study of lactucine derivatives for anticancer properties. *Lett. Drug Des. Discov., 20*: 432–440. doi: 10.2174/1570180820666230224143303.

Assefa, A.D. 2021. Nutritional metabolites of red pigmented lettuce (*Lactuca sativa*) germplasm and correlations with selected phenotypic characters. *Foods, 10*: 2504. doi: 10.3390/foods10102504.

Assefa, A.D., Choi, S., Lee, J.E., Sung, J.S., Hur, O.S., Ro, N.Y., Lee, H.S., Jang, S.W. and Rhee, J.H. 2019. Identification and quantification of selected metabolites in differently pigmented leaves of lettuce (*Lactuca sativa* L.) cultivars harvested at mature and bolting stages. *Chem. Cent. J., 13*: 56. doi: 10.1186/s13065-019-0570-2.

Ayo, V.I., Abah, M.A., Ale, E., Ekele, J.U., Amilo, D., Ochuele, D.A., Ayantse, L.M., Barrah, C., Abdulsalam, I.O., Sani, B., Tibile, S.T., Mohammed, R.I. and Caleb, E.B. 2024. Effect of *Lactuca sativa* supplemented diet on Poloxamer 407 induced hyperlipidemic albino rats. *Asian J. Nat. Prod. Biochem., 21*: 67–78. doi: 10.13057/biofar/f210203.

Aziz, R.A. 2024. Determination of antioxidant and antimicrobial activity of prickly lettuce (*Lactuca serriola* L.) leaves extract. *J. Coll. Basic Educ., 30*: 17–27. doi: 10.35950/cbej.v30i125.12184.

Berríos, D., Nahuelcura, J., González, F., Peña, F., Cornejo, P., Pérez-Navarro, J., Gómez-Alonso, S. and Ruiz, A. 2024. Phenolic compounds and antioxidant response in *Lactuca sativa* inoculated with yeast and nanoparticles. *Plants, 13*: 388. doi: 10.3390/plants13030388.

Bouymajane, A., Filali, F.R., Moujane, S., Majdoub, Y.O.E., Otzen, P., Channaoui, S., Ed-Dra, A., Bouddine, T., Sellam, K., Boughrous, A.A., Miceli, N., Altemimi, A.B. and Cacciola, F. 2024. Phenolic compound, antioxidant, antibacterial, and *in silico* studies of *Lactuca saligna*. *Molecules, 29*: 596. doi: 10.3390/molecules29020596.

Brychkova, G., de Oliveira, C.L., Gomes, L.A.A., de Souza Gomes, M., Fort, A., Esteves-Ferreira, A.A., Sulpice, R., McKeown, P.C. and Spillane, C. 2023. Regulation of Carotenoid Biosynthesis and Degradation in Lettuce (*Lactuca sativa* L.) from Seedlings to Harvest. *Int. J. Mol. Sci., 24*: 10310. doi: 10.3390/ijms241210310.

Buendía-García, A., Lozano-Cavazos, C.J., Rodríguez-Ortiz, J.C., Carballo-Méndez, F.deJ., Moreno-Reséndez, A., M. de los A. Sariñana-Navarrete and Preciado-Rangel, P. 2021. Selenium biofortification improves the nutraceutical quality and antioxidant capacity of lettuce (*Lactuca sativa* L.). *Ecosistemas Recur. Agropec., 8*: e3104. doi: 10.19136/era.a8n3.3104.

Cao, S., Sawettalake, N., Li, P., Fan, S. and Shen, L. 2024. DNA methylation variations underlie lettuce domestication and divergence. *Genome Biol., 25*: 158. doi: 10.1186/s13059-024-03310-x.

Carillo, P., Cozzolino, C., D'Abrosca, B., Nacca, F., DellaGreca, M., Fiorentino, A. and Fuggi, A., 2010. Effects of the Allelochemicals Dihydrodiconiferyl Alcohol and Lariciresinol on Metabolism of *Lactuca sativa*. *Open Bioact. Compd. J., 3*: 18–26.

Ceccanti, C. 2020. *Effect of Pre- and Post-harvest Treatments on Quality, Organoleptic and Nutraceutical Properties of Wild Edible Plants*. PhD Thesis, University of Pisa, Pisa, Italy.

Cheng, D.M., Pogrebnyak, N., Kuhn, P., Poulev, A., Waterman, C., Rojas-Silva, P., Johnson, W.D. and Raskin, I. 2014. Polyphenol-rich Rutgers Scarlet lettuce improves glucose metabolism and liver lipid accumulation in diet-induced obese C57BL/6 mice. *Nutr., 30*: S52–S58. doi: 10.1016/j.nut.2014.02.022.

Cho, E., Gurdon, C., Zhao, R., Peng, H., Poulev, A., Raskin, I. and Simko, I. 2023. Phytochemical and Agronomic Characterization of High-Flavonoid Lettuce Lines Grown under Field Conditions. *Plants, 12*: 3467. doi: 10.3390/plants12193467.

Choi, J., An, J., Lee, H.D., Kim, W.J., Lee, S. and Lee, S. 2023. Comprehensive analysis of phenolic compounds, carotenoids, and antioxidant activities in *Lactuca sativa* var. longifolia in smart farm system. *Processes*, *11*: 2993. doi: 10.3390/pr11102993.

Chon, S.U. 2005. Antioxidant potential and chlorogenic acid level of lettuce. *Korean J. Plant Res.*, *8*: 103–109.

Cofelice, M., Iftikhar, A., Lopez, F. and De Leonardis, A. 2024. Effect of edible coatings on quality parameters and phenol composition of ready-to-eat Salanova lettuce. *Eur. Food Res. Technol.*, *250*: 1–10. doi: 10.1007/s00217-023-04425-4.

Coria-Cayupán, Y.S., Sánchez de Pinto, M.I. and Nazareno, M.A. 2009. Variations in bioactive substance contents and crop yields of lettuce (*Lactuca sativa* L.) cultivated in soils with different fertilization treatments. *J. Agric. Food Chem.*, *57*: 10122–10129. doi: 10.1021/jf903019d.

Cui, X.X., Yang, X., Wang, H.J., Rong, X.Y., Jing, S., Xie, Y.H., Xie, Y.H., Huang, D. and Zhao, C. 2017. Luteolin-7-O-Glucoside Present in Lettuce Extracts Inhibits Hepatitis B Surface Antigen Production and Viral Replication by Human Hepatoma Cells *in Vitro*. *Front. Microbiol.*, *8*: 2425. doi: 10.3389/FMICB.2017.02425.

De Gregorio, M.A., Zengin, G., Alp-Turgut, F.N., Elbasan, F., Ozfidan-Konakci, C., Arikan, B., Yildiztugay, E., Zhang, L. and Lucini, L., 2023. Modulation of lettuce phenolic profile by glutamate and humic acids. *Plants*, *12*: 1822. doi: 10.3390/plants12091822.

de Souza, A.S.N., Schmidt, H.O., Pagno, C., Rodrigues, E., Silva, M.A.S.D., Flôres, S.H. and Rios, A.O. 2022. Influence of cultivar and season on carotenoids and phenolic compounds from red lettuce. *Food Res. Int.*, *155*: 111110. doi: 10.1016/j.foodres.2022.111110.

De Vries, I.M. 1997. Origin and domestication of *Lactuca sativa* L. *Genet. Resour. Crop Evol.*, *44*: 165–174.

Dyląg, A., Smoleń, S., Wisła-Świder, A., Kowalska, I., Sularz, O., Krzemińska, J., Pitala, J. and Koronowicz, A. 2023. Evaluation of the chemical composition and nutritional value of lettuce (*Lactuca sativa* L.) biofortified in hydroponics with iodine in the form of iodoquinolines. *Front. Plant Sci.*, *14*: 1288773. doi: 10.3389/fpls.2023.1288773.

Edziri, H.L., Smach, M.A., Ammar, S., Mahjoub, M.A., Mighri, Z., Aouni, M. and Mastouri, M. 2011. Antioxidant, antibacterial, and antiviral effects of *Lactuca sativa* extracts. *Ind. Crops Prod.*, *34*: 1182–1185.

Erdoğan, M.K., Gundogdu, R., Toy, Y., Halil Gecibesler, I., Yapar, Y., Behcet, L. and Zengin, G. 2024. Comparison of anticancer, antioxidant, and phytochemical contents between edible lettuce (*Lactuca sativa)* and a new wild species (*Lactuca anatolica)*. *Chem. Biodivers.*, *2*: e202400552. doi: 10.1002/cbdv.202400552.

Feás, X., Pacheco, L., Iglesias, A. and Estevinho, L.M. 2014. Use of propolis in the sanitization of lettuce. *Int. J. Mol. Sci.*, *15*: 12243–12257. doi: 10.3390/ijms150712243.

Ferrantelli, V., Vasto, S., Alongi, A., Sabatino, L., Baldassano, D., Caldarella, R., Gagliano, R., L. Di Rosa B.B., Consentino, L. Vultaggio and Baldassano, S. 2023. Boosting plant food polyphenol concentration by saline eustress as supplement strategies for the prevention of metabolic syndrome: An example of randomized interventional trial in the adult population. *Front. Nutr.*, *10*: 112233. doi: 10.3389/fnut.2023.1288064.

Flores, M., Amorós, A. and Escalona, V. 2023. Changes in antioxidant compounds in green and red lettuces. *Horticulturae*, *9*: 677. doi: 10.3390/horticulturae9060677.

[FAOSTAT] Food and Agriculture Organization Statistics 2022. *Lettuce Production Statistics*. FAO, Rome, Italy. Available at: https://www.fao.org/faostat/.

Foudah, A., Alqarni,. M., Alam, A., Salkini, M.A., Ahmed, E.O. and Yusufoglu, H. 2019. Evaluation of in vitro antioxidant and antimicrobial activities of methanol extract of standardized *Lactuca sativa* leaves. *Indo Am. J. Pharm. Sci.*, *6*: 123–130. doi: 10.5281/zenodo.2576069.

Ghanadi, K., Hasanvand, A., Abbaszadeh, S., Heidari-Soureshjani, S. and Suhan, T. 2019. Phytotherapy: An anti-hepatotoxicity and hepatoprotective approach in chemotherapy. *Plant Sci. Today*, *6*: 114–122. doi: 10.14719/PST.2019.6.2.514.

Gómez-López, V.M. (Ed.). 2012. *Decontamination of Fresh and Minimally Processed Produce*. John Wiley & Sons, Hoboken, New Jersey, USA. doi: 10.1002/9781118229187.

Gopal, S.S., Lakshmi, M.J., Sharavana, G., Sathaiah, G., Sreerama, Y.N. and Baskaran, V. 2017. Lactucaxanthin A potential anti-diabetic carotenoid from lettuce (*Lactuca sativa*) inhibits α-amylase and α-glucosidase activity in vitro and in diabetic rats. *Food Func.*, *8*: 1124–1131. doi: 10.1039/C6FO01655C.

Gridling, M., Popescu, R., Kopp, B., Wagner, K.H., Krenn, L. and Krupitza, G. 2010. Anti-leukaemic effects of two extract types of *Lactuca sativa* correlate with the activation of Chk2, induction of p21, downregulation of cyclin D1 and acetylation of alpha-tubulin. *Oncol. Rep.*, *23*: 1145–1151. doi: 10.3892/or_00000744.

Gu, C., Zeng, B., Wang, M., Zhang, Y., Yan, C., Lin, Y., Khan, A., Zeng, R. and Song, Y. 2024. Study on Active Components and Mechanism of Lettuce Latex Against *Spodoptera litura*. *Chem. Biodivers.*, *21*: e202400993. doi: 10.1002/cbdv.202400993.

Gyawali, R., Ghimire, A., Khatiwada, A., Niraula, P., Sharma, U. and Thapa, R. 2020. Anti-inflammatory, anxiolytic, and antioxidant property of *Lactuca sativa* L. and formulation of microspheres loaded sustained release anti-inflammatory gel. *J. Nepal Chem. Soc.*, *41*: 8–15. doi: 10.3126/jncs.v41i1.30371.

Han, A.L., Lee, H.K., Chon, H.S., Pae, H.O., Kim, M.S., Shin, Y.I. and Kim, S. 2022. Evaluation of the effectiveness of fermented soybean–lettuce powder for improving menopausal symptoms. *Nutrients*, *14*: 2878. doi: 10.3390/nu14142878.

Harsha, S.N., Anilakumar, K.R. and Mithila, M.V. 2013. Antioxidant properties of *Lactuca sativa* leaf extract. *Biomed. Prev. Nutr.*, *3*: 367–377. doi: 10.1016/j.bionut.2013.06.003.

Hill, C.R., Liu, H.A., McCahon, L., Zhong, L., Shafaei, A., Balmer, L., Lewis, J.R., Hodgson, J.M. and Blekkenhorst, L.C. 2025. S-methyl cysteine sulfoxide and its potential role in human health: A scoping review. *Crit. Rev. Food Sci. Nutr.*, *65*: 87–100. doi: 10.1080/10408398.2023.2267133.

Huang, X., Lin, K.H., Liu, S.H., Yang, J., Zhao, H., Zheng, X., Tsai, M.T., Chang, C.S., Huang, L. and Weng, C. 2024. Combination of plant metabolites hinders starch digestion and glucose absorption while facilitating insulin sensitivity to diabetes. *Front. Pharmacol.*, *15*: 1362150. doi: 10.3389/fphar.2024.1362150.

Hung, H.C., Joshipura, K.J., Jiang, R., Hu, F.B., Hunter, D., Smith-Warner,, S.A., Colditz, G.A., Rosner, B., Spiegelman, D. and Willett, W.C. 2004. Fruit and vegetable intake and risk of major chronic disease. *J. Natl. Cancer Inst.*, *96*: 1577–1584. doi: 10.1093/jnci/djh296.

Ismail, H. and Mirza, B. 2015. Evaluation of analgesic, anti-inflammatory, anti-depressant, and anti-coagulant properties of *Lactuca sativa* (cv. Grand Rapids) plant tissues and cell suspension in rats. *BMC Complement. Altern. Med.*, *15*: 199. doi: 10.1186/s12906-015-0742-0.

Ismail, H., Dilshad, E., Waheed, M.T. and Mirza, B. 2017. Transformation of Lettuce with rol ABC Genes: Extracts Show Enhanced Antioxidant, Analgesic, Anti-Inflammatory, Antidepressant, and Anticoagulant Activities in Rats. *Appl. Biochem. Biotechnol.*, *181*: 1179–1198. doi: 10.1007/s12010-016-2277-3.

Ivanova, T., Marchev, A., Chervenkov, M., Bosseva, Y., Georgiev, M., Kozuharova, E. and Dimitrova, D. 2023. Catching the green-diversity of ruderal spring plants traditionally consumed in Bulgaria and their potential benefit for human health. *Diversity*, *15*: 435. doi: 10.3390/d15030435.

Jang, J.H., Park, C.Y., Sung, E.G., Song, I.H., Kim, J.Y., Jung, C., Sohn, H.Y. and Lee, T.J. 2021 . Lactucin induces apoptosis through reactive oxygen species-mediated BCL-2 and $CFLAR_L$ downregulation in Caki-1 cells. *Genes and Genomics*, *43*: 1199–1207. doi: 10.1007/s13258-021-01142-8.

Jiangseubchatveera, N., Saechan, C., Petchsomrit, A., Treeyaprasert, T., Leelakanok, N. and Prompanya, C. 2023. Phytochemicals and antioxidant activities of red oak, red coral, and butterhead. *Trop. Life Sci. Res.*, *34*: 1–17. doi: 10.21315/tlsr2023.34.1.1.

Kader, A.A. 1992. Postharvest biology and technology: An overview. *In: Postharvest Technology of Horticultural Crops*. (2nd Edn.), pp. 39–48. University of California, Division of Agriculture and Natural Resources, Oakland, California, USA.

Kim, H.W., Suh, H.J., Choi, H.S., Hong, K.B. and Jo, K. 2019. Effectiveness of the Sleep Enhancement by Green Romaine Lettuce (*Lactuca sativa*) in a Rodent Model. *Biol. Pharm. Bull.*, *42*: 1726–1732. doi: 10.1248/bpb.b19-00454.

Kim, M.J., Moon, Y., Tou, J.C., Mou, B. and Waterland, N.L. 2016. Nutritional value, bioactive compounds, and health benefits of lettuce (*Lactuca sativa* L.). *J. Food Compos. Anal.*, *49*: 19–34. doi: 10.1016/j.jfca.2016.03.004.

Knez, M., Mattas. K., Gurinovic, M., Gkotzamani, A. and Koukounaras, A. 2024. Revealing the power of green leafy vegetables: Cultivating diversity for health, environmental benefits, and sustainability. *Glob. Food Secur.*, *43*: 100816. doi: 10.1016/j.gfs.2024.100816.

Ko, H., Shon, Y.J., Lim, W., Nam, T.G., Kim, W.C., Kim, N.H. and Lim, T.G. 2024. The antimelanogenic activity of heukharang lettuce (*Lactuca sativa* L.) leaf. *J. Food Biochem.*, *48*: e14672. doi: 10.1155/2024/6668222.

Körber, T.T., Sitz, T., Abdalla, M.A., Mühling, K.H. and Rohn, S. 2023. LC-ESI-MS/MS Analysis of Sulfolipids and Galactolipids in Green and Red Lettuce (*Lactuca sativa* L.) as Influenced by Sulfur Nutrition. *Int. J. Mol. Sci.*, *24*: 3728. doi: 10.3390/ijms24043728.

Kwon, M., Hodgins, C.L., Haslam, T.M., Roth, S.A., Nguyen, T.D., Yeung, E.C. and Ro, D.K. 2022. Germacrene A Synthases for Sesquiterpene Lactone Biosynthesis are Expressed in Vascular Parenchyma Cells Neighboring Laticifers in Lettuce. *Plants* (*Basel*). *11*: 1192. doi: 10.3390/plants11091192.

Lafarga, T., Villaró, S., Rivera, A., Bobo, G. and Aguiló-Aguayo, I. 2020. Bioaccessibility of polyphenols and antioxidant capacity of fresh or minimally processed modern or traditional lettuce (*Lactuca sativa* L.) varieties. *J. Food Sci. Technol.*, *57*: 754–763. doi: 10.1007/s13197-019-04108-7.

Lebeda, A., Křístková, E., Doležalová, I., Kitner, M. and Widrlechner, M.P. 2019. Wild *Lactuca* species in North America. In: *North American Crop Wild Relatives* (Vol. 2): *Important Species* (pp. 131—194). Springer International Publishing, Cham, Switzerland. doi: 10.1007/978-3-319-97121-6_5.

Lindqvist, K. 1960. On the origin of cultivated lettuce. *Hereditas*, *46*: 319–350. doi: 10.1111/j.1601-5223.1960.tb03091.x.

Liu, X., Ardo, S., Bunning, M., Parry, J., Zhou, K., Stushnoff, C., Stoniker, F., Yu, L.L. and Kendall, P. 2007. Total phenolic content and DPPH radical scavenging activity of lettuce (*Lactuca sativa* L.) grown in Colorado. *LWT Food Sci. Technol.*, *40*: 552–557. doi: 10.1016/j.lwt.2005.09.007.

Maity, S., Padhy, G.K., Kanthal, L.K. and Pattanayak, S. 2024. Potential of Lactuca and Gyrocarpus Species: A Comprehensive Review. *Int. J. Pharm. Qual. Assur.*, *15*: 454–460. doi: 10.25258/ijpqa.15.1.70.

Malarz, J., Michalska, K. and Stojakowska, A. 2020. Stem lettuce metabolites: Does the variety make any difference? *Foods*, *10*: 59. doi: 10.3390/foods10010059.

Malik, J., Kaur, S., Karan, M. and Choudhary, S. 2022. Neuroprotective effect of standardized extracts of three *Lactuca sativa* Linn. varieties against 3-NP induced Huntington's disease like symptoms in rats. *Nutr. Neurosci.*, *25*: 1173–1187. doi: 10.1080/1028415X.2020.1841500.

Mandal, S.K., Debnath, U., Kumar, A. Thomas, S., Mandal, S.C., Dutta Choudhury, M. and Palit, P. 2020. Natural Sesquiterpene Lactones in the Prevention and Treatment of Inflammatory Disorders and Cancer: A Systematic Study of this Emerging Therapeutic Approach based on Chemical and Pharmacological Aspect. *Lett. Drug Des. Discov.*, *17*: 1102–1116. doi: 10.2174/157018081799 9200421144007.

Marwaha, A. 2022. Lactucaxanthin Regulates the Cascade of Retinal Oxidative Stress, Endoplasmic Reticulum Stress, and Inflammatory Signaling in Diabetic Rats. *J. Med. Food.*, *31*: 320–328. doi: 10.1080/09273948.2022.2027464.

Medina-Lozano, I., Bertolín, J.R. and Díaz, A. 2021. Nutritional value of commercial and traditional lettuce (*Lactuca sativa* L.) and wild relatives: Vitamin C and anthocyanin content. *Food Chem.*, *359*: 129864. doi: 10.1016/j.foodchem.2021.129864.

Michalska, K., Jantas, D., Malarz, J. and Jakubowska, K. 2024. *Lactuca racemosa* as antioxidant source. *Molecules*, *29*: 5975. doi: 10.3390/molecules29245975.

Mohammed, N.J, Salman, W.A.A. and Al-Fahham, A. 2024. The chemical structure, classification, and clinical significance of alkaloids. *Int. J. Hum. Med. Res.*, *3*: 10. doi: 10.58806/ijhmr.2024.v3i10n10.

Mou, B. 2009. Nutrient content of lettuce and its improvement. *Curr. Nutr. Food Sci.*, *5*: 242–248. doi: 10.2174/157340109790218030.

Mou, B. 2011. Mutations in lettuce improvement. *Int. J. Plant Genomics*, *2011*: 723518. doi: 10.1155/2011/723518.

Mughrbi, H.N., Auzi, A. and Maghrbi, H. 2020. Phytochemicals, nutritional value, antioxidant, and anticoagulant activity of *Lactuca sativa* L. leaves and stems. *Borneo J. Pharm.*, *3*: 152–161. doi: 10.33084/bjop.v3i3.1394.

Mulabagal, V., Ngouajio, M., Nair, A., Zhang, Y., Gottumukkala, A.L. and Nair, M.G. 2010. *In vitro* evaluation of red and green lettuce (*Lactuca sativa*) for functional food properties. *Food Chem.*, *118*: 300–306. doi: 10.1016/j.foodchem.2009.04.119.

Naseem, S. and Ismail. H. 2022. *In vitro* and *in vivo* evaluations of antioxidative, anti-Alzheimer, antidiabetic, and anticancer potentials of hydroponically and soil-grown *Lactuca sativa*. *BMC Complement. Med. Ther.*, *22*: 30. doi: 10.1186/s12906-022-03520-5.

Negrão, L.D., Sousa, P.V.L., Barradas, A.M., Brandão, A.C.A.S., Araujo, M.A.M. and Moreira-Araújo, R.S.R. 2021. Bioactive compounds and antioxidant activity of crisphead lettuce (*Lactuca sativa* L.) of three different cultivation systems. *Food Sci. Technol.* doi: 10.1590/fst.04120.

Nisa, N.U. and Tajammul, A. 2022. *In vitro* antimicrobial activity of *Lactuca sativa* leaves against isolated clarithromycin-resistant superbugs. *Proc. Int. Halal Sci. Technol. Conf. 14*: 145–153. doi: 10.31098/ihsatec.v14i1.496.

Noumedem, J.A., Djeussi, D.E., Hritcu, L., Mihasan, M. and Kuete, V. 2017. *Lactuca sativa*. In: Kuete V. (Ed.), *Medicinal Spices and Vegetables from Africa* (pp. 437–449). Academic Press, Cambridge, Massachusetts, USA.

Ogunlade, C.A., Jekayinfa, S.O., Oyedele, O.J., Olajire, A.S. and Oladipo, A.O. 2023. Drying parameters of wild lettuce (*Lactuca taraxaxifolia* L.) affected by different drying methods. *Croat. J. Food Sci. Technol.*, *15*: 115–122. doi: 10.17508/CJFST.2023.15.1.03.

Pandya, J.B. 2025. *Post-Harvest Nutraceuticals: Edible Coatings for Enhanced Quality and Nutrition*. Academic Guru Publishing House, New Delhi, India.

Park, J., Ryu, J.H., Kim, B.Y., Chun, H.S., Kim, M.S. and Shin, Y.I. 2023. Fermented Lettuce Extract Containing Nitric Oxide Metabolites Attenuates Inflammatory Parameters in Model Mice and in Human Fibroblast-Like Synoviocytes. *Nutrients*, *15*: 1106. doi: 10.3390/nu15051106.

Pour, Z.S., Hosseinkhani, A., Asadi, N., Shahraki, H.R., Vafaei, H., Kasraeian, M., Bazrafshan, K. and Faraji, A. 2018. Double-blind randomized placebo-controlled trial on efficacy and safety of *Lactuca sativa* L. seeds on pregnancy-related insomnia. *J. Ethnopharmacol.*, *227*: 176–180. doi: 10.1016/j.jep.2018.08.001.

Qadir, O.K., Seal, C.J., Ashor, A.W., Tassotti, M., Mena, P., Del Rio, D., Siervo, M. and Brandt, K. 2022. Double-blind controlled dietary cross-over intervention with differentially fertilised intact lettuce leaves shows acute reduction in blood pressure in young adults, associated with faster uptake of nitrate than of phenolics. *Eur. J. Nutr.*, *61*: 4191–4203. doi: 10.1007/s00394-022-02961-5.

Qin, X.X., Zhang, M.Y., Han, Y.Y., Hao, J.H., Liu, C.J. and Fan, S.X. 2018. Beneficial Phytochemicals with Anti-Tumor Potential Revealed through Metabolic Profiling of New Red Pigmented Lettuces (*Lactuca sativa* L.). *Int. J. Mol. Sci.*, *19*: 1165. doi: 10.3390/ijms19041165.

Ribas-Agustí, A., Gratacós-Cubarsí, M., Sárraga, C., García-Regueiro, J.A. and Castellari, M. 2011. Analysis of eleven phenolic compounds including novel p-coumaroyl derivatives in lettuce (*Lactuca sativa* L.) by ultra-high-performance liquid chromatography with photodiode array and mass spectrometry detection. *Phytochem. Anal.*, *22*: 555–563. doi: 10.1002/pca.1318.

Roknul, A.S., Zhang, M., Mujumdar, A.S. and Wang, Y. 2014. A comparative study of four drying methods on drying time and quality characteristics of stem lettuce slices (*Lactuca sativa* L.). *Dry. Technol.*, *32*: 657–666. doi: 10.1080/07373937.2013.850435.

Rubatzky, V.E. and Yamaguchi, M. 1997. Lettuce and other composite vegetables: Family: Asteraceae (Compositae). pp. 333–370. In: Rubatzky V.E. and Yamaguchi (2nd Edn.), *World Vegetables: Principles, Production, and Nutritive Values* (pp. 333–370). Springer US, Boston, Massachusetts, USA. doi:10.1007/978-1-4615-6015-9_18.

Sadino, A., Renggana, H., Suwendar, S., Apriani,, R. and Nurhandayani, Y. 2024. Hypnotic-sedative activity of lettuce (*Lactuca sativa* L.) in male white mice swiss webster strain. *J. Ilmiah Farmako Bahari*, *15*: 113–120. doi: 10.52434/jifb.v15i2.3090.

Schoch, C.L., Ciufo, S., Domrachev, M., Hotton, C.L., Kannan, S., Khovanskaya, R., Leipe, D., Mcveigh, R., O'Neill, K., Robbertse, B., Sharma, S., Soussov, V., Sullivan, J.P., Sun, L., Turner, S. and Karsch-Mizrachi, I. 2020. NCBI taxonomy: A comprehensive update on curation, resources, and tools. *Database*, *2020*: baaa062. doi: 10.1093/database/baaa062.

Seo, H. and Jeong, J.B. 2020. Anti-inflammatory effects of lettuce (*Lactuca sativa* L.) extracts through the TLR4-MAPK/NF-κB signalling pathways in RAW264.7 macrophage cells. *Korean J. Plant Res.*, *33*: 183–193. doi: 10.7732/kjpr.2020.33.3.183.

Shi, M., Gu, J., Wu, H., Rauf, A., Emran, T.B., Khan, Z., Mitra, S., Aljohani, A.S.M., Alhumaydhi, F.A., Al-Awthan, Y.S., Bahattab, O., Thiruvengadam, M. and Suleria, H.A.R. 2022. Phytochemicals, nutrition metabolism, bioactivity, and health benefits of lettuce: A comprehensive review. *Antioxidants* (*Basel*) *11*: 1158. doi: 10.3390/antiox11061158.

Shivam, S., Gupta, A.K. and Kumar, S. 2023. Role of antioxidants in mitigating diabetic complications. *J. Nat. Prod.*, *13*: 100421. doi: 10.2174/2210315513666230330082412.

Šimko, I. 2024. Spatio-temporal dynamics of lettuce metabolome: A framework for targeted nutritional quality improvement. *Plants* (*Basel*), *13*: 3316. doi: 10.3390/plants13233316.

Stojanović, M., Savić, S., Delcourt, A., Hilbert, J.L., Hance, P., Maksimović, J.D. and Maksimović, V. 2023. Phenolics and sesquiterpene lactones profile of red and green lettuce: combined effect of cultivar, microbiological fertiliser, and season. *Plants* (*Basel*), *12*: 2616. doi: 10.3390/plants12142616.

Sularz, O., Smoleń, S., Koronowicz, A., Kowalska, I. and Leszczyńska, T. 2020. Chemical composition of lettuce (*Lactuca sativa* L.) biofortified with iodine by KIO_3, 5-iodo-, and 3,5-diiodosalicylic acid in a hydroponic cultivation. *Agronomy*, *10*: 1022. doi: 10.3390/agronomy10071022.

Syed, R.U., Moni, S.S., Lila, A.S.A., Abdallah, M.H., Abouzied, A.S., Banu, H., Alreshidi, K.S.M., Alrashidi, B.M.W., Hadi, M.A., El-Horany, H., Abdelwahab, S.I. and Taha, M.M.E. 2022. Spectral analysis and antiulcer potential of *Lactuca sativa* through the amelioration of proinflammatory cytokines and apoptosis markers. *Life* (*Basel*), *12*: 1641. doi: 10.3390/life12101641.

Tian, Y., Zhang, S., Lv, J., Liu, H., Zhang, X. and Zhang, X., 2025. Structura; characterization of a novel *Lactuca sativa* L. polysaccharide and its adjuvant hypoglycemic effect in HFD/STZ-induced obese C57 BL/6J mice. *Food Biosci.*, *68*: 106458. doi: 10.1016/j.fbio.2025.106458.

Tofiq, M., Salih, M., Rash, N.K., Ali, R., Ahmed, R. and Wahid, H.. 2024. Assessment of toxicological effect of *Lactuca serriola* L. in Sprague-Dawley rat. *Egypt. J. Vet. Sci.*, *55*: 1–13. doi: 10.21608/EJVS.2024.300225.2209.

Turatbekova, A., Masharipova, M., Umarova, F., Khalmuradova, E., Rustamova, R., Abdixoshimov, M. and Teshaboyev, R. 2024. Research into biologically active plant terpenoids and biological activity. *E3S Web Conf.*, *563*: 03076. doi: 10.1051/e3sconf/202456303076.

Ulfah, M., Putro, A.L. and Safitri, E.E. 2019. Antioxidant activity of romaine and curly lettuce extracts. *J. Ilmu Farm. Farmasi Klinik*, *16*: 21–27. doi: 10.31942/jiffk.v16i01.2925.

Utami, N. 2023. Phytochemical analysis, antioxidant activity, and ftir spectroscopic analysis of red leaf lettuce and green leaf lettuce (*Lactuca sativa* L.). *Indian Drugs*, *60*: 50. doi: 10.53879/id.60.05.13378.

Vella, F.M., Calandrelli, R., Cautela, D. and Laratta, B. 2020. Screening of bioactivity in extracts from different varieties of lettuce. In: *The 1st International Electronic Conference on Plant Science*. doi: 10.3390/IECPS2020-08755.

Wang, Y., Li, X., Chen, S., Yang, J., Fang, B., Chen, H., Yao, Y. and Sun, H. 2024. Structure-dependent distribution, metabolism, and toxicity effects of alkyl organophosphate esters in lettuce (*Lactuca sativa* L.). *Environ. Sci. Technol.*, *58*: 17441–17453. doi: 10.1021/acs.est.4c05523.

Wani, S., Tantray, Y., Jan, I., Singhal, V.K., and Gupta, R.C. 2020. *Lactuca* L.; World Distribution and Importance. In: Krüger J. (Ed.), *Lactuca: Cultivation and Uses*. Nova Science Publishers.

Wei, T., van Treuren, R., Liu, X., Zhang, Z., Chen, J., Liu, Y., Dong, S., Sun, P., Yang, T., Lan, T., Wang, X., Xiong, Z., Liu, Y., Wei, J., Lu, H., Han, S., Chen, J.C., Ni, X., Wang, J., Yang, H., Xu, X., Kuang, H., van Hintum, T., Liu, X. and Liu, H. 2021. Whole-genome resequencing of 445 *Lactuca* accessions reveals the domestication history of cultivated lettuce. *Nat. Genet.*, *53*: 1–9. doi: 10.1038/s41588-021-00831-0.

Yang, J., Song, J., Liu, J., Dong, X., Zhang, H. and Jeong, B.R. 2024. Prolonged post-harvest preservation in lettuce (*Lactuca sativa* L.) by reducing water loss rate and chlorophyll degradation regulated through lighting direction-induced morphophysiological improvements. *Plants*, *13*: 2564. doi: 10.3390/plants13182564.

Yang, N., Guo, J., Zhang, J., Gao, S., Xiang, Q., Wen, J., Huang, Y., Rao, C. and Chen, Y. 2024. Toxicological review of alkaloids. *Drug Chem. Toxicol.*, *47*: 1267–1281. doi: 10.1080/01480545.2024.2326051.

Yang, X., Cui, X., Zhao, L., Guo, D., Feng, L., Wei, S., Zhao, C. and Huang, D., 2017. Exogenous glycine nitrogen enhances flavonoids and antioxidant activity in lettuce (*Lactuca sativa* L.). *Front. Plant Sci.*, *8*: 2098. doi: 10.3389/fpls.2017.02098.

Yang, X., Gil, M.I., Yang, Q. and Tomás-Barberán, F.A. 2022. Bioactive compounds in lettuce: Highlighting the benefits to human health and impacts of preharvest and postharvest practices. *Compr. Rev. Food Sci. Food Saf.*, *21*: 4–45. doi: 10.1111/1541-4337.12877.

Yang, X., Wei, S., Liu, B., Guo, D., Zheng, B., Feng, L., Liu, Y., Tomás-Barberán, F.A., Luo, L. and Huang, D. 2018. A novel integrated non-targeted metabolomic analysis reveals significant metabolite variations between different lettuce (*Lactuca sativa.* L.) varieties. *Hortic. Res.*, *5*: 33. doi: 10.1038/s41438-018-0050-1.

Yu, X., Liang, L., Xie, Y., Tang, Y., Tan, H., Zhang, J., Lin, L., Sun, B., Huang, Z., Liu, J., Li, X., Tu, L. and Li, H. 2022. Comparative analysis of Italian lettuce (*Lactuca sativa* L. var. ramose) transcriptome profiles reveal the molecular mechanism on exogenous melatonin preventing cadmium toxicity. *Genes*, *13*: 955. doi: 10.3390/genes13060955.

Zdravković, J., Pavlović, N., Pavlović, R., Mašković, P., Mladenović, J., Đurić, M. and Aćamović-Đoković, G. 2012. Antimicrobial activity of lettuce (*Lactuca sativa* L.) extracts grown in plastic and glasshouses. *Acta Hortic.*, *960*: 299–303. doi: 10.17660/ActaHortic.2012.960.43.

Zekkori, B., Bentayeb, A., Ed-Dra, A., Filali, F.R., Sbai, A., Chlif, N., El Omari, M. and Khallouki, F. 2021. Effect of hydro-alcohol solvent polarity on the antioxidant, antibacterial, and anti-inflammatory activities of four Moroccan lettuce varieties (*Lactuca sativa* L.): A comparative study. *Carpathian J. Food Sci. Technol.*, *13*: 133–149. doi: 10.34302/crpjfst/2021.13.4.11.

Zekkori, B., Khallouki, F., Bentayeb, A., Fiorito, S., Preziuso, F., Taddeo, V.A., Epifano, F. and Genovese, S. 2018. A new phytochemical and anti-oxidant and anti-inflammatory activities of different *Lactuca sativa* L. var. crispa extracts. *Nat. Prod. Commun.*, *13*: 1139–1142. doi: 10.1177/1934578x1801300910.

Zemanová, V., Lhotská, M., Novák, M., Hnilička, F., Popov, M. and Pavlíková, D. 2024. Multicontamination Toxicity Evaluation in the Model Plant *Lactuca sativa* L. *Plants* (*Basel*), *13*: 1356. doi: 10.3390/plants13101356.

Zeng, W., Yang, J., He, Y. and Zhu, Z. 2023. Bioactive compounds in cruciferous sprouts and microgreens and the effects of sulfur nutrition. *J. Sci. Food Agric.*, *103*: 7323–7332. doi: 10.1002/jsfa.12755.

Zha, L., Wei, S., Huang, D. and Zhang, J. 2024. Multi-Omics Analyses of Lettuce (*Lactuca sativa*) Reveals Primary Metabolism Reorganization Supporting Distinct Features of Secondary Metabolism Induced by Supplementing UV-A Radiation. *J. Agric. Food Chem.*, *72*: 15498–15511. doi: 10.1021/acs.jafc.4c00394.

Zhang, K., Yuan, J., Kong, W. and Yang, Z. 2013. Genotype variations in cadmium and lead accumulations of leafy lettuce (*Lactuca sativa* L.) and screening for pollution-safe cultivars for food safety. *Environ. Sci.: Processes Impacts*, *15*: 1245–1255. doi: 10.1039/C3EM00158J.

Zhao, X., Sui, X., Zhao, L., Gao, X., Wang, J., Wen, X. and Li, Y. 2022. Morphological and physiological response mechanism of lettuce (*Lactuca Sativa* L.) to consecutive heat stress. *Scientia Horticulturae*, *301*: 111112. doi: 10.1016/j.scienta.2022.

Zohary, D. 1991. The wild genetic resources of cultivated lettuce (*Lactuca sativa* L.). *Euphytica*, *53*: 31–35. doi: 10.1007/BF00032029.

8 Nutritive Values, Bioactive Compounds, and Biological Activities of Cabbage (*Brassica oleracea* var. *capitata*)

Kadanthottu Sebastian Joseph[1*], Jaya Gangwar[2] and Vijayalaxmi S. Dandin[3]

Introduction

The *Brassica* genus, which belongs to the Brassicaceae family, has a vast range of commercially and nutritionally essential vegetables grown all over the world. The species in this genus, like *Brassica rapa*, *Brassica oleracea*, and *Brassica napus*, are distinguished by their diverse morphologies and compounds that are beneficial to health (Burbulis and Kott, 2013). *Brassica oleracea var. capitata* (Cabbage) is one of the most popular cruciferous vegetables grown globally due to its low cost, availability, and versatility in food preparation. The leaves of these are formed by characteristic cabbage heads with varying shape, size, and leaf texture, leading to a great number of Cullivers, especially white, red, and conical. Many forms of cabbage have been reported, such as *B. oleracea var. capitata f. alba, B. oleracea var. capitata f. conica,* and *B. oleracea var. capitata f. rubra.* Cabbage is a leafy green plant grown for its multilayered leaves. The cabbage seedlings have a tap root system and heart-shaped cotyledons. The first leaves are ovate, and the initial leaves are in rosette form. This is followed by leaves with short petioles, where leaves begin to cup inwards to form the head. The cabbage heads are formed during the first year and flower during the second year of growth. The inflorescence is a terminal raceme that is unbranched. Each flower has four petals and sepals, each set in a perpendicular pattern, with six stamens, and a two-celled ovary containing a single stigma and style. Fruit is a silique that opens at maturity. Seeds are brown to black in color (Dixon, 2007).

Cabbage has an important role in various cuisines and cultures. According to many ancient texts, cabbage has been grown in coastal Europe for nearly 8,000 years (Al-Shehbaz et al., 2006; Al-Shehbaz, 2011). Due to its affordable price and availability in local markets, it is an important source of many phytonutrients. It is

[1,2] Department of Life Sciences, Christ University, Bangalore, 560029, India.
[3] Department of Biology, Alliance Pre-University College, Bangalore 560099, India.
* Corresponding author: joseph.ks@christuniversity.in

DOI: 10.1201/9781003687832-8

consumed raw in salads, boiled, stir-fried, or consumed as a fermented product. It is known as the "medicine of the poor" and has been used for health improvement and treatment of various ailments like inflammation, blood purification (Cavender, 2006; Hatfield, 2004). Juice of cabbage has been used as a laxative by Egyptians for relieving constipation (Hatfield, 2004). In some folk medicine, it is used as an antidote for mushroom poisoning and to cure hangover effects of alcohol. In Europe, it's used for sore throat, melancholy colic, etc. (Hatfield, 2004). The leaves of cabbage have also been reported to reduce swelling and breast engorgement during breastfeeding (Ayers, 2000). It has also been used to relieve rheumatic diseases, neuralgia, healing contusions, and wounds (Marc et al., 2008; Passalacqua et al., 2007).

Cabbage contains high levels of vital nutrients such as vitamins C and K, folate, and dietary fiber (Ağagündüz et al., 2022). Aside from its fundamental nutritional value, cabbage has a diverse range of bioactive substances such as glucosinolates, flavonoids, and anthocyanins, which contribute to its antioxidant, anti-inflammatory, and anticancer activities. It can be ingested raw, cooked, or fermented, with each providing unique health advantages. Cabbage, when fermented into sauerkraut and kimchi, is a functional food that promotes gut health due to its microbial content (Haghighi et al., 2022). This chapter will provide an in-depth overview of cabbage's nutritional composition, phytochemical profile, and biological activity. The medicinal uses of its bioactive constituents are highlighted, as is the impact of environmental and processing conditions on its nutritional quality and position as a functional food. Furthermore, the chapter discusses safety concerns and future research initiatives to support the use of cabbage in dietary interventions and nutraceutical formulations.

Nutritional Composition

Macronutrients

Carbohydrates

Cabbage has a modest quantity of carbohydrates, ranging from 5.0 to 6.0 g per 100 g fresh weight, largely in the form of simple sugars such as glucose and fructose, with trace levels of starch and polysaccharides. It is a non-starchy, fiber-rich vegetable. The resulting carbohydrates contribute to cabbage's moderate sweetness while also providing easily available energy. The cultivar and growing environment might have a minor impact on the amount of carbohydrates. Because of its anthocyanin-related sugar content, red cabbage often contains a little more carbohydrate than green cabbage (Sayas-Barberá et al., 2022).

Proteins

Cabbages provide a small quantity of plant-based protein to the diet, usually 1.2–1.5 g per 100 g, despite not being a high-protein item. These proteins contain modest amounts of important amino acids like valine, isoleucine, leucine, methionine, threonine, histidine, phenylalanine, and tryptophan (Park, Valan Arasu, et al., 2014; Peñas et al., 2017). Research has documented the existence of defense proteins and cysteine-rich peptides, which may also contribute to human health and plant immunity. Most cabbage types have stable protein content, while environmental influences such as soil nitrogen levels can impact overall protein accumulation (Ten Caat et al., 2022).

Fats

The natural low total fat content of cabbage (0.1–0.2 g/100 g) makes it a good vegetable for low-fat diets. Traces of polyunsaturated fatty acids, such as omega-3 (α-linolenic acid) and omega-6 (linoleic acid), are present; nevertheless, and they support cardiovascular health and anti-inflammatory properties. Despite being found in relatively modest levels, these fatty acids have a more important function when cabbage is eaten in large quantities or as part of a balanced plant-based diet that is balanced (Beecher, 1994).

Micronutrients

Vitamins

Cabbage is high in various vitamins, notably vitamin C, which ranges from 30 to 50 mg per 100 g and acts as a powerful antioxidant while also boosting collagen formation and immunological defense (Leja et al., 2010). The cabbage is also a rich source of thiamin (0.061– 0.064 mg/100g), riboflavin (0.040– 0.069 mg/100g), niacin (0.234–0.418 mg/100g), vitamin B6 (0.124–0.209 mg/100g), folate (18–43 μ g/100g), beta-carotene (10–670 μ g/100g) lutein (20–320 μg/100g), and trace amounts of vitamin E and K among different cultivars of cabbage (Moreb et al., 2020). Furthermore, cabbage contains trace levels of vitamin A precursors such as beta-carotene (particularly in red and savoy varieties) and folate (vitamin B9), which are essential for DNA synthesis and fetal development. The inclusion of other B-complex vitamins, such as B6 and thiamine, strengthens its involvement in metabolic activities.

Minerals

Cabbage provides an adequate supply of vital minerals. Potassium is the most abundant, with amounts ranging from 150 to 250 mg per 100 g, and it plays an important role in blood pressure and electrolyte balance. Calcium (35–40 mg/100 g) and magnesium (10–15 mg/100 g) promote bone health and enzymatic function. Small levels of trace elements, including iron, manganese, and zinc, can promote immunological and oxidative enzyme functioning (Hounsome et al., 2008a; Podsędek, 2007). In addition to these cabbages are a rich source of calcium (0.56–105g), iron (53–80 mg) magnesium (0.13–28 mg), phosphorous (0.30–40 mg), Zinc (0.18–22 mg), copper (0.019 mg), manganese (0.16), and trace amount of selenium (Moreb et al., 2020).

Dietary Fiber and Its Role

Cabbage contains around 2.5–3.5 grams of dietary fiber per 100 grams, the majority of which is insoluble fiber such as cellulose and hemicellulose. This fiber promotes bowel regularity, increases hunger, and helps regulate cholesterol (Slavin, 2013). Soluble fibers also help to maintain gut bacteria equilibrium by acting as prebiotics. Cabbage fermentation improves fiber digestibility and boosts the generation of beneficial short-chain fatty acids (SCFAs) in the colon (Knez et al., 2023).

Effects of Cooking and Processing on Nutritional Value

Cabbage's nutritional composition is considerably affected by thermal processing. Heat-sensitive vitamins, particularly vitamin C and several B-complex vitamins, can be lost while boiling or heating for an extended period of time. Steaming and microwaving are excellent techniques for preserving nutritional value (Turkmen et al., 2005). Glucosinolate levels decrease with cooking; however, their breakdown products (e.g., isothiocyanates) may still have biological activity depending on the preparation technique. Fermentation, as seen in sauerkraut and kimchi, not only preserves cabbage but also increases its nutritional value and adds helpful lactic acid bacteria (Park, Arasu, et al., 2014).

Bioactive Compounds

Numerous health-promoting bioactive chemicals, including those that contribute to the antioxidant, anti-inflammatory, anticancer, and cardioprotective qualities of cabbage, have been reported in cabbage.

Glucosinolates and Their Hydrolysis Products

Glucosinolates are secondary metabolites that include sulphur and nitrogen and are most commonly seen in cruciferous vegetables. The main glucosinolates found in cabbage include sinigrin, glucobrassicin, and gluconasturtiin (Cartea and Velasco, 2008; Martinez-Villaluenga et al., 2012; Rosen et al., 2005). Their content varies according to cultivar, maturity, environmental circumstances, and post-harvest processing (Cartea et al., 2008; Fahey et al., 2001; Hanschen et al., 2017; Kusznierewicz et al., 2008; Peñas et al., 2011; Sarvan et al., 2013) (Figure 1).

When tissue is injured, the enzyme myrosinase catalyzes the hydrolysis of glucosinolates, producing physiologically active compounds such as isothiocyanates, nitriles, epithionnitriles, oxazolidines, and indoles (Agerbirk et al., 2009; Bones and Rossiter, 2006; Martinez-Ballesta and Carvajal, 2015). However, allyl isothiocyanate, iberin, and indol-3-carbinol are the immediate breakdown products of sinigrin, glucoiberin, and glucobrassicin (Agerbirk et al., 2009) (Figure 1). Isothiocyanates, such as sulforaphane and phenethyl isothiocyanate, are known for their chemo-preventive and detoxification-enhancing properties, largely through their activation of Phase II detoxifying enzymes (Dinkova-Kostova and Kostov, 2012).

Polyphenols and Flavonoids

Polyphenols are prevalent in cabbage, which contributes to its antioxidant properties. Cabbage's primary flavonoids include quercetin and kaempferol, which have anti-inflammatory and free radical-scavenging activities. Red cabbage has a high concentration of anthocyanins, notably cyanidin-3-glucoside, which contributes to its red-purple color and has antioxidant and anticancer properties (Chun et al., 2004; Cuong et al., 2022).

Carotenoids and Vitamins

Carotenoids, including β-carotene, lutein, and zeaxanthin, are found in variable levels in cabbage. According to the USDA 2011, these chemicals boost provitamin A activity and improve ocular and immunological health by inhibiting singlet oxygen and regulating cellular signaling pathways (Podsędek, 2007; Podsędek et al., 2017).

Sulfur-containing Compounds

Cabbage contains sulfur compounds such as S-methylcysteine sulfoxide in addition to glucosinolates, which give them their distinctive smell and may have anticarcinogenic and cholesterol-lowering properties. In addition, 20 sulfur compounds are reported, of which five each of isothiocyanates and sulfides, nine are disulfides, one trisulfide, two additional isothiocyanates, and one trisulfide (Bailey et al., 1961; Rosa et al., 2001)

Ascorbic Acid and Other Antioxidants

Cabbage contains a high content of ascorbic acid (vitamin C), ranging from 30 to 50 mg/100 g in raw form. This vitamin is essential for maintaining redox equilibrium, collagen formation, and immunological function. Along with vitamin E and polyphenols, ascorbic acid improves overall antioxidant defense (Hounsome et al., 2008b; Leja et al., 2010).

Biological and Pharmacological Activities

Antioxidant Activity

Natural antioxidants, including vitamin C, polyphenols, and glucosinolates, which are abundant in cabbage, are essential for scavenging reactive oxygen

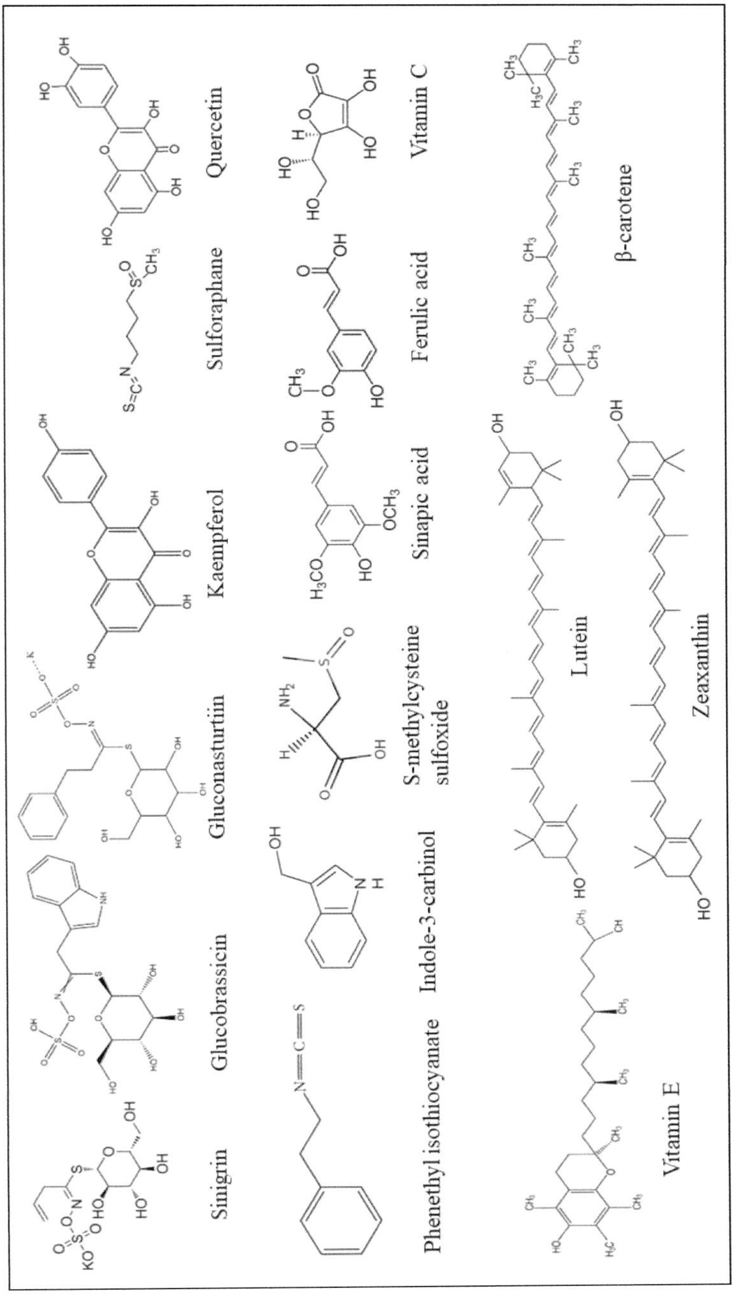

Fig. 1 Bioactive compounds present in cabbage.

species (ROS) and preventing oxidative damage. Red cabbage, in particular, has a greater anthocyanin concentration, which contributes to its increased antioxidant properties. Studies revealed 69.82, 87.23%, and 0.53% antioxidant activities by DPPH, ABTS, and FRAP assays (Liang et al., 2019). According to many other studies, the phenolic chemicals in cabbage help the body's enzymatic antioxidant defenses, such as catalase and superoxide dismutase, and scavenge free radicals (Chun et al., 2004; Guan et al., 2023; Parada et al., 2023) (Table 1).

Table 1 Antioxidant activities of cabbage leaves.

Extracts	*Assay Type*	*Effect*	*Reference*
Aqueous extract isolated at different boiling temperatures	DPPH and ORAC Assays	The effect was significant, but showed a significant difference at different drying temperatures compared to fresh samples	Vega-Galvez et al. (2023)
Methanolic extracts of different cabbage varieties	TAC by Phosphomolybdate method, TEAS, ABTS, DPPH, and FRAP assays	The results showed significant antioxidant activities with Chinese variety, ABTS = 5.72 µmol TE/g fresh weight; Green, highest DPPH ≈ 91.2 µmol TE/g fw and Red, highest FRAP ≈ 80.8 µmol TE/g fw	Rokayya et al. (2013)
Water and lipid-soluble antioxidant fractions	DPPH and ABTS assays	Evaluated both water- and lipid-soluble antioxidants and reported cabbage has appreciable antioxidant capacity	Podse dek et al. (2006)
Fresh-cut red cabbage with storage treatments in different packaging films	DPPH, ABTS (radical scavenging), T-AOC (total antioxidant capacity); enzymatic antioxidants (SOD, POD, CAT)	PVDC packaging best preserved phenolics or anthocyanins and maintained higher DPPH, ABTS, T-AOC and higher activities of SOD, POD, CAT compared to other films	Guan et al. (2023)
Different forms of cabbage	Measurements of antioxidant substances (total phenolics, flavonoids, anthocyanins/glucosinolates) and standard antioxidant assays	Comparative difference among forms with higher antioxidant substances and higher activity among different forms	Yue et al. (2024)

Contd...

Table 1 *Contd...*

Extracts	*Assay Type*	*Effect*	*Reference*
Different drying methods and methanolic extracts	DPPH and ORAC Assays	Different drying methods affected the retention of polyphenols and thus antioxidant potentials	Mejías et al. (2024)
Brassica microgreens grown in different LED lights	ABTS and DPPH assays	Red LED light increased the bioactive compound and improved the antioxidant potential	Ntsoane et al. (2024)
Fresh tissue extracts prepared with 80% methanol for phenolic extraction	DPPH, FRAP, and ABTS assays along with TPC, TFC and phenolic acid profiling using HPLC	Antioxidant activity of DPPH ≈ 69.82% scavenging, ABTS ≈ 87.23% scavenging, FRAP ≈ 0.53 scavenging	Liang et al. (2019)
Extracted using ethanol and portioned using different solvents	DPPH and H_2O_2 scavenging assays	Methanolic extracts showed the highest DPPH and H_2O_2 scavenging activities	Archana et al. (2018)
Methanolic extracts	DPPH scavenging assays	Showed strong dose dependent antioxidant activity with $IC_{50} \approx$ 750 μg/mL	Tajalli et al. (2020)
Acetone extracts	TEAC and FRAP assays	Antioxidant activity of TEAC≈ 1.73 μmol Trolox/g fw, FRAP ≈ 1.56 μmol Fe^{2+}/g	Bahorun et al. (2004)

Anti-inflammatory Effects

Cabbage components, especially indole-3-carbinol and sulforaphane, have anti-inflammatory properties via regulating inflammatory signaling pathways, including NF-κB and decreasing pro-inflammatory cytokines (Dinkova-Kostova and Kostov, 2012). Similarly, methanolic extracts of different cabbage heads showed different anti-inflammatory effects (Rokayya et al., 2013). The methanolic extracts of cabbage lowered the levels of tumor necrosis factor-alpha, interferon-gamma, interleukin-6, and monocyte chemotactic protein-1 in inflamed tissues of mice (Y. Lee et al., 2018). Such effects have been shown in both *in-vitro* cell lines and animal models, indicating their potential for treating chronic inflammatory disorders (Kasarello et al., 2022; Kwak et al., 2020; Lin et al., 2008; Mizgier et al., 2016).

Anticancer Activity

Cabbage extracts, particularly those derived from cruciferous vegetables, have demonstrated cytotoxicity against cancer cell lines such as breast, colon, and prostate malignancies. The anticancer property of the cabbage is due to the presence of phytochemicals like vitamin C, Polyphenols, carotenoids, glycosinolates, etc. The ability to alter the expression of Phase I and II detoxification enzymes, prevent DNA damage in cells, and regulate the cell cycle and apoptosis are the mechanisms underlying the anticancer activity of glucosinolate breakdown products. Glucosinolates and their breakdown products (such as isothiocyanates) decrease cancer cell growth and trigger death by caspase activation and cell cycle arrest (Ahmad Jan, 2018; Anubhuti et al., 2016; Drozdowska et al., 2020; Farag and Motaal, 2010; Hafidh et al., 2013; Mandrich and Caputo, 2020; Tajalli et al., 2020; Verhoeven et al., 1996). Epidemiological studies indicate a negative relationship between cruciferous vegetable consumption and cancer incidence, notably colorectal and lung cancers. Animal studies have also shown cabbage's chemopreventive benefits by modulating Phase I and II detoxifying enzymes and suppressing tumor development (Higdon et al., 2007). Similarly, breast cancer research suggested that higher intake of Chinese cabbage reduced the risk of breast cancer development (Fowke et al., 2003; Lin et al., 2017). Similar studies on prostate cancer risk suggested that consumption of cabbage reduced the risk of cancer (Cohen et al., 2000; Kristal and Lampe, 2002). Studies suggested that consumption of more than three portions of cabbage reduced the risk of cancer by 41% in males when compared to others with a lower rate of consumption (Cohen et al., 2000).

Antimicrobial Activity

The antimicrobial properties of cabbage are due to the presence of bioactive compounds like glucosinolates, isothiocyanates, flavonoids, and phenols (Table 2). These natural compounds can inhibit the growth of bacteria, fungi, and other microbes. Studies revealed that the extracts of cabbage have a significant effect on dental pathogens (Kavishri et al., 2024). Diethyl ether extracts of cabbage showed high growth inhibition against pathogenic bacteria and fungi (Arrais et al., 2022). Chloroform extracts of cabbage showed improvement of quality and shelf-life of meat and potential effect on food-borne pathogens (Rubab et al., 2020). Similarly, several studies showed potential effects of cabbage extracts on various food-borne pathogens (Brito et al., 2021; Muntean et al., 2021; Vale et al., 2015).

Table 2 Anti-microbial activities of cabbage leaves.

Extract	*Test Organism*	*Effect*	*Reference*
Methanolic, ethanolic, and Aqueous extracts	E. coli, B. subtilis, C. Albicans, P. vulgeris A. Niger	Showed inhibition zone of 6– 12 mm	Waghulde et al. (2018)
Aqueous extracts	E. coli, P. aeruginosa, S. enterica, S. aureus, S. mutans, B. cereus	Showed inhibition zone of 9– 18 mm	Rahmani et al. (2024)

Contd...

Table 2 *Contd...*

Extract	*Test Organism*	*Effect*	*Reference*
Diethyl ether	C. albicans, C. glabrata, S. aureus, P. aeruginosa, K. pneumoniae	MIC results showed higher inhibitory effect against the two Candida species S. aureus	Arrais et al. (2022)
Chloroform, dichloromethane, ethyl ether, toluene, ethanol, and methanol extracts	7 bacterial strains and 4 Fungal strains from ATCC and KCTC	Potential dose dependent activity and preservation of meat quality	Rubab et al. (2020)
Aqueous extracts	S. mutans, E. faecalis, S.aureus, C. albicans	Showed inhibition zone for 9–15 mm	Kavishri et al. (2024)
Aqueous extracts	E. coli, S. typhimurium, L monocytogenes, B. cereus, S. aureus	MIC results showed significant inhibition of growth of food pathogens	Vale et al. (2015)
Aqueous and ethanol extracts	S. epidermidis, A. baumannii	80% ethanol showed inhibition	Abdullah and Mohammed (2024)

Cardiovascular Protection

Cabbage promotes heart health by reducing oxidative stress, enhancing endothelial function, and lowering cholesterol levels. The extracts protect against H_2O_2-induced cell death, suppress ROS production, and increase antioxidant expression (Yang, 2018). Cabbage extracts also inhibited aortic inflammation in hyperlipidemic apolipoprotein E-deficient mice (Joo et al., 2018). Similar studies in cabbage show cardioprotective activities (Doha et al., 2021; Sankhari et al., 2012). Anthocyanins, particularly in red cabbage, have been associated with lower blood pressure and better lipid profiles. Sulforaphane in cabbage also reduces vascular inflammation (Nallasamy et al., 2014). Several other studies have suggested the role of cabbage in the reduction of low-density lipoproteins, triacylglycerol, and hepatic cholesterol levels (Bacchetti et al., 2014; Huang et al., 2016; Järvi et al., 2016)

Antidiabetic and Anti-Obesity Effects

Cabbage is good for metabolic health since it is high in fiber while being low in calories. Cabbage extract has been found in animal experiments to enhance insulin sensitivity, lower blood glucose levels, and alter adipokines associated with fat metabolism. These properties are attributed to the bioactive compounds present in cabbage leaves. These compounds help in regulating blood glucose, insulin regulation, and delaying carbohydrate digestion and absorption (Uuh-Narvaez and Segura-Campos, 2021). (Buko et al., 2018; Kataya and Hamza, 2008; Mehmood et al., 2023). Pancreatic β-cells may be further shielded from oxidative damage by the antioxidant activity (Islam and Choi, 2009; Sahai and Kumar, 2020).

Neuroprotective and Hepatoprotective Roles

Bioactive chemicals found in cabbage, particularly flavonoids and glucosinolates, may have neuroprotective properties by lowering oxidative stress in neurons and controlling neuroinflammation (Heo and Lee, 2006; K.-J. Lee et al., 2002; Vega-Galvez et al., 2023; Zhang and Jing, 2023). Hepatoprotective benefits have been seen in animal models by increasing liver antioxidant enzymes and mitigating chemically induced liver damage (Al-Dosari, 2014; Ashfaq et al., 2020; Morales-López et al., 2017; Nugroho et al., 2023).

Gastrointestinal Anti-Hypercholesterolemic Activities

Cabbage has a long history of treatment for gastrointestinal disorders. Animal model studies have shown aqueous cabbage extracts have potential in the recovery of acute gastric ulcers (Carvalho et al., 2011). The study showed the potential of cabbage extracts in managing gastric disorders. The study was further supported by similar animal models (Oguwike et al., 2014). Studies have shown the anti-hypercholesterolemic effect of cabbage extracts. Significant decrease in serum cholesterol level and increase in cholesterol 7-alpha-hydroxylase activity were seen in hepatoma-bearing rats administered with cabbage aqueous extracts (Komatsu et al., 1998). Similarly, in hypercholesterolemic human patients, total cholesterol and LDL cholesterol decreased when a vegetable and fruit beverage containing cabbage was used (Suido et al., 2002).

Functional and Medicinal Food Potential

Cabbage is not only a healthy vegetable, but it also plays an important role in the field of functional and therapeutic foods due to its high concentration of bioactive chemicals. It is gaining popularity for its role in illness prevention, intestinal health, and metabolic control, notably through fermentation and nutraceutical advancements.

Fermentation improves cabbage's functional value by stimulating the development of beneficial lactic acid bacteria (LAB), increasing nutrient bioavailability, and producing extra bioactive chemicals. Sauerkraut, a type of fermented cabbage from Europe, has high quantities of probiotics, organic acids, and vitamins C and K. It improves gut microbial balance, immunity, and digestion (Polak-Berecka et al., 2018; Siddeeg et al., 2022; Viander et al., 2003; Wiander and Ryhnen, 2005; S. Zhang et al., 2023). Fermentation also results in the partial breakdown of glucosinolates, which produces bioactive metabolites such as isothiocyanates and indoles, both of which have proven health advantages (Knez et al., 2023)

Factors Influencing Bioactive Compound Content

The content and composition of bioactive chemicals in cabbage are greatly influenced by a variety of internal and external variables. These include genetics, ambient circumstances, agricultural methods, and postharvest management. Understanding these aspects is critical for maximizing the health benefits of cabbage.

Cultivar and Genetic Variation

The genetic makeup of cabbage cultivars impacts their ability to produce and store beneficial chemicals such as glucosinolates, flavonoids, and anthocyanins. Red cabbage types have more anthocyanins and ascorbic acid than green types (Leja et al., 2010).

Agronomic Practices and Environmental Factors

Agronomic techniques like fertilizer use, irrigation, plant density, and organic vs conventional farming all have an impact on secondary metabolites production. Nitrogen levels can influence glucosinolate biosynthesis, whereas sulfur availability stimulates the formation of sulfur-containing compounds (Gu et al., 2012). Environmental variables such as light intensity, temperature, and soil type all influence the content of phenolic compounds and vitamins (Oloyede et al., 2021).

Harvest Time and Maturity Stage

The level of maturity upon harvest has a significant impact on cabbage's nutritional composition. Young cabbage plants frequently have greater glucosinolate concentrations, but complete maturation results in enhanced fibre and antioxidant levels (Rosa, 1997). Delayed harvesting can cause a decrease in vitamin C and phenolic content owing to over-ripening.

Storage and Postharvest Handling

Temperature, humidity, and storage length all have a direct impact on bioactive component stability after harvest. Vitamin C is particularly vulnerable to long-term storage and high temperatures (Lee and Kader, 2000). Modified environment packing and cold storage can help preserve glucosinolates and polyphenols, but incorrect handling causes oxidative destruction. Fermentation, blanching, and drying all affect the concentration and bioavailability of these chemicals.

Safety, Toxicity, and Antinutritional Aspects

Cabbage, like other cruciferous vegetables, contains goitrogens, principally glucosinolates and their hydrolysis products (e.g., thiocyanates and isothiocyanates), which may interfere with iodine absorption by the thyroid. Typical cabbage consumption is unlikely to cause hypothyroidism in healthy persons with appropriate iodine levels. Cooking or fermenting dramatically decreases goitrogenic chemicals, henceforth lowering risk.

Cabbage is deemed safe and well-tolerated when taken as part of a healthy diet. The WHO and national dietary standards require at least 400 g of fruits and vegetables per day, including cruciferous vegetables. A regular portion of cabbage (75– 100 g raw or cooked) contains vital nutrients without exceeding safety limits. Long-term excessive consumption, particularly uncooked, may contribute to gas, bloating, and mineral absorption problems due to the fiber and phytate content. Vulnerable groups, such as those with thyroid issues or on certain drugs, should seek expert advice on recommended intake levels.

Future Perspectives and Conclusion

Cabbage contains a high concentration of bioactive chemicals with known health advantages, but numerous elements require more exploration to fully realize its potential as a functional food. Future studies should concentrate on genetic improvement using precision breeding and current biotechnological technologies like CRISPR/Cas9 to increase the concentration of glucosinolates, flavonoids, and other phytochemicals. Although the biological activities of compounds derived from cabbage have been highlighted by several *in vitro* and *in vivo* studies, well-designed human clinical trials are still needed to confirm their effectiveness, especially regarding cancer prevention, metabolic health, and neurodegenerative ailments. Studies that focus on the gut microbiota and metabolomics might provide important new information on the absorption and metabolic fate of such substances in the human body.

To improve stability and targeted release, new cabbage-based nutraceuticals, fermented foods, and delivery methods such as nano-encapsulation should be developed using food technology advancements. To encourage sustainable agriculture techniques, the impact of climate change and environmental stresses on phytochemical production should be investigated. In order to bridge the gap between laboratory results and the application of public health, interdisciplinary research integrating plant science, nutrition, and clinical care is crucial.

Future studies should concentrate on combining eco-friendly agricultural practices, clinical assessments, and omics technology to fully realize cabbage's nutraceutical potential. Plant scientists, dietitians, and clinical researchers must work together across academic boundaries to convert laboratory results into practical health applications.

References

Ağagündüz, D., Şahin, T.Ö., Yılmaz, B., Ekenci, K.D., Duyar Özer, Ş. and Capasso, R. 2022. Cruciferous Vegetables and Their Bioactive Metabolites: From Prevention to Novel Therapies of Colorectal Cancer. *Evid.-based Complement. Altern. Med., 2022*: 1–20.

Agerbirk, N., De Vos, M., Kim, J.H. and Jander, G. 2009. Indole glucosinolate breakdown and its biological effects. *Phytochem. Rev.*, *8* (1): 101–120. https://doi.org/10.1007/s11101-008-9098-0.

Ahmad Jan, S. 2018. Antioxidant and anticancer activities of *Brassica rapa*: A review. *MOJ Biol. Med.*, *3* (4): 175–178. https://doi.org/10.15406/mojbm.2018.03.00094.

Al-Dosari, M.S. 2014. Red Cabbage (*Brassica oleracea* L.) Mediates Redox-Sensitive Amelioration of Dyslipidemia and Hepatic Injury Induced by Exogenous Cholesterol Administration. *Am. J. Chin. Med.*, *42* (01): 189–206. https://doi.org/10.1142/S0192415X1450013X.

Al-Shehbaz, I.A. 2011. Brassicaceae (Mustard Family). In: *Encyclopedia of Life Sciences*. Wiley.

Al-Shehbaz, I.A. Beilstein, M.A. and Kellogg, E.A. 2006. Systematics and phylogeny of the Brassicaceae (Cruciferae): An overview. *Plant Syst. Evol.*, *259* (2–4): 89–120. https://doi.org/10.1007/s00606-006-0415-z.

Anubhuti, S., Ashok, S., Prashant, Y. and Dhiraj, S. 2016. Isothiocyanates in Brassica: Potential Anti Cancer Agents. *Asian Pac. J. Cancer Prev.*, *17* (9): 4507–4510.

Archana, J., Kusuma, M. and Vijayabhargavi, C. 2018. Antioxidant and Anti Cholinesterase Potential of Red Cabbage (*Brassica oleracea* var. *capitata* f. rubra). *Eur. J. Med. Plants*, *22* (1): 1–7.

Arrais, A., Testori, F., Calligari, R., Gianotti, V., Roncoli, M., Caramaschi, A., Todeschini, V., Massa, N. and Bona, E. 2022. Extracts from Cabbage Leaves: Preliminary Results towards a "Universal" Highly-Performant Antibacterial and Antifungal Natural Mixture. *Biology, 11* (7): 1080. https://doi.org/10.3390/biology11071080.

Ashfaq, F., Butt, M.S., Bilal, A. and Suleria, H.A.R. 2020. Hepatoprotective Effects of Red Cabbage in Hypercholesterolemic Diet-induced Oxidative Stressed Rabbits. *Curr. Bioact. Compd.*, *16* (4): 469–480. https://doi.org/10.2174/1573407215666190124113738.

Ayers, J.F. 2000. The Use of Alternative Therapies in the Support of Breastfeeding. *J. Hum. Lact.*, *16* (1): 52–56. https://doi.org/10.1177/089033440001600111.

Bacchetti, T., Tullii, D., Masciangelo, S., Gesuita, R., Skrami, E., Brugè, F., Silvestri, S., Orlando, P., Tiano, L. and Ferretti, G. 2014. Effect of black and red cabbage on plasma carotenoid levels, lipid profile, and oxidized low density lipoprotein. *J. Funct. Foods*, *8*: 128–137. https://doi.org/10.1016/j.jff.2014.02.020.

Bahorun, T., Luximon-Ramma, A., Crozier, A. and Aruoma, O.I. 2004. Total phenol, flavonoid, proanthocyanidin, and vitamin C levels and antioxidant activities of Mauritian vegetables. *J. Sci. Food Agri.*, *84* (12): 1553–1561.

Bailey, S.D., Bazinet, M.L., Driscoll, J.L. and McCarthy, A.I. 1961. The Volatile Sulfur Components of Cabbage. *J. Food Sci.*, *26* (2): 163–170.

Beecher, C., 1994. Cancer preventive properties of varieties of *Brassica oleracea*: A review. *Am. J. Clin. Nutr.*, *59* (5): 1166S–1170S. https://doi.org/10.1093/ajcn/59.5.1166S.

Bones, A. and Rossiter, J. 2006. The enzymic and chemically induced decomposition of glucosinolates. *Phytochemistry, 67* (11): 1053–1067.

Brito, T.B.N., Lima, L.R.S., Santos, M.C.B., Moreira, R.F.A., Cameron, L.C., Fai, A.E.C. and Ferreira, M.S.L.M. 2021. Antimicrobial, antioxidant, volatile, and phenolic profiles of cabbage-stalk and pineapple-crown flour revealed by GC-MS and UPLC-MSE. *Food Chem.*, *339*: 127882. https://doi.org/10.1016/j.foodchem.2020.127882.

Buko, V., Zavodnik, I., Kanuka, O., Belonovskaya, E., Naruta, E., Lukivskaya, O., Kirko, S., Budryn, G., Żyżelewicz, D., Oracz, J. and Sybirna, N. 2018. Antidiabetic effects and erythrocyte stabilization by red cabbage extract in streptozotocin-treated rats. *Food Funct.*, *9* (3): 1850–1863.

Burbulis, N. and Kott, L.S. 2013. Application of doubled haploid technology in breeding of Brassica napus. In: *From Plant Genomics to Plant Biotechnology*. Elsevier, pp. 183–203. https://doi.org/10.1533/9781908818478.183.

Cartea, M.E. and Velasco, P. 2008. Glucosinolates in Brassica foods: Bioavailability in food and significance for human health. *Phytochem. Rev.*, *7* (2): 213–229. https://doi.org/10.1007/s11101-007-9072-2.

Cartea, M.E., Velasco, P., Obregón, S., Padilla, G. and De Haro, A.A. 2008. Seasonal variation in glucosinolate content in *Brassica oleracea* crops grown in northwestern Spain. *Phytochemistry, 69* (2): 403–410.

Carvalho, C.A.D., Fernandes, K.M., Matta, S.L.P., Silva, M.B.D., Oliveira, L.L.D. and Fonseca, C.C. 2011. Evaluation of antiulcerogenic activity of aqueous extract of *Brassica oleracea* var. *capitata* (cabbage) on Wistar rat gastric ulceration. *Arq. Gastroenterol., 48* (4): 276–282. https://doi.org/10.1590/s0004-28032011000400011.

Cavender, A. 2006. Folk medical uses of plant foods in southern Appalachia, United States. *J. Ethnopharmacol., 108* (1): 74–84. https://doi.org/10.1016/j.jep.2006.04.008.

Chun, O.K., Smith, N., Sakagawa, A. and Lee, C.Y. 2004. Antioxidant properties of raw and processed cabbages. *Int. J. Food Sci. Nutr., 55* (3): 191–199.

Cohen, J.H., Kristal, A.R. and Stanford, J.L. 2000. Fruit and Vegetable Intakes and Prostate Cancer Risk. *J. Natl. Cancer Inst., 92* (1): 61–68. https://doi.org/10.1093/jnci/92.1.61.

Cuong, D.M., Kim, H.Y., Keshawa Ediriweera, M. and Cho, S.K. 2022. Evaluation of Phytochemical Content and the Antioxidant and Antiproliferative Potentials of Leaf Layers of Cabbage Subjected to Hot Air and Freeze-Drying. *J. Food Qual., 2022*: 1–13.

Dinkova-Kostova, A.T. and Kostov, R.V. 2012. Glucosinolates and isothiocyanates in health and disease. *Trends Mol. Med., 18* (6): 337–347.

Dixon, G.R. 2007. *Vegetable Brassicas and Related Crucifers*. Wallingford, UK; Cambridge, MA: CABI Pub.

Doha, M., Hoda, M., Sherein, A. and Hagar, E. 2021. Cardioprotective potency of anthocyanin-rich extract of red cabbage against isoproterenol-induced myocardial infarction in experimental animals. *J. Appl. Pharm. Sci., 11*: 22– 30. https://doi.org/10.7324/JAPS.2021.110804.

Drozdowska, M., Leszczyńska, T., Koronowicz, A., Piasna-Słupecka, E. and Dziadek, K. 2020. Comparative study of young shoots and the mature red headed cabbage as antioxidant food resources with antiproliferative effect on prostate cancer cells. *RSC Adv., 10* (70): 43021–43034.

Fahey, J.W., Zalcmann, A.T. and Talalay, P. 2001. The chemical diversity and distribution of glucosinolates and isothiocyanates among plants. *Phytochemistry, 56* (1): 5–51.

Farag, M.A. and Motaal A.A.A. 2010. Sulforaphane composition, cytotoxic, and antioxidant activity of crucifer vegetables. *J. Adv. Res., 1* (1): 65–70.

Fowke, J.H., Chung, F.-L., Jin, F., Qi, D., Cai, Q., Conaway, C., Cheng, J.-R., Shu, X.-O., Gao, Y.-T. and Zheng, W. 2003. Urinary isothiocyanate levels, brassica, and human breast cancer. *Cancer Res., 63* (14): 3980–3986.

Gu, Z., Guo, Q. and Gu, Y. 2012. Factors Influencing Glucoraphanin and Sulforaphane Formation in Brassica Plants: A Review. *J. Integrat. Agri., 11* (11): 1804–1816.

Guan, Y., Ji, Y., Yang, X., Pang, L., Cheng, J., Lu, X., Zheng, J., Yin, L. and Hu, W. 2023. Antioxidant activity and microbial safety of fresh-cut red cabbage stored in different packaging films. *LWT, 175*: 114478. https://doi.org/10.1016/j.lwt.2023.114478.

Hafidh, R.R., Abdulamir, A.S., Abu Bakar, F., Jalilian, F.A., Jahanshiri, F., Abas, F. and Sekawi, Z. 2013. Novel anticancer activity and anticancer mechanisms of *Brassica oleracea* L. var. *capitata* f. rubra. *Eur. J. Integrat. Med., 5* (5): 450–464.

Haghighi, M., Barzegar Sadeghabad, A. and Abolghasemi, R. 2022. Effect of exogenous amino acids application on the biochemical, antioxidant, and nutritional value of some leafy cabbage cultivars. *Sci. Rep., 12* (1): 17720.

Hanschen, F.S., Klopsch, R., Oliviero, T., Schreiner, M., Verkerk, R. and Dekker, M. 2017. Optimizing isothiocyanate formation during enzymatic glucosinolate breakdown by adjusting pH value, temperature, and dilution in Brassica vegetables and Arabidopsis thaliana. *Sci. Rep., 7* (1): 40807.

Hatfield, G. 2004. *Encyclopedia of Folk Medicine: Old World and New World Traditions*. Santa Barbara, Calif: ABC-CLIO.

Heo, H.J. and Lee, C.Y. 2006. Phenolic phytochemicals in cabbage inhibit amyloid β protein-induced neurotoxicity. *LWT - Food Sci. Technol., 39* (4): 331–337.

Higdon, J., Delage, B., Williams, D. and Dashwood, R. 2007. Cruciferous vegetables and human cancer risk: Epidemiologic evidence and mechanistic basis. *Pharm. Res., 55* (3), 224–236.

Hounsome, N., Hounsome, B., Tomos, D. and Edwards-Jones, G. 2008a. Plant Metabolites and Nutritional Quality of Vegetables. *J. Food Sci.. 73* (4): R48– 65. https://doi.org/10.1111/j.1750-3841.2008.00716.x.

Hounsome, N., Hounsome, B., Tomos, D. and Edwards-Jones, G.G. 2008b. Plant Metabolites and Nutritional Quality of Vegetables. *Journal of Food Science, 73* (4): R48– 65.

Huang, H., Jiang, X., Xiao, Z., Yu, L., Pham, Q., Sun, J., Chen, P., Yokoyama, W., Yu, L.L., Luo, Y.S. and Wang, T.T.Y. 2016. Red Cabbage Microgreens Lower Circulating Low-Density Lipoprotein (LDL), Liver Cholesterol, and Inflammatory Cytokines in Mice Fed a High-Fat Diet. *J. Agri. Food Chem., 64* (48): 9161–9171.

Islam, Md.S. and Choi, H.H. 2009. Antidiabetic Effect of Korean Traditional Baechu (Chinese Cabbage) Kimchi in a Type 2 Diabetes Model of Rats. *J. Med. Food, 12* (2): 292–297.

Järvi, A., Karlström, B., Vessby, B. and Becker, W. 2016. Increased intake of fruits and vegetables in overweight subjects: Effects on body weight, body composition, metabolic risk factors, and dietary intake. *Br. J. Nutr., 115* (10): 1760–1768.

Joo, H., Choi, S., Lee, Y., Lee, E., Park, M., Park, K., Kim, C.-S., Lim, Y., Park, J.-T. and Jeon, B. 2018. Anthocyanin-Rich Extract from Red Chinese Cabbage Alleviates Vascular Inflammation in Endothelial Cells and Apo E−/− Mice. *Int. J. Mol. Sci., 19* (3): 816.

Kasarello, K., Köhling, I., Kosowska, A., Pucia, K., Lukasik, A., Cudnoch-Jedrzejewska, A., Paczek, L., Zielenkiewicz, U. and Zielenkiewicz, P. 2022. The Anti-Inflammatory Effect of Cabbage Leaves Explained by the Influence of bol-miRNA172a on FAN Expression. *Front. Pharmacol., 13*: 846830.

Kataya, H.A.H. and Hamza, A.A. 2008. Red Cabbage (*Brassica oleracea*) Ameliorates Diabetic Nephropathy in Rats. *Evid.-based Compl. Alter. Med., 5* (3): 281–287.

Kavishri, S., Panneer Selvam, S., Shanmugam, R., Ramadoss, R., Sundar, S. and Ramani, P. 2024. Exploring the Antimicrobial Potential and Cytotoxic Effects of Different *Brassica oleracea* Varieties. *Cureus, 16*: e59613. https://doi.org/10.7759/cureus.59613.

Knez, E., Kadac-Czapska, K. and Grembecka, M. 2023. Effect of Fermentation on the Nutritional Quality of the Selected Vegetables and Legumes and Their Health Effects. *Life, 13* (3): 655.

Komatsu, W., Miura, Y. and K. Yagasaki. 1998. Suppression of hypercholesterolemia in hepatoma-bearing rats by cabbage extract and its component, S-methyl-L-cysteine sulfoxide. *Lipids, 33* (5): 499–503.

Kristal, A.R. and Lampe, J.W. 2002. Brassica Vegetables and Prostate Cancer Risk: A Review of the Epidemiological Evidence. *Nutr. Cancer, 42* (1): 1–9.

Kusznierewicz, B., Bartoszek, A., Wolska, L., Drzewiecki, J., Gorinstein, S. and Namieśnik, J. 2008. Partial characterization of white cabbages (*Brassica oleracea* var. *capitata* f. alba) from different regions by glucosinolates, bioactive compounds, total antioxidant activities, and proteins. *LWT - Food Sci. Technol., 41* (1), 1–9.

Kwak, J.H., Kim, Y., Ryu, S.I., Lee, M., Lee, H., Lim, Y.P. and J.K. Paik. 2020. Anti-inflammatory effect from extracts of Red Chinese cabbage and Aronia in LPS-stimulated RAW 264.7 cells. *Food Sci. Nutr., 8* (4): 1898–1903.

Lee, K.-J., Sok, D.-E., Kim, Y.-B. and Kim, M.R. 2002. Protective effect of vegetable extracts on oxidative stress in brain of mice administered with NMDA. *Food Res. Int., 35* (1): 55–63.

Lee, S.K. and Kader, A.A. 2000. Preharvest and postharvest factors influencing vitamin C content of horticultural crops. *Postharvest Biology and Technology, 20* (3), 207–220.

Lee, Y., Kim, S., Yang, B., Lim, C., Kim, J.-H., Kim, H. and Cho, S. 2018. Anti-inflammatory effects of *Brassica oleracea* var. *capitata* L. (Cabbage) methanol extract in mice with contact dermatitis. *Pharmacogn. Mag.*, *14* (54): 174.

Leja, M., Kamińska, I. and Kołton, A. 2010. Phenolic compounds as the major antioxidants in red cabbage. *Folia Hortic.*, *22* (1): 19–24.

Liang, Y., Li, Y., Zhang, L. and Liu, X. 2019. Phytochemicals and antioxidant activity in four varieties of head cabbages commonly consumed in China. *Food Prod. Process. Nutr.*, *1* (1): 3.

Lin, J.-Y., Li, C.-Y. and Hwang, I.-F. 2008. Characterisation of the pigment components in red cabbage (*Brassica oleracea* L. var.) juice and their anti-inflammatory effects on LPS-stimulated murine splenocytes. *Food Chem.*, *109* (4): 771–781.

Lin, T., Zirpoli, G.R., McCann, S.E., Moysich, K.B., Ambrosone, C.B. and Tang, L. 2017. Trends in Cruciferous Vegetable Consumption and Associations with Breast Cancer Risk: A Case-Control Study. *Curr. Devel. Nutr.*, *1* (8), e000448.

Mandrich, L. and Caputo, E. 2020. Brassicaceae-Derived Anticancer Agents: Towards a Green Approach to Beat Cancer. *Nutrients*, *12* (3): 868.

Marc, E.B., Nelly, A., Annick, D.-D. and Frederic, D. 2008. Plants used as remedies antirheumatic and antineuralgic in the traditional medicine of Lebanon. *J. Ethnopharmacol.*, *120* (3): 315–334.

Martinez-Ballesta, M.D.C. and Carvajal, M. 2015. Myrosinase in Brassicaceae: The most important issue for glucosinolate turnover and food quality. *Phytochem. Rev.*, *14* (6): 1045–1051.

Martinez-Villaluenga, C., Peñas, E., Sidro, B., Ullate, M., Frias, J. and Vidal-Valverde, C. 2012. White cabbage fermentation improves ascorbigen content, antioxidant, and nitric oxide production inhibitory activity in LPS-induced macrophages. *LWT - Food Sci. Technol.*, *46* (1): 77–83.

Mehmood, A., Zeb, A. and Ateeq, M.K. 2023. *In vivo* antidiabetic effects of phenolic compounds of spinach, mustard, and cabbage leaves in mice. *Heliyon*, *9* (6): e16616.

Mejías, N., Vega-Galvez, A., Gomez-Perez, L.S., Pasten, A., Uribe, E., Cortés, A., Valenzuela-Barra, G., Camus, J., Delporte, C. and Bernal, G. 2024. Health-Promoting Properties of Processed Red Cabbage (*Brassica oleracea* var. *capitata* f. rubra): Effects of Drying Methods on Bio-Compound Retention. *Foods*, *13* (6): 830.

Mizgier, P., Kucharska, A.Z., Sokół-Łętowska, A., Kolniak-Ostek, J., Kidoń, M. and Fecka, I. 2016. Characterization of phenolic compounds and antioxidant and anti-inflammatory properties of red cabbage and purple carrot extracts. *J. Funct. Foods*, *21*: 133–146.

Morales-López, J., Centeno-Álvarez, M., Nieto-Camacho, A., López, M.G., Pérez-Hernández, E., Pérez-Hernández, N. and Fernández-Martínez, E. 2017. Evaluation of antioxidant and hepatoprotective effects of white cabbage essential oil. *Pharm. Biol.*, *55* (1): 233–241.

Moreb, N., Murphy, A., Jaiswal, S. and Jaiswal, A.K. 2020. Cabbage. In: *Nutritional Composition and Antioxidant Properties of Fruits and Vegetables*. Elsevier, pp. 33–54.

Muntean, D., Ştefănuţ, M., Căta, A., Buda, V., Danciu, C., Bănică, R., Pop, R., Licker, M. and Ienaşcu, I. 2021. Symmetrical Antioxidant and Antibacterial Properties of Four Romanian Cruciferous Extracts. *Symmetry, 13* (5): 893.

Nallasamy, P., Si, H., Babu, P.V.A., Pan, D., Fu, Y., Brooke, E.A.S., Shah, H., Zhen, W., Zhu, H., Liu, D., Li, Y. and Jia, Z. 2014. Sulforaphane reduces vascular inflammation in mice and prevents TNF-α-induced monocyte adhesion to primary endothelial cells through interfering with the NF-κB pathway. *J. Nutr. Biochem.*, *25* (8): 824–833.

Ntsoane, M.L.L., Manhivi, V.E., Shoko, T., Seke, F., Sultanbawa, Y. and Sivakumar, D., 2024. Brassica microgreens cabbage (*Brassica oleracea*), radish (*Raphanus sativus*) and rocket (*Eruca vesicaria*) (L.) Cav: Application of red-light emitting diodes lighting during postharvest storage and *in vitro* digestion on bioactive compounds and antioxidant activity. *Int. J. Food Sci. Technol.*, *59* (3): 1432–1442.

Nugroho, D., Surya, R., Nurkolis, F., Surya, E., Thinthasit, A., Kamal, N., Oh, J.-S. and Benchawattananon, R. 2023. Hepatoprotective effects of ethnic cabbage dishes: A comparison study on kimchi and pao cai. *J. Ethn. Foods*, *10* (1): 31.

Oguwike, F.N., C.C, O., A.N, N. and S.O, E. 2014. Evaluation of Efficacy of Cabbage Juice (*Brassica Oleracea* Linn.) As Potential Antiulcer Agent and Its Effect on the Haemostatic Mechanism of Male Albino Wistar Rats. *IOSR J. Dental Med. Sci.*, *13* (1): 92–97.

Oloyede, O.O., Wagstaff, C. and Methven, L. 2021. Influence of Cabbage (*Brassica oleracea*) Accession and Growing Conditions on Myrosinase Activity, Glucosinolates, and Their Hydrolysis Products. *Foods*, *10* (12): 2903.

Parada, R.B., Marguet, E., Campos, C. and Vallejo, M. 2023. Improved antioxidant capacity of three Brassica vegetables by two-step-controlled fermentation using isolated autochthone strains of the genus *Leuconostoc* spp. and *Lactiplanti bacillus* spp. *Food Chem.: Mol. Sci.*, *6*: 100163.

Park, S., Arasu, M.V., Jiang, N., Choi, S.-H., Lim, Y.P., Park, J.-T., Al-Dhabi, N.A. and Kim, S.-J. 2014. Metabolite profiling of phenolics, anthocyanins, and flavonols in cabbage (*Brassica oleracea* var. *capitata*). *Ind. Crops Prod.*, *60*: 8–14.

Park, S., Valan Arasu, M., Lee, M.-K., Chun, J.-H., Seo, J.M., Lee, S.-W., Al-Dhabi, N.A. and Kim, S.-J. 2014. Quantification of glucosinolates, anthocyanins, free amino acids, and vitamin C in inbred lines of cabbage (*Brassica oleracea* L.). *Food Chem. 145*: 77–85.

Passalacqua, N.G., Guarrera, P.M. and De Fine, G. 2007. Contribution to the knowledge of the folk plant medicine in Calabria region (Southern Italy). *Fitoterapia*, *78* (1): 52–68.

Peñas, E., Frias, J., Martínez-Villaluenga, C. and Vidal-Valverde, C. 2011. Bioactive Compounds, Myrosinase Activity, and Antioxidant Capacity of White Cabbages Grown in Different Locations of Spain. *J. Agric. Food Chem.*, *59* (8): 3772–3779.

Peñas, E., Martinez-Villaluenga, C. and Frias, J. 2017. Sauerkraut. In: *Fermented Foods in Health and Disease Prevention*. Elsevier, pp. 557–576.

Podse dek, A., Sosnowska, D., Redzynia, M. and Anders, B. 2006. Antioxidant capacity and content of *Brassica oleracea* dietary antioxidants. *Int. J. Food Sci. Technol.*, *41* (S1): 49–58.

Podsędek, A. 2007. Natural antioxidants and antioxidant capacity of Brassica vegetables: A review. *LWT - Food Sci. Technol.*, *40* (1): 1–11.

Podsędek, A., Majewska, I. and Kucharska, A.Z. 2017. Inhibitory Potential of Red Cabbage against Digestive Enzymes Linked to Obesity and Type 2 Diabetes. *J. Agric. Food Chem.*, *65* (33): 7192–7199.

Polak-Berecka, M., Kubik-Komar, A., Gustaw, K., Michalak, M., Kazimierczak, W. and Waśko, A. 2018. Functional traits of *Lactobacillus plantarum* from fermented *Brassica oleracea* var. *capitata* L. in view of multivariate statistical analysis. *Eur. Food Res. Technol.*, *244* (10): 1719–1727.

Rahmani, M., Esmaeili. A., Taherkhani. M., 2024. Antimicrobial potential of Brassica oleracea extracts (White and Broccoli) and their resistance compared to doxycycline against gram positive and gram negative bacteria. *Food Sci. Nutr*. 12: 7492–7503.

Rokayya, S., Li, C.-J., Zhao, Y., Li, Y. and Sun, C.-H. 2013. Cabbage (Brassica oleracea L. var. capitata) Phytochemicals with Antioxidant and Anti-inflammatory Potential. *Asian Pac. J. Cancer Prev.*, *14* (11): 6657–6662.

Rosa, E. 1997. Daily Variation in Glucosinolate Concentrations in the Leaves and Roots of Cabbage Seedlings in Two Constant Temperature Regimes. *J. Sci. Food Agric.*, *73* (3): 364–368.

Rosa, E., David, M. and Gomes, M.H. 2001. Glucose, fructose and sucrose content in broccoli, white cabbage, and Portuguese cabbage grown in early and late seasons. *J. Sci. Food Agric.*, *81* (12): 1145–1149.

Rosen, C.J., Fritz, V.A., Gardner, G.M., Hecht, S.S., Carmella, S.G. and Kenney, P.M. 2005. Cabbage Yield and Glucosinolate Concentrations as Affected by Nitrogen and Sulfur Fertility. *HortScience*, *40* (5): 1493–1498.

Rubab, M., Chelliah, R., Saravanakumar, K., Kim, J.-R., Yoo, D., Wang, M.-H. and Oh, D.-H. 2020. Phytochemical characterization, and antioxidant and antimicrobial activities of white cabbage extract on the quality and shelf-life of raw beef during refrigerated storage. *RSC Adv.*, *10* (68): 41430–41442.

Sahai, V. and Kumar, V., 2020. Anti-diabetic, hepatoprotective, and antioxidant potential of *Brassica oleracea* sprouts. *Biocatal. Agric. Biotechnol.*, *25*: 101623.

Sankhari, J.M., Thounaojam, M.C., Jadeja, R.N., Devkar, R.V. and Ramachandran, A.V. 2012. Anthocyanin-rich red cabbage (*Brassica oleracea* L.) extract attenuates cardiac and hepatic oxidative stress in rats fed an atherogenic diet. *J. Sci. Food Agric.*, *92* (8): 1688–1693.

Sarvan, I., Valerio, F., Lonigro, S.L., De Candia, S., Verkerk, R., Dekker, M. and Lavermicocca, P. 2013. Glucosinolate content of blanched cabbage (*Brassica oleracea* var. *capitata*) fermented by the probiotic strain Lactobacillus paracasei LMG-P22043. *Food Res. Int.*, *54* (1): 706–710.

Sayas-Barberá, E., Pérez-Álvarez, J.A., Navarro-Rodríguez De Vera, C., Fernández-López, M., Viuda-Martos, M. and Fernández-López, J. 2022. Sustainability and Gender Perspective in Food Innovation: Foods and Food Processing Coproducts as Source of Macro- and Micro-Nutrients for Woman-Fortified Foods. *Foods*, *11* (22): 3661.

Siddeeg, A., Afzaal, M., Saeed, F., Ali, R., Shah, Y.A., Shehzadi, U., Ateeq, H., Waris, N., Hussain, M., Raza, M.A. and Al-Farga, A. 2022. Recent updates and perspectives of fermented healthy super food sauerkraut: A review. *International J. Food Prop.*, *25* (1): 2320–2331.

Slavin, J. 2013. Fiber and Prebiotics: Mechanisms and Health Benefits. *Nutrients*, *5* (4); 1417–1435.

Suido, H., Tanaka, T., Tabei, T., Takeuchi, A., Okita, M., Kishimoto, T., Kasayama, S. and Higashino, K. 2002. A Mixed Green Vegetable and Fruit Beverage Decreased the Serum Level of Low-Density Lipoprotein Cholesterol in Hypercholesterolemic Patients. *J. Agric. Food Chem.*, *50* (11): 3346–3350.

Tajalli, F., Saeedi, M. and Vafaei Malekabadi, A. 2020. Anticancer and antioxidant effects of red cabbage on three cancerous cell lines and comparison with a normal cell line (HFF-3). *J. Genes Cells*, *6* (1): 12.

Ten Caat, N., Tenpierik, M. and Van Den Dobbelsteen, A. 2022. Towards a More Sustainable Urban Food System: Carbon Emissions Assessment of a Diet Transition with the FEW print Platform. *Sustainability*, *14* (3): 1797.

Turkmen, N., Sari, F. and Velioglu, Y. 2005. The effect of cooking methods on total phenolics and antioxidant activity of selected green vegetables. *Food Chem.*, *93* (4): 713–718.

Uuh-Narvaez, J.J. & M.R. Segura-Campos. 2021. Cabbage (*Brassica oleracea* var. *capitat*a): A food with functional properties aimed to Type 2 diabetes prevention and management. *J. Food Sci.*, *86* (11): 4775–4798.

Vale, A.P., Santos, J., Melia, N., Peixoto, V., Brito, N.V. and Oliveira, M.B.P.P. 2015. Phytochemical composition and antimicrobial properties of four varieties of *Brassica oleracea* sprouts. *Food Cont.*, *55*: 248–256.

Vega-Galvez, A., Gomez-Perez, L.S., Zepeda, F., Vidal, R.L., Grunenwald, F., Mejías, N., Pasten, A., Araya, M. and Ah-Hen, K.S. 2023. Assessment of Bio-Compounds Content, Antioxidant Activity, and Neuroprotective Effect of Red Cabbage (*Brassica oleracea* var. *capitata f. rubra*) Processed by Convective Drying at Different Temperatures. *Antioxidants*, *12* (9): 1789.

Verhoeven, D.T., Goldbohm, R.A., van Poppel, G., Verhagen, H. and P.A. van den Brandt, P.A., 1996. Epidemiological studies on brassica vegetables and cancer risk. *Cancer Epidemiol. Biomarkers Prev.*, *5* (9): 733–748.

Viander, B., Mäki, M. and Palva, A. 2003. Impact of low salt concentration, salt quality on natural large-scale sauerkraut fermentation. *Food Microbiol.*, *20* (4): 391–395.

Waghulde, S., Abid Khan, N., Gorde, N., Kale, M., Naik, P., and Yewale, R.P., 2018. Comparative antimicrobial activity study of *Brassica oleceracea*. *Proceedings*, 9: 64.

Wiander, B. and Ryhnen, E.-L. 2005. Laboratory and large-scale fermentation of white cabbage into sauerkraut and sauerkraut juice by using starters in combination with mineral salt with a low NaCl content. *Eur. Food Res. Technol.*, *220* (2): 191–195.

Yang, D.K., 2018. Cabbage (*Brassica oleracea* var. *capitata*) Protects against H_2O_2-Induced Oxidative Stress by Preventing Mitochondrial Dysfunction in H9c2 Cardiomyoblasts. *Evid.-Bas. Complement. Alter. Med.*, *2018* (1): 2179021.

Yue, Z., Zhang, G., Wang, J., Wang, J., Luo, S., Zhang, B., Li, Z. and Liu, Z. 2024. Comparative study of the quality indices, antioxidant substances, and mineral elements in different forms of cabbage. *BMC Plant Biol.*, *24* (1): 187.

Zhang, N. and Jing, P. 2023. Red Cabbage Anthocyanins Attenuate Cognitive Impairment by Attenuating Neuroinflammation and Regulating Gut Microbiota in Aging Mice. *J. Agric. Food Chem.*, *71* (41): 15064–15072.

Zhang, S., Zhang, Y., Wu, L., Zhang, L. and Wang, S. 2023. Characterization of microbiota of naturally fermented sauerkraut by high-throughput sequencing. *Food Sci. Biotechnol.*, *32* (6): 855–862.

9 Bioactive Compounds and Biological Activities of Amaranth (*Amaranthus* spp.) Leaves

Lecholocholo N.[1] and Sivakumar D.[2,3*]

Introduction

Food insecurity and the overall scarcity of calorie intake, mainly in developing countries, have led to malnutrition, affecting approximately 735 million people currently facing hunger, compared to 613 million in 2019 (WHO, 2023). Among them, about 20 million suffer from hidden hunger caused by mineral and vitamin deficiencies. While main foods are considered the primary sources of energy, they can sometimes lack iron, zinc, iodine, vitamin E, ascorbic acid, and carotenoids. Amaranth leaves contain significantly more calcium (20 times), vitamin C (13 times), iron (7 times), and beta-carotene (18 times) compared to lettuce (Sarker et al., 2024). Amaranth (*Amaranthus* spp.) is known as a highly versatile and resilient crop, holding substantial economic and nutritional value, especially in tropical and subtropical regions (Mukuwapasi et al., 2024).

This quickly growing and affordable vegetable has a C4 pathway and versatile uses, including cultivating amaranth as grains, ornamentals, or vegetables (Sarker et al., 2020). The leafy parts of amaranth are rich in protein, digestible fibre, carotenoids, vitamin C, and various elements like phosphorus, magnesium, copper, sulfur, calcium, zinc, potassium, sodium, manganese, boron, molybdenum, and iron (Sarker et al., 2024). Among red and green amaranthus (Figure 1), it also contains abundant colourants such as betalains, betaxanthins, carotenoids, chlorophylls, and betacyanins, along with bioactive compounds including vitamin C, phenols, their acids, and flavonoids (Sarker et al., 2022). These pigments and phytochemicals have a significant ability to neutralize free radicals, offer potential health benefits, and greatly influence the food industry (Tufail et al., 2024; Sarker et al., 2024). Li et al. (2015) also pointed out that highly pigmented vegetables,

[1] Department of Crop Science, Tshwane University of Technology, Pretoria, South Africa.
[2] South African Research in Phytochemical Food Network, Tshwane University of Technology, Pretoria, South Africa.
[3] Centre for Nutrition and Food Sciences, Queensland Alliance for Agriculture and Food Innovation, The University of Queensland, QLD 4108, Queensland, Australia.
* Corresponding author: d.sivakumar@uq.edu.au; SivakumarD@tut.ac.za

DOI: 10.1201/9781003687832-9

such as Amaranthus, typically have higher nutritional value than other species of the same plant. Most Amaranthus plants display a bright red-violet colour, and their pigments are generally attributed to the maroon pigment. Amaranth leaves contribute to food security due to their rich nutritional content, especially protein, vitamins (A, C, folate), and antioxidants, which combat malnutrition and provide essential micronutrients, particularly in developing regions (Mukuwapasi et al., 2024).

The crop is an annual or short-cycle perennial, monoecious, dicotyledonous plant species commonly consumed as leafy vegetables (Sharma et al., 2025). It belongs to the Amaranthaceae family, which was previously classified under Chenopodium, within the Caryophyllales order. This family includes roughly 70 species of flowering plants and over 400 varieties (Iamonico et al., 2023). It has been rediscovered among the few edible crops with multipurpose potential worth exploring. This ancient plant offers a wide range of edible and non-edible products, making it valuable to households and industries (Mukuwapasi et al., 2024). Every part of the amaranth plant, from its seeds to its leaves, can be utilized, providing profitable opportunities for smallholder farmers and contributing to food security. As a result, it has gained recognition as a super crop in the context of climate-smart agriculture (Olusanya et al., 2024). Unlike most cultivated plants, amaranth is widely recognized as a dual-purpose crop, providing both pseudocereal grains and leafy vegetables (Sharma et al., 2025). The crop's adaptability, drought tolerance, and short growth cycle make it increasingly appealing in regions affected by climate variability (Yadav and Yadav, 2024). Nonetheless, the genetic diversity and nutritional richness of amaranth, characterized by high protein content, essential amino acids, and micronutrients, highlight its promise as a sustainable and valuable crop for the future (Selamoglu et al., 2024).

The most prevalent species include *Amaranthus thunbergii*, *A. greazicans*, *A. spinosus, A. deflexus, A. hypochondriacus, A. viridis,* and *A. hybridus* (Olusanya et al., 2024). They have a deep root system, efficient water use, heat tolerance, and a quick growth cycle, ensuring reliable yields even in unpredictable conditions, making it a valuable asset for sustainable food production and an ideal candidate for climate-smart agriculture (Yadav and Yadav, 2024). Despite their significance, there is limited scientific data available on the bioactive properties of Amaranth leaves. Nambafu et al. (2024) reported that obstacles to the commercialization of Indigenous leafy vegetables such as Amaranthus, include the long distances to markets, their perishable nature, price fluctuations due to varying seasons, and market opportunism that prevents farmers from making profits. Thus, this chapter focuses specifically on the bioactive compounds and biological activities of amaranth leaf genotypes. It examines the diversity of phytochemicals found in amaranth leaves, their associated biological activities, the influence of genotype variation, and the potential applications of leafy amaranth in functional foods and nutraceuticals, which underline its role as an underutilized but valuable crop in promoting human health and nutrition security.

Fig. 1 Amaranth genotypes released from the Agricultural Research Council of South Africa, which are widely consumed.

Taxonomic Background

Amaranthus exhibits remarkable morphological diversity, both within and across species (Sharma et al., 2025). As a member of the *Amaranthaceae* family within the Caryophyllales order shown in Figure 1 (Xu et al., 2024). *Amaranthus* species are annual or short-lived perennials with extraordinary variation in growth forms. Plants can range from low, prostrate forms to erect bushes exceeding 2 meters in height. Stems are often fleshy, smooth, or pubescent, and some species, such as *Amaranthus spinosus*, develop sharp axillary spines. The leaves are simple and alternate, varying in shape from oval to lanceolate or rhomboid, with entire to undulating margins. Leaf colouration ranges from green to purple or variegated forms (Figure 2). Inflorescences are dense spikes or panicles of tiny, unisexual flowers, often adorned with showy bracts. Flower colours also vary, with some species bearing greenish or yellow-green flowers, while others, like *Amaranthus cruentus* and *Amaranthus caudatus*, display vibrant red or purple tassels.

Traditional Uses and Modern Uses

Amaranthus is widely marketed as a vegetable in Africa due to its high economic potential (Akoth, 2021). This versatile plant is valued for its nutritious leaves, which are consumed as a vegetable or used as animal feed and fodder (Sharma et al., 2025). Its ornamental appeal, attributed to the vibrant hues of its inflorescence, further enhances its utility (Sharma et al., 2025). The uses are summarized in Figure 3.

Culinary Uses

Amaranthus leaves are integral to traditional African dishes, often boiled or steamed and combined with ingredients like groundnuts, onions, or tomatoes, and served with maize meal in Africa. Young, tender leaves can also be eaten raw in salads or blended into smoothies, while dried and powdered leaves are used to enrich flours, porridges, or sauces (Baraniak and Kania-Dobrowolska, 2022). Fresh leaves are sometimes juiced or infused into herbal drinks for traditional healing purposes (Chandrashekhar, 2018).

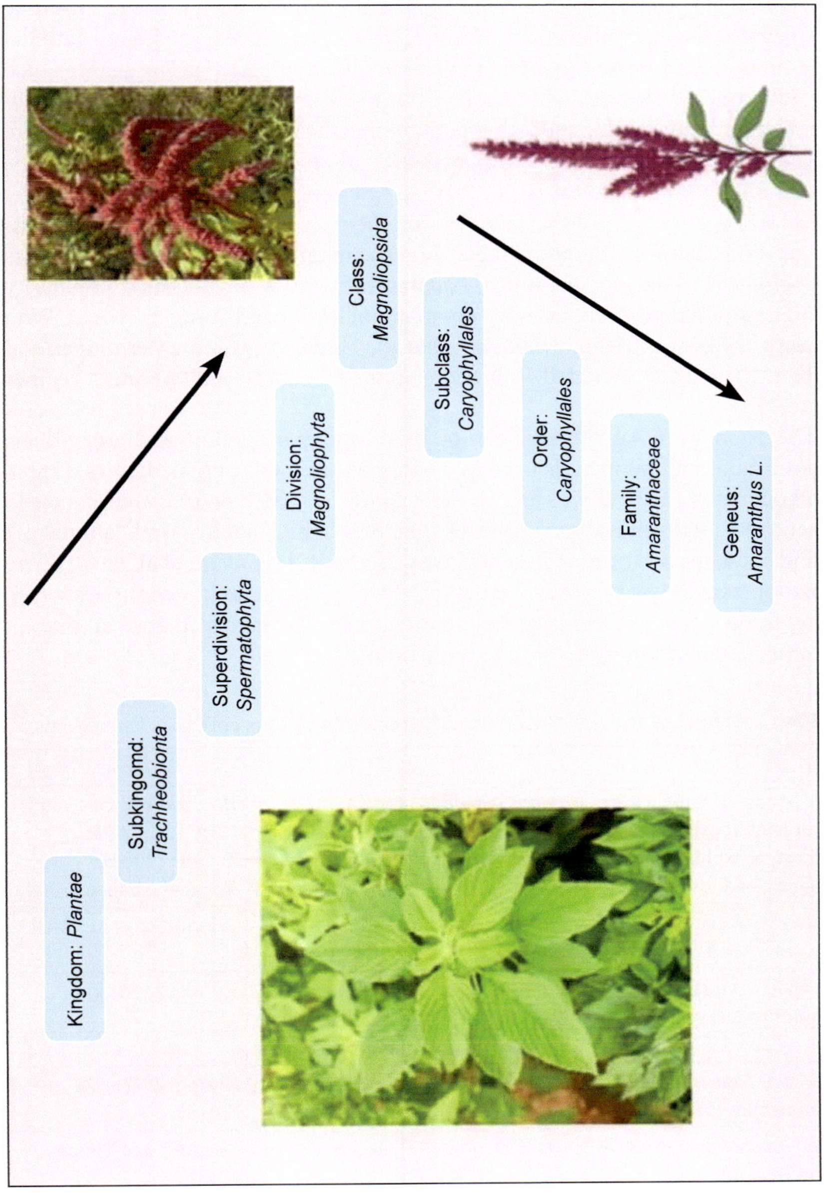

Fig. 2 Classification of *Amaranthus* spp.

Medicinal and Cultural Uses

The leaves of Amaranthus are particularly beneficial for young children, lactating mothers, and individuals suffering from constipation, fever, haemorrhage, anaemia, or kidney complaints (Sharma et al., 2025). It is also recognized as a diuretic, aiding in increased urine production (Peter and Ghandi, 2017). The plant is also used traditionally for various medicinal purposes, including as a diuretic, anti-snake venom, antipyretic, antidiabetic, antigonorrheal, and antileprotic (Ganjare and Raut, 2019). The sap from the plant is used as an eyewash to treat ophthalmia and convulsions in children (Rao, 2019). Amaranthus leaves are particularly beneficial for young children, lactating mothers, and patients with conditions such as constipation, haemorrhage, fever, anaemia, or kidney complaints (Alegbejo, 2013). In modern times, Amaranthus leaves have gained popularity beyond traditional farming, entering commercial health and culinary spaces. With increasing awareness of their health benefits, they are now promoted as a superfood in urban diets, often included in green smoothies, salads, and wellness recipes (Figure 3).

The pigment in Amaranthus can be used to prepare yellow and green dyes, while red amaranth pigment is used as a colourant in foods and medicines. (Table 1) Anthocyanins and betalains are the most commonly used natural alternatives to synthetic reds, with anthocyanins performing best in high acidity applications (pH < 4) and betalains performing best at pH levels above 4 (Howard et al., 2022). The fascination with the purple-red pigments produced by Amaranthus continues today, leading to the selection of many different cultivars displaying intense coloration and intricate patterning from around the world.

Table 1 Medicinal uses of Amaranthus leaves across various countries and regions.

Uses	*Region*	*References*
Many people in India use *Amaranthus tricolor* to treat jaundice, intestinal and urinary discharge disorders, as well as blood conditions, scrofula, stranguria, and fever.	India	Borthakur & Goswami (1995); Gupta & Tandon (2004); Yoganarasimhan (2000)
The leaves of *Amaranthus cruentus* are used as a tapeworm expellant.	Ethiopia	Aderibigbe et al. (2022)
"Lal bhaji" is a famous dish made from Amaranthus leaves and used to wash faces.	Chhattisgarh & Kondagoan region in India	Rastogi & Shukla (2013)
Traditional Amaranthus dish called 'Mchincha' is used to cure low red blood cell count.	East Africa	Rastogi & Shukla (2013)
Heated *Amaranthus* leaves were traditionally used to treat tumours.	Gabon	Grubben and Sloten (1981)
The crushed leaves are effective in helping treat dermatitis, psoriasis, and rashes.	Philippines	Kumar et al. (2008)

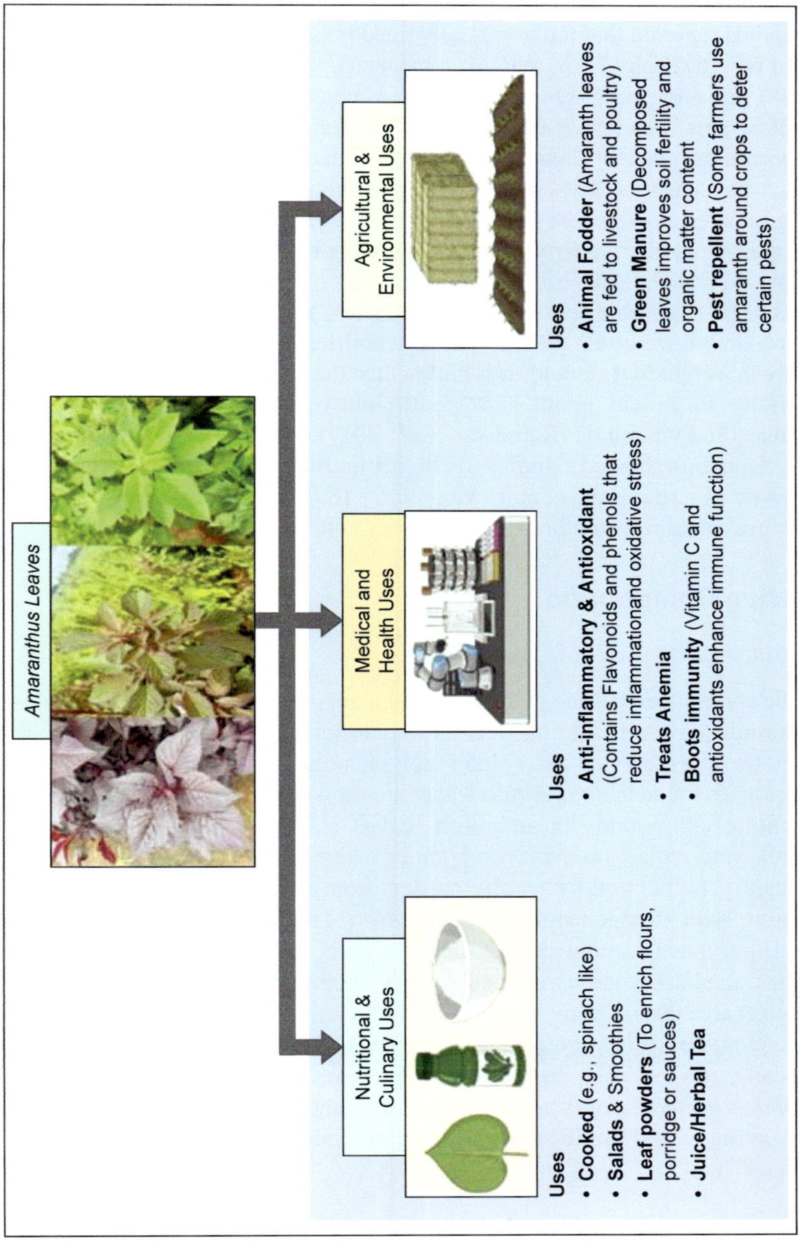

Fig. 3 Different uses of leaves of *Amaranthus* spp.

Nutritional Composition

Amaranthus leaves are high in protein and micronutrients. Element analysis of Amaranthus reported that the leaves contained potassium (54.20), sodium (7.43), calcium (44.15), Iron (13.58 mg/100), magnesium (231.22 mg/100), zinc (3.80 mg/100), and phosphorus (34.91 mg/100) (Akubugwo et al., 2007) on dry weight basis. Raw amaranth leaves contain 2.1 g to over 62 g per 100 g of dry weight for *Amaranthus tricolor* (Jahan et al., 2023). Amaranthus contains 17.5 – 38.3% protein, which consists of 5% lysine and sulphur-containing amino acids, which are lacking in cereals and tubers (Kour and Gupta, 2022). Additionally, Amaranthus leaves are rich in fiber, β-carotene (3.29 mg/100 g DW), thiamine (2.75 mg/100 g DW), pyridoxine (2.33 mg/100 g DW), riboflavin (4.24 mg/100 g DW), niacin (1.54 mg/100 g DW), ascorbic acid (25.40 mg/100 g DW), and tocopherol (0.50 mg/100 g DW) (Akubugwo et al., 2007). A study confirms that the nutrient content per 100 g fresh weight of Amaranthus is higher than that of cabbage, with Amaranthus being richer in protein, about 47.96%; calcium was 81.39%, and iron 79.74%, β-carotene, and vitamin C (Ramdwar et al., 2017). Similarly, when compared to lettuce, Amaranth showed significantly higher nutritional content, including seven times more iron, three times more vitamin C, 18 times more vitamin A, and 20 times more calcium than other similar plants (Ruth et al., 2021).

Bioactive Compounds

Phenolic Acids

Phenolic compounds are one of the most abundant groups of bioactive compounds in amaranth leaves, including phenolic acids like gallic, caffeic, ferulic, and p-coumaric acids, as well as complex polyphenols (Karamač et al., 2019). The phenolic acids found in *Amaranthus* leaves are shown in Table 2. The concentration of phenolic compounds in amaranth leaves is highly genotype-dependent. Red-leaf amaranth (*Amaranthus tricolor*) has been shown to accumulate significantly higher total phenolic content compared to green-leaf varieties, correlating with stronger antioxidant activity (Managa and Nemadodzi, 2023). In food, polyphenols contribute to taste, colour, and stability, and diets high in plant polyphenols are associated with protection against numerous diseases (Manach et al., 2004). The main polyphenols identified in *Amaranthus* spp. include hydroxycinnamic acids, hydroxybenzoic acids, flavonols, and their glycosides (Kongdang et al., 2021). In the leaves of *Amaranthus hypochondriacus* and *Amaranthus tricolor*, gallic, p-coumaric, ferulic, and salicylic acids are prominent, sinapic, syringic, and vanillic acids are the most abundant phenolic acids (Khanam and Oba, 2013).

Table 2 Phenolic acids identified in the leaves of Amaranthus species.

Compounds	*Amaranthus Species*	*References*
Hydroxycinnamic acids		
Caffeic acid	*A. hybridus*	Oboh et al. (2019)
	A. hypochondriacus	Khanam and Oba (2012)
	A. spinosus	Rjeibi et al. (2019)
	A. tricolor	Sarker & Oba (2020b), Khanam and Oba (2012)
2-Caffeoylisocitric acid	*A. hybridus*	Schroter et al. (2019)
Chlorogenic acid	*A. hybridus*	Oboh et al. (2016)
	A. tricolor	Sarker & Oba (2020b), Khanam and Oba (2012)
	A. virdis	Kumari et al. (2018)
Cinnamic acid	*A. spinosus*	Rjeibi et al. (2019)
	A. tricolor	Sarker & Oba (2020b)
m-Coumaric acid	*A. hypochondriacus*	Khanam and Oba (2012)
	A. spinosus	Sarker & Oba (2020b); Khanam and Oba (2012)
p-Coumaric acid (4′ -hydroxycinnamic acid)	*A. hypochondriacus*	Khanam and Oba (2012)
	A. spinosus	Rjeibi et al. (2019)
Ferulic acid	*A. hypochondriacus*	Khanam and Oba (2012)
	A. spinosus	Rjeibi et al. (2019)
	A. tricolor	Sarker & Oba (2020b), Khanam and Oba (2012)
	A. virdis	Kumari et al. (2018)
Hydroxybenzoic acids		
Gallic acid	*A. cruentus*	Klimczak et al. (2002), Pa´sko et al. (2008)
	A. spinosus	Rjeibi et al. (2016)
	A. hybridus	Oboh et al. (2016)
	A. hypochondriacus	Khanam and Oba (2012)
	A. tricolor	Sarker & Oba (2020b); Khanam and Oba (2012)
2,4-Dihydroxy benzoic acid	*A. tricolor*	Sarker & Oba (2020b)
Ellagic acid	*A. hybridus*	Oboh et al. (2016)
	A. hypochondriacus	Khanam and Oba (2012)
	A. tricolor	Sarker & Oba (2020b); Khanam and Oba (2012)

Contd...

Table 2 *Contd...*

Compounds	*Amaranthus Species*	*References*
Vanillic acid	*A. hypochondriacus*	Khanam and Oba (2012)
	A. tricolor	Sarker & Oba (2020b); Khanam and Oba (2012)
Syringic acid	*A. hypochondriacus*	Khanam and Oba (2012)
	A. tricolor	Sarker & Oba (2020b); Khanam and Oba (2012)
p-Hydroxybenzoic acid (4-hydroxybenzoic acid)	*A. hypochondriacus*	Khanam and Oba (2012)
	A. tricolor	Sarker & Oba (2020b); Khanam and Oba (2012)

Flavonoids

Flavonoids are a major class of phytochemicals found in amaranth leaves, including compounds such as quercetin, rutin, kaempferol, and isoquercitrin (Barba et al., 2019). Rutin and quercetin are regarded as predominant flavonoids in Amaranths leaves (Sarker et al., 2022. Flavonoid content varies across amaranth genotypes, with pigmented species typically exhibiting higher concentrations (Sarker et al., 2022). They are responsible for the colour, fragrance, and flavour characteristics of many plants and are part of the larger family of phenolic compounds or polyphenols, which include more than 6,000 different structures (Dias et al., 2021). Studies report total phenolic content on the order of 200–300 mg gallic acid equivalents (GAE) per 100 g (dry leaves), and total flavonoids 150 mg quercetin equivalents/100 g (dry weight) (Belew et al., 2025). In *Amaranthus hypochondriacus,* a ferulic acid was detected only in green leaves (0.7 g/g), whereas p-coumaric acid was only found in red leaves. Flavonoid derivatives identified in the leaves of *Amaranthus* species are shown in Table 3. The specific types and amounts of flavonoids can vary depending on the amaranth cultivar and growing conditions, with some genotypes showing exceptionally high flavonoid (Barba et al., 2019).

Table 3 Flavonoid derivatives identified in the leaves of Amaranthus species.

Compounds	*Amaranthus Species*	*Biological Activities*	*Reference*
Hyperoside (Quercentin-3-ogalactoside)	*A. hypochondriacus*	Antioxidant activity	Khanam and Oba (2012)
Isoquercetin (Quercetin-3-O-glucoside)	*A. hypochondriacus*	Antioxidant activity	Khanam and Oba (2012)
	A. tricolor	Antioxidant activity	Sarker & Oba (2020b); Sarker & Oba (2020a); Khanam and Oba (2012)

Contd...

Table 3 *Contd...*

Compounds	*Amaranthus Species*	*Biological Activities*	*Reference*
Rutin (Quercetin-3-O-rutinoside)	*A. caudatus*	n.d.	Kalinova & Dadakova (2009)
	A. hybridus	ACE and cisplatin-induced lipid peroxidation inhibitory activities	Kalinova & Dadakova (2009); Oboh et al. (2016)
	A. hypochondriacus	Antioxidant activity	Kalinova & Dadakova (2009); Khanam and Oba (2012)
	A. retroflexus	n.d.	Kalinova & Dadakova (2009)
	A. spinosus	Antioxidant and antinociceptive activities	Rjeibi et al. (2019)
	A. tricolor	Antioxidant activity	Kalinova & Dadakova (2009); Khanam and Oba (2012); Sarker & Oba (2020a, b)
Quercetin	*A. caudatus*	n.d.	Khanam and Oba (2012)
	A. tricolor	Antioxidant activity	Medoua & Oldewage-Theron (2014)
	A. hypochondriacus	n.d.	Kalinova & Dadakova (2009)
	A. spinosus	Antioxidant and antinociceptive activities	Rjeibi et al. (2019)
	A. tricolor	Antioxidant activity	Sarker & Oba (2020b); Sarker & Oba (2020a)
(+)-Catechin	*A. tricolor*	Antioxidant activity	Sarker & Oba (2020b)
	A. spinosus	Antioxidant and antinociceptive activities	Rjeibi et al. (2019)

Contd...

Table 3 *Contd...*

Compounds	*Amaranthus Species*	*Biological Activities*	*Reference*
	A. viridis	Antioxidant activity	Kumari et al. (2018)
(−)-Epicatechin	*A. viridis*	Antioxidant activity	Kumari et al. (2018)
Flavones			
Apigenin	*A. hybridus*	ACE and cisplatin-induced lipid peroxidation inhibitory activities	Oboh et al. (2016)
	A. tricolor	Antioxidant activity	Sarker & Oba (2020b)
	A. viridis	Antioxidant activity	Kumari et al. (2018)
Luteolin	*A. spinosus*	Antioxidant and antinociceptive activities	Rjeibi et al. (2019)
Vitexin (Apigenin-8-C-dglucopyranoside)	*A. hybridus*	ACE and cisplatin-induced lipid peroxidation inhibitory activities	Oboh et al. (2016)
Flavanones			
Naringenin	*A. tricolor*	Antioxidant activity	Sarker & Oba (2020b)

Betalains and Anthocyanins

Betalains are water-soluble nitrogen-containing pigments unique to the *Caryophyllales* order, to which *Amaranthus* belongs (Gengatharan et al., 2015). They are divided into betacyanins (red-violet pigments such as *amaranthin* and *isoamaranthin*) and betaxanthins (yellow pigments) (Pratiwi et al., 2025). The concentration of betalains is strongly influenced by leaf pigmentation. Red and purple-leaf genotypes of *Amaranthus tricolor* and *Amaranthus cruentus* exhibit high betalain levels as observed by Hussain et al. (2018). Betalains can be found in all tissues of amaranth and are typically the most concentrated in the leaves or inflorescence tissue (Howard et al., 2022). *Amaranthus tricolor* leaves are various colours, including violet, red, and purple. Previous studies reported that concentrations of amaranthine/isoamarathine are higher than those of betanin/isobetanin in *Amaranthus tricolor* seedlings (Sarker and Oba, 2021). Total leaf betacyanins and betaxanthins were determined in four Bangladeshi vegetable amaranth genotypes, namely: Rocto ranga, Rocto alta, Baromashi, and Alto pati. The concentrations of betaxanthin were higher than those of betacyanins in the Baromashi and Alto pati genotypes (Sarker et al., 2022). Several authors (Arunachalam et al., 2022; Howard et al., 2022; Khandaker et al., 2009) also previously reported differences in the amounts of betacyanins among cultivars. Amaranths replace red beets because of the natural source of betalains (Sarker et al., 2022).

These plant-derived natural pigments are currently gaining acceptance for use as natural colourants in the beverage, pharmacology, and food industries owing to their physiological functions, including antioxidative (Mishra et al., 2025), anti-cancer (Masyita et al., 2025), anti-lipidemic, and antimicrobial (Gengatharan et al., 2015). This may be beneficial for eliminating certain disorders related to oxidative stress (Magalhães et al., 2024). Betalains are extensively utilized in the food industry because of their natural colourant properties, and because there are no toxic effects at higher concentrations (Martinez et al., 2024).

Carotenoids

Carotenoids are organic plant pigments (Hermanns et al., 2020) and play a crucial role as provitamin A, which is converted into retinol by the enzymes oxygenase and reductase. The number of proteins involved regulates the efficiency of vitamin A and intestinal carotenoid absorption (Harrison, 2012). Key carotenoids identified in amaranth leaves include lutein, β-carotene, α-carotene, neoxanthin, and violaxanthin (Nyonje et al., 2021). These compounds play vital roles in vision (e.g., prevention of age-related macular degeneration), immune function, and protection against oxidative damage. A study by Peter and Gandhi (2017) concluded that *Amaranthus cruentus* is a potentially good dietary source of the pro-vitamin A carotenoid (β-carotene). The carotenoids present in the amaranth leaves, particularly β-carotene and lutein, help prevent age-related eye diseases. β-carotene serves as a pro-vitamin A source, supporting normal vision and preventing conditions associated with vitamin A deficiency, while Lutein, together with its isomer zeaxanthin, lowers the risk of age-related macular degeneration and other degenerative disorders (Chandrika et al., 2010). The level of β-carotene (28.5 mg/100 g) in *Amaranthus cruentus* was seven times higher than in tomatoes, making it a potential aid in treating anaemia in African countries, as reported by Peter and Gandhi (2017).

Chlorophyll

Chlorophylls are the most widely distributed and important pigments, playing essential roles in photosynthesis (Saini et al., 2024). Green amaranth (*Amaranthus viridis*) is an excellent source of phytochemicals, including chlorophylls.

Amaranthus leaves have been reported to produce 1504 mg/kg of total chlorophyll (Mohd-Amin et al., 2023). Green-leaf amaranth genotypes tend to have higher carotenoid and chlorophyll concentrations, whereas pigmented genotypes may exhibit a trade-off with higher betalains but lower carotenoids (Sarker et al., 2022).

Biological Activities

Amaranth leaves are not only nutrient-dense but also a rich source of bioactive compounds that exhibit various biological activities, including antioxidant, anti-inflammatory, antidiabetic, antihypertensive, anticancer, and antimicrobial effects (Singh et al., 2024). These activities are primarily attributed to phenolic acids, flavonoids, betalains, carotenoids, vitamins, and saponins, whose concentrations vary across genotypes and environmental conditions (Segura-Jiménez et al., 2025). Understanding these biological activities is crucial for supporting the role of amaranth leaves as functional foods and nutraceutical candidates (Kongdang et al., 2021). One of the most widely reported biological activities of amaranth leaves is their strong antioxidant capacity. Extracts from both green and pigmented genotypes show significant radical scavenging ability in assays such as 2,2-diphenyl-1-picrylhydrazyl (DPPH), 2,2′-azino-bis (3-ethylbenzothiazoline-6-sulfonic acid (ABTS), and ferric ion reducing antioxidant potential (FRAP) (Karamač et al., 2019). The antioxidant effects are mainly attributed to phenolic acids, flavonoids, vitamin C, carotenoids, and betalains (Sarker and Oba, 2021); Red and purple-leaf genotypes, especially *Amaranthus tricolor*, often exhibit superior antioxidant activity due to their high betalain and total phenolic contents. Antioxidants from amaranth leaves help prevent oxidative stress-related cellular damage, which is implicated in the pathogenesis of cardiovascular diseases, neurodegenerative disorders, diabetes, and cancer (Ullah et al., 2020).

Anti-inflammatory Properties

Amaranth leaf extracts have shown anti-inflammatory effects by downregulating the expression of pro-inflammatory cytokines, such as Tumour necrosis factor (TNF-α) and Interleukin-6 (IL-6), and enzymes like cyclooxygenase-2 (COX-2) (Montoya-Rodríguez et al. 2014). Flavonoids, including quercetin and rutin, along with betalain pigments, are the main contributors to these activities. They work by modulating signalling pathways such as (nuclear factor) NF-kB and Mitogen-Activated Protein Kinase pathway (MAPK), thereby reducing the production of inflammatory mediators (Al-Khayri et al., 2022). These properties suggest that amaranth leaves may be beneficial in preventing or managing conditions such as arthritis, metabolic syndrome, and inflammatory bowel diseases (Baraniak and Kania-Dobrowolska, 2022).

Antidiabetic Effects

Amaranth leaves may play a role in glycaemic control by inhibiting the activity of carbohydrate-hydrolyzing enzymes, such as α-amylase and α-glucosidase, which slow down starch digestion and reduce postprandial blood glucose spikes (Kaur et al., 2025; Yang et al., 2020). A previous study done by Yang et al. (2020) also highlighted that the flavonoids and phenolic acids from amaranth leaves improve insulin sensitivity and glucose uptake. Additionally, carotenoids and anthocyanins help reduce oxidative stress, a key factor in the progression of diabetes and its complications (Sharma et al., 2023). Extracts prepared from red amaranth exhibited marked anti-glycative, anti-α-amylase, and anti-α-glucosidase activities (Nkobole et al., 2021). Hence, amaranth varieties developed as functional foods for diabetes prevention and/or mitigation (Yang et al., 2020).

Antihypertensive and Cardio-protective Roles

Amaranth leaves contain bioactive peptides and other compounds that contribute to cardio-protective and antihypertensive roles (Segura-Jiménez and Jacobo-Velázquez, 2025). These peptides, which can be released through digestion or fermentation, work by inhibiting enzymes in the Renin-Angiotensin-Aldosterone System (RAAS), thereby regulating blood pressure (Nardo et al., 2020). Protein hydrolysates from amaranth leaves have shown angiotensin-converting enzyme (ACE) inhibitory activity, which helps regulate blood pressure as reported by Ramírez-Torres et al. (2017). Additionally, the high potassium and magnesium content in amaranth leaves supports cardiovascular health by promoting vasodilation and electrolyte balance (Singh et al., 2024). The antioxidant properties of flavonoids and betalains also protect vascular tissues from oxidative stress, reducing the risk of hypertension and atherosclerosis (Barba-Ostria et al., 2025).

Anticancer Potential

Despite advancements in technology and medicine over time, cancer remains a global concern (Tripathi and Hariwal, 2023). It is the second major cause of death worldwide and accounts for thousands of fatalities annually. *Amaranthus spinosus* can induce a cytotoxic action against various cancer cell lines and initiate cell death (Rajasekaran et al., 2014); consumption of the parts of these plants or ingestion may impart anticancer effects. Phenolic compounds (Liu et al., 2023), flavonoids (Mutha et al., 2021), and betalains (Martínez-Rodríguez et al., 2025) are believed to exert these effects through mechanisms such as inducing apoptosis, inhibiting cell proliferation, and modulating oxidative stress pathways. Betacyanins, like amaranthin, which are abundant in red-leaf genotypes, have shown cytotoxicity against tumour cells while sparing normal cells, suggesting their potential as natural chemopreventive agents (Quiroga et al., 2015). However, most evidence remains at the laboratory stage, and further clinical studies are needed to validate these effects in humans.

Antimicrobial Properties

The biologically active compounds contained in amaranth plants are characterized by significant antimicrobial activity and cytotoxicity (Al-Mamun et al., 2016).

Extracts from amaranth leaves have also demonstrated antimicrobial activities against a range of bacterial and fungal pathogens (Spórna-Kucab et al., 2023). Research conducted by Jamiołkowska et al. (2023) demonstrated that the strongest antifungal activity was recorded for extracts from *Amaranthus cruentus*, *Amaranthus hypohondriacus*, and *Amaranthus hybridus*, of which *Amaranthus hybridus* was characterized by the highest polyphenol contents. Phenolic compounds and flavonoids disrupt microbial cell membranes, interfere with enzyme systems, and inhibit nucleic acid synthesis (De Rossi et al., 2025). Leaf extracts have been particularly effective against Gram-positive bacteria such as *Staphylococcus aureus* and *Bacillus subtilis*, as well as fungal strains like *Candida albicans* (Jimoh et al., 2020). These antimicrobial properties support the potential use of amaranth as a natural food preservative and therapeutic agent in combating infections (Fouad et al., 2024).

Conclusion

Amaranthus species represent a unique and underutilized plant group with significant potential in human nutrition and health promotion, primarily due to their diverse profile of bioactive compounds. These bioactive constituents include phenolic acids, flavonoids, carotenoids, betalains, saponins, and ascorbic acid, all of which exhibit strong antioxidant properties. Collectively, these compounds contribute to the reduction of oxidative stress by scavenging free radicals and mitigating cellular damage, thereby lowering the risk of chronic non-communicable diseases such as cardiovascular disorders, diabetes, and cancer.

The functional roles of these compounds extend beyond antioxidant activity. For instance, phenolic compounds and flavonoids possess anti-inflammatory and antimicrobial properties, while carotenoids are associated with improved eye health and immune modulation. Saponins contribute to cholesterol-lowering effects, and vitamin C (ascorbic acid) enhances immune function and collagen synthesis. This rich chemical diversity positions *Amaranthus* not only as a dietary staple but also as a promising candidate for the development of functional foods, nutraceuticals, and pharmaceutical products. Furthermore, the adaptability of *Amaranthus* to diverse agro-ecological conditions, its rapid growth cycle, and high yield of leaves make it an attractive crop for food security programmes, particularly in regions facing malnutrition and resource constraints. Incorporating Amaranthus into diets can help address micronutrient deficiencies, promote health, and support sustainable food systems.

References

Aderibigbe, O.R., Ezekiel, O.O., Owolade, S.O., Korese, J.K., Sturm, B. and Hensel, O. 2022. Exploring the potentials of underutilized grain amaranth (*Amaranthus* spp.) along the value chain for food and nutrition security: A review. *Crit. Rev. Food Sci. Nutr.*, *62*: 656– 669. https://doi.org/10.1080/10408398.2020.1825323.

Akoth, B., 2021. *Potential of Amaranthus in Improving Urban Farmers' Livelihoods in Kampal*a. Doctoral dissertation, Stellenbosch: Stellenbosch University, South Africa.

Akubugwo, I.E., Obasi, N.A., Chinyere, G.C. and Ugbogu, A.E. 2007. Nutritional and chemical value of *Amaranthus hybridus* L. leaves from Afikpo, Nigeria. *Afr. J. Biotechnol.*, *6*: 2833– 2889. https://doi.org/10.5897/AJB2007.000-2452.

Alegbejo, J.O. 2013. Nutritional value and utilization of Amaranthus (*Amaranthu*s spp.): A review. *Bayero J. Pure Appl. Sci.*, *6*: 136–143. http://dx.doi.org/10.4314/bajopas.v6i1.27.

Al-Khayri, J.M., Sahana, G.R., Nagella, P., Joseph, B.V., Alessa, F.M. and Al-Mssallem, M.Q. 2022. Flavonoids as potential anti-inflammatory molecules: A review. *Molecules*, *27*: 2901. https://doi.org/10.3390/molecules27092901.

Al-Mamun, M.A., Husna, J., Khatun, M., Hasan, R., Kamruzzaman, M., Hoque, K.M.F., Reza, M.A. and Ferdousi, Z. 2016. Assessment of antioxidant, anticancer, and antimicrobial activity of two vegetable species of *Amaranthus* in Bangladesh. *BMC Complement. Altern, Med.*, *16*: 157. https://doi.org/10.1186/s12906-016-1130-0.

Arunachalam, V., Salgaonkar, D.C., Kevat, N.V., Walawalkar, B.V. and Das B. 2022. Quantification of betacyanin content variation of amaranth varieties by an Android App, colourimeter, and infrared spectroscopy. *Chin. J. Anal. Chem.*, *50*: 100145. https://doi.org/10.1016/j.cjac.2022.100145.

Baraniak, J. and Kania-Dobrowolska, M. 2022. The dual nature of amaranth: Functional food and potential medicine. *Foods*, *11*: 618. https://doi.org/10.3390/foods11040618.

Barba de la Rosa, A.P., León-Rodríguez, A.D., Laursen, B. and Fomsgaard, I.S. 2019. Influence of the growing conditions on the flavonoids and phenolic acids accumulation in amaranth (*Amaranthus hypochondriacus* L.) leaves. *Terra Latinoamericana*, *37*: 449–457. https://doi.org/10.28940/terra.v37i4.541.

Barba-Ostria, C., Guamán-Bautista, J., Tosi-Vélez, A.A., Puente-Pineda, J.A., Cedeño-Zambrano, M.A., Teran, E. and Guamán, L.P. 2025. Recent advances in the therapeutic potential of bioactive molecules from Plants of Andean Origin. *Nutrients*, *17*: 1749. https://doi.org/10.3390/nu17111749.

Belew, A.A., Hanan, G.G.M.W., Meshesha, D.S. and Akele, M.L. 2025. Evaluation of total phenolic, flavonoid contents, antioxidant, and antibacterial activity of leaf extracts from *Rhus vulgaris*. *Discov. Plants*, *2*: 141. https://doi.org/10.1007/s44372-025-00222-3.

Borthakur, S.K. and Goswami, N. (1995). Herbal remedies from Dimasa of Kamrup district of Assam in North-Eastern India. *Ancient Science of Life*, 14(4), 258–267.

Chandrashekhar, K. 2018. A review on Tanduliyaka (*Amaranthus spinosus* L.): A weed, a vegetable, and a medicinal plant. *Int. J. Ayurvedic Med.*, *9*: 231–238. https://doi.org/10.47552/ijam.v9i4.1169.

Chandrika, U.G., Basnayake, B.M.L.B., Athukorala, I., Colombagama, P.W.N.M. and Goonetilleke, A. 2010. Carotenoid content and *in vitro* bioaccessibility of lutein in some leafy vegetables popular in Sri Lanka. *J. Nutrition. Sci. Vitaminol.*, *56*: 203–207. https://doi.org/10.3177/jnsv.56.203.

De Rossi, L., Rocchetti, G., Lucini, L. and Rebecchi, A. 2025. Antimicrobial potential of polyphenols: Mechanisms of action and microbial responses: A narrative review. *Antioxidants*, *14*: 200. https://doi.org/10.3390/antiox14020200.

Dias, M.C., Pinto, D.C. and Silva, A.M. 2021. Plant flavonoids: Chemical characteristics and biological activity. *Molecules*, *26*: 5377. https://doi.org/10.3390/molecules26175377.

Fouad, M.S., Ghareeb, M.A., Hamed, A.A., Aidy, E.A., Tabudravu, J., Sayed, A.M., Tammam, M.A. and Mwaheb, M.A. 2024. Exploring the antioxidant, anticancer, and antimicrobial potential of *Amaranthus viridis* L. collected from Fayoum depression: Phytochemical, and biological aspects. *S. Afr. J. Bot.*, *166*: 297–310. https://doi.org/10.1016/j.sajb.2024.01.047.

Gandhi, P., Samarth, R.M. and Peter, K. 2021. Bioactive compounds of amaranth (Genus *Amaranthus*). pp.39–74. In: Murthy, H.N., Paek, K.Y. (Eds.), *Bioactive Compounds in Underutilized Vegetables and Legumes*. Spinger, Cham, https://doi.org/10.1007/978-3-030-57415-4_3.

Gengatharan, A., Dykes, G.A. and Choo, W.S. 2015. Betalains: Natural plant pigments with potential application in functional foods. *LWT-Food Sci. Technol.*, *64*: 645–649. https://doi.org/10.1016/j.lwt.2015.06.052.

Gupta, A.K., Sharma, M., and Tandon N. (2004). Reviews on Indian Medicinal Plants: Vol. 2. Indian Council of Medical Research, New Delhi, India.

Grubben, G.J.H. and Van Stolen, D.H. (1981). Genetic resources and Amaranths. International Board for Plant Genetic Resources, FAO, Rome.

Harrison, E.H. 2012. Mechanisms involved in the intestinal absorption of dietary vitamin A and pro-vitamin A carotenoids. *Biochim. Biophys. Acta.*, *1821*: 70–77. https://doi.org/10.1016/j.bbalip.2011.06.002

Hermanns, A.S., Zhou, X., Xu, Q., Tadmor, Y. and Li, L. 2020. Carotenoid Pigment Accumulation in Horticultural Plants. *Hortic. Plant J.*, *6*: 343–360. https://doi.org/10.1016/j.hpj.2020.10.002.

Howard, J.E., Villamil, M.B. and Riggins, C.W. 2022. Amaranth as a natural food colourant source: Survey of germplasm and optimization of extraction methods for betalain pigments. *Front. Plant Sci.*, *13*: 932440. https://doi.org/10.3389/fpls.2022.932440.

Hussain, E.A., Sadiq, Z. and M. Zia-Ul-Haq. 2018. *Betalains: Biomolecular Aspects*. Springer, London. https://doi.org/10.1007/978-3-319-95624-4.

Iamonico, D., Hussain, A.N., Sindhu, A., Saradamma Anil Kumar, V.N., Shaheen, S., Munir, M. and Fortini, P. 2023. Trying to understand the complicated taxonomy in *Amaranthus* (Amaranthaceae): Insights on seeds micromorphology. *Plants*, *12*: 987. https://doi.org/10.3390/plants12050987.

Jahan, N., Sarker, U., Saikat, M.M.H., Hossain, M.M., Azam, M.G., Ali, D., Ercisli, S. and Golokhvast, K.S. 2023. Evaluation of yield attributes and bioactive phytochemicals of twenty amaranth genotypes of Bengal floodplain. *Heliyon*, *9*: e19644. https://doi.org/10.1016/j.heliyon.2023.e19644.

Jamiołkowska, A., Skwaryło-Bednarz, B., Kowalski, R., Yildirim, I. and Patkowska, E. 2023. Antifungal potency of amaranth leaf extract: An *in vitro* study. *Plants*, *12*: 1723. https://doi.org/10.3390/plants12081723.

Jimoh, M.O., Afolayan, A.J. and Lewu, F.B. 2020. Toxicity and antimicrobial activities of *Amaranthus caudatus* L. (Amaranthaceae) harvested from formulated soils at different growth stages. *J. Evid. Based Integr. Med.*, *25*: 2515690X20971578. https://doi.org/10.1177/2515690X20971578.

Kalinova, J. and Dadakova, E. 2009. Rutin and total quercetin content in amaranth (*Amaranthus* spp.). *Plant Foods Hum Nutr.*, *64*: 68–74. https://doi.org/10.1007/s11130-008-0104-x.

Kalinova, J., and Dadakova, E. (2009). Rutin content in amaranth (Amaranthus spp.) at different stages of blossoming. *Czech Journal of Food Sciences*, 27(Special Issue), S410-S412.

Karamać, M., Gai, F., Longato, E., Meineri, G., Janiak, M.A., Amarowicz, R. and Peiretti, P.G. 2019. Antioxidant activity and phenolic composition of amaranth (*Amaranthus caudatus*) during plant growth. *Antioxidants*, *8*: 173. https://doi.org/10.3390/antiox8060173.

Kaur, S., Gadpayale, D., Kumari, A., Kaur, G., Kumar, A., Seem, K., Bhardwaj, R. and Riar, A. 2025. Antidiabetic Potential of Underutilized Crops: Nutritional, Phytochemical Insights, and Prospects for Diabetes Management. *Appl. Food Res.*, *5*: 101127. https://doi.org/10.1016/j.afres.2025.101127.

Khanam, U.K.S., Oba, S. 2012. Phenotypic plasticity of vegetable amaranth, *Amaranthus tricolor* L. under a natural climate. *Plant Prod. Sci.* 17: 166–172.

Khanam, U.K.S. and Oba, S. 2013. Bioactive substances in leaves of two amaranth species, *Amaranthus tricolor*, and *A. hypochondriacus*. *Can. J. Plant Sci.*, *93*: 47–58. https://doi.org/10.4141/cjps2012-117.

Khandaker, L., Ali, M.B. and Oba, S. 2009. Influence of cultivar and growth stage on pigments and processing factors on betacyanins in red amaranth (*Amaranthus tricolor* L.). *Food Sci. Technol. Int.*, *15*: 259–265. https://doi.org/10.1177/1082013209341119.

Kongdang, P., Dukaew, N., Pruksakorn, D. and Koonrungsesomboon, B. 2021. Biochemistry of *Amaranthus* polyphenols and their potential benefits on gut ecosystem: A comprehensive review of the literature. *J. Ethnopharmacol.*, *281*: 114547. https://doi.org/10.1016/j.jep.2021.114547.

Kour, D.P. and Gupta, N. 2022. Health benefits and uses of Amaranth. *Just Agric.*, *2*: 1–6.

Kumari, S., Elancheran, R., Devi, R. 2018. Phytochemical screening, antioxidant, antityrosinase, and antigenotoxic potential of Amaranthus virids extract. *Indian J Phamacol*, 50: 130–138.

Kumar, S., Korra, T., Thakur, R., Arutselvan, R., Kashyap, A.S., Nehela, Y., Chaplygin, V., Minkina, T. and Keswani, C. 2023. Role of plant secondary metabolites in defence and transcriptional regulation in response to biotic stress. *Plant Stress*, *8*: 100154. https://doi.org/10.1016/j.stress.2023.100154.

Li, H., Deng, Z., Liu, R., Zhu, H., Draves, J., Marcone, M., Sun, Y. and Tsao, R. 2015. Characterization of phenolics, betacyanins, and antioxidant activities of the seed, leaf, sprout, flower, and stalk extracts of three *Amaranthus* species. *J. Food Compos. Anal.*, *37*: 75–81. https://doi.org/10.1016/j.jfca.2014.09.003.

Liu, W., Cui, X., Zhong, Y., Ma, R., Liu, B. and Xia, Y. 2023. Phenolic metabolites as therapeutic in inflammation and neoplasms: Molecular pathways explaining their efficacy. *Pharmacol. Res.*, *193*: 6812. https://doi.org/10.1016/j.phrs.2023.106812.

Magalhães, D., Gonçalves, R., Rodrigues, C.V., Rocha, H.R., Pintado, M. and Coelho, M.C. 2024. Natural pigments recovery from food by-products: Health benefits towards the food industry. *Foods*, *13*: 2276. https://doi.org/10.3390/foods13142276.

Manach, C., Scalbert, A., Morand, C., Rémésy, C. and Jiménez, L. 2004. Polyphenols: Food sources and bioavailability. *Am. J, Clin. Nutr.*, *79*: 727–747. https://doi.org/10.1093/ajcn/79.5.727.

Managa, G.M. and Nemadodzi, L.E. 2023. Comparison of agronomic parameters and nutritional composition on red and green amaranth species grown in open field versus greenhouse environment. *Agriculture*, *13*: 685. https://doi.org/10.3390/agriculture13030685.

Martinez, R.M., Melo, C.P., Pinto, I.C., Mendes-Pierotti, S., Vignoli, J.A., Verri, W.A. and Casagrande, R.R. 2024. Betalains: A narrative review on pharmacological mechanisms supporting the nutraceutical potential towards health benefits. *Foods*, *13*: 3909. https://doi.org/10.3390/foods13233909.

Martínez-Rodríguez, P., Malo-García, M., Hernández-García, S., Youssefi, S., Nateri, A.S., Guerrero-Rubio, M.A. and Gandia-Herrero, F. 2025. Betalain-containing foods and their antitumoral potential: A comprehensive review of *in vitro* and *in vivo* evidence. *Trends Food Sci. Technol.*, *163*: 105150. https://doi.org/10.1016/j.tifs.2025.105150.

Masyita, A., Hardinasinta, G., Astuti, A.D., Firdayani, F., Mayasari, D., Hori, A., Nisha, I.N.A., Nainu, F. and Kuraishi, T. 2025. Natural pigments: Innovative extraction technologies and their potential application in health and food industries. *Front. Pharmacol.*, *15*: 1507108. https://doi.org/10.3389/fphar.2024.1507108.

Medoua, G.N., and Oldewage-Theron, W.H. (2014). Effect of drying and cooking on nutritional value and antioxidant capacity of morogo (Amaranthus hybridus) a traditional leafy vegetable grown in South Africa. *Journal of Food Science and Technology*, 51, 736–742. https://doi.org/10.1007/s13197-011-0560-4 [DOI] [PMC free article] [PubMed] [Google Scholar].

Mishra, G., Dash, S.P., Mahapatra, S.K., Swain, D. and Rout, G.R. 2025. Deeper insights into the physiological and metabolic functions of the pigments in plants and their applications: Beyond natural colourants. *Physiol. Plant.*, *177*: e70168. https://doi.org/10.1111/ppl.70168.

Mohd Amin, S.F., Karim, R., Yusof, Y.A. and Muhammad, K. 2023. Effects of metal concentration, pH, and temperature on the chlorophyll derivative content, green colour, and antioxidant activity of Amaranth (*Amaranthus viridis*) purees. *Appl. Sci.*, *13*: 1344. https://doi.org/10.3390/app13031344.

Montoya-Rodríguez, A., de Mejía, E.G., Dia, V.P., Reyes-Moreno, C. and Milán-Carrillo, J. 2014. Extrusion improved the anti-inflammatory effect of amaranth (*Amaranthus hypochondriacus*) hydrolysates in LPS-induced human THP-1 macrophage-like and mouse RAW 264.7 macrophages by preventing activation of NF-κ B signalling. *Mol. Nutr. Food Res.*, *58*: 1028– 1041. https://doi.org/10.1002/mnfr.201300764.

Mukuwapasi, B., Mavengahama, S. and Gerrano, A.S. 2024. Grain amaranth: A versatile untapped climate-smart crop for enhancing food and nutritional security. *Discov. Agric.*, *2*: 44. https://doi.org/10.1007/s44279-024-00057-8.

Mutha, R.E., Tatiya, A.U. and Surana, S.J. 2021. Flavonoids as natural phenolic compounds and their role in therapeutics: An overview. *Future J. Pharm. Sci.*, *7*: 25. https://doi.org/10.1186/s43094-020-00161-8.

Nambafu, S.N., Bett, H. and Sibiko, K.W., 2024. Determinants of commercialization of African Indigenous Vegetables among smallholder farmers in Bungoma County, Kenya. *Discov. Agric.*, *2*: 14. https://doi.org/10.1007/s44279-024-00013-6.

Nardo, A.E., Suárez, S., Quiroga, A.V. and Añón, M.C. 2020. Amaranth as a source of antihypertensive peptides. *Front. Plant Sci.*, *11*: 578631. https://doi.org/10.3389/fpls.2020.578631.

Nkobole, N., Bodede, O. and Hussein, A.A. 2021. *In vitro* α-glucosidase and α-amylase activities of wild and cultivated Amaranthus spp. and isolated compounds. *Pharmacog. J.*, *13*: 1614–1620. https://doi.org/10.5530/pj.2021.13.208.

Nyonje, W.A., Schafleitner, R., Abukutsa-Onyango, M., Yang, R.Y., Makokha, A. and Owino, W. 2021. Precision phenotyping and association between morphological traits and nutritional content in Vegetable Amaranth (*Amaranthus* spp.). *J. Agric. Food Res.*, *5*: 100165. https://doi.org/10.1016/j.jafr.2021.100165.

Oboh, G., Oyeleye, S.I., Ademiluyi, A.O. 2019. The food and medicinal values of indigenous leafy vegetables. *Acta Hortic*. 1238: 137–156.

Olusanya, R.N., Kolanisi, U., Ngobese, N.Z. and Mayashree, C. 2024. Dynamics of Amaranthus in Urban and Rural Value Chains in Communities of KwaZulu-Natal Province, South Africa. *Resources*, *13*: 161. https://doi.org/10.3390/resources13110161.

Peter, K. and Gandhi, P. 2017. Rediscovering the therapeutic potential of *Amaranthus* species: A review. *Egypt. J. Basic Appl Sci.*, *4*: 196–205. https://doi.org/10.1016/j.ejbas.2017.05.001

Pratiwi, R., Maharani, D.S. and Redjeki, S.G. 2025. Betalain Pigments: Isolation and application as reagents for colourimetric methods and biosensors. *Biosensors*, *15*: 349. https://doi.org/10.3390/bios15060349.

Quiroga, A.V., Barrio, D.A. and Añón, M.C. 2015. Amaranth lectin presents potential antitumour properties. *LWT-Food Sci. Technol.*, *60*: 478–485. https://doi.org/10.1016/j.lwt.2014.07.035.

Rajasekaran, S., Dinesh, M.G., Kansrajh, C. and Baig, F.H.A. 2014. *Amaranthus spinosus* leaf extracts and its anti-inflammatory effects on cancer. *Indian J. Res. Pharm. Biotechnol.*, *2*: 1058.

Rjeibi, I., Saad, A.B., Sdayria, J., Feriani, A., Ncib, S., Allagui, M.S., Hfaiedh, N., Souid, S. 2019. HPLC-DAD identification of polyphenols from ethyl acetate extract of Amaranthus spinosus leaves and determination of their antioxidant and antinociceptive effects. *Inflammopharmacology*, 27: 975–984.

Ramdwar, M.N., Chadee, S.T. and Stoute, V.A. 2017. Estimating the potential consumption level of amaranth for food security initiatives in Trinidad, West Indies. *Cogent Food Agric.*, *3*: 1321475. https://doi.org/10.1080/23311932.2017.1321475.

Ramírez-Torres, G., Ontiveros, N., Lopez-Teros, V., Ibarra-Diarte, J.A., Reyes-Moreno, C., Cuevas-Rodríguez, E.O. and Cabrera-Chávez, F. 2017. Amaranth protein hydrolysates efficiently reduce systolic blood pressure in spontaneously hypertensive rats. *Molecules*, *22*: 1905. https://doi.org/10.3390/molecules22111905.

Rastogi, A., Shukla, S. 2013. Amaranth: a new millennium crop of nutraceutical values. *Crti. Rev. Food Sci. Nutr*. 53: 109–125.

Rao, P.P. 2019. Essential leafy vegetables for healthy eyes. *Int. J. Herb. Med.*, *4*: 120–123.

Ruth, O.N., Unathi, K., Nomali, N. and Chinsamy, M. 2021. Underutilization versus nutritional-nutraceutical potential of the *Amaranthus* food plant: A mini-review. *Appl. Sci.*, *11*: 6879. https://doi.org/10.3390/app11156879.

Saini, D.R., Prakash, P. and Kumar, S. 2024. Chlorophyll: unveiling its dual role in plant vitality and human health. *Krishi Sci.*, *4*: 23–28.

Sarker, U. and Oba, S. 2021. Color attributes, betacyanin, and carotenoid profiles, bioactive components, and radical quenching capacity in selected *Amaranthus gangeticus* leafy vegetables. *Sci. Rep.*, *11*: 11559. https://doi.org/10.1038/s41598-021-91157-8.

Sarker, U., Hossain, M.M. and Oba, S. 2020a Nutritional and antioxidant components and antioxidant capacity in green morph Amaranthus leafy vegetable. *Sci. Rep.*, *10*: 1336. https://doi.org/10.1038/s41598-020-57687-3.

Sarker, U., Hossain, M.N., Iqbal, M.A. and Oba, S. 2020b. Bioactive components and radical scavenging activity in selected advance lines of salt-tolerant vegetable amaranth. *Front. Nutr.*, *7*: 587257. https://doi.org/10.3389/fnut.2020.587257.

Sarker, U., Lin, Y.P., Oba, S., Yoshioka, Y. and Hoshikawa, K. 2022. Prospects and potentials of underutilized leafy Amaranths as vegetable use for health-promotion. *Plant Physiol. Biochem.*, *182*: 104–123. https://doi.org/10.1016/j.plaphy.2022.04.011.

Sarker, U., Oba, S., Ullah, R., Bari, A., Ercisli, S., Skrovankova, S., Adamkova, A., Zvonkova, M. and Mlcek, J. 2024. Nutritional and bioactive properties and antioxidant potential of *Amaranthus tricolor*, *A. lividus*, *A viridis*, and *A. spinosus* leafy vegetables. *Heliyon*, *10*: e30453. https://doi.org/10.1016/j.heliyon.2024.e30453.

Sarker, U., Rabbani, M.G., Oba, S., Eldehna, W.M., Al-Rashood, S.T., Mostafa, N.M. and Eldahshan, O.A. 2022. Phytonutrients, colourant pigments, phytochemicals, and antioxidant potential of orphan leafy *Amaranthus* species. *Molecules*, *27*: 2899. https://doi.org/10.3390/molecules27092899.

Segura-Jiménez, M.E. and Jacobo-Velázquez, D.A. 2025. *Amaranthus* spp.: A Multifunctional Crop at the Nexus of Nutrition, Health, and Sustainable Innovation. *J. Agric. Food Res.*, *24*: 102322. https://doi.org/10.1016/j.jafr.2025.102322.

Selamoglu, Z., Naeem, M.Y., Rakhmetova, Y. and Gaipov, T. 2024. The Nutritional and Biotechnological Potential of Amaranth: A Versatile Plant for Health and Sustainability. *J. Biores. Manag.*, *11*:12.

Sharma, A.K., Dwivedi, S.V., Devi, J., Gupta, N., Rai, N., Behera, T.K. and Sagar, V. 2025. Botany, ethnomedicine, phytochemistry, and pharmacology of *Amaranthus* spp.: A review. *S. Afr. J. Bot.*, *178*: 198–216. https://doi.org/10.1016/j.sajb.2025.01.030.

Sharma, S., Pandita, G. and Bhosale, Y.K. 2023. Anthocyanin: Potential tool for diabetes management and different delivery aspects. *Trends Food Sci. Technol.*, *140*: 104170. https://doi.org/10.1016/j.tifs.2023.104170.

Singh, N., Samarth, R.M., Vashishth, A. and Pareek, A. 2024. Amaranthus as a potential dietary supplement in sports nutrition. *CyTA-J. Food*, *22*: 2375253. https://doi.org/10.1080/19476337.2024.2375253

Spórna-Kucab, A., Tekieli, A., Kisiel, A., Grzegorczyk, A., Skalicka-Woźniak, K., Starzak, K. and Wybraniec, S. 2023. Antioxidant and antimicrobial effects of baby leaves of *Amaranthus tricolor* L. Harvested as vegetable in correlation with their phytochemical composition. *Molecules*, *28*: 1463. https://doi.org/10.3390/molecules28031463.

Tripathi, G.K. and Harıwal, M. 2023. Anticancer activity of Amaranthus spinosus Linn. (Tanduliya): A Review. *Curr. Persp. Med. Aroma. Plants*, *6*: 66–77. https://doi.org/10.38093/cupmap.1324101.

Tufail, T., Bader Ul Ain, H., Noreen, S., Ikram, A., Arshad, M.T. and Abdullahi, M.A. 2024. Nutritional benefits of lycopene and beta-carotene: A comprehensive overview. *Food Sci. Nutr.*, *12*: 8715–8741. https://doi.org/10.1002/fsn3.4502.

Ullah, A., Munir, S., Badshah, S.L., Khan, N., Ghani, L., Poulson, B.G., Emwas, A.H. and Jaremko, M. 2020. Important flavonoids and their role as a therapeutic agent. *Molecules*, *25*: 5243. https://doi.org/10.3390/molecules25225243.

World Health Organization (WHO), 2023. 122 million more people pushed into hunger since 2019 due to multiple crises. Reveals UN Report.

Xu, H., Guo, Y., Xia, M., Yu, J., Chi, X., Han, Y., Li, X. and Zhang, F. 2024. An updated phylogeny and adaptive evolution within Amaranthaceae s.l. inferred from multiple phylogenomic datasets. *Food Ecol. Evol.*, *14*: e70013.https://doi.org/10.1002/ece3.70013.

Yadav, A. and Yadav, K. 2024. From humble beginnings to nutritional powerhouse: The rise of amaranth as a climate-resilient superfood. *Tropical Plants*, *3*: e037. https://doi.org/10.48130/tp-0024-0037.

Yadav, S.K. and Sehgal, S. 2002. Effect of domestic processing and cooking methods on total, HCl extractable iron and *in vitro* availability of iron in spinach and amaranth leaves. *Nutr. Health*, *16*: 113– 120. https://doi.org/10.1177/026010600201600205.

Yang, Y.C., Mong, M.C., Wu, W.T., Wang, Z.H. and Yin, M.C. 2020. Phytochemical profiles and anti-diabetic benefits of two edible *Amaranthus* species. *CyTA-J. Food*, *18*: 94–101. https://doi.org/10.1080/19476337.2020.1716850.

Yoganarasimthan, S.N. (2000). Medicinal Plants of Tamil Nadu, Vol. 2. www.eap.mcgill.ca/CPAT 1.htm. Nov. 2008. www.wikepedia.com. USDA National Research Council,

10 Nutritive Value, Bioactive Compounds, and Biological Activities of Spinach (*Spinacia oleracea* L.)

Guggalada Govardhana Yadav[1,2*] and Hosakatte Niranjana Murthy[2-4*]

Introduction

A leaf is a determinate lateral organ of the shoot with a dorsiventral bilateral symmetry. The leaf has three zones, fundamentally, viz., the distal lamina or blade with dorsiventral symmetry, the middle petiole or stalk, and the proximal leaf base (Kaplan, 2022). The basic functions of a leaf are photosynthesis, gaseous exchange, and transpiration. In addition to these fundamental duties, leaves are modified morphologically and anatomically to perform other functions that are crucial in a plant's adaptability, survival, and reproduction. The leaves are modified into spines which protect against herbivores and reduce transpiration (*Cactus* spp.), store water in succulent plants (*Bryophyllum* spp.), store food materials (*Allium* spp.), tendrils for support and climbing (*Pisum* spp.), trap insects and provide nutrition in insectivorous plants (*Nepenthes*, *Drosera*, *Utricularia* and *Dionaea*), and produce plantlets on the margins which help in vegetative propagation (*Bryophyllum* spp.). Further, leaves are the chemical factories of the plant, which synthesize and accumulate a variety of bioactive compounds, such as phenolics, flavonoids, tannins, alkaloids, and terpenoids, which are being used in curing a variety of ailments (De et al., 2023).

The leaves of many plant species have been used as vegetables globally since ancient times. The leafy vegetables have a high nutrient profile and serve as a source of essential components of a diet, such as minerals, vitamins, and antioxidants. *Amaranthus* spp., *Ipomoea* spp., *Moringa oleifera*, *Basella alba*,

[1] Department of Botany, A. D. B. First Grade College, Harapanahalli, India 583131.
[2] P. G. Department of Studies in Botany, Karnatak University, Dharwad, India 580003.
[3] Department of Horticultural Science, Chungbuk National University, Cheongju 28644, Republic of Korea.
[4] Department of Life Sciences, CHRIST University, Bengaluru 5600039, India.
* Corresponding author: govardhanyadavgs@gmail.com[1]; hnmurthy60@gmail.com[2]; hosakatte.niranjanamurthy@christuniversity.in[2]

DOI: 10.1201/9781003687832-10

Hibiscus sabdariffa, *Trigonella*, and *Portulaca oleracea* are some of the major leafy vegetables that possess great nutritional benefits (Sarkar et al., 2023). These underutilized leafy vegetables play a vital role in the nutrition security of the ever-increasing population by providing essential nutrients as mentioned above, and help in combating malnutrition and hidden hunger (Gowthami et al., 2019). The underutilized species are well adapted to local environmental conditions, and hence can withstand climate change and have the potential to contribute significantly to the diversification and resilience of agriculture systems. They are easily accessible to rural and marginalized communities and also play a prominent role in generating income for them (Padulosi et al., 2011).

Spinacia oleracea L. is one such underutilized leafy vegetable (Figure 1A) that belongs to the family Amaranthaceae, which includes other important vegetable crops, such as beetroot (*Beta vulgaris* L.) and amaranth (*Amaranthus* spp.). This species is native to central Asia and Western Pakistan and has been introduced to other parts of the world, including India (POWO, 2025). *Spinacia oleracea* is well-known in India for its nutritional quality and known with many regional names, such as Spinach in English, Palak or Panakana Palle in Kannada, Palak in Hindi and Marathi, Vachalaikkirai in Tamil, and Dumpabachhali or Matturbachhali in Telugu (Flowers of India, 2025). *Spinacia oleracea* has two recognized subspecies: *S. oleracea* subsp. *oleracea* and *S. oleracea* subsp. *turkestanica*. Spinach is cultivated worldwide, with global production reaching 34.1 million tons in 2023. China alone produced 31.7 million tons, accounting for 92.98% of the total, followed by the United States, Turkey, and Japan. In global trade, spinach leaf exports generated 846.72 million USD, with China contributing 356.99 million USD, followed by the United States, Italy, and Mexico (FAO, 2024). Thus, cultivation and export of spinach provide great opportunities in India for both providing nutrition security to the growing population and generating income for the farmers.

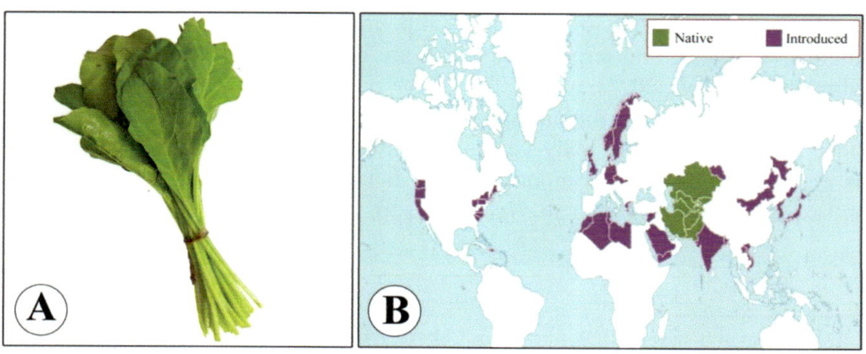

Fig. 1 A. Leaves of *Spinacia oleracea* L.; B. Distribution of *Spinacia oleracea* L. (Source of distribution map: POWO website, available at: https://powo.science.kew.org/taxon/urn:lsid:ipni.org:names:167384-1)

In addition, this leafy vegetable is also used for its beneficial role as a medicine, such as stomach and colon-related disorders, lung cancer, dyspepsia, and preventing arteriosclerosis in the traditional medicine system (Salehi et al., 2019). Several pharmacological studies proved the efficiency of this plant as an antioxidant, antimicrobial, anticancer, anti-inflammatory, anti-obesity, hypoglycemic, and hypolipidemic agent. These activities of the spinach leaves may be attributed to the wide array of bioactive compounds it possesses. Spinach is rich in flavonoids, which are regarded as natural antioxidants, along with phenolics, carotenoids, ecdysteroids, lignans, terpenoids, xanthines, and other potent compounds (Roberts and Moreau, 2016; Salehi et al., 2019). The present chapter focuses on the distribution, traditional uses, bioactive compounds isolated, and biological and pharmacological activities of extracts and isolated compounds of *Spinacia oleracea* leaves.

Distribution and Botanical Description

Spinacia oleracea is native to Central Asia, including the countries of Afghanistan, Iran, Kazakhstan, Kirgizstan, Pakistan, Tadzhikistan, Turkmenistan and Uzbekistan, and introduced to India, Bangladesh, United States, Czechia-Slovakia, Denmark, Dominican Republic, Germany, Britain, Greece, Mongolia, Japan, Lebanon-Syria, Yemen, and other parts of the world (Figure 1B; POWO, 2025). Spinach is an annual or biennial dioecious erect herb, reaching up to 1 m in height. Tap roots are reddish in colour, rarely white; the stem is simple or with few branches, hollow; leaves are light green, simple, ovate to triangular-hastate, base cuneate, margin entire or dentate, lower ones are long-petiolate. Male flowers are in panicles, with perianth segments usually 4, filiform filaments, flattened, and anthers without an appendage; Female flowers are in dense axillary sessile clusters, each separated and falling so in the fruit; fruit is an urticle, ovoid or sub-globose, compressed, and about 2.5 mm in diameter, pericarp brown (World Flora Online, 2025).

Traditional Uses

Spinacia oleracea has long been valued not only as a nutritious vegetable, but also for its medicinal properties. Compounds, such as folic acid and ferric salts present in spinach exhibit strong anti-proliferative activity, helping to combat cancers of the stomach, colon, and lungs, and are also beneficial in treating dyspepsia (Maeda et al., 2005). Traditionally, the whole plant has been employed as a remedy for urinary calculi, while the leaves are specifically used to alleviate inflammation of the lungs and intestines, as well as disorders of the kidney, bladder, and liver. Spinach also contains choline and inositol, which play a role in preventing arteriosclerosis (hardening of arteries), and it serves as a rich source of chlorophyll that supports digestion (Salehi et al., 2019).

Ethnomedicinally, spinach is recommended for ailments, such as phlegm, tuberculosis, rickets, halitosis, and weight loss (Ahmed, 2016). Its leaves are known for their cooling, emollient, nutritious, antipyretic, diuretic, laxative, and anthelmintic properties. They are traditionally used in treating urinary calculi, lung and bowel inflammation, sore throat, joint pain, colds, sneezing, eye infections, skin conditions like ringworm, scabies, and leukoderma, as well as vomiting, biliousness, flatulence, and febrile illnesses. Seeds, on the other hand, are valued for managing fever, leucorrhoea, urinary discharges, lumbago, and disorders of the brain and heart. Being laxative and cooling in nature, seeds are also prescribed for respiratory distress, liver inflammation, and jaundice (Salehi et al., 2019). It has also been used in preventing neurological, hepatic, and respiratory disorders and is considered safe for both children and pregnant women (Maeda et al., 2005). Furthermore, its nutritional components—such as saponins, dietary fibre, ferric salts, and folic acid—contribute to its effectiveness in treating anemia and constipation. Spinach also supports bone health by reducing osteoporosis-related bone loss, exerts anti-inflammatory effects, improves cardiovascular function, and is recognized for its energy-boosting benefits (Subhash et al., 2010).

Nutritional Properties

Spinach, a widely consumed leafy vegetable, is well recognized for its high nutritional density and medicinal properties. Its nutritional composition encompasses proximate nutrients, vitamins, minerals, pigments, fatty acids, amino acids, organic acids, and certain antinutritional factors. The nutritional composition of spinach leaves is provided in Table 1.

Table 1 Nutrient composition of Spinach leaves (Longvah et al., 2017; Naseem et al., 2023).

Nutrient components	*Composition*
Proximate composition (g/100 g FW)	
Moisture	90.31±0.46
Carbohydrate	2.05±0.31
Fat	0.64±0.03
Protein	2.14±0.14
Fibre	2.38±0.24
Ash	2.47±0.38
Energy (KJ/100 g)	102.0±7.0
Vitamins (per 100 g FW)	
Vitamin B1 (Thiamine)	0.16±0.02 mg
Vitamin B2 (Riboflavin)	0.10±0.01 mg
Vitamin B3 (Niacin)	0.33±0.03 mg
Vitamin B5 (Pantothenic acid)	0.22±0.03 mg

Contd...

Table 1 *Contd...*

Nutrient components	*Composition*
Vitamin B6 (Total)	0.15±0.01 mg
Vitamin B7 (Biotin)	4.14±0.27 µg
Vitamin B9 (Total folates)	142.0±10.3 µg
Vitamin C (Ascorbic acid)	30.28±4.71 mg
Vitamin D2 (Ergocalciferol)	0.26 µg
Vitamin E (Tocopherols)	1.39±0.03 mg
Vitamin K1 (Phylloquinones)	325.0±2.3 µg
Pigments (mg/100 g FW)	
Lutein	3.85±9.26
Zeaxanthin	0.017±0.01
Beta-carotene	2.60±0.52
Total carotenoids	9.55±1.49
Minerals (mg/100 g FW)	
Aluminium	1.51±0.47
Calcium	82.29±6.51
Copper	0.17±0.05
Iron	2.95±0.68
Magnesium	86.97±8.58
Manganese	1.12±0.42
Phosphorous	32.59±8.08
Potassium	625.0±5.90
Selenium	2.09±0.22
Sodium	42.55±4.12
Zinc	0.46±0.09
Fatty acid profile (mg/100 g FW)	
Myristic acid (C14:0)	17.01±0.72
Palmitic acid (C16:0)	147.0±6.3
Stearic acid (C18:0)	9.54±1.17
Arachidic acid (C20:0)	1.29±0.25
Behenic acid (C22:0)	4.13±0.29
Lignoceric acid (C24:0)	3.99±0.39
Palmitoleic acid (C16:1)	8.77±0.84
Oleic acid (C18:1n9)	32.87±1.74
Linoleic acid (C18:2n6)	75.06±4.22
α-Linolenic acid (C18:3n3)	220.0±8.3

Contd...

Table 1 *Contd...*

Nutrient components	*Composition*
TSFA (Total saturated fatty acids)	183.0±8.1
TMUFA (Total monounsaturated fatty acids)	42.83±2.54
TPUFA (Total polyunsaturated fatty acids)	295.0±12.2
Amino acid profile (g/100 g protein)	
Alanine	6.78±0.42
Arginine	5.41±0.36
Aspartic acid	9.24±0.71
Cystine	0.84±0.06
Glutamic acid	14.77±1.57
Glycine	5.70±0.25
Histidine	2.18±0.16
Isoleucine	4.31±0.29
Leucine	8.70±0.59
Lysine	4.68±0.13
Methionine	1.33±0.05
Phenylalanine	5.38±0.46
Proline	4.70±0.59
Serine	4.40±0.22
Threonine	4.70±0.26
Tryptophan	1.53±0.11
Tyrosine	3.95±0.32
Valine	5.83±0.38
Organic acids (mg/100 g FW)	
Citric acid	67.34±14.43
Fumaric acid	5.44±0.79
Malic acid	15.79±2.95
Quinic acid	254.0±7.9
Succinic acid	1.64±0.25
Antinutrient components (mg/100 g DW)	
Oxalate	0.07±0.01
Phytate	1.97±0.08
Alkaloids	6.45±0.02
Saponins	10.56±0.481
Glucosinolates (μmol/g DW)	201.20±2.47

Spinach is composed predominantly of water (90.31 g/100 g FW), making it a hydrating and low-energy food. Despite being low in calories (102 kJ/100 g), it is nutritionally rich, containing 2.14 g protein, 2.05 g carbohydrates, 0.64 g fat, 2.38 g fibre, and 2.47 g ash per 100 g of fresh weight (Longvah et al., 2017; Table 1). The moderate protein content is notable for a leafy vegetable, while the fibre fraction enhances digestive health and contributes to satiety. The high ash content reflects the abundance of minerals present in spinach. Spinach is a powerhouse of vitamins, particularly vitamin K1 (phylloquinone, 325 μg/100 g), which plays an essential role in blood clotting and bone metabolism. The vegetable is also an excellent source of folates (142 μg/100 g), crucial for DNA synthesis and cell division, thereby making spinach especially valuable during pregnancy (Krishnaswamy and Nair, 2001; Longvah et al., 2017). It provides considerable amounts of vitamin C (30.28 mg/100 g) and Vitamin E (1.39 mg/100 g), which are known for their strong antioxidant activities and other health benefits. Spinach also supplies small but important amounts of the B-complex group (thiamine, riboflavin, niacin, pantothenic acid, pyridoxine, and biotin), which are vital for energy metabolism. Interestingly, spinach contains vitamin D2 (0.26 μg/100 g), rare in plant sources, enhancing its nutritional uniqueness (Longvah et al., 2017).

Carotenoid pigments are among the most valuable bioactive compounds in spinach. The leaves contain lutein (3.85 mg/100 g), β-carotene (2.60 mg/100 g), and Zeaxanthin (0.017 mg/100 g) with total carotenoids reaching 9.55 mg/100 g. These pigments function as antioxidants, protecting cells from oxidative damage, and play a crucial role in eye health by preventing age-related macular degeneration (Longvah et al., 2017). Spinach is particularly rich in minerals, many of which are essential for human health. Potassium (625 mg/100 g) is the most abundant mineral, followed by magnesium (86.97 mg/100 g) and Calcium (82.29 mg/100 g). The iron content was 2.95 mg/100 g, which is crucial for haemoglobin synthesis and the prevention of anaemia. Other important elements include manganese (1.12 mg/100 g), zinc (0.46 mg/100 g), copper (0.17 mg/100 g), and selenium (2.09 mg/100 g). Moderate sodium content (42.55 mg/100 g) makes spinach suitable even for low-sodium diets (Longvah et al., 2017).

Though spinach is naturally low in fat, the quality of its lipid fraction is nutritionally significant. It is particularly rich in polyunsaturated fatty acids (PUFAs, 295 mg/100 g), dominated by α-linolenic acid (220 mg/100 g), an essential omega-3 fatty acid with cardioprotective and anti-inflammatory properties. Linoleic acid (75 mg/100 g), an omega-6 fatty acid, complements the fatty acid profile. Monounsaturated fatty acids (MUFAs), mainly oleic acid (32.87 mg/100 g), are present in smaller amounts, whereas saturated fatty acids (183 mg/100 g) are relatively low, dominated by palmitic acid (147 mg/100 g) (Longvah et al., 2017). Spinach protein has a balanced amino acid composition. Glutamic acid is present in prominent composition with a value of 14.77 g/100 g protein, followed by aspartic acid, leucine, alanine, and valine, with contents of 9.24, 8.7, 6.78, and 5.83 g/100 g protein, respectively. They are the most abundant, contributing to flavour and neurotransmitter precursors. The essential amino acids, such as leucine, valine, lysine (4.68 g/100 g protein), and isoleucine (4.31 g/100 g protein), enhance the nutritional value of spinach, making it an important supplement in vegetarian diets where essential amino acid intake may be limited. Spinach contains diverse organic acids, with quinic acid (254 mg/100 g) as the most abundant, followed by citric acid (67.34 mg/100 g) and malic acid (15.79 mg/100 g) (Longvah et al., 2017).

Like many leafy greens, spinach contains certain antinutritional compounds. Saponins (10.56 mg/100 g DW) and glucosinolates (201.20 μmol/g DW) are present in significant amounts. The oxalate, phytate, and alkaloids are present at a concentration of 0.07, 1.97, and 6.45 mg/100 g DW of the leaves (Longvah et al., 2017). While these anti-nutrients may reduce mineral bioavailability, they also provide some health benefits. Also, these antinutrient compounds can be lowered by adopting suitable processing techniques, such as soaking and boiling (Murthy et al., 2024). The nutritional profile of spinach underscores its role as a low-calorie, nutrient-dense functional food.

Bioactive Compounds

Medicinal plants are the storehouse of bioactive compounds, which are responsible for curing various ailments. Globally, around 70–80% of the human population relies on medicinal plants for primary healthcare (Yadav and Murthy, 2025). These bioactive compounds help the plant in different ways, such as adapting to surrounding environmental conditions, protection against herbivores, microbial attack, and UV rays, attracting pollinating and seed dispersing agents, and responding to the varied biotic and abiotic stress (Guerriero et al., 2018). Though spinach is regarded as a vegetable, it is rich in bioactive compounds. A total of 133 bioactive compounds, belonging to the prominent chemical groups, such as flavonoids, phenolic acids, carotenoids, coumarins, ecdysteroids, lignans, and xanthines were identified in this plant. The bioactive compounds isolated or identified from the leaves of spinach have been listed in Table 2 and detailed in the following sections.

Flavonoids

Flavonoids are one of the major groups of phytochemicals, synthesized via the phenylpropanoid pathway, and possess 15 carbon atoms arranged in three rings. Plants are the major source of these polyphenolic compounds, which are reported to possess a wide range of biological activities, including antioxidant, anti-inflammatory, anticancer, and antimutagenic properties. These compounds have numerous pharmacological applications (Chen et al., 2023). *Spinacia oleracea* is rich in flavonoids, consisting of a total of 82 flavonoid compounds (Table 2; Figure 2). The major flavonoid compounds of spinach include, 5,3',4'-trihydroxy-3-methoxy-6:7-methylendioxyflavone-4'-glucuronide; 5,3',4'-trihydroxy-3-methoxy-6:7-methylen-dioxyflavone-

4'-β-D-(2'-O-feurloyl-glucuronide); 5,4'-dihydroxy-3,3'-dimethoxy -6:7-methylendioxyflavone-4′-β-D-glucuronide; 5,4'-dihydroxy-3-methoxy-6:7-methylendioxyflavone-4′-glucuronide; apigenin; catechin; dihydrokaempferol; genistein; isorhamnetin; jaceidin; jaceidin 4'-glucuronide; kaempferol; luteolin; myrecetin; patuletin; patuletin 3-O-β-D-(2"feruloylglucopyranosyl)(1→6)-[β-D-apiofuranosyl(1→2)]-β-D-glucopyranoside; patuletin 3-O-β-D-glucopyranosyl; patuletin-3-O-β-D-glucopyranosyl-(1 → 6)-β-D-glucopyranoside; patuletin-3-O-β-D-(2"-feruloylglucopyranosyl-(1→6)-β-D-glucopyranoside; patuletin-3-O-β-D-(2"-β-coumaroylglucopyranosyl)-(1→6)-[β-D-apiofuranosyl-(1→2)]-β-D-glucopyranoside; patuletin-3-O-β-D-glucopyranosyl-(1 → 6)-[β-D-apiofuranosyl-(1 → 2)]-β-D-glucopyranoside; quercetin; spinacetin; spinacetin 3-O-β-D-(2″feruloylglucopyranosyl)(1 → 6)-[β-D-apiofuranosyl(1 → 2)]-β-D-glucopyranoside; spinacetin 3-O-β-D-(2"-feruloylglucopyranosyl)(1 → 6)-β-D-glucopyranoside; spinacetin-3-O-β-D-(2″-coumaroylglucopyranosyl)-(1→6)-β-D-glucopyranoside; spinacetin-3-O-β-D-glucopyranosyl-(1→6)-[β-D-apiofuranosyl-(1→2)]-β-D-glucopyranoside; and spinacetin-3-O-β-D-glucopyranosyl-(1→6)-β-D-glucopyranoside (Ferreres et al., 1997; Chu et al., 2000; Nuutila et al., 2002; Aehle et al., 2004; Sultana and Anwar, 2008; Dehkharghanian et al., 2010; Morishita et al., 2015; Singh, 2016; Khalid et al., 2017; Barkat et al., 2018; Kasote et al., 2019; Yuk et al., 2019; Grace et al., 2022; Lee et al., 2022; Vujčić Bok et al., 2022; Cerulli et al., 2024; Dzakovich et al., 2025).

Table 2 List of bioactive compounds isolated/identified from Spinach leaves.

Sl. No.	*Compound group*	*Name of the Compound*	*References*
1.	Flavonoids	5,3,4 -Trihydroxy-3-methoxy-6:7 methylendioxyflavone-4-β-D-glucuronide	Kasote et al. (2019)
2.		5,3,4 -Trihydroxy-3-methoxy-6:7-methylendioxyflavone	Dzakovich et al. (2025)
3.		5,3',4'-Trihydroxy-3-methoxy-6:7-methylendioxyflavone-4'-glucuronide	Ferreres et al. (1997); Aehle et al. (2004); Morishita et al. (2015); Barkat et al. (2017); Yuk et al. (2019); Grace et al. (2022); Lee et al. (2022); Cerulli et al. (2024); Dzakovich et al. (2025)
4.		5,3',4'-Trihydroxy-3-methoxy-6:7-methylen-dioxyflavone-4'-β-D-(2'-O-feurloyl-glucuronide)	Dzakovich et al. (2025)

Contd...

Table 2 *Contd...*

Sl. No.	*Compound group*	*Name of the Compound*	*References*
5.		5,4 -Dihydroxy-3,3 -dimethoxy-6:7-methylendioxyflavone-4-β-D-glucuronide	Dzakovich et al. (2025)
6.		5,4 -Dihydroxy-3-methoxy-6:7-methylendioxyflavone-4-β-D-glucuronide	Dzakovich et al. (2025)
7.		5,4'-Dihydroxy-3,3'-dimethoxy-6,7-methylenedioxyflavone	Dzakovich et al. (2025)
8.		5,4'-dihydroxy-3,3'-dimethoxy-6:7-methylen-dioxyflavone-4' -β-D-(2'-O-feurloyl-glucuronide)	Dzakovich et al. (2025)
9.		5,4'-Dihydroxy-3-3'-dimethoxy-6:7-methylenedioxyflavone 4'-glucuronide	Dehkharghanian et al. (2010)
10.		5,4'-Dihydroxy-3-methoxy-6,7-methylenedioxyflavone	Dzakovich et al. (2025)
11.		5,4'- Dihydroxy-3,3'-dimethoxy -6,7-methylendioxyflavone-4′-β-D-glucuronide	Ferreres et al. (1997); Aehle et al. (2004); Barkat et al. (2017); Yuk et al. (2019); Grace et al. (2022); Lee et al. (2022); Cerulli et al. (2024)
12.		5,4'-Dihydroxy-3-methoxy-6,7-methylendioxyflavone-4′-glucuronide	Barkat et al. (2017); Grace et al. (2022); Yuk et al. (2019); Cerulli et al. (2024)
13.		5,7,3',4'-Tetrahydroxy-3,6-dimethoxyflavone 4'-glucuronide	Aehle et al. (2004)
14.		5,7,4'-Trihydroxy-3,6,3'-trimethoxyflavone 4'-glucuronide	Aehle et al. (2004)
15.		5,7-Dihydroxy-3,6-dimethoxy-flavone-4'- glucuronide	Dzakovich et al. (2025)
16.		6-Methoxykaempferol	Dehkharghanian et al. (2010)
17.		6-Metoxyapigenin-5-methylether	Cerulli et al. (2024)
18.		Apigenin	Dehkharghanian et al. (2010)
19.		Apigenin-C-glucoside	Singh (2016)

Contd...

Table 2 *Contd...*

Sl. No.	*Compound group*	*Name of the Compound*	*References*
20.		Axillarin-4'-β-glucuronide	Morishita et al. (2015)
21.		Cassiaside	Khalid et al. (2017)
22.		Catechin	Singh (2016)
23.		Daidzein	Singh (2016)
24.		Dihydrokaempferol	Singh (2016)
25.		Dihydroquercetin	Singh (2016)
26.		Epicatechin	Singh (2016)
27.		Genistein	Singh (2016)
28.		Glycitein	Singh (2016)
29.		Isorhamnetin	Singh (2016); Cerulli et al. (2024)
30.		Isorhamnetin-3-gentiobioside-7-glucoside	Khalid et al. (2017)
31.		Isorhamnetin-3-O-rutinoside	Singh (2016)
32.		Jaceidin	Dzakovich et al. (2025)
33.		Jaceidin 4'-glucuronide	Ferreres et al. (1997); Morishita et al. (2015); Barkat et al. (2017); Yuk et al. (2019); Grace et al. (2022); Cerulli et al. (2024)
34.		Jaceidin-4'-β-D-(2'-O-glucuronopyranosyl-glucuronide)	Dzakovich et al. (2025)
35.		Jaceidin-4-β-D-glucuronide	Dzakovich et al. (2025)
36.		Jaceosidin	Dzakovich et al. (2025)
37.		Kaempferol	Chu et al. (2000); Nuutila et al. (2002); Sultana & Anwar (2008); Dehkharghanian et al. (2010); Khalid et al. (2017)
38.		Kaempferol-3-O-rutinoside	Singh (2016)

Contd...

Table 2 *Contd...*

Sl. No.	*Compound group*	*Name of the Compound*	*References*
39.		Kaempferol-rutinosideacetylhexoside	Singh (2016)
40.		Kaempferol-rutinoside-hexoside-rhamnoside	Singh (2016)
41.		Luteolin	Khalid et al. (2017)
42.		Myrecetin	Chu et al. (2000); Sultana & Anwar (2008)
43.		Naringenin	Singh (2016)
44.		Patuletin	Ferreres et al. (1997); Dzakovich et al. (2025)
45.		Patuletin 3-O-β-D-(2”feruloylglucopyranosyl)(1→6)-[β-D-apiofuranosyl(1→2)]-β-D-glucopyranoside	Ferreres et al. (1997); Yuk et al. (2019); Grace et al. (2022); Cerulli et al. (2024)
46.		Patuletin 3-O-β-D-glucopyranosyl	Dehkharghanian et al. (2010); Dzakovich et al. (2025)
47.		Patuletin 3-O-β-gentiobioside	Dehkharghanian et al. (2010)
48.		Patuletin-3-[apiosyl-(1→2)]-glucoside	Dzakovich et al. (2025)
49.		Patuletin-3-methoxy-4′-glucuronoide	Grace et al. (2022)
50.		Patuletin-3-O-(2”-coumaroylglucosyl)-(1→6)-glucoside	Yuk et al. (2019); Grace et al. (2022)
51.		Patuletin-3-O-β-D glucopyranosyl-(1 → 6)-β-D-glucopyranoside	Aehle et al. (2004); Singh (2016); Barkat et al. (2017); Kasote et al. (2019); Yuk et al. (2019); Grace et al. (2022); Dzakovich et al. (2025)
52.		Patuletin-3-O-β-D-(2”-feruloylglucopyranosyl-(1→6)-β-D-glucopyranoside	Barkat et al. (2017); Yuk et al. (2019); Grace et al. (2022); Cerulli et al. (2024)

Contd...

Table 2 *Contd...*

Sl. No.	*Compound group*	*Name of the Compound*	*References*
53.		Patuletin-3-O-β-D-(2"-β-coumaroylglucopyranosyl)-(1→6)-[β-D-apiofuranosyl-(1→2)]-β-D-glucopyranoside	Barkat et al. (2017); Yuk et al. (2019); Grace et al. (2022); Cerulli et al. (2024)
54.		Patuletin-3-O-β-D-(2-feruloylglucopyranosyl)-(1→6)-[β-D-apiofuranosyl-(1 → 2)]-β-D-glucopyranoside	Dzakovich et al. (2025)
55.		Patuletin-3-O-β-D-(2-feruloylglucopyranosyl)-(1→6)-β-D-glucopyranoside	Dzakovich et al. (2025)
56.		Patuletin-3-O-β-D-(2-ρ-coumaroylglucopyranosyl-(1 → 6)-[β-D-apiofuranosyl-(1 → 2)]-β-D-glucopyranoside	Dzakovich et al. (2025)
57.		Patuletin-3-O-β-D-(2-ρ-coumaroylglucopyranosyl-(1 → 6)-β-D-glucopyranoside)	Dzakovich et al. (2025)
58.		Patuletin-3-O-β-D-glucopyranosyl-(1 → 6)-[β-D apiofuranosyl-(1 → 2)]-β-D-glucopyranoside	Ferreres et al. (1997); Aehle et al. (2004); Singh (2016); Barkat et al. (2017); Kasote et al. (2019); Yuk et al. (2019); Grace et al. (2022); Cerulli et al. (2024); Dzakovich et al. (2025)
59.		Patuletin-3-O-β-D-glucopyranosyl-(1→5)-[β-D-apiofuranosyl-(1→2)]-β-D glucopyranoside	Grace et al. (2022); Cerulli et al. (2024)
60.		Patuletin-3-β-D-glycopyranosyl-(1→2)-β-D-glucopyranoside	Aehle et al. (2004)
61.		Quercetin	Chu et al. (2000); Nuutila et al. (2002); Dehkharghanian et al. (2010); Singh (2016); Khalid et al. (2017); Vujčić Bok et al. (2022)
62.		Quercetin-3-gentiobioside-7-glucoside	Khalid et al. (2017)

Contd...

Table 2 *Contd...*

Sl. No.	*Compound group*	*Name of the Compound*	*References*
63.		Quercetin-3-O-glucoside	Dzakovich et al. (2025)
64.		Quercetin-3-O-rutinoside	Singh (2016)
65.		Rhamnetin-3-O-glucosyl-(1→5)-[apiosyl-(1→2)]-glucoside	Yuk et al. (2019)
66.		Satuletin-3-methoxy-4′-O-β-D-glucuronide	Cerulli et al. (2024)
67.		Spinacetin	Dzakovich et al. (2025)
68.		Spinacetin 3-[p-coumaroyl-(1 → 2)-glucosyl-(1 → 6)-[apiosyl-(1 → 2)]-glucoside]	Dzakovich et al. (2025)
69.		Spinacetin 3-gentiobioside	Ferreres et al. (1997)
		Spinacetin 3-O-β-D-(2″feruloylglucopyranosyl)(1 → 6)-[β-D-apiofuranosyl(1 → 2)]-β-D-glucopyranoside	Ferreres et al. 1997; Yuk et al. 2019; Grace et al. 2022; Cerulli et al. (2024)
70.		Spinacetin 3-O-β-D-(2”-feruloylglucopyranosyl)(1 → 6)-β-D-glucopyranoside	Ferreres et al. (1997); Singh (2016); Barkat et al. (2017); Yuk et al. (2019); Grace et al. (2022); Cerulli et al. (2024)
71.		Spinacetin 3-O-β-D-(2″-p-coumaroylglucopyranosyl)(1 → 6)-[β-D-apiofuranosyl(1 → 2)]-β-D-glucopyranoside	Ferreres et al. (1997); Yuk et al. (2019); Grace et al. (2022); Cerulli et al. (2024)
72.		Spinacetin-3-[apiosyl-(1→2)]-glucoside	Dzakovich et al. (2025)
73.		Spinacetin-3-glucoside	Dzakovich et al. (2025)
74.		Spinacetin-3-O- β-D-(glucopyranosyl)-(1→6)-[β-glucopyranosyl (1→ 2)]-β-D-glucopyranoside	Dzakovich et al. (2025)
75.		Spinacetin-3-O-β-D-(2″-coumaroylglucopyranosyl)-(1→6)-β-D-glucopyranoside	Grace et al. (2022); Yuk et al. (2019); Cerulli et al. (2024); Dzakovich et al. (2025)

Contd...

Table 2 *Contd...*

Sl. No.	*Compound group*	*Name of the Compound*	*References*
76.		Spinacetin-3-O-β-D-(2-feruloylglucopyranosyl) -(1→6)-[β-glucopyranosyl (1 → 2)]-β-D-glucopyranoside	Dzakovich et al. (2025)
77.		Spinacetin-3-O-β-D-(2-feruloylglucopyranosyl)-(1 → 6)-[β-D-apiofuranosyl-(1 → 2)]-β-D-glucopyranoside	Dzakovich et al. (2025)
78.		Spinacetin-3-O-β-D-(2-feruloylglucopyranosyl)-(1 → 6)-β-D-glucopyranoside	Dzakovich et al. (2025)
79.		Spinacetin-3-O-β-D-glucopyranosyl-(1→6)-[β-D-apiofuranosyl-(1→2)]-β-D-glucopyranoside	Ferreres et al. (1997); Singh (2016); Barkat et al. (2017); Kasote et al.(2019); Yuk et al. (2019); Grace et al. (2022); Cerulli et al. (2024); Dzakovich et al. (2025)
80.		Spinacetin-3-O-β-D-glucopyranosyl-(1→6)-β-D-glucopyranoside	Aehle et al. (2004); Singh (2016); Barkat et al. (2017); Yuk et al. (2019); Grace et al. (2022); Cerulli et al. (2024); Dzakovich et al. (2025)
81.		Spinatoside	Barkat et al. (2017); Yuk et al. (2019); Dzakovich et al. (2025)
82.		Spinatoside-4′-β-D-(2′-O-feruloyl-glucuronide)	Dzakovich et al. (2025)
83.	Phenolic acids	4′-Geranyloxyferulic acid	Fiorito et al. (2019)
84.		5-p-Coumaroylquinic acid	Khalid et al. (2017)
85.		Boropinic acid	Fiorito et al. (2019)
86.		Caffeic acid	Khalid et al. (2017)
87.		Caffeoylmalic acid	Khalid et al. (2017)
88.		Caffeoyltartaric acid	Khalid et al. (2017)
89.		Chlorogenic acid	Khalid et al. (2017)
90.		Cinnamic acid	Singh (2016)

Contd...

Table 2 *Contd...*

Sl. No.	*Compound group*	*Name of the Compound*	*References*
91.		Coumaroyltartaric acid	Nemzer et al. (2021)
92.		Ferulic acid	Bunea et al. (2008); Khalid et al. (2017); Vujčić Bok et al. (2022)
93.		Feruloyltartaric acid	Nemzer et al. (2021)
94.		Gallic acid	Singh (2016)
95.		Methyl o-coumarate	Ewais et al. (2023)
96.		o-Coumaric acid	Bunea et al. (2008)
97.		p-Coumaric acid	Bunea et al. (2008); Khalid et al. (2017); Vujčić Bok et al. (2022)
98.	Carotenoids	Lutein	Bunea et al. (2008)
99.		Neoxanthin	Bunea et al. (2008)
100.		Violaxanthin	Bunea et al. (2008)
101.		β-Carotene	Bunea et al. (2008)
102.	Coumarins	7-Isopentenyloxycoumarin	Fiorito et al. (2019)
103.		Auraptene	Fiorito et al. (2019)
104.		Umbelliprenin	Fiorito et al. (2019)
105.	Glycoglycerolipids	Monogalactosyl-diacylglycerol	Shiota et al. (2010)
106.		Digalactosyl-diacylglycerol	Shiota et al. (2010)
107.		Sulfoquinovosyl-diacylglycerol	Shiota et al. (2010)
108.	Lignans	Benzoylgomisin H	Khalid et al. (2017)
109.		Interiotherin A	Khalid et al. (2017)
110.		Phillyrin	Khalid et al. (2017)
111.		Schisantherin D	Khalid et al. (2017)
112.		Lariciresinol	Milder et al. (2005)
113.		Pinoresinol	Milder et al. (2005)
114.		Secoisolariciresinol	Milder et al. (2005)
115.	Diterpene alcohol	Phytol	Ewais et al. (2023)
116.	Ecdysteroid	20-Hydroxyecdysone	Grebenok et al. (1991); Bakrim et al. (2008); Hunyadi et al. (2016); Lee et al. (2022); Cerulli et al. (2024)

Contd...

Table 2 *Contd...*

Sl. No.	*Compound group*	*Name of the Compound*	*References*
117.		Polypodine B	Grebenok et al. (1991)
118.	Saponin	Spinasaponin B	Cerulli et al. (2024)
119.	Sterol	3,7,12-Trihhdroxycholan-24-oic acid	Ewais et al. (2023)
120.		Stigmasterol	Ewais et al. (2023)
121.		α-Spinasterol	Majeed et al. (2022)
122.	Tannins	Trimethylellagic acid	Cerulli et al. (2024)
123.	Terpenoid	Betulin	Ewais et al. (2023)
124.	Xanthines	(8′R)-Neochrome	Rotimi et al. (2019)
125.		(8′S)-Neochrome	Rotimi et al. (2019)
126.	Other compounds	2-Methoxy-4-vinylphenol	Ewais et al. (2023)
127.		5-Azecanol	Ewais et al. (2023)
128.		Allantoin	Cerulli et al. (2024)
129.		Benzoic acid	Singh (2016)
130.		Chorismic acid	Singh (2016)
131.		Coumaran	Ewais et al. (2023)
132.		Desulphosinigrin	Ewais et al. (2023)
133.		Quinic acid	Singh (2016)

Phenolic Acids

Phenolic acids are a large group of compounds having a benzene ring and one or more hydroxyl groups; they are reported to have prominent roles in the plant's physiological activities, including flavouring, antioxidation, colouring, and resistance against stress (Zhang et al., 2022). Spinach leaves are rich in phenolic compounds, and a total of 15 compounds have been identified (Table 2; Figure 2). The major phenolic acid compounds of the spinach include 4′-geranyloxyferulic acid; 5-p-coumaroylquinic acid; caffeic acid; chlorogenic acid; ferulic acid; gallic acid; and p-coumaric acid (Bunea et al., 2008; Singh, 2016; Khalid et al., 2017; Fiorito et al., 2019; Vujčić Bok et al., 2022).

Carotenoids

The carotenoids are C40 tetraterpenoids and consist of eight isoprene units. Carotenoids are brightly coloured compounds that are present in chloroplasts and act as accessory pigments along with the chlorophylls. Carotenoids are reported to possess antidiabetic, anticancer, skin-protective, and vision-improving activities (Crupi et al., 2023). Vegetables and fruits are excellent sources of these valuable compounds, and spinach possesses an impressive amount of carotenoids. Four carotenes, viz., lutein, neoxanthin, violaxanthin, and β-carotene (Table 2; Figure

2; Bunea et al., 2008). Total xanthophyll content of the spinach leaves ranged from 5.21 mg/100 g dw obtained through light petroleum ether solvent to 37.22 mg/100 g in dichloromethane solvent. The total carotenoid content of the leaves ranged between 10.90 (through hexane, acetone, and methanol in a 1:1:1 ratio) and 21.71 (in diethyl ether) mg/100 g DW (Turkiewicz et al., 2025). The lutein content of spinach ranged between 41.9–92.4 mg/100 g DW, whereas the β-carotene ranged from 26.2 to 54.4 mg/100 g dw. The drying procedures influenced the extracted content of these compounds and their bioaccessibility (Grace et al., 2022).

Lignans

Lignans are the prominent group of phenolic compounds, formed through the oxidative coupling of phenylpropanoid units. They are mainly used by plants in their defence, such as antimicrobial, insecticidal, and antifeeding activities, and possess many pharmacological activities (Singh et al., 2024). The spinach is rich in lignan content, and seven lignans are identified, viz., benzoylgomisin H; interiotherin A; phillyrin; schisantherin D; lariciresinol; pinoresinol; and secoisolariciresinol (Milder et al., 2005; Khalid et al., 2017).

Other Compounds

Spinacia oleracea possesses many other phytochemicals, apart from the groups detailed in the previous sections. Coumarins are one of the major secondary metabolite groups having a benzopyrane ring as the primary structure, possessing anti-inflammatory, cytotoxic, antimicrobial, antioxidant, and antidiabetic activities (Nan et al., 2025). Spinach possesses three coumarin compounds: 7-isopentenyloxycoumarin, auraptene, and umbelliprenin (Fiorito et al., 2019). Ecdysteroids are the hormones that are involved in the metamorphosis of insects by stimulating larval molting and transformation. The analogues (phytoecdysteroids) of these hormones are found in certain plants, and spinach is one among them (Bajguz et al., 2015). Spinach majorly consists of 20-hydroxyecdysone and polypodine B compounds from this category, and they are reported to show many useful biological activities (Grebenok et al., 1991; Bakrim et al., 2008; Hunyadi et al., 2016; Lee et al., 2022; Cerulli et al., 2024). Three biologically active glycoglycerolipids were reported from spinach, viz., monogalactosyl-diacylglycerol, digalactosyl-diacylglycerol, and sulfoquinovosyl-diacylglycerol (Shiota et al., 2010). Further, spinach also consists of a saponin, spinasaponin B (Cerulli et al., 2024), three sterols, 3,7,12-trihydroxycholan-24-oic acid, stigmasterol, and α-spinasterol (Ewais et al., 2023), a tannin, trimethylellagic acid (Cerulli et al., 2024), a terpenoid, betulin (Ewais et al., 2023), and two xanthines, viz., (8′R)-neochrome and (8′S)-neochrome (Rotimi et al., 2019). Additionally, there are some other important compounds also identified from the spinach, and that include phytol, 2-methoxy-4-vinylphenol, 5-azecanol; allantoin, benzoic acid; chorismic acid; coumaran; desulphosinigrin; and quinic acid (Singh, 2016; Ewais et al., 2023; Cerulli et al., 2024).

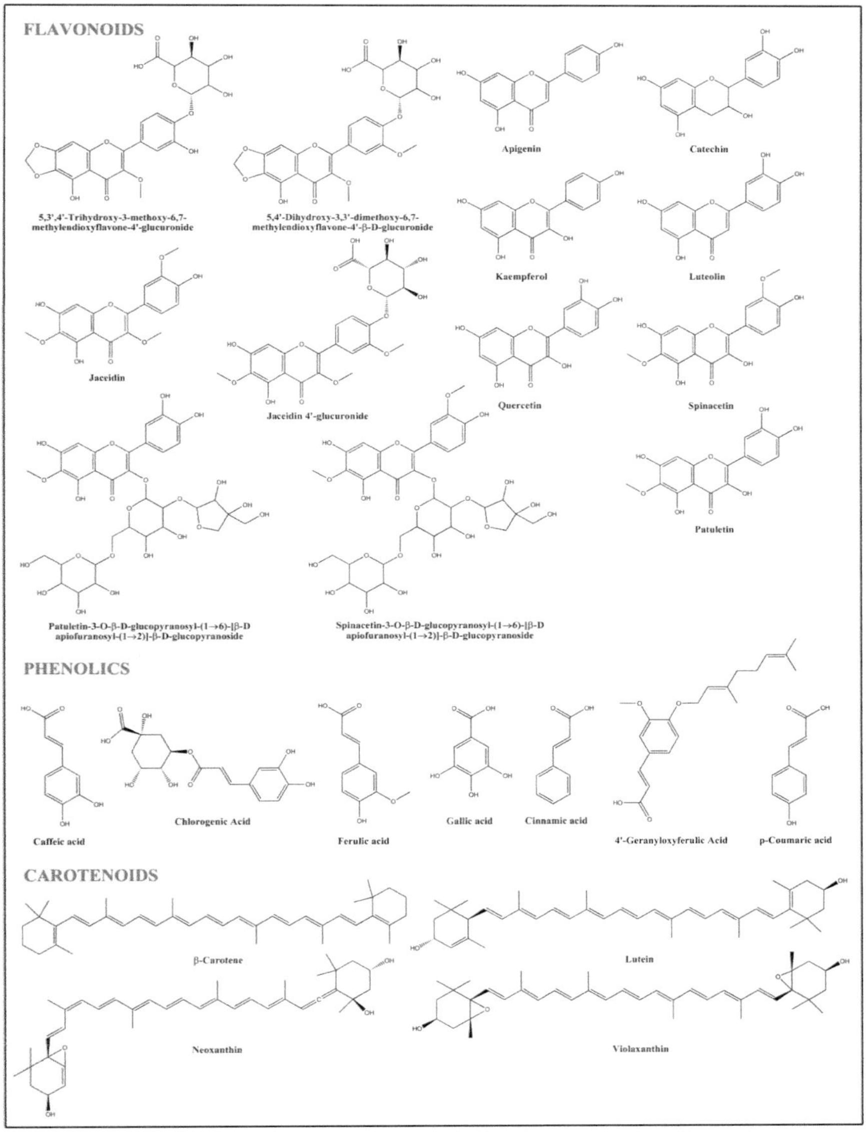

Fig. 2 Some important bioactive compounds of *Spinacia oleracea* L.

Biological Activities

Antioxidant Activity

Oxidation is an inevitable process during the metabolic activities in the cell, even in normal conditions, and some unfavourable conditions, such as biotic and abiotic stress, nutrient deficiency, and high temperature, generate elevated levels of oxidants, such as reactive oxygen species (ROS), reactive nitrogen species, and free radicals. These oxidant molecules damage the cell, and the plants produce a variety of bioactive compounds, such as phenolics, flavonoids, and tocopherols, to neutralize these oxidant molecules (Chaves et al., 2020). In humans, oxidative stress leads to many health complications, including Alzheimer's and Parkinson's disease, inflammation, cancer, cardiovascular disease, stroke, and DNA damage (Nwozo et al., 2023). Spinach is one of the best leafy vegetables, showing impressive antioxidant capacity due to the presence of natural antioxidant compounds, such as flavonoids and phenolics. The antioxidant activity of aqueous extract of spinach was assessed through *in vitro* methods, such as DPPH and ABTS radical scavenging activity (Rao et al., 2015). The aqueous extract was able to inhibit 42.63% of DPPH radicals and 69.55% of ABTS radicals at a concentration of 1.8 mg/mL. Similarly, Pellegrini et al. (2003) analysed the antioxidant activity of water extract of spinach leaves by using TEAC (tronjlox equivalent antioxidant capacity), TRAP (total radical-trapping antioxidant parameter), and FRAP (ferric reducing antioxidant power). The extract showed 8.49 and 5.79 mM trolox equivalent/kg fresh weight activity in TEAC and TRAP assays, respectively, and 26.94 mM Fe2+ equivalent/kg fresh weight activity in FRAP assay. These activities were very impressive and more as compared to the other beverages, oils, and vegetables, such as radish, pumpkin, pepper, lettuce, fennel, avocado, beet, cabbage, and carrot. The antioxidant activity of spinach leaves obtained from different genotypes was assessed by extracting the leaves with methanol, water, and formic acid mixture solvent (in the ratio 60:37:3) (Cho et al., 2008). The extract showed good activity in ORAC (oxygen radical absorbing capacity) and PCL (photochemiluminescence) assay, with an activity ranging between 48.7–84.4 mM trolox equivalent (TE)/kg and between 9.0 to 14.0 mM TE/kg, respectively. Additionally, they isolated three flavonoids, viz., patuletin-3-O-β-D-glucopyranosyl-(1 → 6)-[β-D-apiofuranosyl-(1 → 2)]-β-D-glucopyranoside, spinacetin-3-O-β-D-glucopyranosyl-(1 → 6)-[β-D-apiofuranosyl-(1 → 2)]-β-D-glucopyranoside, and 5,3′,4′-trihydroxy-3-methoxy-6:7methylendioxyflavone-4′-β-D-glucuronide and assessed their antioxidant activity by ORAC and PCL assays. The compound's activity was impressive in the PCL assay, and it outperformed the standards Trolox and ascorbic acid. Bergman et al. (2001) separated hydrophilic and hydrophobic fractions from the spinach water extract and assessed their antioxidant activity by using the thiobarbituric acid (TBA) assay and ferrous oxidation in the presence of xylenol orange (FOX) assay. The flavonoid and hydrophilic fraction and p-coumaric acid showed impressive antioxidant activity with IC_{50} values of 15, 120, and 15 μM, respectively. The activity of these fractions and a compound was higher compared to a standard butylated hydroxytoluene, which had an IC_{50} value of 100 μM. Similarly, many studies proved the efficacy of spinach extracts against the free radicals by using various *in vitro* methods (Yosefi et al., 2010; Raghavendra et al., 2013; Ko et al., 2014; Jaiswal et al., 2017; Barkat et al., 2018; Machado et al., 2018).

In addition, several *in vivo* studies also proved the efficacy of spinach extracts in lowering oxidative stress. The effect of spinach powder on reducing oxidative stress was studied in rats by Ko et al. (2014). The rats were fed with a high-fat-cholesterol diet for 6 weeks, and which induced oxidative stress, indicated by the increase of blood and liver lipid profiles, plasma conjugated diene concentration, and liver thiobarbituric acid reactive substances (TBARS) level, and reduction in manganese superoxide dismutase (Mn-SOD) activity compared with the normal diet-fed rats. The addition of 5% spinach powder to the high-fat-cholesterol diet was able to reduce the liver TBARS level and DNA damage in leukocytes and increase plasma conjugated dienes and Mn-SOD activity. Ewais et al. (2023) studied the effect of spinach on the reduction of oxidative stress in chickens caused by the infection of a protozoan, *Eimeria tenella*. In chickens, the infection of *E. tenella* caused a lowering of body weight, goblet cells, superoxide dismutase (SOD), catalase (CAT), and glutathione peroxidase (GPX), and an increase of malondialdehyde (MDA) and nitric oxide (NO). When the chickens were administered 50 mg/kg body weight of spinach ethanolic extract, the effects were reversed and resulted in the restored body weight and goblet cells, upregulation of SOD, CAT, and GPX, and downregulation of MDA and NO. In another study, the oxidative stress induced by the treatment of lipopolysaccharide was effectively reduced by the spinach leaf water extract in Male BALB/c Mice (Sharma et al., 2014). The above reports suggest that spinach is an excellent antioxidant source, and this can be attributed to its bioactive compounds, such as flavonoids and phenolics. Further, the spinach extract is very effective against hyperlipidemia, inflammation, and neural imbalance through its antioxidant activities and which are discussed in the following sections.

Anti-inflammatory Activity

Inflammation is the body's natural defence mechanism against injury and infection, which enables the living cells to protect themselves. The inflammation response may include swelling, redness, loss of functioning of cells, pain, and heat. These responses may be acute or chronic, and during this process, the inflammation indicators, such as ROS, RNS (reactive nitrogen species), interleukins (ILs), tumour necrosis factor-α (TNF-α), cyclooxygenase-1 (COX-1), and cyclooxygenase-2 (COX-2) are produced (Gonfa et al., 2023). Bioactive compounds and extracts of plant origin are gaining importance over synthetic drugs for their anti-inflammatory activities. Spinach is reported to have impressive anti-inflammatory activities. Shiota et al. (2010) studied the anti-inflammatory effect of glycoglycerolipid extract of spinach leaves in 5-fluorouracil-induced intestinal mucosal injury in rats. The treatment of 5-fluorouracil increased the expression of IL-1α and TNF-α genes significantly, compared to the control groups. Administration of spinach extract at 20 mg/kg body weight/ day inhibited the expression of both IL-1α and TNF-α genes. In addition, the extract also reduced the oxidative stress caused by the PMA, a phorbol ester in Caco-2 cells. In another study, the spinach polyphenol extract at 10 mg/kg body weight/day was found to be very effective in reducing the production of nitric oxide synthase and COX-2 significantly in the liver cells of rats, which were injured by the treatment of lipopolysaccharide. In fact, the activity of spinach polyphenol surpassed the activity of apocynin, a well-known antioxidant flavonoid (Lomnitski et al., 2000a). In a related study, Lomnitski et al. (2000b) investigated the protective effects of spinach polyphenol extract against lipopolysaccharide (LPS)-induced cellular damage in rats. LPS administration led to several pathological changes, including multifocal hepatocellular necrosis with acute inflammation, lymphoid necrosis in the thymus and spleen, retinal hemorrhage with acute endophthalmitis, and vacuolation or necrosis in the adrenal medulla along with reduced cytoplasmic vacuolation in the adrenal cortex. Treatment with spinach extract significantly reduced these LPS-induced lesions, particularly liver inflammation and necrosis, lymphoid necrosis in the thymus and spleen, and degenerative alterations in the adrenal cortex and medulla. Cartford et al. (2002) studied the effect of spinach on the inhibition of TNF-α and TNF-β in the cerebellum, which in turn improves the cerebellar-dependent delay of classical eyeblink learning in 18-month-old Fischer 344 Rats. When the rats were fed with a diet consisting of spinach (2% w/w dry spinach) for six weeks, they learned delay eyeblink conditioning significantly faster than the rats without spinach in their diet; also, TNF-α and TNF-β in the cerebelli of spinach-diet fed rats were significantly higher than those of rats that had a normal diet. This shows the cognitive improvement effect of spinach is through its anti-inflammatory activity. Further, Sharma et al. (2014) assessed the effect of spinach water extract on the levels of IL-1β and TNF-α in the lipopolysaccharide (LPS) induced brain cell-inflammation in mice. The levels of IL-1β and TNF-α increased significantly when they were treated with LPS, and treating 50 mg/kg body weight to the mice effectively lowered the IL-1β and TNF-α expression. However, contrasting results were reported in another study, in which the effect of spinach on the suppression of inflammation was assessed as indicated by the levels of C-reactive protein in humans (Graydon et al., 2012). The subjects (healthy males and females aged between 18 and 75 years) took 10.4 g of dried spinach daily for eightr weeks, and the result showed that the spinach had no significant effect on the reduction of inflammation. Thus, further work concentrating on the isolation of active bioactive compounds from spinach and assessing their anti-inflammatory potential is needed.

Neuroprotective Activity

Excessive oxidative stress and chronic neuroinflammation lead to cellular damage, protein misfolding, and neuronal death, contributing to the onset and progression of numerous nervous system disorders, including Alzheimer's, Parkinson's, and Huntington's diseases, multiple sclerosis, epilepsy, and schizophrenia. Neuroprotection is the body's strategy to protect the neuronal structure and function against the injuries mentioned above by a range of actions, such as reducing oxidative stress, enhancing nerve impulse transmission, and repair of damaged neurons (Rehman et al., 2019; Shalini et al., 2021). In a study by Wang et al. (2005), the neuroprotective effect of spinach in the focal ischemic brain was assessed in adult male Sprague–Dawley rats. The rats were induced to reperfusional injury in the right middle cerebral artery through a ligation with a 10-O suture for 60 min. The rats fed with a diet consisting of 2% spinach significantly reduced the volume of infarction in the cerebral cortex and increased the post-stroke locomotor activity. In another study, the effect of spinach water extract on age-related neuronal signal-transduction and cognitive behavioral deficits was assessed (Joseph et al., 1998). The results showed that feeding 6.4 g/kg dry water extract improved the oxotremorine-enhanced striatal dopamine (OX-K 1-ERDA), GABA (Gamma-Aminobutyric Acid), and Morris water maze performance. This shows the beneficial effect of spinach in the inhibition of age-related neural damage. An experiment on rats revealed that treatment with *S. oleracea* extract at a dose of 400 mg/kg body weight elevated serotonin levels while reducing dopamine and norepinephrine concentrations (Das and Guha, 2008). The study also documented reduced locomotor activity, prolonged pentobarbitone-induced sleep, and significant alterations in seizure activity induced by pentylenetetrazole. Additionally, another experiment demonstrated that pretreating 36 rats with fresh young spinach leaves for 14 days before amygdala-kindled (AMK) epileptogenesis resulted in notable reductions in serotonin and monoamine levels, along with modulated norepinephrine and dopamine activity in the brains of AMK rats. Based on these findings, the authors suggested that spinach extract delayed the onset of AMK epilepsy by influencing both neurochemical balance and behavioural outcomes (Das and Guha, 2011). Similarly, Sharma et al. (2014) observed that the spinach leaf extract at 50 mg/kg body weight administered Intraperitoneally for 21 days was able to reverse the Parkinson's disease like symptoms, such as impaired motor functions, total locomotor activity, catalepsy, and rotarod, increased oxidative burden (MDA, GSH, SOD), enhanced mRNA expression of proinflammatory cytokines (TNF-α and IL-1β) and decreased dopamine turnover in striatum. Cartford et al. (2002) reported an improvement in the cognitive functioning and reduction of neural inflammation of rats when they are fed with a diet containing 2% w/w dry spinach. The above reports prove spinach as an excellent neuroprotective agent, and that action may be attributed to the antioxidant compounds present in it. However, further studies concentrating on the neuroprotective effects of bioactive compounds of spinach are needed in order to explore this plant against neural disorders.

Anti-hyperlipidemic Activity

Hyperlipidemia is a lipid metabolic disorder that leads to abnormally increased levels of lipids, such as high cholesterol, triglycerides, and low-density lipoproteins in plasma and a reduction in high-density lipoproteins. Hyperlipidemia is associated with a series of health complications, such as obesity, atherosclerosis, coronary artery disease, heart attack, and stroke (Ezeh and Ezeudemba, 2021; Wei et al., 2024). Medicinal plants, their extracts, and bioactive compounds isolated from them serve as excellent sources to treat hyperlipidemia. Spinach is one of the plants used traditionally in the treatment of hyperlipidemia, and there are many pharmacological reports that support this view. Spinach thylakoids are very effective in suppressing the high-calorie food intake in rats and humans, which in turn lowers the body weight (Köhnke et al., 2009b; Montelius et al., 2013; Rebello et al., 2015). Further, Köhnke et al. (2009a) proved the appetite-suppressing activity of thylakoids obtained from spinach leaves in humans. In that study, the healthy individuals were offered a high-fat meal containing thylakoids obtained from spinach leaves, and the blood samples were analysed for satiety and hunger hormones (CCK, leptin, and ghrelin), insulin, and blood metabolites (glucose and free fatty acids). The spinach thylakoids reduced the appetite-stimulating hormone ghrelin and enhanced the satiety hormones CCK (cholecystokinin) and leptin. In addition, glucose and free fatty acids were lowered after the meal with spinach thylakoids. This report proves the ability of spinach to induce satiety hormones, thereby reducing the energy-rich foods. Similar results were obtained when the spinach thylakoids were taken along with a high-carbohydrate meal (Stenblom et al., 2013). The report by Montelius et al. (2014a) demonstrated the weight loss effect of spinach in women when spinach thylakoids were given 5 g per day for 12 weeks. The body weight of women who took the spinach with their diet was reduced by 35% more than that of women with a control diet. The spinach also reduced the total and LDL-cholesterol and increased postprandial release of GLP-1 (glucagon-like peptide-1), and suppressed the urge for sweet and chocolate. The hypolipidemic effects of spinach-derived phytochemicals have been investigated in animal models. In one study, alloxan-induced type 1 diabetic Wistar rats received a daily oral dose of 70% ethanolic spinach extract (100 mg/kg body weight). The treatment resulted in a 62.3% reduction in serum triglyceride levels compared to untreated diabetic controls, and plasma triglycerides were restored to concentrations similar to those of non-diabetic rats (83 mg/dL) (Kumar and Loganathan, 2010). Another experiment using Sprague-Dawley rats fed a high-fat, cholesterol-enriched diet (9.5% fat) supplemented with 5% freeze-dried spinach powder for six weeks demonstrated an 11% decline in serum triglycerides, though the reduction was not statistically significant (Ko et al., 2014). The above studies suggest that the thylakoids of spinach have the capability of anti-hyperlipidemic activity and anti-obesity activity, both by modulating hunger and satiety-related hormones and biochemical parameters. Hence, spinach could be explored as an anti-hyperlipidemic product effectively.

Antidiabetic Activity

Diabetes mellitus is a serious metabolic disorder in the human population globally, which is characterized by an increased level of glucose in the blood serum due to either the body producing lower levels of insulin or the inefficient use of produced insulin or both. Increased oxidative stress, pancreatic beta-cell dysfunction, insulin resistance, inflammation, mitochondrial dysfunction, and selenium deficiency trigger the development of diabetes, which leads to various complications, such as increased appetite, polydipsia, dysuria, weight loss, vision problems, cardiovascular disease, kidney damage, and nervous system impairment (Antar et al., 2023). Plant extracts and their secondary metabolites have been proven to be efficient in treating diabetes through various mechanisms and bring back the blood glucose level to normal (Rahman et al., 2022). Spinach is one such plant that has the potential to cure diabetes. α-Glucosidase inhibition is one of the major mechanisms to reduce the blood glucose level. 50 μL of 50 mg/mL concentrated spinach leaf juice and ethanolic extract inhibited the α-glucosidase activity by 5.2% and 19.6%, respectively. The spinach also promoted the cells to differentiate into adipocytes (Park et al., 2012). The spinach thylakoid-enriched diet increased the blood insulin level in rats, while unaltering the glucose level (Montelius et al., 2013). It was assumed that the decreased insulin level might be due to higher insulin sensitivity or the lipid and glucose homeostasis action of microbiota. In their continued study, Montelius et al. (2014b) studied the effect of spinach thylakoids in suppressing blood glucose and modulating insulin levels in a pig model. The pigs were on a high-fat diet for one month before the oral glucose tolerance test. When the pigs were fed 0.5 g/kg body weight of thylakoids along with 1 g/kg body weight of 20% glucose solution, blood glucose was reduced in the first hour, and increased plasma insulin level in the first 30 minutes. Further, Montelius et al. (2011) proved that spinach thylakoids lowered the methyl-glucose uptake significantly in a dose-dependent manner than the control, and trypsin-treated thylakoids exhibited an even stronger inhibition of the methyl-glucose uptake in the rat intestine. The thylakoids also reduced the intestinal permeability of macromolecules, such as FITC-dextran and ovalbumen, up to 64.6% and 65%, respectively, in a dose-dependent manner. The authors suggested that the attachment of thylakoids to the mucosal surface subsequently hinders the uptake of macromolecules. In another study (Kumar and Loganathan, 2010), when the alloxan-induced rats were administered orally the 100 mg/kg body weight of 70% ethanolic extract of spinach once a day for 10 days, it effectively lowered serum glucose level. The lowering of serum glucose level began on the 6th day, and on the 9th day, it was able to reduce 54% glucose, compared to the control diabetic rats. The report by Montelius et al. (2014a) demonstrated the lowering of glucose and insulin levels after 15 minutes of food intake, which was preceded by 5 g of spinach thylakoids, compared to the control, in humans. In another study (Stenblom et al., 2013), the effect of spinach thylakoids in moderating the levels of glucose and insulin was assessed in moderately overweight females. However, the results obtained were contrasting, in which the spinach-supplemented meal increased the glucose and insulin levels after 90 minutes, compared to the control. In the study conducted by Maruyama et al. (2013), the spinach did not show any effect on the reduction of glucose level. In summary, while numerous *in vitro* and animal studies demonstrate the anti-hyperglycaemic potential of spinach, evidence from human trials remains inconclusive. Most research has focused on the activity of spinach thylakoids; therefore, further investigations using spinach extracts and bioactive compounds are essential to fully establish its effectiveness as an antidiabetic agent.

Anticancer Activity

Cancer is a condition in which the cells divide uncontrollably and invade other parts of the body. The cancer may develop due to the mutations in the genes that control normal cell division and create oncogenes, inactivation of tumour suppressor genes, or disabling the DNA repair genes. Both genetic and environmental factors, such as viral infections, toxic substances, chemicals, oxidative stress, and inflammation, lead to cancer initiation (Amjad et al., 2022; Sanjai et al., 2024). Due to unwarranted side effects associated with synthetic drugs in cancer treatment, natural products are being preferred over them. The natural products or plant extracts exert anticancer activity by damaging DNA, inhibiting cell division, boosting the immune system, and inhibiting the growth of blood vessels that supply tumors (Kumar et al., 2018). Several *in vitro* and *in vivo* studies have reported the anticancer potential of spinach through various mechanisms. Spinach extract was very effective against HepG2 liver cancer lines *in vitro* with an EC_{50} of 42.51 mg/mL (Chu et al., 2002). This activity was highest when compared with the other vegetables, such as cabbage, red pepper, onion, broccoli, lettuce, cucumber, and carrot. The methanolic extract of spinach leaf showed a good inhibitory effect against HeLa and MCF7 cancer cells, with IC_{50} values of 45.9 and 47.9 μg/mL, respectively, as proved by *in vitro* MTT assay (Umamaheswari and Nishanthini, 2017). Nyska et al. (2003) assessed the effect of natural antioxidants of spinach rich in polyphenols on the prevention of cancer in the TRAMP (Transgenic Adenocarcinoma of the Mouse Prostate) model and human prostatic cancer cell lines DU145 and PC3. The results revealed that the 200 mg/kg spinach extract exerted a significant effect in reducing hyperplasia at 9 and 13 weeks in TRAMP and inhibited the proliferation of DU145 and PC3 cells, and reduced ROS levels. In another study, Bakshi et al. (2004) assessed the mechanism of action of spinach antioxidants and their compounds (two each of p-coumaric acid derivatives and flavonoid derivatives) suppression of prostate cancer by using human PC3 cell lines. The spinach extract and constituents showed anticancer activity by suppressing the proliferation of cells by cell-cycle prolongation, and the activity was dose and time-dependent. The extract and compounds downregulated the expression of ppRb and E2F-1 proteins and which led to cell arrest at G1 phase and inhibition of the expression of Cyclin A and CDK-2 in the PC3 cells. Rotimi et al. (2019) assessed the binding affinities of two terpenoid compounds of spinach, (8′R)-neochrome and (8′S)-neochrome, with GUCY1A2 proteins, a protein majorly responsible for the development of castration resistant prostate cancer (CRPC). The molecular docking studies revealed that the compounds mentioned have the highest binding affinities with the GUCY1A2 proteins than the other compounds assessed. The authors suggested neochrome as a lead compound that could be explored in the treatment of CRPC. It is also proven that the neoxanthin, one of the major carotenoids of spinach, is partially converted into (R/S)-neochrome by intragastric acidity before intestinal absorption and that (R/S)-neochrome exhibits an antiproliferative effect on PC-3 cells by the induction of cytostasis (Asai et al., 2004). Dietary heme increases the risk of colon cancer because the heme converts into a cytotoxic metabolite, which in turn causes cytotoxicity and cancer risk. When the rats were fed a diet supplemented with 0.5 mM heme/kg and spinach (1.2 mM/kg chlorophyll content), the spinach chlorophylls were able to lower the colonic content cytotoxicity and improve the exfoliation of colonocytes. This effect was due to inhibition of detrimental, cytotoxic, and hyperproliferative colonic effects of dietary haem by spinach (de Vogel et al., 2005). Lutein, another carotenoid obtained from spinach, showed selective cytotoxicity in human colon adenocarcinoma cells (HT-29), but not on normal colon epithelial cells (CCD 841 CoTr) (Grudzinski et al., 2018). Interestingly, individual glycolipid components exhibited varying levels of effectiveness. In human gastric carcinoma cells (NUGC-3), neither digalactosyldiacylglycerol (DGDG) nor sulfoquinovosyldiacylglycerol (SQDG) suppressed growth, whereas monogalactosyldiacylglycerol (MGDG) and its lipase-derived monoacylglycerol form (MGMG) strongly reduced cell viability in a dose-dependent manner, with IC_{50} values of 49 and 40 μg/mL, respectively (Murakami et al., 2003). Beyond *in vitro* studies, the anticancer and chemopreventive activities of spinach-derived glycolipids have been confirmed in animal models. In Greene's melanoma-bearing golden hamsters, subcutaneous administration of a glycolipid fraction containing MGMG, DGDG, and SQMG effectively suppressed solid tumour growth (Maeda et al., 2005). Similarly, oral administration of 20 mg/kg of spinach glycolipid extract for two weeks to BALB/c mice implanted with Colon-26 carcinoma reduced tumor volume by 56.1% without signs of toxicity, including body weight loss or organ damage (Maeda et al., 2008). Mechanistic studies revealed that the treatment inhibited angiogenesis and downregulated tumour cell proliferation markers, such as Ki-67, PCNA, cyclin E, and mitotic index (Maeda et al., 2011). In another model, ICR mice implanted with S-180 sarcoma cells and treated with 70 mg/kg of the glycolipid fraction for 27 days showed a significant reduction in tumour growth compared with untreated controls (Maeda et al., 2009). These *in vitro* and *in vivo* studies clearly demonstrate the beneficial effects of spinach-derived extracts, fractions, and bioactive compounds against various types of cancer. Therefore, spinach holds significant promise as a natural source of anticancer agents and warrants further investigation for its therapeutic and preventive potential.

Conclusions

Spinach is a globally consumed leafy vegetable, originally native to Central Asia and now cultivated widely, including in India. Beyond its role as a common food crop, it is recognized as a functional food owing to its rich nutritional and medicinal profile. It contains high levels of protein, fibre, vitamins, minerals, pigments, unsaturated fatty acids, essential amino acids, and organic acids. Moreover, spinach is abundant in bioactive compounds such as flavonoids, phenolics, xanthines, carotenoids, and lignans, which contribute to its diverse biological activities, including antioxidant, anti-inflammatory, neuroprotective, antihyperlipidemic, antidiabetic, and anticancer effects. While extensive research has examined the biological activities of spinach extracts, fewer studies have focused on the isolated compounds themselves. Therefore, future research should emphasize the isolation, characterization, and evaluation of these bioactive constituents to better understand and utilize spinach's full therapeutic potential.

Acknowledgments

Hosakatte Niranjana Murthy is thankful for the "Brain Pool" fellowship (Grant No. RS-2025-25425249) of the National Research Foundation, Republic of Korea.

Disclosure Statement

No potential conflict of interest was reported by the authors.

Author contribution HNM: conceptualization; HNM, GGY: investigation, data curation, and formalization; HNM, GGY: resources and validation; HNM, GGY: writing, review, and editing. All the authors have read and agreed to the published version of this manuscript.

References

Aehle, E., Raynaud-Le Grandic, S., Ralainirina, R., Baltora-Rosset, S., Mesnard, F., Prouillet, C., Mazière, J.C. and Fliniaux, M.A. 2004. Development and evaluation of an enriched natural antioxidant preparation obtained from aqueous spinach (*Spinacia oleracea*) extracts by an adsorption procedure. *Food Chem.*, *86*: 579–585. https://doi.org/10.1016/j.foodchem.2003.10.006.

Ahmed, H.M. 2016. Ethnopharmacobotanical study on the medicinal plants used by herbalists in Sulaymaniyah Province, Kurdistan, Iraq. *J. Ethnobiol. Ethnomed.*, *12*: 8. https://doi.org/10.1186/s13002-016-0081-3.

Amjad, E., Sokouti, B. and Asnaashari, S. 2022. A systematic review of anti-cancer roles and mechanisms of kaempferol as a natural compound. *Cancer Cell Int.*, *22*: 260. https://doi.org/10.1186/s12935-022-02673-0.

Antar, S. A., Ashour,, N.A., Sharaky, M., Khattab, M., Ashour, N.A., Zaid, R.T., Roh, E.J., Elkamhawy, A. and Al-Karmalawy, A.A. 2023. Diabetes mellitus: Classification, mediators, and complications; a gate to identify potential targets for the development of new effective treatments. *Biomed. Pharmacother.*, *168*: 115734. https://doi.org/10.1016/j.biopha.2023.115734.

Asai, A., Terasaki, M. and Nagao, A. 2004. An epoxide–furanoid rearrangement of spinach neoxanthin occurs in the gastrointestinal tract of mice and *in vitro*: Formation and cytostatic activity of neochrome stereoisomers. *J. Nutr.*, *134*: 2237–2243. https://doi.org/10.1093/jn/134.9.2237.

Bajguz, A., Bąkała, I. and Talarek, M.M. 2015. Ecdysteroids in plants and their pharmacological effects in vertebrates and humans. In: F.R.S. Atta-ur-Rahman (Ed.), *Studies in Natural Products Chemistry* (Vol. 45), pp. 121–145. Elsevier. https://doi.org/10.1016/B978-0-444-63473-3.00005-8.

Bakrim, A., Maria, A., Sayah, F., Lafont, R. and Takvorian, N. 2008. Ecdysteroids in spinach (*Spinacia oleracea* L.): Biosynthesis, transport, and regulation of levels. *Plant Physiol. Biochem.*, *46*: 844–854. https://doi.org/10.1016/j.plaphy.2008.06.002.

Bakshi, S., Bergman, M., Dovrat, S. and Grossman, S. 2004. Unique natural antioxidants (NAOs) and derived purified components inhibit cell cycle progression by downregulation of ppRb and E2F in human PC3 prostate cancer cells. *FEBS Lett.*, *573*: 31–37. https://doi.org/10.1016/j.febslet.2004.06.101.

Barkat, N., Singh, J., Jayaprakasha, G.K. and Patil, B.S. 2018. Effect of harvest time on the levels of phytochemicals, free radical-scavenging activity, α-amylase inhibition, and bile acid-binding capacity of spinach (*Spinacia oleracea*). *J. Sci. Food Agric.*, *98*: 3468–3477. https://doi.org/10.1002/jsfa.8862.

Bergman, M., Varshavsky, L., Gottlieb, H.E. and Grossman, S. 2001. The antioxidant activity of aqueous spinach extract: Chemical identification of active fractions. *Phytochem.*, *58*: 143–152. https://doi.org/10.1016/S0031-9422(01)00137-6.

Bunea, A., Andjelkovic, M., Socaciu, C., Bobis, O., Neacsu, M., Verhé, R. and van. Camp, J.. 2008. Total and individual carotenoids and phenolic acids content in fresh, refrigerated, and processed spinach (*Spinacia oleracea* L.). *Food Chem.*, *108*: 649–656. https://doi.org/10.1016/j.foodchem.2007.11.056.

Cartford, M.C., Gemma, C. and Bickford, P.C. 2002. Eighteen-month-old Fischer 344 rats fed a spinach-enriched diet show improved delay classical eyeblink conditioning and reduced expression of tumor necrosis factor α (TNFα) and TNFβ in the cerebellum. *J. Neurosci.*, *22*: 5813–5816. https://doi.org/10.1523/JNEUROSCI.22-14-05813.2002.

Cerulli, A., Polcaro, L.M., Masullo, M. and Piacente, S. 2024. *Spinacia oleracea* L. baby leaves as a source of bioactive principles: The chemical profiling of eco-sustainable extracts by using LC-ESI/HRMS- and 1H NMR-based metabolomics. *Foods*, *13*: 3699. https://doi.org/10.3390/foods13223699.

Chaves, N., Santiago, A. and Alías, J.C. 2020. Quantification of the antioxidant activity of plant extracts: Analysis of sensitivity and hierarchization based on the method used. *Antioxidants*, *9*; 76. https://doi.org/10.3390/antiox9010076.

Chen, S., Wang, X., Cheng, Y., Gao, H. and Chen, X. 2023. A review of classification, biosynthesis, biological activities, and potential applications of flavonoids. *Molecules*, *28*: 4982. https://doi.org/10.3390/molecules28134982.

Cho, M.J., Howard, L.R., Prior, R.L. and Morelock, T. 2008. Flavonoid content and antioxidant capacity of spinach genotypes determined by high-performance liquid chromatography/mass spectrometry. *J. Sci. Food Agric.*, *88*: 1099–1106. https://doi.org/10.1002/jsfa.3206.

Chu, Y. H., Chang, C.L.and Hsu, H.F. 2000. Flavonoid content of several vegetables and their antioxidant activity. *J. Sci. Food Agric.*, *80*: 561–566.

Chu, Y.-F., Sun, J., X. Wu and Liu, R.H. 2002. Antioxidant and antiproliferative activities of common vegetables. *J. Agric. Food Chem.*, *50*: 6910–6916. https://doi.org/10.1021/jf020665f.

Crupi, P., Faienza, M.F., Naeem, M.Y., Corbo, F., Clodoveo, M.L. and Muraglia, M. 2023. Overview of the potential beneficial effects of carotenoids on consumer health and well-being. *Antioxidants*, *12*: 1069. https://doi.org/10.3390/antiox12051069.

Das, S. and Guha, D. 2008. CNS depressive role of aqueous extract of *Spinacia oleracea* L. leaves in adult male albino rats. *Indian J. Exp. Biol.*, *46*: 185–190.

Das, S. and Guha, D. 2011. *Spinacia oleracea* retards the development of Amygdala kindled epilepsy in rats. *Al Ameen J. Med. Sci.*, *4*(2): 175–190.

de Vogel, J., Jonker-Termont, D.S.M.L., van Lieshout, E.M.M., Katan, M.B. and van der Meer, R.. 2005. Green vegetables, red meat, and colon cancer: Chlorophyll prevents the cytotoxic and hyperproliferative effects of haem in rat colon. *Carcinogenesis*, *26*: 387–393. https://doi.org/10.1093/carcin/bgh331.

De, M., Sharma, L. and Acharya, C.K. 2023. A comprehensive chemical characterization of leaves of five potential medicinal plants in Paschim Medinipur District, W. B., India. *Int. J. Exp. Res. Rev.*, *36*: 20–36. https://doi.org/10.52756/ijerr.2023.v36.002.

Dehkharghanian, M., Adenier, H. and Vijayalakshmi, M.A.. 2010. Study of flavonoids in aqueous spinach extract using positive electrospray ionisation tandem quadrupole mass spectrometry. *Food Chem.*, *121*: 863–870. https://doi.org/10.1016/j.foodchem.2010.01.007.

Dzakovich, M.P., Le, E.A., Tak, A.L. and Chacko, S.K. 2025. A comprehensive UHPLC-MS/MS and extraction method reveals flavonoid profile and concentration are diverse in spinach (*Spinacia oleracea* L.). *Front. Plant Sci.*, *16*: 1496200. https://doi.org/10.3389/fpls.2025.1496200.

Ewais, O., Abdel-Tawab, H., El-Fayoumi, H., Aboelhadid, S.M., Al-Quraishy, S., Falkowski, P. and Abdel-Baki, A.A.S.. 2023. Administration of ethanolic extract of *Spinacia oleracea* rich in omega-3 improves oxidative stress and goblet cells in broiler chickens infected with *Eimeria tenella*. *Molecules*, *28*: 6621. https://doi.org/10.3390/molecules28186621.

Ezeh, K.J. and Ezeudemba, O. 2021. Hyperlipidemia: A review of the novel methods for the management of lipids. *Cureus*, *13*: e16412. https://doi.org/10.7759/cureus.16412.

FAO. 2024. *Food and Agriculture Organization Statistics*. https://www.fao.org/faostat/en/#data. Accessed on 25th August 2025.

Ferreres, F., Castañer, M. and Tomás-Barberán, F.A. 1997. Acylated flavonol glycosides from spinach leaves (*Spinacia oleracea*). *Phytochem.*, *45*: 1701–1705. https://doi.org/10.1016/S0031-9422(97)00244-6.

Fiorito, S., Preziuso, F., Epifano, F., Scotti, L., Bucciarelli, T., Taddeo, V.A. and Genovese, S. 2019. Novel biologically active principles from spinach, goji, and quinoa. *Food Chem.*, *276*: 262–265. https://doi.org/10.1016/j.foodchem.2018.10.018.

Flowers of India. 2025. *Spinacia oleracea L.* https://www.flowersofindia.net/catalog/slides/Spinach.html. Accessed on 25th August 2025.

Gonfa, Y.H., Tessema, F.B., Bachheti, A., Rai, N., Tadesse, M.G., Singab, A.N., Chaubey, K.K. and Bachheti, R.K. 2023. Anti-inflammatory activity of phytochemicals from medicinal plants and their nanoparticles: A review. *Curr. Res. Biotechnol.*, *6*: 100152. https://doi.org/10.1016/j.crbiot.2023.100152.

Gowthami, R., Prakash, B.G., Raghavendra,, K.V., Brunda, S.M. and Kumara, B.N.. 2019. Role of underutilised leafy vegetables to attain nutrition security. *Acta Hortic.*, *1241*: 583–588. https://doi.org/10.17660/ActaHortic.2019.1241.86.

Grace, M.H., Hoskin, R.T., Hayes, M., Iorizzo, M., Kay, C., Ferruzzi, M.G. and Lila, M.A. 2022. Spray-dried and freeze-dried protein-spinach particles; effect of drying technique and protein type on the bioaccessibility of carotenoids, chlorophylls, and phenolics. *Food Chem.*, *388*: 133017. https://doi.org/10.1016/j.foodchem.2022.133017.

Graydon, R., Hogg, R.E., Chakravarthy, U., Young, I.S. and Woodside, J.V. 2012. The effect of lutein- and zeaxanthin-rich foods *v.* supplements on macular pigment level and serological markers of endothelial activation, inflammation, and oxidation: Pilot studies in healthy volunteers. *Br. J. Nutr.*, *108*: 334–342. https://doi.org/10.1017/S0007114511005599.

Grebenok, R.J., Ripa, P.V. and Adler, J.H. 1991. Occurrence and levels of ecdysteroids in spinach. *Lipids*, *26*: 666–668. https://doi.org/10.1007/BF02536433.

Grudzinski, W., Piet, M., Luchowski, R., Reszczynska, E., Welc, R., Paduch, R. and Gruszecki, W.I. 2018. Different molecular organization of two carotenoids, lutein, and zeaxanthin, in human colon epithelial cells and colon adenocarcinoma cells. *Spectrochim. Acta A Mol. Biomol. Spectrosc.*, *188*: 57–63. https://doi.org/10.1016/j.saa.2017.06.041.

Guerriero, G., Berni, R., Muñoz-Sanchez, J.A., Apone, F., Abdel-Salam, E.M., Qahtan, A.A., Alatar, A.A., et al. 2018. Production of plant secondary metabolites: Examples, tips, and suggestions for biotechnologists. *Genes*, *9*: 309. https://doi.org/10.3390/genes9060309.

Hunyadi, A., Herke, I., Lengyel, K., Báthori, M., Kele, Z., Simon, A., Tóth, G. and Szendrei, K. 2016. Ecdysteroid-containing food supplements from *Cyanotis arachnoidea* on the European market: Evidence for spinach product counterfeiting. *Sci. Rep.*, *6*: 37322. https://doi.org/10.1038/srep37322.

Jaiswal, S.G., Patel, M., Saxena, D.K. and Naik, S. 2017. Comparison of measurements of antioxidant activity in the selected leafy vegetables depending on extraction solvent. *J. Hortic. Res.*, *25*: 75–80. https://doi.org/10.1515/johr-2017-0023.

Joseph, J.A., Shukitt-Hale, B., Denisova, N.A., Prior, R.L., Cao, G., Martin, A., Taglialatela, G. and Bickford, P.C. 1998. Long-term dietary strawberry, spinach, or vitamin E supplementation retards the onset of age-related neuronal signal-transduction and cognitive behavioral deficits. *J. Neurosci.*, *18*: 8047–8055. https://doi.org/10.1523/JNEUROSCI.18-19-08047.1998.

Kaplan, D.R. 2022. *Kaplan's Principles of Plant Morphology*. In: C. D. Specht (Ed.), 1st Edn. CRC Press.

Kasote, D.M., Jayaprakasha, G.K. and Patil, B.S. 2019. Leaf disc assays for rapid measurement of antioxidant activity. *Sci. Rep.*, *9*: 1884. https://doi.org/10.1038/s41598-018-38036-x.

Khalid, M., Hassani, D., Bilal, M., Asad, F. and Huang, D. 2017. Influence of bio-fertilizer containing beneficial fungi and rhizospheric bacteria on health promoting compounds and antioxidant activity of *Spinacia oleracea* L. *Bot. Stud.*, *58*: 35. https://doi.org/10.1186/s40529-017-0189-3.

Ko, S.H., Park, J.H., Kim, S.Y., Lee, S.W., Chun, S.S. and Park, E. 2014. Antioxidant effects of spinach (*Spinacia oleracea* L.) supplementation in hyperlipidemic rats. *Prev. Nutr. Food Sci.*, *19*: 19–26.

Köhnke, R., Lindbo, A., Larsson, T., Lindqvist, A., Rayner, M., Emek, S.C., Albertsson, P.-Å., Rehfeld, J.F., Landin-Olsson, M. and Erlanson-Albertsson, C. 2009a. Thylakoids promote release of the satiety hormone cholecystokinin while reducing insulin in healthy humans. *Scand. J. Gastroenterol.*, *44*: 712–719. https://doi.org/10.1080/00365520902803499.

Köhnke, R., Lindqvist, A., Göransson, N., Emek, S.C., Albertsson, P., Rehfeld, J.F., Hultgårdh-Nilsson, A. and Erlanson-Albertsson, C. 2009b. Thylakoids suppress appetite by increasing cholecystokinin resulting in lower food intake and body weight in high-fat fed mice. *Phytother. Res.*, *23*: 1778–1783. https://doi.org/10.1002/ptr.2855.

Krishnaswamy, K. and Nair, K.M. 2001. Importance of folate in human nutrition. *Br. J. Nutr.*, *85*: S115–S124. https://doi.org/10.1079/BJN2000303.

Kumar, B., Singh, S., Skvortsova, I. and Kumar, V. 2018. Promising targets in anti-cancer drug development: Recent updates. *Curr. Med. Chem.*, *24*: 4729–4752. https://doi.org/10.2174/0929867324666170331123648.

Kumar, N.J. and Loganathan, P. 2010. Hypoglycemic effect of *Spinacia oleracea* in alloxan induced diabetic rat. *Global J. Biotechnol. Biochem.*, *5*: 87–91.

Lee, B.S., So, H.M., Kim, S., Kim, J.K., Kim, J.C., Kang, D.M., Ahn, M.J., Ko, Y.J. and Kim, K.H. 2022. Comparative evaluation of bioactive phytochemicals in *Spinacia oleracea* cultivated under greenhouse and open field conditions. *Arch. Pharm. Res.*, *45*: 795–805. https://doi.org/10.1007/s12272-022-01416-z.

Lomnitski, L., Nyska, A., Ben-Shaul, V., Maronpot, R.R., Haseman, J.K., Harrus, T.L., Bergman, M. and Grossman, S. 2000b. Effects of antioxidants apocynin and the natural water-soluble antioxidant from spinach on cellular damage induced by lipopolysaccharide in the rat. *Toxicol. Pathol.*, *28*: 580–587. https://doi.org/10.1177/019262330002800412.

Lomnitski, L., Foley, J.F., Grossman, S., Ben Shaul, V., Maronpot, R.R., Moomaw, C.R., Carbonatto, M. and Nyska, A. 2000a. Effects of apocynin and natural antioxidant from spinach on inducible nitric oxide synthase and cyclooxygenase-2 induction in lipopolysaccharide-induced hepatic injury in rat. *Pharmacol. Toxicol.*, *87*: 18–25. https://doi.org/10.1111/j.0901-9928.2000.870104.x.

Longvah, T., Ananthan, R., Bhaskarachary, K. and Venkaiah, K. 2017. *Indian Food Composition Tables*. National Institute of Nutrition, Indian Council of Medical Research, Hyderabad, India.

Machado, R.M.A., Alves-Pereira, I. and Ferreira, R.M.A. 2018. Plant growth, phytochemical accumulation, and antioxidant activity of substrate-grown spinach. *Heliyon*, *4*: e00751. https://doi.org/10.1016/j.heliyon.2018.e00751.

Maeda, N., Matsubara, K., Yoshida, H. and Mizushina, Y. 2011. Anti-cancer effect of spinach glycoglycerolipids as angiogenesis inhibitors based on the selective inhibition of DNA polymerase activity. *Mini Rev. Med. Chem.*, *11*: 32–38. https://doi.org/10.2174/138955711793564042.

Maeda, Naoki, Hada, T., Murakami-Nakai, C., Kuriyama, I., Ichikawa, H., Fukumori, Y., Hiratsuka, J., Yoshida, H., Sakaguchi, K. and Mizushina, Y. 2005. Effects of DNA polymerase inhibitory and antitumor activities of lipase-hydrolyzed glycolipid fractions from spinach. *J. Nutr. Biochem.*, *16*: 121–128. https://doi.org/10.1016/j.jnutbio.2004.08.005.

Maeda, Naoki, Kokai, Y., Ohtani, S., Sahara, H., Kumamoto-Yonezawa, Y., Kuriyama, I., Hada, T., Sato, N., Yoshida, H. and Mizushina, Y. 2008. Anti-tumor effect of orally administered spinach glycolipid fraction on implanted cancer cells, colon-26, in mice. *Lipids*, *43*: 741–748. https://doi.org/10.1007/s11745-008-3202-5.

Maeda, Naoki, Kokai, Y., Ohtani, S., Hada, T., Yoshida, H. and Mizushina, Y. 2009. Inhibitory effects of preventive and curative orally administered spinach glycoglycerolipid fraction on the tumor growth of sarcoma and colon in mouse graft models. *Food Chem.*, *112*: 205–210. https://doi.org/10.1016/j.foodchem.2008.05.059.

Majeed, M., Ahmad, F., Mundkur, L. and Appian, S. 2022. Pharmacology of α-spinasterol, a phytosterol with nutraceutical values: A review. *Phytother. Res.*, *36*: 3681–3690. https://doi.org/10.1002/ptr.7560.

Maruyama, C., Kikuchi, N., Masuya, Y., Hirota, S., Araki, R. and Maruyama, T. 2013. Effects of green-leafy vegetable intake on postprandial glycemic and lipidemic responses and α-tocopherol concentration in normal weight and obese men. *J. Nutr. Sci. Vitaminol.*, *59*: 264–271. https://doi.org/10.3177/jnsv.59.264.

Milder, I.E.J., Arts, I.C.W., van de Putte, B., Venema, D.P. and Hollman, P.C.H. 2005. Lignan contents of Dutch plant foods: A database including lariciresinol, pinoresinol, secoisolariciresinol, and matairesinol. *Br. J. Nutr.*, *93*: 393–402. https://doi.org/10.1079/BJN20051371.

Montelius, C., Erlandsson, D., Vitija, E., Stenblom, E.L., Egecioglu, E. and Erlanson-Albertsson, C. 2014a. Body weight loss, reduced urge for palatable food, and increased release of GLP-1 through daily supplementation with green-plant membranes for three months in overweight women. *Appetite*, *81*: 295–304. https://doi.org/10.1016/j.appet.2014.06.101.

Montelius, C., Szwiec, K., Kardas, M., Lozinska, L., Erlanson-Albertsson, C., Pierzynowski, S., Rehfeld, J.F. and Weström, B. 2014b. Dietary thylakoids suppress blood glucose and modulate appetite-regulating hormones in pigs exposed to oral glucose tolerance test. *Clinic. Nutr.*, *33*: 1122–1126. https://doi.org/10.1016/j.clnu.2013.12.009.

Montelius, C., Gustafsson, K., Weström, B., Albertsson, P.Å., Emek, S.C., Rayner, M. and Erlanson-Albertsson, C. 2011. Chloroplast thylakoids reduce glucose uptake and decrease intestinal macromolecular permeability. *Br. J. Nutr.*, *106*: 836–844. https://doi.org/10.1017/S0007114511001267.

Montelius, C., Osman, N., Weström, B., Ahrné, S., Molin, G., Albertsson, P.Å. and Erlanson-Albertsson, C. 2013. Feeding spinach thylakoids to rats modulates the gut microbiota, decreases food intake, and affects the insulin response. *J. Nutr. Sci.*, *2*: e20. https://doi.org/10.1017/jns.2012.29.

Morishita, Y., Saito, E., Takemura, E., Fujikawa, R., Yamamoto, R., Kuroyanagi, M., Shirota, O. and Muto, N. 2015. Flavonoid glucuronides isolated from spinach inhibit IgE-mediated degranulation in basophilic leukemia RBL-2H3 cells and passive cutaneous anaphylaxis reaction in mice. *Integr. Mol. Med.*, *2*: 99–105. https://doi.org/10.15761/IMM.1000119.

Murakami, C., Kumagai, T., Hada, T., Kanekazu, U., Nakazawa, S., Kamisuki, S., Maeda, N. et al,. 2003. Effects of glycolipids from spinach on mammalian DNA polymerases. *Biochem. Pharmacol.*, *65*: 259–267. https://doi.org/10.1016/S0006-2952(02)01483-1.

Murthy, H.N., Yadav, G.G., Kadapatti, S.S., Lamani, S., Desai, A.S., Sumbad, M.M., Dewir, Y.H. and Magyar-Tábori, K. 2024. Nutritional and oil characterization of *Erythrina stricta* Roxb. seeds:

A potential resource for functional foods. *Cogent Food Agric.*, *10*: 2337770. https://doi.org/10.1080/23311932.2024.2337770.

Nan, Z.D., Shang, Y., Zhu, Y.D., Zhang, H., Sun, R.R., Tian, J.J., Jiang, Z.B., Ma, X.L. and Bai, C. 2025. Systematic review of natural coumarins in plants (2019–2024): Chemical structures and pharmacological activities. *Phytochem.*, *235*: 114480. https://doi.org/10.1016/j.phytochem.2025.114480.

Naseem, A., Akhtar, S., Ismail, T., Qamar, M., Sattar, D., Saeed, W., Esatbeyoglu, T., Bartkiene, E. and Rocha, J.M. 2023. Effect of growth stages and lactic acid fermentation on anti-nutrients and nutritional attributes of spinach (*Spinacia oleracea*). *Microorganisms*, *11*: 2343. https://doi.org/10.3390/microorganisms11092343.

Nemzer, B., Al-Taher, F. and Abshiru, N. 2021. Extraction and natural bioactive molecules characterization in spinach, kale, and purslane: A comparative study. *Molecules*, *26*: 2515. https://doi.org/10.3390/molecules26092515.

Nuutila, A.M., Kammiovirta, K. and Oksman-Caldentey, K.M. 2002. Comparison of methods for the hydrolysis of flavonoids and phenolic acids from onion and spinach for HPLC analysis. *Food Chem.*, *76*: 519–525. https://doi.org/10.1016/S0308-8146(01)00305-3.

Nwozo, O.S., Effiong, E.M., Aja, P.M. and Awuchi, C.G. 2023. Antioxidant, phytochemical, and therapeutic properties of medicinal plants: A review. *Int. J.Food Prop.*, *26*: 359–388. https://doi.org/10.1080/10942912.2022.2157425.

Nyska, A., Suttie, A., Bakshi, S., Lomnitski L., Grossman S., Bergman, M., Ben-Shaul, V. et al., 2003. Slowing tumorigenic progression in tramp mice and prostatic carcinoma cell lines using natural antioxidant from spinach, NAO: A comparative study of three antioxidants. *Toxicol. Pathol.*, *31*: 39–51. https://doi.org/10.1080/01926230390173833.

Padulosi, S., Heywood, V., Hunter, D. and Jarvis, A. 2011. Underutilized species and climate change: Current status and outlook. In: S.S. Yadav, R.J. Redden, J.L. Hatfield, H. Lotze-Campen and A. E. Hall (Eds.), *Crop Adaptation to Climate Change*, pp. 507–-521. Wiley. https://doi.org/10.1002/9780470960929.ch35.

Park, J.H., Kim, R.Y. and Park, E. 2012. Antidiabetic activity of fruits and vegetables commonly consumed in Korea: Inhibitory potential against α-glucosidase and insulin-like action *in vitro*. *Food Sci. Biotechnol.*, *21*: 1187–1193. https://doi.org/10.1007/s10068-012-0155-5.

Pellegrini, N., Colombi, B., Del Rio, D., Salvatore, S., Bianchi, M., Brighenti, F., and Serafini, M. 2003. Total antioxidant capacity of plant foods, beverages, and oils consumed in Italy assessed by three different *in vitro* assays. *The Journal of Nutrition*, *133*: 2812–2819. https://doi.org/10.1093/jn/133.9.2812.

POWO. 2025. *Spinacia oleracea L.* Plants of The World Online. https://powo.science.kew.org/taxon/urn:lsid:ipni.org:names:167384-1. Accessed on 25th August 2025.

Raghavendra, M., Reddy, A.M., Yadav, P.G., Raju, A.S. and Kumar, L.S. 2013. Comparative studies on the *in vitro* antioxidant properties of methanolic leafy extracts from six edible leafy vegetables of India. *Asian J. Pharm. Clin. Res.*, *6*: 96–99.

Rahman, Md.M., Dhar, P.S., Sumaia, F.A., Ahmed, L., Islam, Md.R., Sultana, N.A., Cavalu, S., Pop, O. and Rauf, A. 2022. Exploring the plant-derived bioactive substances as antidiabetic agent: An extensive review. *Biomed. Pharmacother.*, *152*: 113217. https://doi.org/10.1016/j.biopha.2022.113217.

Rao, G.N., Rao, P.G.P., Sulochanamma, G. and Satyanarayana, A. 2015. Physico-chemical amino acid composition, fatty acid profile, functional and antioxidant properties of *Spinacia oleracea* L. leaf. *J. Food Pharmaceutic. Sci.*, *3*: 27–37.

Rebello, C.J., Chu, J., Beyl, R., Edwall, D., Erlanson-Albertsson, C. and Greenway, F.L. 2015. Acute effects of a spinach extract rich in thylakoids on satiety: A randomized controlled crossover trial. *J. Am. Coll. Nutr.*, *34*: 470–477. https://doi.org/10.1080/07315724.2014.1003999.

Rehman, M.U., Wali, A.F., Ahmad, A., Shakeel, S., Rasool, S., Ali, R., Rashid, S.M., Madkhali, H., Ganaie, M.A. and Khan, R. 2019. Neuroprotective strategies for neurological disorders by natural products: An update. *Curr. Neuropharmacol.*, *17*: 247–267. https://doi.org/10.2174/1570159X16666180911124605.

Roberts, J.L. and Moreau, R. 2016. Functional properties of spinach (*Spinacia oleracea* L.) phytochemicals and bioactives. *Food Function*, *7*: 3337–3353. https://doi.org/10.1039/C6FO00051G.

Rotimi, S.O., Rotimi, O.A., Salako, A.A., Jibrin, P., Oyelade, J. and Iweala, E.E.J. 2019. Gene expression profiling analysis reveals putative phytochemotherapeutic target for castration-resistant prostate cancer. *Front. Oncol.*, *9*: Article 714. https://doi.org/10.3389/fonc.2019.00714.

Salehi, B., Tumer, T.B., Ozleyen, A., Peron, G., Dall'Acqua, S., Rajkovic, J., Naz, R., et al. 2019. Plants of the genus *Spinacia*: From bioactive molecules to food and phytopharmacological applications. *Trends Food Sci. Technol.*, *88*: 260–273. https://doi.org/10.1016/j.tifs.2019.03.028.

Sanjai, C., Hakkimane, S.S., Guru, B.R. and Gaonkar, S.L. 2024. A comprehensive review on anticancer evaluation techniques. *Bioorg. Chem.*, *142*: 106973. https://doi.org/10.1016/j.bioorg.2023.106973.

Sarkar, T., Salauddin, M., Roy, S., Chakraborty, R., Rebezov, M., Shariati, M.A., Thiruvengadam, M. and Rengasamy, K.R.R. 2023. Underutilized green leafy vegetables: Frontier in fortified food development and nutrition. *Crit. Rev. Food Sci. Nutr.*, *63*: 11679–11733. https://doi.org/10.1080/10408398.2022.2095555.

Shalini, V.T., Neelakanta, S.J. and Sriranjini, J.S. 2021. Neuroprotection with *Bacopa monnieri*: A review of experimental evidence. *Mol. Biol. Rep.*, *48*: 2653–2668. https://doi.org/10.1007/s11033-021-06236-w.

Sharma, N., Kapoor, M. and Nehru, B. 2014. LPS induced oxidative stress, inflammatory response and ensuing biochemical, neurochemical, and neurobehavioral impairment in mice. *Int. J. Pharm. Pharm. Sci.*, *6*: 203–210.

Shiota, A., Hada, T., Baba, T., Sato, M., Yamanaka-Okumura, H., Yamamoto, H., Taketani, Y. and Takeda, E. 2010. Protective effects of glycoglycerolipids extracted from spinach on 5-fluorouracil induced intestinal mucosal injury. *J. Med. Investig.*, *57*: 314–320. https://doi.org/10.2152/jmi.57.314.

Singh, R., Iqbal, N., Umar, S. and Ahmad, S. 2024. Lignan enhancement: An updated review on the significance of lignan and its improved production in crop plants. *Phyton*, *93*: 3237–3271. https://doi.org/10.32604/phyton.2024.057189.

Singh, S. 2016. Enhancing phytochemical levels, enzymatic and antioxidant activity of spinach leaves by chitosan treatment and an insight into the metabolic pathway using DART-MS technique. *Food Chem.*, *199*; 176–184. https://doi.org/10.1016/j.foodchem.2015.11.127.

Stenblom, E.L., Montelius, C., Östbring, K., Håkansson, M., Nilsson, S., Rehfeld, J.F. and Erlanson-Albertsson, C. 2013. Supplementation by thylakoids to a high carbohydrate meal decreases feelings of hunger, elevates CCK levels, and prevents postprandial hypoglycaemia in overweight women. *Appetite*, *68*: 118–123. https://doi.org/10.1016/j.appet.2013.04.022.

Subhash, G.P., Virbhadrappa, S.R. and Vasant, O.K. 2010. *Spinacia oleracea* Linn: A pharmacognostic and pharmacological overview. *Int. J. Res. Ayurveda Pharma.*, *1*: 74–84.

Sultana, B. and Anwar, F. 2008. Flavonols (kaempferol, quercetin, myricetin) contents of selected fruits, vegetables, and medicinal plants. *Food Chem.*, *108*: 879–884. https://doi.org/10.1016/j.foodchem.2007.11.053.

Turkiewicz, I.P., Tkacz, K. and Wojdyło, A. 2025. Phytonutrient profiling of spinach leaves using LC-ESI-QTOF/MS: Linking polar and non-polar extracts to enzyme inhibitory properties. *Food Chem.*, *487*: 144745. https://doi.org/10.1016/j.foodchem.2025.144745.

Umamaheswari, G. and Nishanthini, A. 2017. *In vitro* anticancer activity of *Spinacia oleracea* against various mammalian cell lines. *World J. Pharm. Res.*, *6*: 1723–1730. https://doi.org/10.20959/wjpr20177-8910.

Vujčić Bok, V., Šola, I. and Rusak, G. 2022. Lemon juice formulations modulate *in vitro* digestive recovery of spinach phytochemicals. *Food Technol. Biotechnol.*, *60*: 293–307. https://doi.org/10.17113/ftb.60.03.22.7104.

Wang, Y., Chang, C.F., Chou, J., Chen, H.L., Deng, X., Harvey, B.K., Cadet, J.L. and Bickford, P.C. 2005. Dietary supplementation with blueberries, spinach, or spirulina reduces ischemic brain damage. *Exp. Neurol.*, *193*: 75–84. https://doi.org/10.1016/j.expneurol.2004.12.014.

Wei, L., Wu, H. Wang, X., Wen, L., Cui, B. and Cheng, Y. 2024. Comprehensive review of plant-derived anti-hyperlipidemia peptides: Production, anti-hyperlipidemia mechanism, and structure-activity relationship study. *Food Chem.*, *461*: 140715. https://doi.org/10.1016/j.foodchem.2024.140715.

World Flora Online. 2025. *Spinacia oleracea L.* https://www.worldfloraonline.org/taxon/wfo-0000437543. Accessed on 25th August 2025.

Yadav, G.G. and Murthy, H.N. 2025. Phytochemicals and biological activities of *Lumnitzera racemosa* Willd. pp. 267-295. In: H. N. Murthy (Ed.), *Bioactive Compounds in Mangroves and their Associates, Reference Series in Phytochemistry* (1st Edn.), pp. 267–295. Springer Nature Switzerland AG. https://doi.org/10.1007/978-3-031-63920-3_12-1.

Yosefi, Z., Tabaraki, R., Gharneh, H.A.A. and Mehrabi, A.A. 2010. Variation in antioxidant activity, total phenolics, and nitrate in spinach. *Int. J. Veg. Sci.*, *16*: 233–242. https://doi.org/10.1080/19315260903577278.

Yuk, H.J., Ryu, H.W., Kim, D.Y., Park, M.H., Seo, W.D., Jeong, S.H. and Oh, S.R. 2019. Comparison of flavonoid and policosanol profiles in Korean winter spinach (*Spinacia oleracea* L.) cultivated in different regions. *Food Chem.*, *279*: 202–208. https://doi.org/10.1016/j.foodchem.2018.11.143.

Zhang, Y., Cai, P. G. Cheng and Zhang, Y. 2022. A brief review of phenolic compounds identified from plants: Their extraction, analysis, and biological activity. *Nat. Prod. Commun.*, *17*. https://doi.org/10.1177/1934578X211069721.

11 Nutritive Value, Bioactive Compounds, and Biological Activities of *Hyssopus* spp. Leaves

Yana K. Levaya[1*], Karakoz Zh. Badekova[2] and Gayane A. Atazhanova[3]

Introduction

Hyssopus spp., particularly *Hyssopus officinalis* L., is a small genus of perennial, aromatic, and medicinally significant plants belonging to the Lamiaceae family. Native to the dry and rocky slopes of Southern Europe, Central Asia, and parts of the Middle East, *Hyssopus* has adapted well to a variety of ecological conditions, ranging from temperate steppes to mountainous and semi-arid regions (Nistor et al., 2025; Sharifi-Rad et al., 2022a). Among its species, *H. officinalis* is the most widely cultivated and studied, having been naturalized in several regions outside its native range, including North America and East Asia. This species has historically been cultivated in monasteries and home gardens for its medicinal, aromatic, and culinary properties (Fatemeh Fathiazad, 2011). Within the species *H. officinalis*, modern taxonomic concepts distinguish four officially recognized subspecies: subsp. *officinalis* (type autonym), subsp. *aristatus* (Godr.) Nyman, subsp. *canescens* (DC.) Nyman, and subsp. *montanus* (Jord. & Fourr.) Briq. (Guerrini et al., 2021). Each of these taxa exhibits unique geographical ranges and morphological characteristics: subsp. *aristatus* is distributed in the western and central Mediterranean, subsp. *canescens* in the southwestern part of the basin, subsp. montanus inhabits southern Europe (including France, Ukraine, and southern European Russia), while subsp. *officinalis* covers a wide area from Europe to Iran (WFO, 2025). In addition to the main subspecies, subsp. *austro oranensis* Maire is also taxonomically recognized as a characteristic representative of North African flora (Algeria, Morocco). These subspecies, being the most widespread and structurally justified within the species *H. officinalis*, reflect the geographical and ecological plasticity of the species, its adaptation to various climatic and soil conditions. In addition, important *Hyssopus* species are: *H. cuspidatus* Boriss., *H. ambiguus* (Trautv.) Iljin, *H. seravschanicus* (Lipsky) Pazij, *H. latilabiatus* Gontsch., *H. macranthus* Boriss. These species are distributed from Eastern Europe (Ukraine, Romania) through the Caucasus, Iran,

[1-3] School of Pharmacy, Karaganda Medical University, Gogol Street, 40, Karaganda 100017, Kazakhstan.

* Corresponding author: yaninka_25@mail.ru

DOI: 10.1201/9781003687832-11

and Central Asia (Kazakhstan, Uzbekistan, Kyrgyzstan), extending into Western China and Mongolia. *H. officinalis* has also been introduced to North America and cultivated widely for medicinal and ornamental purposes.

As perennial subshrubs or small shrubs typically reaching heights between 20–80 cm, species of the *Hyssopus* genus are characterized by woody or semi-woody stems, narrow linear to lanceolate leaves, and dense spikes of bilabiate flowers that range in color from blue and violet to pink (Mićović et al., 2023). These plants are distinguished by their high content of essential oils (EOs), especially concentrated in their leaves and flowers, which contribute to their distinctive aromatic profile and notable pharmacological properties. Adapted to thrive in dry, rocky, and calcareous soils, *Hyssopus* spp. exhibit remarkable drought tolerance and resistance to pests, making them well-suited for sustainable cultivation in low-input agricultural systems and dryland agroecologies (Kizil et al., 2016).

Despite its long history of traditional use and modern application in herbal medicine, *Hyssopus* remains underutilized in global agriculture and pharmacology (Jankovský and Landa, 2002). In traditional systems of medicine across Europe, Central Asia, and the Middle East, hyssop has been used for the treatment of respiratory ailments, digestive disorders, and infections, as well as a general tonic and stimulant. In modern phytotherapy, it is increasingly recognized for its antioxidant (Moulodi et al., 2018; Vlase et al., 2014; Kizil et al., 2010), antimicrobial (Mićović et al., 2023; Moulodi et al., 2018; Vlase et al., 2014), anti-inflammatory (Mićović et al., 2022; Ghasempour et al., 2022), and spasmolytic properties (Aćimović et al., 2021), attributed largely to its EOs and a diverse array of phenolic compounds (Skrypnik et al., 2022).

In recent decades, increasing scientific interest has been directed toward the nutritional and phytochemical composition of *Hyssopus* spp. leaves, which are rich not only in EOs but also in bioactive flavonoids, phenolic acids, and terpenoids (Atazhanova et al., 2024). These compounds have demonstrated significant biological activities *in vitro* and *in vivo*, suggesting potential therapeutic roles in managing oxidative stress (Huyut et al., 2017), metabolic disorders, microbial infections (Sharma et al., 2025; Gupta et al., 2022; Carević et al., 2023), and inflammation (Chagas et al., 2022; Almasoudi et al., 2024). The growing evidence base supports the inclusion of hyssop in functional food formulations, nutraceutical development, and botanical drug research.

Furthermore, the adaptability of *Hyssopus* to marginal and degraded lands, coupled with its ecological value as a pollinator-friendly plant, positions it as a promising candidate for sustainable agriculture, biodiversity conservation, and land restoration efforts, particularly in arid and semi-arid regions (Sharifi-Rad et al., 2022). Its role as a component in agroecological systems—whether as an intercrop, hedgerow, or medicinal cash crop—aligns with broader global efforts to diversify cropping systems, reduce input dependency, and enhance resilience in the face of climate change.

However, despite its potential, comprehensive evaluations of the nutritional value, phytochemical diversity, and biological activities of *Hyssopus* leaves remain fragmented across the scientific literature. Moreover, standardized data on its composition, biological efficacy, toxicity, and postharvest management are limited. Therefore, this chapter aims to provide a comprehensive review of the current state of knowledge concerning the nutritive value, bioactive constituents, and biological effects of *Hyssopus spp.* leaves. Special emphasis is placed on their role in promoting health, their pharmacological potential, and possible applications in food, pharmaceutical, and functional products.

Taxonomic Background

The genus *Hyssopus* belongs to the family Lamiaceae. It comprises aromatic perennial herbaceous plants native to the Mediterranean region, Central Asia, and parts of Eastern Europe. The most well-known and widely studied species is *H. officinalis*, commonly referred to as hyssop. The name "Hyssopus" is derived from the Greek word ὕσσωπος (*hyssopos*), which itself may have Semitic origins (Hebrew: אזוב, *ezov*). In ancient texts, hyssop was mentioned as a sacred herb used in purification rituals, most notably in the Bible (Fleisher et al., 1998) and writings of Hippocrates, Dioscorides, and Galen. Medieval herbalists recognized it for its medicinal, culinary, and symbolic properties. The formal scientific description of *H. officinalis* was provided by Carl Linnaeus in Species Plantarum (1753), where it was given its current binomial name. Since then, *Hyssopus* spp. have become an important focus of phytochemical, pharmacological, and agricultural research due to their EOs profile, ethnomedicinal relevance, and ecological adaptability. References to hyssop can also be found in the works of Avicenna, the great medieval philosopher and healer. In the Middle Ages, it was an indispensable ingredient in all herbal mixtures, potions, and elixirs. The Jews considered it a sacred plant, and in ancient times, hyssop began to be valued for its medicinal properties. Towards the 14th century in Europe, this herb began to be widely used not only for medical purposes, but also for cosmetic purposes. "The herb called hyssop cleanses the chest of phlegm. Hyssop is beneficial for the lungs when boiled with honey, and it is said that Hyssop is beneficial for the lungs when boiled together with honey, and it is said to give the face an excellent complexion," wrote the famous Spanish physician Arnold of Villanova in his *Salerno Code of Health*. Nowadays, this remarkable plant is also called blue St. John's wort, and there are more than 50 species of it. The taxonomic classification for *Hyssopus* spp. is given in Table 1.

Table 1 Taxonomic classification of *Hyssopus* spp.

Level	*Classification*
Kingdom	Plantae Haeckel
Subkingdom	Pteridobiotina Britton & Brown
Division	Angiosperms

Contd...

Table 1 *Contd...*

Level	*Classification*				
Order	Lamiales Bromhead				
Family	Lamiaceae Martinov				
Genus	*Hyssopus* L.				
Species	*Hyssopus cuspidatus* Boriss.	*Hyssopus ambiguus* (Trautv.) Iljin ex Prochorov. & Lebel	*Hyssopus seravschanicus* (Dubj.) Pazij	*Hyssopus latilabiatus* C.Y.Wu & H.W.Li	*Hyssopus macranthus* Boriss.
	Hyssopus officinalis				
Subspecies	subsp. *officinalis*, subsp. *aristatus* (Godr.) Nyman, subsp. *canescens* (DC.) Nyman, subsp. *montanus* (Jord. & Fourr.) Briq. subsp. *austro-oranensis* Maire				

The most widely known and studied species is *H. officinalis*, commonly referred to as common or garden hyssop. This species has been cultivated since ancient times for its aromatic, culinary, and medicinal properties. Its use dates back to the Greco-Roman and Biblical periods, where it was valued for purification, digestion, and respiratory ailments. Taxonomically, *H. officinalis* exhibits considerable intraspecific variation across its wide geographic range. Worldwide *Hyssopus* spp. are distributed: Africa, Northern Africa, Algeria, Morocco, Asia-Temperate Caucasus, North Caucasus, Transcaucasus, China, Xinjiang, Middle Asia, Kazakhstan, Kirgizistan, Tadzhikistan, Uzbekistan, Mongolia, Siberia, Altay, Buryatiya, West Siberia, Western Asia, Afghanistan, Iran, Turkey, Asia-Tropical Indian Subcontinent, Pakistan, Europe, Eastern Europe, Belarus, Central European Russia, Krym, South European Russia, Ukraine, Middle Europe, Austria, Czechoslovakia, Hungary, Switzerland, Southeastern Europe, Albania, Bulgaria, Greece, Italy, Yugoslavia, Southwestern Europe, France, Spain (Figure 1).

Hyssop is valued both as an ornamental plant and for its practical use as a garden hedge due to its robust, upright growth habit and attractive flowers. Propagation can be achieved through seeds, cuttings, or root divisions. Optimal cultivation requires full sun to partial shade exposure and well-drained, dry soils with a pH around 6.7. Other species within the genus tend to prefer drier, rockier, and more calcareous substrates (Simon et al., 1984). Hyssop thrives in USDA hardiness zones 4–5 and is typically planted at intervals of approximately one foot. Pruning is generally limited to the removal of spent flower heads to encourage new growth and cutting back the plant in autumn to protect against damage from strong winds. Following transplantation of seedlings into field conditions, two years are required before the initial harvest, which is ideally conducted immediately after the first bloom of the season. Subsequent harvests can be maintained for up to ten years through regular cuttings. For maximum medicinal efficacy, harvesting of the aerial parts—specifically the tops and leaves—should occur in August. The prolonged flowering period of hyssop, combined with the availability of cultivars

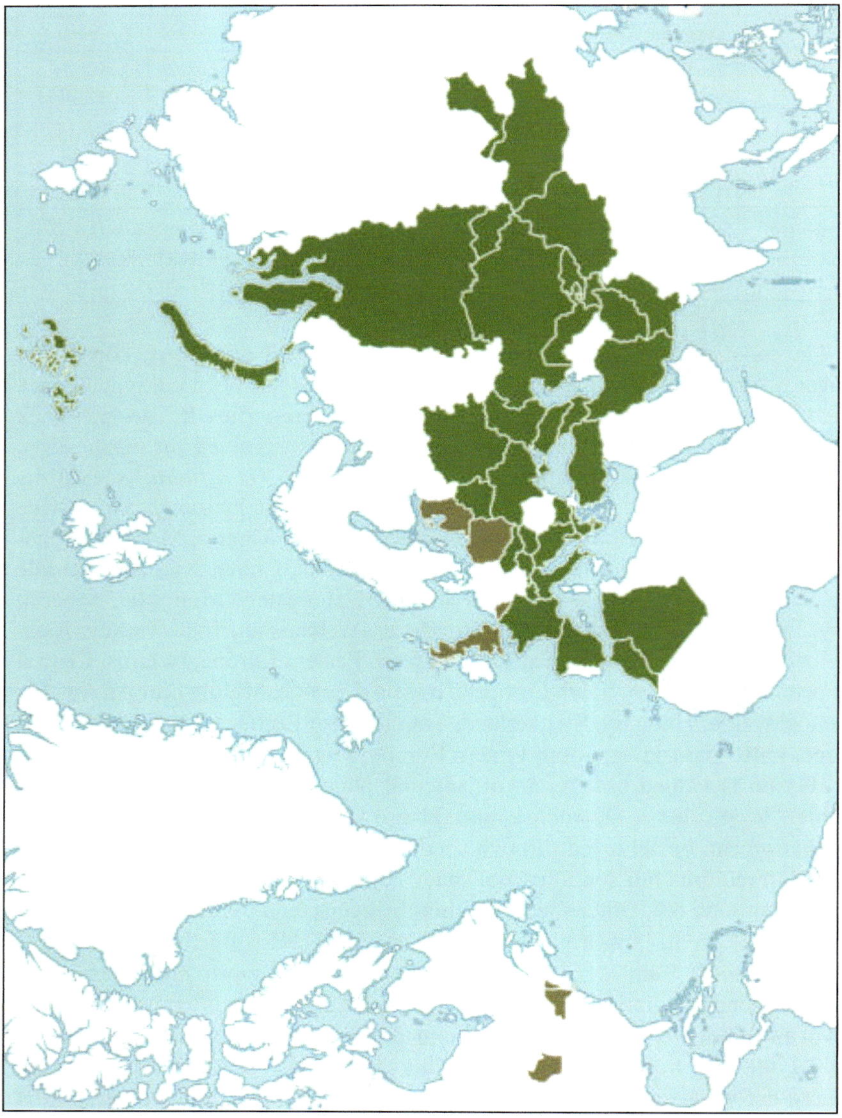

Fig. 1 Distribution of *Hyssopus* spp. According to WFO (2025).

exhibiting diverse flower colors, makes it a valuable filler plant in ornamental horticulture. Additionally, hyssop is well-suited for container cultivation, allowing for versatility in its use within both garden and indoor environments. *Hyssopus* spp. introduced into Europe, Eastern Europe, Baltic States, Middle Europe, Belgium, Poland, Northern Europe, Great Britain, Northern America, North-Central U.S.A., Wisconsin, Northeastern U.S.A., New York, Vermont.

More precisely, several subspecies have been described based on morphological traits and geographical distribution:

- subsp. *officinalis* – the typical subspecies widespread in Central and Southern Europe, representing cultivated and wild populations.
- subsp. *aristatus* – distributed across the western and central Mediterranean, including France, Spain, Algeria, and Morocco.
- subsp. *canescens* – occurring in southwestern Mediterranean areas such as Spain and southern France.
- subsp. *montanus* – native to mountainous regions of Southern and Eastern Europe, including France, Ukraine, and parts of Russia.
- subsp. *austro-oranensis* – a rare taxon confined to southern Algeria (Oran region).

In addition to *H. officinalis*, the genus comprises several lesser-known but ecologically and phytochemically important species, such as:

- *H. cuspidatus* and *H. macranthus* – endemic to Central Asia and adapted to high-altitude, rocky terrains.
- *H. ambiguus* and *H. seravschanicus* – typically found in arid Central Asian zones, often on calcareous or gypsum-rich soils (Lakomkina et al., 2022).
- *H. latilabiatus* and *H. cretaceus* – localized endemics of Central Asia and southeastern Europe with distinct floral morphologies and ecological preferences.

Representatives of the genus *H. officinalis* are perennial semi-shrubs that develop from a woody rhizome. The shoots are erect or ascending, 20–80 cm tall, tetrahedral, varying from almost glabrous to shortly pubescent, often with aromatic glands (Missouri Botanical Garden and Harvard University Herbaria, 2005). The leaves are opposite, sessile or almost sessile, narrow, from linear to lanceolate or oblanceolate, 1–4 cm long and 2–7 mm wide; the edges are entire, often slightly curved downward (revolute), the leaf surface may be glabrous, sparsely pubescent, or covered with glands (Gleason, 1991). Inflorescences are formed in the upper part of the shoots in the form of spike-like or paniculate formations consisting of 3–14 flowers gathered in axillary false whorls (Jarvis, 2007). The flowers are sessile or on very short pedicels, with a tubular or funnel-shaped calyx 6–8 mm long, with triangular-lanceolate teeth and glandular pubescence. The corolla is 7–12 mm long, most often blue-violet, less often white, bilabiate; the upper lip is two-parted, the lower lip is three-lobed, with a strongly expanded middle lobe. The stamens usually protrude beyond the corolla. The fruit is a four-seeded capsule, oblong, finely papillate, and mucilaginous when wet.

H. cuspidatus Borris stems reach 30–60 cm in height, thickened, with a woody, twisted brown base that exfoliates with age and branches profusely from the bottom. Branches are numerous, ranging from nearly glabrous to slightly hairy, with a purplish tint near the base. Leaves are sessile, linear, 1.5–4.5 cm long and 2–4 mm wide, generally exceeding the length of the internodes (Missouri Botanical Garden and Harvard University Herbaria, 2002). They are glabrous, narrowly tapering at the base, with finely strigose margins and slightly incurved edges; the apex is sharply pointed (cuspidate), about 2 mm long, and often deciduous. Inflorescences are 10-flowered verticillasters, arranged on short peduncles (1–2 mm), forming one-sided spikes 3–8 cm long. Bracts and bracteoles are linear, 4–8 mm, longer than the pedicels, with tips extending 2–3 mm. The calyx is tubular, around 1 cm long, slightly widened at the throat, sparsely covered with yellow glands; veins and teeth are minutely hairy (puberulent). Calyx teeth are triangular-lanceolate, about 4 mm long, with a cuspidate apex. The corolla is purple to white, around 1.2 cm long, puberulent and glandular, with a slightly curved tube that gradually widens toward the apex. The upper lip is straight, about 3 mm long, with sharply pointed lobes; the lower lip is about 4 mm, featuring an obcordate central lobe with a notched apex, and two broadly ovate lateral lobes around 1.5 mm wide. Nutlets are brown, three-angled, oblong (ca. 2.5 × 0.7 mm), with a rounded tip, glandular surface, and basal white areolae.

H. latilabiatus (C.Y. Wu and H.W. Li.) Subshrubs. Stems glabrous, base woody, brown, twisted, branches numerous from base, wandlike, ca. 20 cm, bark exfoliating, internodes longer than leaves. Leaves sessile, linear, 1–2 cm × 1–2 mm, glabrous, impressed glandular, base cuneate, margin ± involute, apex obtuse (Missouri Botanical Garden and Harvard University Herbaria, 2002). Verticillasters ± secund, pedunculate, in terminal spikes 5–8 cm, lower spikes 1–4-flowered, upper 6–8-flowered; peduncle ca. 2 cm; floral leaves similar, 3–10 × less than 1 mm; bracts and bracteoles linear-subulate, shorter than pedicels. Pedicel 2–4 mm. Calyx tubular-campanulate, tinged purple, ca. 6 mm, pubescent, glandular; teeth triangular, ca. 2 mm, apex acute. Corolla purple, ca. 1.3 cm, pubescent, sometimes glandular toward apex; upper lip straight, oblong, ca. 5 mm, lobes ovate; lower lip spreading, ca. 7 mm wide; middle lobe 1 cm wide, much wider than lateral lobes, obcordate, abruptly constricted at base, 2-lobulate, undulate; lateral lobes ovate, ca. 2 mm wide. Ovary apex pubescent. Flowering in August.

Species of the genus *Hyssopus* exhibit both common features and pronounced morphological differences. They are all perennial subshrubs with characteristic tetrahedral stems, linear or lanceolate leaves, sessile or almost sessile, and spike-like inflorescences consisting of false whorls with 3–14 flowers. The flowers are usually violet-blue or purple, bilabiate, with a pronounced lower lip. The differences lie in the details of the leaves and inflorescences: *H. cuspidatus* is distinguished by its long, pointed leaves and twisted woody stem; *H. seravschanicus* is shorter, with broad, pubescent leaves and short inflorescences; *H. ambiguus* has the longest and narrowest leaves and compact spikelets; *H. macranthus* is characterized by the largest flowers with a divided lower lip; *H. latilabiatus* is distinguished by a broad lower lip of the corolla; while *H. officinalis* is a typical species with a classic habit and moderately expressed characteristics. Thus, despite their morphological proximity and biological similarity, each species has a set of distinctive features that allow it to be reliably identified.

Uses

Traditional uses. *H. officinalis* has been utilized in traditional medicine for millennia. Historical sources, including Hippocrates, document its use in treating respiratory ailments such as bronchitis and pleurisy. During the 17th and 18th centuries (Kowalchik and Hylton, 1987), preparations such as teas and tinctures derived from hyssop flowers were employed to reduce excessive perspiration and to manage conditions like jaundice and dropsy. Traditionally, hyssop has been applied to treat a range of ailments, including asthma, congestion, rheumatism, sore throats, wounds, ulcers, and tumors (Hartwell, 1971; Lewis et al., 1977). Contemporary research has further elucidated the pharmacological properties of hyssop. Experimental studies have demonstrated its stimulatory effects and potential benefits in rheumatic conditions. The herb is also noted for its mucosal soothing properties, which underpin its efficacy in managing bronchitis, pleurisy, and asthma (Reader's Digest Association, 1986). Moreover, hyssop has been reported to reduce blood pressure by enhancing circulatory function. Its antioxidant capacity has recently garnered attention for potential use as a dietary supplement in individuals with hyperglycemia (Miyazaki et al., 2003). Notably, a polysaccharide isolated from hyssop has shown promise as a therapeutic agent in the context of human immunodeficiency virus (HIV)-1 infection (Gollapudi et al., 1995). Hyssop continues to be recommended in herbal medicine, with leaf infusions used externally as compresses to relieve muscle pain and insect bites. Although its strong, spicy, and bitter flavor limits culinary use, hyssop leaves and flowers are occasionally incorporated into salads, pasta dishes, soups, teas, and alcoholic beverages such as absinthe, Benedictine, and Chartreuse. Historically, hyssop was used by the Romans to produce herbal wines and by medieval monks as a soup seasoning (Reader's Digest Association, 1979). Additionally, the plant served as a valuable source of nectar for honey production, with bees capable of yielding up to 60 pounds of honey per acre of hyssop. Industrial applications of hyssop EO are diverse, encompassing use as a bittering agent in perfumes, liquors, soaps, colognes, and creams, as well as in laundry and bathing rinses. In Hungary, dedicated plantations cultivate hyssop specifically for oil extraction, with optimal yield achieved by harvesting flowers immediately post-bloom. However, instances of adulteration of hyssop oil with substances such as lavender, rosemary, or camphor oils have been reported, the latter linked to toxicity manifested as epileptic and neuromuscular symptoms in animal models (Millet et al., 1981). The antifungal efficacy of hyssop is attributed to the volatilization of active oil components from the leaf surface, reducing their concentration (Letessier et al.,

2001). Finally, dried hyssop leaves and flowers are utilized for their aesthetic and aromatic qualities in potpourri formulations.

Modern uses. *Hyssopus* spp. continues to find diverse applications in contemporary medicine, agriculture, and industry due to its rich phytochemical profile. Modern research highlights its potential as a source of bioactive compounds, including EOs, flavonoids, and phenolic acids, which contribute to its pharmacological activities. The leaves contain oil of hyssop, widely used in the cosmetic and fragrance industry due to its refreshing, herbal, and slightly sweet aroma (Judžentienė, 2016). Its volatile constituents, particularly pinocamphone and isopinocamphone, are responsible for its characteristic scent, which is utilized in perfumes, soaps, and skin care products (Atazhanova et al., 2024). Hyssop EO exhibits strong antimicrobial activity, making it a valuable natural preservative and active ingredient in skin care formulations (Michalak, 2023), particularly those targeting acne-prone or irritated skin. In aromatherapy, hyssop oil is applied for its mood-stabilizing and mentally clarifying properties, contributing to emotional balance and stress reduction. These properties also make it a popular additive in wellness-focused cosmetic formulations. Hyssop extracts and EOs exhibit significant antimicrobial, antiviral, antioxidant, anti-inflammatory, and expectorant properties. These attributes have led to its use as a complementary treatment for respiratory conditions such as bronchitis, asthma, and pleurisy (Ma et al., 2014; Daneshfard et al., 2024; Amini et al., 2022). Hyssop's attractiveness to pollinators such as bees supports biodiversity and enhances honey production, making it valuable in sustainable agriculture. Its ability to grow in marginal and dry soils positions it as a candidate for land reclamation and ecological restoration projects, particularly in arid and semi-arid regions. Its modern uses are its strong flavor limits widespread culinary application. Hyssop is occasionally used for flavoring meats, fish, vegetables, salads, sweets, and such liqueurs as absinthe, Benedictine, and Chartreuse (Samany et al., 2022).

Nutritional Composition

During the study, changes in the accumulated content of macro- and microelements (N, K, P, Ca, Mg, Cu, Zn, Pb) in *H. officinalis* from Iran (Table 2), as well as ash indicators (total ash — TA, acid-insoluble ash — AA, water-insoluble ash — WA) depending on plant height. The results showed that moving from the top of the shoots to the base, the content of Mg (by 22.67%), Ca (43.74%), Cu (12.87%), Zn (39.02%), and Pb (85.04%) increased. At the same time, there was a decrease in the content of N (by 50.16%) and K (6.41%), while the content of P increased by 29.06%. In the stem residues, when moving from the upper shoots to the base, there was a decrease in the content of Mg (by 1.01%), Ca (21.03%), Cu (9.11%), Zn (17.02%), and Pb (51.06%), while the content of N and P increased by 60.59% and 3.15%, respectively. An increase in the height of the cut led to an increase in TA, AA, and WA by 33.48%, 27.03%, and 18.25%, respectively. In the residues, these indicators also increased, but less significantly: TA by 11.44%, AA by 6.35%, and WA by 5.22% (Saebi et al., 2021).

Table 2 The dynamics of macro- and micro-element accumulation in *Hyssopus officinalis*.

Indicator	*Change from Top to Base of Shoot (%)*	*Change in Stem Residues (%)*
Nitrogen (N)	50.16%	60.59%
Potassium (K)	6.41%	—
Phosphorus (P)	29.06%	34.74%
Magnesium (Mg)	22.67%	1.01%
Calcium (Ca)	43.74%	21.03%
Copper (Cu)	12.87%	9.11%
Zinc (Zn)	39.02%	17.02%
Lead (Pb)	85.04%	51.06%
Total ash (TA)	33.48%	11.44%
Acid-insoluble ash (AA)	27.03%	6.35%
Water soluble ash (WA)	18.25%	5.22%

Analysis of ash indicators and extractive substances in *H. cuspidatus* Boriss from Xinjiang, China, showed that TA is 4.18% (w/w). At the same time, the AA value was significantly lower—0.28% (w/w), which indicates high solubility of mineral components in an acidic environment and, consequently, low content of silicates and sand-like impurities. The WA content was 1.56% (w/w), which indicates the presence of a significant amount of mineral substances that are easily soluble in water. At the same time, the content of water-soluble extractives was very high—11.47% (w/w), which may indicate that the plant is rich in bioactive compounds such as phenolic compounds, sugars, and other polar metabolites. In contrast, the values for ethanol- and ether-soluble extractive substances were relatively low—2.50% (w/w) and 1.79% (w/w), respectively, indicating a limited amount of less polar components such as lipids, terpenoids, and steroids (Zhou et al., 2010). The authors (Shanayda, 2015) studied the amino acid composition of the aerial parts of *H. officinalis* harvested during flowering when cultivated in the conditions of Western Podolia, Ukraine (Table 3).

Table 3 Amino acids from *Hyssopus officinalis*.

Amino Acid	*Content (mg/100 g)*
Aspartic acid	6.8
Glutamic acid	46.8
4-hydroxyproline	8.6
Asparagine	40.2
Glutamine	5.4
Serine	8.0

Contd...

Table 3 *Contd...*

Amino Acid	*Content (mg/100 g)*
Arginine	57.2
Glycine	24.6
Threonine	41.0
Alanine	16.8
Proline	469.9
Aminobutyric acid	25.7
Valine	47.3
Methionine	3.1
Isoleucine	8.8
Leucine	8.9
Phenylalanine	15.1
Cysteine	51.7
Histidine	22.6
Lysine	11.3
Tyrosine	20.5
Total number	940.6

The amino acid profile of the plant sample studied is characterized by a high total concentration of 940.6 mg/100 g, with proline (469.9 mg/100 g) predominating, accounting for about 50% of the total content, which may indicate its structural and stress-protective function. Significant amounts of arginine (57.2 mg), valine (47.3 mg), glutamic acid (46.8 mg), and threonine (41.0 mg) indicate active protein and nitrogen metabolism. The relatively high level of cystine (51.7 mg) suggests the presence of stable disulfide bonds, while the low methionine content (3.1 mg) is typical for many plants and may limit protein biosynthesis. A wide range of amino acids was found, including both proteinogenic and non-proteinogenic ones (e.g., γ-aminobutyric acid), confirming a balanced and functionally significant amino acid composition.

Bioactive Compounds

Hyssopus spp. possess a wide variety of different chemical groups of secondary metabolites. Species of this genus are known for their high content of biologically active substances of various classes: polyphenols, terpenoids, alkaloids, as well as sulfur-containing compounds. The wide range of these substances determines the anti-inflammatory, antioxidant, antimicrobial, and other pharmacological properties of hyssop, making it a promising subject for further research in the field of phytotherapy and pharmacognosy.

Polyphenols

Plants of the genus *Hyssopus*, including *H. officinalis* and *H. cuspidatus*, are characterized by a rich polyphenolic composition, covering the following classes: phenolic acids and their esters, flavonoids, glycosides of phenolic compounds, lignans, coumarins, and tannins. The total phenol content (TPC) of the ethanol leaves extract of *H. officinalis* from Iran was reported as 117.43 ± 9.22 mg gallic acid equivalent (GAE)/100 g (Rezaei Savadkouhi et al., 2020). Compared to another study, the amount of TPC is higher than TPC of the ethanolic extract obtained from *H. officinalis L.* from Romania resulted in 77.72 mg/g (Vlase et al., 2014); such differences may be due to the extraction method. The results indicated that the total flavonoid content (TFC) is higher than TPC were about 2.06% and 4.48%, respectively (Zhao et al., 2020). TPC was found to be higher than TFC in *H. officinalis* L. subsp. *angustifolius* leaves aqueous extracts were 1.3% and 4.7%, respectively (Hatipoğlu et al., 2013). Authors showed that the TPC of the methanolic extract obtained from Iranian *H. officinalis L.* subs. *angustifolius* was 90 mg GAE /g. TPC of the n-butanol and ethylacetate extracts of the *H. officinalis* aerial parts from Iran was found to be 246 mg GAE/g and 51 mg GAE/g, respectively (Fathiazad et al., 2011). In another study, TPC of six samples of *H. officinalis* from Montenegro were analyzed (Benea et al., 2022). It was concluded that the highest TPC was found in *H. officinalis* with blue flowers, 39.056 mg GAE/g dry extract. Furthermore, the authors assessed the total content of hydroxycinnamic acids, expressed as caffeic acid equivalents, in both the aerial parts and dry extracts of *H. officinalis*. The highest levels were observed in plants with white flowers (1.484 mg/g in the aerial parts and 3.014 mg/g in the dry extract), followed by those with pink flowers (1.190 mg/g and 2.915 mg/g, respectively) and blue flowers (1.015 mg/g and 2.851 mg/g, respectively). TPC of *H. officinalis* from Romania expressed as rosmarinic acid equivalent (RAE) is 1.80 ± 0.19 g RAE/100g (Benedec et al., 2017).

Among the various polyphenolic compounds identified in *Hyssopus*, flavonoids (such as quercetin, luteolin, apigenin, and their glucosides) and phenolic acids (including *p*-hydroxybenzoic acid, protocatechuic acid, ferulic acid, caffeic acid, and chlorogenic acid) represent the two major classes (Figures 2, 3, and Table 4). These groups constitute the predominant polyphenols contributing to the plant's antioxidant and therapeutic potential.

HPLC-DAD analysis identified three phenolic acids in *H. cuspidatus*: caffeic acid **1** (0.04–0.06%), ferulic acid **2** (0.01–0.08%), and rosmarinic acid **3** (0.12–0.13%), with rosmarinic acid being the most abundant (Zhao et al., 2020). A total of eighteen phenolic compounds (Shomirzoeva et al., 2020) were identified in the ethyl acetate fraction of *H. cuspidatus*, including methyl rosmarinate **4**, rosmarinic acid **3**, 4-*O*-caffeoylquinic acid methyl ester **5**, 3-*O*-caffeoylquinic acid methyl ester **6**, oresbiusin A **7**, caffeic acid **1**, esculetin **8**, luteolin **9**, salvigenin **10**, apigenin-4′-*O*-β-*d*-glucopyranoside **11**, quercetin-3-*O*-β-*d*-glucopyranoside **12**, icariin **13**, epimedoside A **14**, epimedin C **15**, glochidioboside [(7S,8R)-

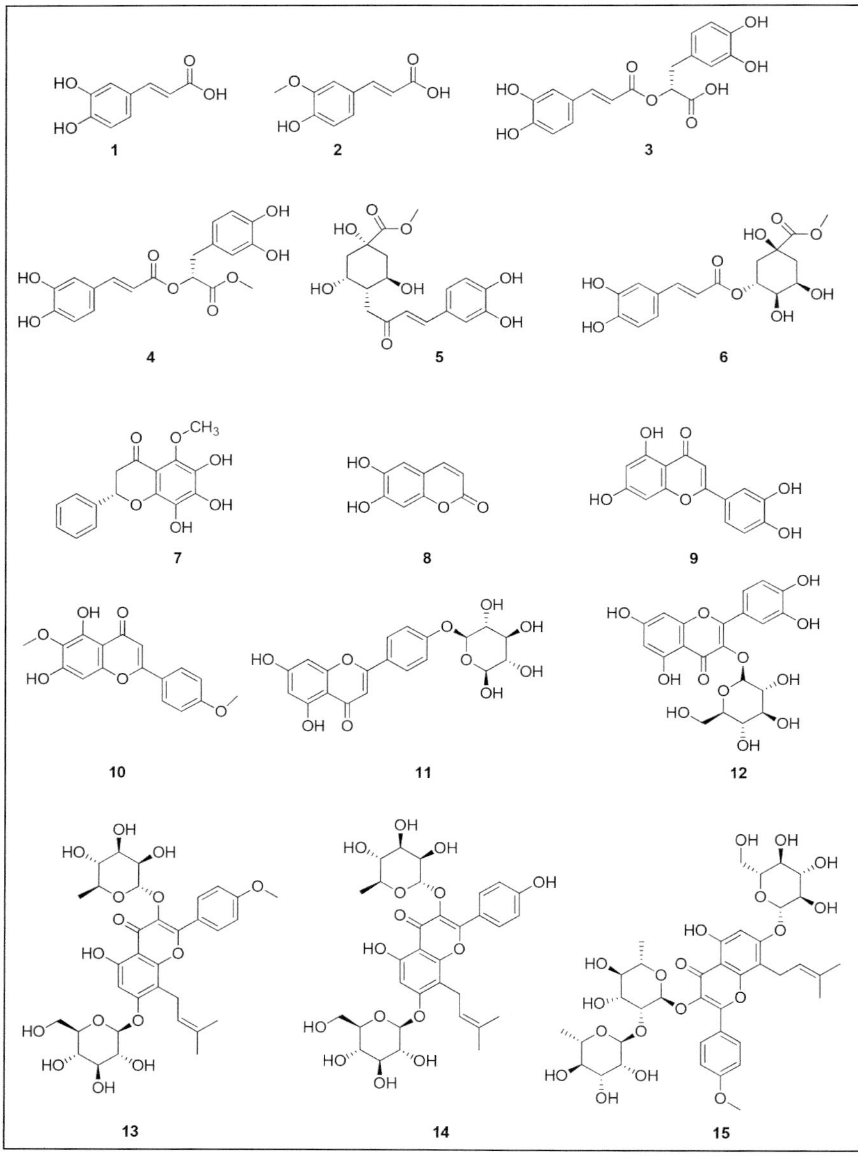

Fig. 2 Phenolic compounds from *Hyssopus* spp. Part 1.

dihydrodehydrodiconiferyl alcohol 9′-*O*-β-*d*-glucopyranoside] **16**, daucosterol **17**, ursolic acid **18**, and 2α,3β,24-trihydroxy-12-en-28-ursolic acid **19**. Among them, compounds **5**, **7**, **8**, **11**, **12–16**, and **19** were isolated from the *Hyssopus* genus for the first time, while compound **4** was newly reported from *H. cuspidatus*. Methyl

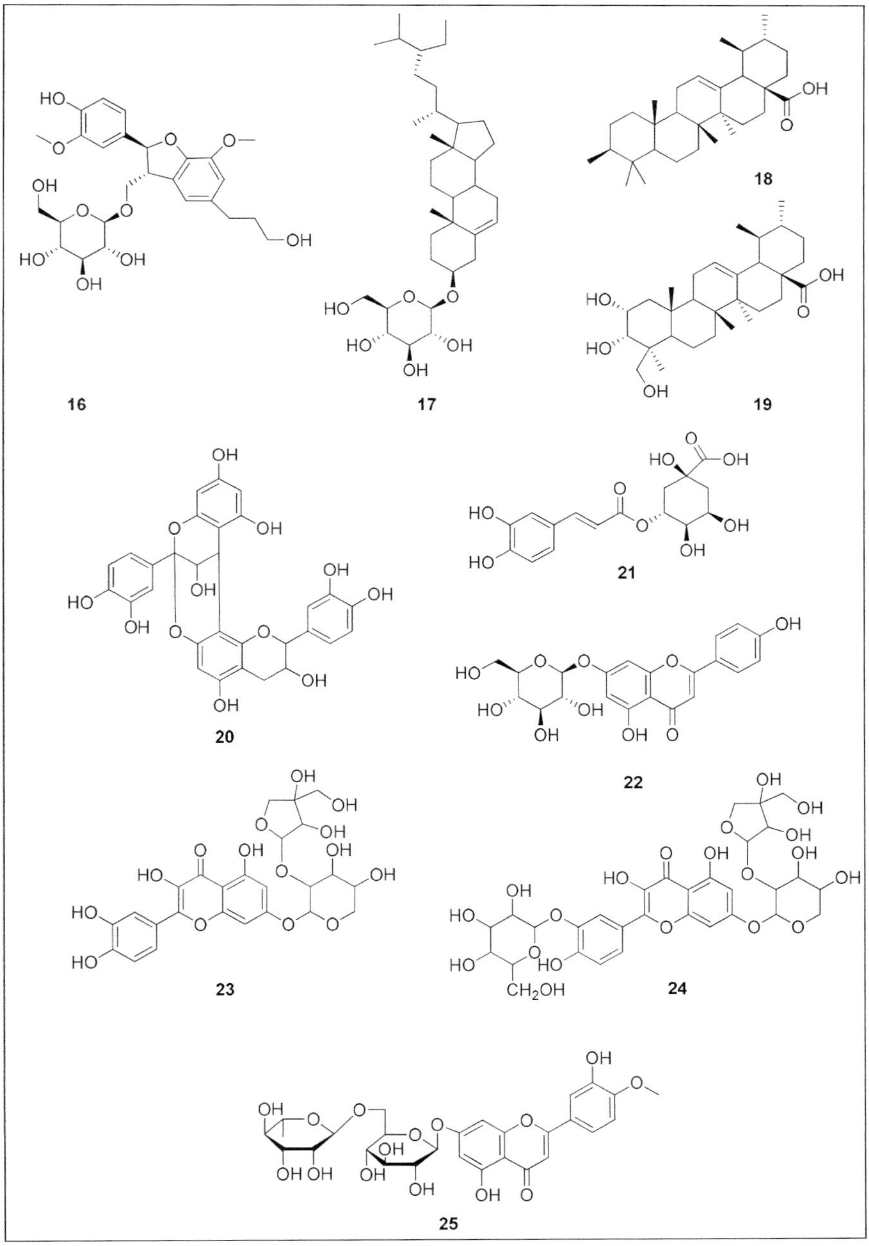

Fig. 3 Phenolic compounds from *Hyssopus* spp. Part 2.

rosmarinate **4** and rosmarinic acid **3** were the most abundant components in the ethyl acetate fraction of *H. cuspidatus* (14.19 mg/g and 24.75 mg/g, respectively). Proanthocyanidin **20** was detected at notably high concentrations in the aqueous, chloroform, and hexane leaf extracts of *H. officinalis*, with the highest level found in the chloroform extract (10,250 mg/L) (Hatipoğlu et al., 2013). Additionally, HPLC analysis revealed in the methanolic leaf extract the presence of caffeic acid **1** (111.09 μg/g) and chlorogenic acid **21** (166.21 μg/g).

From the hydromethanolic extract of the aerial parts, the major flavonoid identified was apigenin-7-*O*-β-*D*-glucuronide **22** (Fathiazad et al., 2011). Wang and Yang (2010) reported the isolation of two previously unidentified flavonoid glycosides from the whole herb of *H. officinalis* grown in China. These compounds were characterized as quercetin-7-*O*-β-*D*-apiofuranosyl-(1→2)-β-*D*-xylopyranoside-3′-*O*-β-*D*-glucopyranoside **23** and quercetin-7-*O*-β-*D*-apiofuranosyl-(1→2)-β-*D*-xylopyranoside **24**. In the aqueous-ethanolic extract of *H. officinalis*, the content of hydroxycinnamic acids was quantified as follows: caffeic acid **1** — 10.6 ± 0.3 mg per 100 g of raw material, and rosmarinic acid **2** — 169.2 ± 0.6 mg per 100 g (Grebennikova et al., 2017). Chlorogenic acid **21** content ranged from 23.35 to 33.46 mg/g, while rosmarinic acid **3** was detected in lower concentrations, between 3.53–17.98 mg/g (Benea et al., 2022). The presence of diosmin **25**, a major biologically active flavonoid, was confirmed in the leaves of *H. officinalis* L. using 1H NMR spectroscopy (Sheremeta et al., 2023).

Table 4 presents a summary of various *Hyssopus* spp. and the identified phenolic compounds found in them, including phenolic acids, flavonoids, glycosides, lignans, tannins, and related classes. The table includes the main compound types, notable quantitative data (where available), and references to the corresponding studies.

Terpenoids

The EOs of *Hyssopus* species exhibit significant chemical diversity influenced by species, subspecies, and geographic origin. Major components typically include oxygenated monoterpenes, sesquiterpenes, diterpenes, and triterpenes (Figure 4, Table 5). Variations in volatile profiles have been reported across different regions, including China, Kazakhstan, Iran, and Europe, reflecting the complexity of their phytochemical composition. Understanding these variations is crucial for their application in pharmacology, aromatherapy, and taxonomy.

The volatile composition of *H. cuspidatus* EO from China includes a diverse array of terpenoid and hydrocarbon compounds (Zhou et al., 2010). Among the oxygenated terpenes (66.33%), monoterpenes (26.14%), oxygenated sesquiterpenes (1.25%). The primary components identified in the sample included berbenone **26** (23.84%), β-pinene **27** (19.76%), pinocamphone **28** (17.95%), 1,8-cineole **29** (7.16%), myrtenol **30** (7.06%), and l-trans-pinocarveol **31** (3.00%). The EO of *H. cuspidatus* from Xinjiang, China, was found to contain various terpenoids as major constituents, including germacrene D **32** (18.67%), germacrene B **33** (15.61%),

Table 4 Phenolic Compounds Identified in *Hyssopus* Species.

Species / Variety	*Identified Phenolic Compounds*	*Compound Type*	**Content / Notes**	*References*
H. officinalis (Iran)	Caffeic acid **1**, chlorogenic acid **2**, apigenin-7-O-β-D-glucuronide **22**, luteolin, quercetin, apigenin, diosmin **25**	Phenolic acids, flavonoids, glycosides	TPC: 117.43 ± 9.22 mg GAE/100g; Caffeic acid: 111.09 µg/g; Chlorogenic acid: 166.21 µg/g; Diosmin confirmed via ^{1}H NMR	Sheremeta et al. (2023; Hatipoğlu et al. (2013); Fathiazad et al. (2011)
H. officinalis (Romania)	Rosmarinic acid **3**, chlorogenic acid **21**, caffeic acid **1**, syringic, p-hydroxybenzoic, ferulic, protocatechuic acids	Phenolic acids	TPC: 77.72 mg/g; RAE: 1.80 ± 0.19 g/100g	Vlase et al. 014; Benedec et al. (2017)
H. officinalis (Iran, subs. *angustifolius*)	Caffeic acid **1**, rosmarinic acid **3**	Phenolic acids	TPC: 90 mg GAE/g; Caffeic acid: 10.6 ± 0.3 mg/100g, Rosmarinic acid: 169.2 ± 0.6 mg/100g	Grebennikova et al. (2017)
H. officinalis (Montenegro)	Rosmarinic acid **3**, chlorogenic acid **21**, caffeic acid **1**	Phenolic acids	Chlorogenic acid: 23.35–33.46 mg/g; Rosmarinic acid: 3.53–17.98 mg/g; TPC: 39.056 mg GAE/g (blue flowers)	Benea et al. (2022)

Contd...

Table 4 *Contd...*

Species / Variety	*Identified Phenolic Compounds*	*Compound Type*	**Content / Notes**	*References*
H. officinalis (China)	Quercetin-7-O-β-D-apiofuranosyl-(1→2)-β-D-xylopyranoside-3′-O-β-D-glucopyranoside **23**, quercetin-7-O-β-D-apiofuranosyl-(1→2)-β-D-xylopyranoside **24**	Flavonoid glycosides	New flavonoid glycosides were first isolated from *H. officinalis*	Wang & Yang (2010)
H. officinalis (subs. *angustifolius*)	Proanthocyanidin **20**	Tannin	Very high: up to 10,250 mg/L in chloroform extract	Hatipoğlu et al. (2013)
H. cuspidatus (China)	Rosmarinic acid **3**, methyl rosmarinate **4**, 3-O-caffeoylquinic acid methyl ester **6**, 4-O-caffeoylquinic acid methyl ester **5**, caffeic acid **1**, ferulic acid **2**, esculetin **8**, luteolin **9**, salvigenin **10**, apigenin glucoside **11**, quercetin glucoside **12**, icariin **13**, epimedoside A **14**, epimedin C **15**, glochidioboside **16**, daucosterol **17**, ursolic acid **18**, trihydroxy-ursolic acid **19**	Phenolic acids, flavonoids, glycosides, lignans	Rosmarinic acid: 0.12–0.13%; Methyl rosmarinate: 14.19 mg/g; Caffeic acid: 0.04–0.06%; Ferulic acid: 0.01–0.08%	Zhao et al. (2020); Shomirzoeva et al. (2020)

(+)-trans-caryophyllene **34** (8.04%), and spathulenol **35** (4.11%) (Ablizl et al., 2009). *H. seravschanicus* EO from Tadjikistan contains mainly pinocamphone **28** (57%–89%), β-pinene **27** (up to 6%), 1,8-cineole **29** (2%–4%), camphor **36** (up to 4%), and spathulenol **36** (up to 5%) (Sharopov et al., 2015). EO from the aerial parts of Iranian *H. officinalis* were found to contain major terpenoid constituents such as β-pinene **27**, isopinocamphone **37**, and pinocamphone **28**, along with myrtenyl acetate **38**, camphor **37,** and spathulenol **36** (Fathiazad et al., 2011). Additionally, the chloroform-soluble fraction of the dichloromethane-ethyl acetate extract of *H. officinalis* cell cultures yielded pentacyclic triterpenes with oleanane and ursane skeletons, including oleanolic acid **39**, ursolic acid **18**, and their hydroxylated derivatives, some of which were isolated as acetates or bromolactone acetates (Skrzypek and Wysokińska, 2003). The EO of *H. ambiguus* from Kazakhstan was found to contain 48 identified compounds, with the major constituents being 1,8-cineole **29** (26%), cis-verbenol **40** (18.2%), pinocarvone **41** (14%), β-pinene **27** (10.5%) (Suleimen et al., 2015). Another study (Lakomkina et al., 2023) revealed that the EO samples of *H. ambiguus* (Trautv.) Iljin collected from the Karkaralinsk area and the Sortau (Kazakhstan) settlement differ in both qualitative and quantitative composition. Specifically, the sample from the first location contained 3-carene **42**, terpinen-4-ol **43**, and germacrene D **33**, while the sample from the second site was characterized by the presence of β-pinene **27**, o-cymene **44**, and β-myrcene **45**. GC and GC-MS analyses of *H. officinalis* EOs from Spain (Vallejo et al., 1995), Turkey (Ozer et al., 2004), and the French Alps (Salvatore et al., 1998) reveal significant variation in composition, with Spain's oil dominated by 1,8-cineole **29** (52.89%) and β-pinene **27** (16.82%), Turkey's by pinocarvone **41** (36.3%), pinocamphone **28** (19.6%), and β-pinene **27** (10.6%), and the French Alps' by linalool **46** (49.6%), 1,8-cineole **29** (13.3%), and limonene **47** (5.4%), reflecting geographic influences on the oil's chemical profile. EO from *H. officinalis* in Poland contained 52 compounds, with the main terpenoid components being isopinocamphone **37** (22.53–28.74%), pinocamphone **28** (11.41–17.99%), β-pinene **27** (6.69–12.01%), elemol **48** (5.02–7.57%), germacrene D **33** (3.14–6.98%), and bicyclogermacrene **49** (2.55–4.36%) (Zawislak et al., 2016).

Garg et al. (1999) reported that key volatile terpenoids in *H. officinalis* oils from various locations included β-pinene **27**, limonene **47**, β-phellandrene **50**, 1,8-cineole **29**, pinocamphone **28**, isopinocamphone **37**, pinocarvone **41**, germacrene D **33**, and methyleugenol **51**, noting variation across subspecies and populations. In Egyptian samples, Said-Al Ahl et al. (2015) found major terpenoids such as cis-pinocamphone **52** (26.85%), β-pinene **27** (20.43%), trans-pinocamphone **53** (15.97%), α-elemol **54** (7.96%), durenol **55** (3.11%), β-phellandrene **50** (2.41%), caryophyllene **56** (2.34%), bicyclogermacrene **49** (1.83%), myrtenol **30** (1.73%), germacrene D **33** (1.68%), limonene **47** (1.56%), γ-eudesmol **57** (1.36%), and linalool **46** (1.08%) (Said-Al Ahl et al., 2015). EO content of *H. officinalis* genotypes with pink (ruber), white (albus), and blue (cyaneus) corollas was evaluated (Gonceariuc and Balmu, 2013). The cyaneus form had 51.77% cis-pinocamphone **52** and 6.70% isopinocamphone **37**; the ruber form contained 66.94% pinocamphone **28**, almost equally split between isopinocamphone **37** (33.31%) and cis-pinocamphone **52** (33.63%); and the albus form was richest in isopinocamphone **37** (61.1%) with only 2.15% cis-pinocamphone **52**. In all genotypes, pinocamphone **28** was followed by β-pinene **27** (8.49% in cyaneus, 7.38% in albus, and 4.15% in ruber) and β-phellandrene **50** at varying concentrations. *H. officinalis* plants were collected from six locations in East Lithuania, and their EOs obtained by hydrodistillation were analyzed using GC and GC/MS (Bernotienė and Butkienė, 2010). Pinocarvone **41** was the major component in four of the oils, ranging from 21.1% to 28.1%, while isopinocamphone **37** dominated in two oils with 16.8% to 33.6%. In addition to pinocarvone **41**, significant amounts of isopinocamphone **37** (11.5–15.9%), β-pinene **27** (7.0–11.4%), germacrene D **33** (3.7–5.5%), and hedycaryol **58** (4.1–4.8%) were found in four oils. The other two oils had main constituents including pinocarvone **41** (9.0–13.6%), β-pinene **27** (7.2–8.6%), and hedycaryol **58** (4.0–9.1%). Compounds with a pinane carbon skeleton constituted 43.6–56.9% of the oils, with oxygenated monoterpenes being the most characteristic components, accounting for 36.0–50.7%. The EO content in three genotypes from Serbia f. *albus* Alef, f. *cyaneus* Alef, f. *ruber* Mill. identified that pinocamphone **28** in both cis and trans forms is the major component across all genotypes (Kiprovski and Rat, 2019). *H. officinalis* subsp. *angustifolius* (Bieb) Arcangeli from Turkey's North East and South Anatolia identified pinocarvone **41** (27.1%), β-pinene (19.0%), and isopinocamphone **37** (13.6%) as the main components of the EO (Kürkçüoğlu, 2016).

Six previously undescribed monoterpenoids, along with twelve known compounds, were isolated and identified from *H. cuspidatus* Boriss using NMR and MS, and further confirmed by electronic circular dichroism calculations and single-crystal X-ray diffraction. Among the identified compounds were 3β-hydroxy-7,8-dihydro-β-ionone **59**, oleanolic acid **39**, as well as several other monoterpenoids and triterpenoids (Aihaiti et al., 2023). The major components of the *H. officinalis* ethanolic extract cultivated in Iran were isopinochampone **37** (10.3%), linalool **46** (6.5%), 1,8-cineole **29** (6.3%), pinocamphone **28** (6.28%), and pinocarvone **41** (5.4%) (Rastkari et al., 2007). Pinane-type monoterpenoid, a menthane-type monoterpenoid, eremophilane-type sesquiterpenoid, abietane-type diterpenoid, and four phenethyl glucosides, together with fourteen known compounds, were isolated from the whole herb of *H. cuspidatus* (Furukawa et al., 2011). The aqueous ethanolic extract of *H. officinalis* is characterized by a high content of monoterpene ketones (81.6%) and monoterpenes (7.3%), with pinocamphone **28** (78.2%), isopinocamphone **37** (3.2%), and β-pinene **27** (4.2%) being the dominant components—consistent with known profiles of *H. officinalis* EO (Grebennikova et al., 2017). Minor constituents include β-phellandrene **50** (1.3%) and myrtenal **60** (1.4%), while other compound classes such as sesquiterpenes (2.1%), monoterpene alcohols (1.6%), sesquiterpene alcohols (1.8%), esters (0.9%), and organic acids (0.8%) are present in trace amounts.

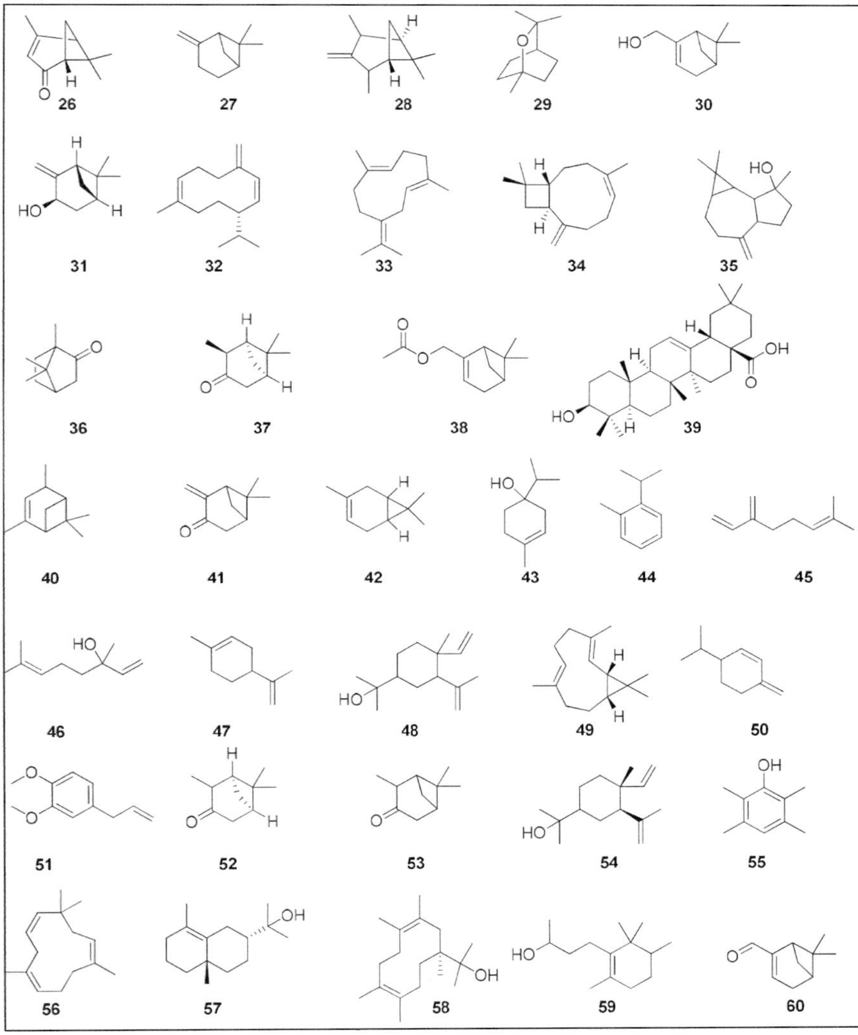

Fig. 4 Terpenoids from *Hyssopus* spp.

Table 5 presents a summary of various *Hyssopus* species and the major volatile components identified in their EOs across different geographic regions. The table includes the main terpenoid compounds—such as monoterpenes, sesquiterpenes, ketones, and alcohols—along with their relative concentrations (where available) and references to the corresponding studies. This comparative overview highlights the chemical variability influenced by species, subspecies, genotype, and environmental factors.

Table 5 Terpenoids Identified in *Hyssopus* Species.

Species	*Region*	*Main Components (and Content, %)*	*References*
H. cuspidatus EO	China	Berbenone (23.84%), β-Pinene (19.76%), Pinocamphone (17.95%), 1,8-Cineole (7.16%), Myrtenol (7.06%), l-trans-Pinocarveol (3.00%)	Zhou et al. (2010)
H. cuspidatus EO	Xinjiang, China	Germacrene D (18.67%), Germacrene B (15.61%), (+)-trans-Caryophyllene (8.04%), Spathulenol (4.11%)	Ablizl et al. (2009)
H. seravschanicus EO	Tajikistan	Pinocamphone (57–89%), β-Pinene (up to 6%), 1,8-Cineole (2–4%), Camphor (up to 4%), Spathulenol (up to 5%)	Sharopov et al. (2015)
H. officinalis EO	Iran	β-Pinene, Isopinocamphone, Pinocamphone, Myrtenyl acetate, Camphor, Spathulenol	Fathiazad et al. (2011)
H. officinalis extract	Iran	Isopinocamphone (10.3%), Linalool (6.5%), 1,8-Cineole (6.3%), Pinocamphone (6.28%), Pinocarvone (5.4%)	Rastkari et al. (2007)
H. ambiguus EO	Kazakhstan	1,8-Cineole (26%), cis-Verbenol (18.2%), Pinocarvone (14%), β-Pinene (10.5%)	Suleimen et al. (2015)
H. ambiguus EO	Kazakhstan, Karkaraly	3-Carene, Terpinen-4-ol, Germacrene D	Lakomkina et al. (2023)
H. ambiguus EO	Kazakhstan, Sortau	β-Pinene, o-Cymene, β-Myrcene	Lakomkina et al. (2023)
H. officinalis EO	Spain	1,8-Cineole (52.89%), β-Pinene (16.82%)	Vallejo et al. (1995)
H. officinalis EO	Turkey	Pinocarvone (36.3%), Pinocamphone (19.6%), β-Pinene (10.6%)	Ozer et al. (2004)
H. officinalis EO	French Alpes	Linalool (49.6%), 1,8-Cineole (13.3%), Limonene (5.4%)	Salvatore et al. (1998)

Contd...

Table 5 *Contd...*

Species	*Region*	*Main Components (and Content, %)*	*References*
H. officinalis EO	Poland	Isopinocamphone (22.53–28.74%), Pinocamphone (11.41–17.99%), β-Pinene (6.69–12.01%), Elemol (5.02–7.57%), Germacrene D, Bicyclogermacrene	Zawislak et al. (2016)
H. officinalis EO	India	β-Pinene, Limonene, β-Phellandrene, 1,8-Cineole, Pinocamphone, Isopinocamphone, Pinocarvone, Germacrene D, Methyleugenol	Garg et al. (1999)
H. officinalis EO	Egypt	Cis-Pinocamphone (26.85%), β-Pinene (20.43%), Trans-Pinocamphone (15.97%), α-Elemol, Durenol, Germacrene D, Limonene, Linalool	Said-Al Ahl et al. (2015)
H. officinalis f. *ruber* EO	Moldova	Pinocamphone (66.94%), Isopinocamphone (33.31%), β-Pinene (4.15%)	Gonceariuc & Balmuù (2013)
H. officinalis f. *albus* EO	Moldova	Isopinocamphone (61.1%), β-Pinene (7.38%), Pinocamphone (2.15%)	Gonceariuc & Balmuù (2013)
H. officinalis f. *cyaneus* EO	Moldova	Cis-Pinocamphone (51.77%), Isopinocamphone (6.70%), β-Pinene (8.49%)	Gonceariuc & Balmuù (2013)
H. officinalis EOs	Lithuania	Pinocarvone (up to 28.1%), Isopinocamphone (up to 33.6%), β-Pinene (до 11.4%), Germacrene D, Hedycaryol	Bernotienė & Butkienė (n.d.)
H. officinalis EO 3 genotypes	Serbia	Pinocamphone	Kiprovski & Rat (2019)
H. officinalis subs. *angustifolius* EO	Turkey	Pinocarvone (27.1%), β-Pinene (19.0%), Isopinocamphone (13.6%)	Mine Kürkçüoğlu (2016)
H. cuspidatus extract	China	3β-Hydroxy-7,8-dihydro-β-ionone, Oleanolic acid	Aihaiti et al. (2023)
H. officinalis extract	Crimea	Pinocamphone (78.2%), Isopinocamphone (3.2%), β-Pinene (4.2%), β-Phellandrene (1.3%), Myrtenal (1.4%)	Grebennikova et al. (2017)

Other Secondary Metabolites

Even though extensive groups of phenolic compounds (phenolic acids, flavonoids, glycosides) and various sesquiterpenes have been identified in the genus *Hyssopus*, recent studies show that the composition of hyssop metabolites is not limited to these classes. For example, A new macrocyclic spermidine alkaloid from the aerial part of *H. cuspidatus* Boriss. novel macrocyclic spermidine alkaloid, hyssopusizine **61** (Figure 5), was isolated from the aerial parts of *H. cuspidatus*, together with sixteen previously identified compounds (Aihaiti et al., 2023). However, the profile featured a variety of non-terpenoid hydrocarbons, including straight- and branched-chain alkanes and cycloalkanes such as octane, methylated octanes, nonane, various dimethylheptanes, and multiple isomers of cyclohexane and cyclopentane (Atazhanova et al., 2024).

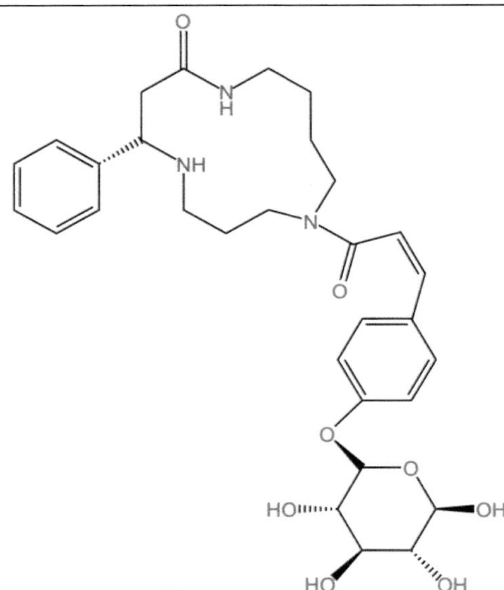

61

Fig. 5 New macrocyclic spermidine from *Hyssopus* spp.

Although extensive phytochemical studies of *Hyssopus* spp. have revealed a rich diversity of secondary metabolites, including terpenoids, flavonoids, phenolic acids, and triterpenes, there is currently no evidence in the scientific literature confirming the presence of sulfur-containing compounds such as glucosinolates, thionins, alliin, or glutathione in this genus. The absence of such compounds in published studies may suggest either their true absence in *Hyssopus* species or a lack of targeted investigations focused on sulfur-containing metabolites. Further specialized phytochemical analyses employing sulfur-specific detection methods may help clarify whether such compounds are present in trace amounts or under specific ecological or developmental conditions.

Biological Activities

Antioxidant Activities

The antioxidant potential of plant extracts is commonly assessed using assays such as 2,2-diphenyl-1-picrylhydrazyl (DPPH) and 2,2′-azino-bis(3-ethylbenzothiazoline-6-sulfonic acid) (ABTS). In the case of hyssop EO, a relatively low antioxidant capacity was recorded in the DPPH assay, with a value of 1.21 µmol Trolox equivalents (TE)/g. However, a markedly higher activity—approximately tenfold greater—was observed in the ABTS assay, reaching 10.6 µmol TE/g (Venditti et al., 2015). Another study confirmed that among the various radical scavenging assays applied to evaluate the antioxidant potential of *Hyssopus* extracts, the highest activity was observed against the ABTS assay. *H. cuspidatus* extract exhibited a notably low IC_{50} value of 0.0245 mg/mL, indicating strong free radical scavenging efficiency (Zhao et al., 2020). Among the 17 tested compounds by Aihaiti et al. (2023) p-hydroxycinnamic acid exhibited the strongest antioxidant activity against ABTS radicals with an IC_{50} of 15.6 µM. Compounds glochidioboside, dihydrodehydrodiconiferyl alcohol, (7R, 8S)-4,3′,9-trihydroxyl-3-methoxyl-7,8-dihydro-benzofuran-10-propyl-neo-lignan-9′-O-β-D-glucopyranoside, and 2-butanone showed moderate scavenging effects against both DPPH (IC_{50}: 49.2–85.8 µM) and ABTS radicals (IC_{50}: 27.2–45.5 µM). Additional studies conducted on hyssop extracts from Serbia and Turkey revealed IC_{50} values of 156.60 mg/mL and 16.37 mg/mL, respectively (Kizil et al., 2010). These discrepancies in antioxidant activity are attributed to genetic variability, environmental influences, and the geographical origin of the plant material (Mićović et al., 2021). The antioxidant evaluation revealed that among the tested compounds extracted from *H. cuspidatus* ethyl acetate fraction, rosmarinic acid (IC_{50} = 23.08 ± 2.14 µM) and Caffeic acid (IC_{50} = 24.68 ± 2.14 µM) exhibited the strongest free radical scavenging activities, followed by oresbiusin A (46.73 ± 3.23 µM). 4-O-Caffeoylquinic acid methyl ester showed weak activity (135.87 ± 7.69 µM), while compounds salvigenin and daucosterol demonstrated no antioxidant effect. The ethyl acetate fraction showed moderate activity (IC_{50} = 26.53 ± 2.32 mg/mL) when compared to the standard L-ascorbic acid (5.34 ± 0.42 mg/mL; 30.32 ± 2.38 µM). This suggests that specific isolated compounds possess significantly higher antioxidant potential than the crude fraction (Shomirzoeva et al., 2020).

The EO and methanolic extract of *H. officinalis* L. var. *angustifolius* have been investigated for *in vitro* antioxidant properties. According to Özer et al. (Ozer et al., 2006), the methanolic extract demonstrated moderate antioxidant activity, inhibiting 50% of DPPH radicals at a concentration of 117.0 µg/mL and exhibiting 40% inhibition in the linoleic acid oxidation system at 2 g/L. The methanolic extract from the aerial parts of *H. officinalis* var. *angustifolius* demonstrated moderate to strong activity in assays measuring reducing power, DPPH radical scavenging, and Fe^{2+} chelation. However, its performance in nitric oxide and hydrogen peroxide scavenging, as well as the ferric thiocyanate assay, was comparatively weak (Ebrahimzadeh et al., 2010). Authors investigated the antioxidant potential of supercritical fluid *H. officinalis* extract, which exhibited considerably lower antioxidant activity compared to other Lamiaceae species (Babovic et al., 2010). In a study conducted by Vlase et al. (2014), ethanol extracts of *H. officinalis* showed substantial antioxidant activity, as evidenced through multiple evaluation methods, including DPPH assay IC_{50} values 125.44 ± 4.70 μg/mL, TE 57.39 ± 13.68 μmol Trolox/mg plant material, and physiologically relevant enzymatic assay (HAPX) 16.17 ± 3.58 % (Vlase et al., 2014). All three *H. officinalis* genotypes showed strong antioxidant activity in the DPPH assay: blue-flowered (IC_{50} = 38.091 mg/mL), pink-flowered (IC_{50} = 34.172 mg/mL), and white-flowered (IC_{50} = 34.774 mg/mL), although their activity was lower compared to the standard antioxidant Trolox (IC_{50} = 3.15 mg/mL) (Benea et al., 2022). The antioxidant potential of *H. officinalis* extracts was assessed using the DPPH radical scavenging assay. Among the tested extracts, the deodorized aqueous extract showed the highest activity with an EC_{50} value of 0.54 mg/mL, followed by the deodorized methanol extract at 0.82 mg/mL and the deodorized ethyl acetate extract at 2.97 mg/mL (Dzamic et al., 2013). The antioxidant mechanism of hyssop EO is believed to involve its capacity to chelate transition metals, contributing to its overall bioactivity. Such antioxidative effects support its potential use in food preservation, particularly in products like ground beef, where it may inhibit the proliferation of pathogenic microorganisms (Michalczyk et al., 2012).

Antimicrobial Activities

The EO of *H. officinalis* contains a variety of bioactive compounds, including pinocarvone, β-pinene, terpinene-4-ol, isopinocamphone, pinocamphone, carvacrol, and *p*-cymene. These constituents are recognized for their antimicrobial properties and hold potential for the development of novel therapeutic agents aimed at preventing or treating infectious diseases in humans. Although *H. officinalis* exhibits relatively weak antimicrobial activity *in vitro* against both Gram-positive and Gram-negative bacteria, it also demonstrates notable antiviral, antifungal, insecticidal, and antioxidant properties.

The EO and methanolic extract of *H. officinalis* L. subsp. *angustifolius* has been investigated *in vitro*. The EO showed pronounced antimicrobial activity against eight bacterial strains, ten fungal species, and the yeast Candida albicans, with minimum inhibitory concentrations (MICs) ranging between 15.625– 250 μL/mL. *H. officinalis* extract exhibited moderate antibacterial activity against *S. aureus* (inhibition zone: 16 mm) and weak activity against *L. monocytogenes*, *E. coli*, and *S. typhimurium* (inhibition zones: 10–15 mm). It also showed antifungal activity against *C. albicans*, with inhibition zones ranging from 16 to 22 mm (Vlase et al., 2014). Nonetheless, the EO did not exhibit notable antioxidant potential (Ozer et al., 2006). Moreover, EO derived from *H. officinalis* collected in Southeastern Anatolia (Turkey) demonstrated strong antibacterial activity against *S. aureus*, *S. pyogenes*, *E. coli*, and *C. albicans* in a disc diffusion assay using 5 μL and 10 μL volumes. However, it showed no activity against *P. aeruginosa*

(Oalđe et al., 2021). The EO of *H. officinalis* was also evaluated for its antifungal properties, specifically against *M. perniciosa*, a major fungal pathogen in cultivated mushrooms in Serbia. The oil exhibited significant antifungal activity, with a minimum inhibitory concentration of 5 μL/mL and a minimum fungicidal quantity between 15–20 μL/mL (Glamoclija et al., 2005). Several compounds isolated from *H. cuspidatus* Boriss. showed moderate inhibitory activities against *S. aureus*, *C. albicans*, and *E. coli.* In this study, the discovery of anti-inflammatory, antibacterial, and antitumor components from *H. cuspidatus* could benefit further development and utilization of this plant (Aihaiti et al., 2023). Several compounds isolated from *H. cuspidatus* exhibited antimicrobial activity and inhibited the growth of *C. albicans*, *E. coli*, and *S. aureus* (Aihaiti et al., 2023). The antibacterial screening revealed that caffeic acid demonstrated the broadest spectrum of activity, exhibiting inhibition against *C. albicans*, *E. coli*, and *S. aureus*. The ethyl acetate fraction showed the strongest overall antibacterial activity against *C. albicans, E. coli, and S. aureus,* suggesting a synergistic effect of multiple constituents (Shomirzoeva et al., 2020). *H. seravschanicus* EO exhibited the strongest antimicrobial activity against *S. aureus* (MIC as low as 156 μg/mL) and *B. cereus*, while showing lower efficacy against *P. aeruginosa*, *E. coli*, and fungal strains such as *C. albicans* and *A. niger* (Sharopov et al., 2015). Authors tested 12 EOs from Mediterranean aromatic plants, including hyssop, against four phytopathogenic fungi: *B. cinerea*, *P. italicum*, *P. citrophthora*, and *R. stolonifer* (Camele et al., 2010). Based on *H. ambigus* and Marshall's thyme EOs composition active against *S. aureus*, *B. subtilis*, *E. coli*, and *C. albicans* fungi were developed and studied (Lakomkina, Atazhanova et al., 2023).

Anticancer Activities

H. officinalis has gained attention for its potential anticancer activity. Several *in vitro* studies have demonstrated that its EOs and extracts possess cytotoxic and apoptotic effects against various human cancer cell lines, including colon (HCT116), melanoma (A375), and breast cancer (MDA-MB-231) cells (Kumar et al., 2023). Using the MTT assay to assess cell viability, researchers observed concentration-dependent cytotoxicity, with IC_{50} values of 35.16 μg/mL (A375), 29.91 μg/mL (HCT116), and 62.66 μg/mL (MDA-MB-231) (Mićović et al., 2021). Further experiments confirmed similar activity on MRC-5, HeLa, and MDA-MB-231 cell lines, with *H. officinalis* subsp. *aristatus* showing slight cytotoxicity toward A549 lung cancer cells (Guerrini et al., 2021). In another study, the EO sample CIV17-EO reduced cell viability by 12% at a concentration of 200 μg/mL compared to the control ($p < 0.01$) (Sharifi-Rad et al., 2022), and (4*S*)-*p*-menth-l-ene-7,8-diol 8-*O*-*β*-D-glucopyranoside showed antitumor activities against MCF-8 and HT-29 cell lines with IC_{50} values of 93.39 ± 3.69 and 71.89 ± 2.94 μM, respectively. Oleanolic acid showed moderate antitumor activity against HT-29 cell lines with an IC_{50} value of 52.62 ± 1.63 μM (Aihaiti et al. 2023). The methanol extract of *H. officinalis* demonstrated significant anti-cancer effects against glioblastoma both *in vitro* and *in vivo*. It increased apoptosis, reduced tumor cell proliferation, decreased tumor volume, prolonged survival, upregulated p53 and p21 expression, and reduced tumor invasion and inflammation in treated rats (Khaksar et al., 2022). The methanol extract of *H. officinalis* flowers significantly reduced the viability of MCF-7 breast cancer cells at concentrations above 500 μg/mL, lowering cell survival from 100% to 60% after 72 hours. In contrast, the ethanol extract of hyssop leaves showed minimal cytotoxic effects. These results suggest that the methanol flower extract has promising anti-cancer potential and warrants further investigation for future therapeutic use (Fallahian et al., 2019).

Antidiabetic Activities

H. officinalis has demonstrated significant antidiabetic potential in both *in vitro* and *in vivo* experimental models (Miyazaki et al., 2003). Studies using aqueous methanolic extracts of dried leaves at doses of 100 mg/kg and 300 mg/kg body weight in mice showed notable α-glucosidase inhibitory activity (Miyazaki et al., 2003). Additional investigations suggest that *H. officinalis* may be effective in reducing blood sugar levels; however, caution is advised for individuals with liver conditions, as its use may not be appropriate in such cases (Muhammed Akram, 2013). Further evidence of its antidiabetic properties was obtained through a comprehensive study involving physicochemical and phytochemical characterization, quantification via high-performance thin-layer chromatography (HPTLC), and *in vitro* assays (Srivastava et al., 2018). The alpha-amylase inhibitory potential was evaluated using a modified starch-iodine method, yielding an IC_{50} value of 0.8366 mg/mL, indicating moderate inhibition. Additionally, the plant extract demonstrated strong antioxidant activity in the DPPH radical scavenging assay, with an IC_{50} value of 0.50 μg/mL. These findings support the hypothesis that *H. officinalis* may help in managing diabetes by inhibiting key carbohydrate-digesting enzymes such as α-amylase and α-glucosidase. Treatment with *H. officinalis* extract in diabetic rats led to a significant upregulation of ABCA1 and ABCG1 gene expression, which are involved in cholesterol efflux (Fallahian et al., 2019). Aqueous methanol extracts of *H. officinalis* leaves demonstrated alpha-glucosidase inhibitory activity. Two active compounds were isolated and structurally identified as (7S, 8S)-syringoylglycerol-9-O-(6'-O-cinnamoyl)-β-D-glucopyranoside and (7S, 8S)-syringoylglycerol 9-O-β-D-glucopyranoside. These findings highlight the potential of hyssop leaf constituents in managing blood glucose levels through inhibition of alpha-glucosidase (Matsuura et al., 2004).

Anti-inflammatory Activities

H. officinalis L., commonly known as hyssop, is a well-established medicinal plant with a long history of traditional use in treating inflammatory conditions. However, scientific validation is essential to confirm its pharmacological potential. One study evaluated the anti-inflammatory activity of an ethanolic extract prepared from the aerial parts of *H. officinalis* (CABI Digital Library, 2025). Using a laboratory rat model of acute inflammation, the extract's effects were compared with those of the non-steroidal anti-inflammatory drug diclofenac. The study included four groups of ten animals each, and results indicated that the extract significantly reduced phagocytic activity, nitrite/nitrate levels, and the Phagocytic Index. Overall, the extract exhibited notable anti-inflammatory efficacy, showing a comparable effect to diclofenac in suppressing acute inflammation (Ghasempour et al., 2022). In another investigation, *H. officinalis* derived from the Uygur region was evaluated for its immunomodulatory effects in a murine model of chronic asthma (Ma et al., 2014). The extract modulated inflammatory responses by decreasing the levels of key pro-inflammatory cytokines such as interleukin-4 (IL-4), IL-6, IL-17, and interferon-gamma (IFN-γ). Additionally, there was a notable downregulation of pro-apoptotic (GSK-3β) and inflammatory (NF-κB, TNF-α) markers, indicating the extract's potential anti-inflammatory and anti-apoptotic effects (Fallahian et al., 2019). Furthermore, it inhibited eosinophilic infiltration in lung tissues, reduced immunoglobulin E (IgE) levels, and helped regulate immune system responses. Some compounds from *H. cuspidatus*, 3*β*-hydroxy-7,8-dihydro-*β*-ionone, oleanolic acid, and acetylpleamolic acid showed strong anti-inflammatory activity against IL-6 and TNF-*α* in lipopolysaccharide (LPS) stimulated RAW 264.7 cells (Aihaiti et al. 2023). *H. cuspidatus* Boriss extract demonstrated significant anti-inflammatory effects in an ovalbumin (OVA)-induced asthma mouse model. Treatment with SXCF partly restored the disturbed metabolic network of lipid mediators (LMs) in lung tissue and serum in a dose-dependent manner, increasing anti-inflammatory metabolites while reducing pro-inflammatory lipids (Ling-Fei et al., 2023). This was accompanied by a downregulation of IgE and pro-inflammatory cytokines in serum and bronchoalveolar lavage fluid (BALF), confirming its anti-inflammatory potential. Hyssopuside, a phenolic glycoside from *H. cuspidatus*, demonstrated strong anti-inflammatory activity by inhibiting nitric oxide production and reducing pro-inflammatory cytokines (TNF-α, IL-6, IL-1β) in LPS-stimulated macrophages. HY suppressed NF-κB activation by preventing IκB degradation, leading to decreased expression of iNOS and COX-2, while having minimal effect on MAPK pathways (Liu et al., 2021). *In silico* analysis supported COX-2 as a key target. These findings suggest that HY exerts its anti-inflammatory effects primarily through the NF-κB signaling pathway, making it a promising candidate for treating inflammation-related diseases. These findings suggest that *Hyssopus* spp. may be beneficial in the management of inflammatory and immune-mediated respiratory conditions, such as asthma.

Anti-Ulcer Activity

H. officinalis has demonstrated notable anti-ulcer activity, likely attributed to its rich composition of polyphenolic compounds such as tannins, flavonoids, and phenolic acids. *In vitro* analysis revealed a high urease inhibition rate of 92.67%, suggesting the potential of the extract in managing urease-associated gastrointestinal disorders, including ulcers. However, its inhibitory effect on α-chymotrypsin was relatively low at 19.6%, indicating enzyme-specific activity (Saini and Sharma, 2012). In a separate *in vivo* study, albino rats were pretreated with ethanolic extracts of *H. officinalis* at doses of 100 mg/kg and 125 mg/kg body weight, administered one hour before ethanol-induced ulceration (Paun et al., 2014). The pre-treatment significantly reduced gastric damage by decreasing nitric oxide levels and reactive oxygen species (ROS) generation. It also enhanced gastric mucosal integrity and increased mucus secretion. These outcomes support the plant's traditional use in folk medicine for treating gastrointestinal ailments and highlight its combined antioxidant and anti-ulcer potential.

Other Reported Activities

H. officinalis and *H. cuspidatus* exhibit a broad range of biological effects beyond their anti-inflammatory and antioxidant properties. Extracts have shown insecticidal activity against *Spodoptera littoralis* larvae, anti-Leishmanial effects comparable to glucantime, and preliminary anti-HIV activity, including inhibition of reverse transcriptase and HIV antigen expression. Studies also report α-glucosidase inhibition, suggesting potential in regulating postprandial hyperglycemia, although evidence remains insufficient for clinical application in diabetes.

The insecticidal activity of *H. officinalis* was assessed using third-instar larvae of the Egyptian cottonworm (*Spodoptera littoralis*). Among eight tested medicinal plant extracts, the methanolic extract of *H. officinalis* at a concentration of 10% (w/v) significantly affected larval growth parameters, indicating notable toxicity. The extract exhibited a median lethal concentration (LC_{50}) of 1.7874, with a 95% confidence interval ranging from 1.66 to 1.82 (Pavela, 2005). Alcoholic extracts of *H. officinalis* demonstrated significant *in vitro* anti-Leishmanial activity against *Leishmania major* promastigotes, with effectiveness confirmed through dose-dependent inhibition and IC_{50} determination. The results, supported by MTT assay, showed that the plant extracts produced comparable inhibitory effects to the reference drug glucantime ($P < 0.05$), indicating their potential as natural anti-Leishmanial agents (Tabatabaie et al., 2014). Although preliminary studies suggest that extracts of *H. officinalis* may possess anti-HIV activity. *In vitro* investigations have demonstrated that hyssop leaf extracts can inhibit syncytia formation, HIV reverse transcriptase activity, and expression of HIV antigens (p17 and p24), without exhibiting toxicity toward uninfected Molt-3 cells (Kreis et al., 1990). Additionally, isolated compounds such as caffeic acid and a novel polysaccharide (MAR-10) have also demonstrated inhibitory effects on HIV-1 replication (Gollapudi et al., 1995). These findings inspired at least one patent application proposing the therapeutic use of *H. officinalis* extracts for AIDS treatment (Toppo, 1995). There is also preliminary evidence indicating that hyssop extracts may reduce intestinal carbohydrate absorption. In *ex vivo* and *in vivo* models, hyssop extract inhibited maltose- and sucrose-induced increases in glucose levels, implicating α-glucosidase inhibition; similar administration in carbohydrate-loaded mice significantly reduced postprandial hyperglycemia (Miyazaki et al., 2003). Despite such observations, the current evidence is not robust enough to substantiate claims that hyssop effectively treats diabetes or promotes weight loss. The muscle-relaxing (myorelaxant) properties of hyssop have also been documented, primarily attributed to the compound isopinocamphone, which has demonstrated the ability to relax the smooth muscles of the intestines and blood vessels (Sharifi-Rad et al., 2022c). The ethanolic extract of *H. officinalis* leaves showed significant gastroprotective effects in indomethacin-induced gastric ulcer in rats and increased antioxidant levels (SOD, GSH, catalase), enhanced the mucus barrier, and reduced ulcer index and lipid peroxidation (TBARS). Histological analysis confirmed that the ethanolic extract of *H. officinalis* leaves protected gastric tissue from ulceration, with effects comparable to the standard drug ranitidine (Tahir et al., 2019). The primary active components of hyssop oil, cis- and trans-3-pinanones, act as GABA receptor antagonists, inhibiting [^{3}H]EBOB binding in mouse brain membranes (IC_{50} = 35–64 μM). This antagonistic activity was associated with tonic-clonic seizures in mice, which could be reversed by diazepam. Metabolism studies revealed that cis-3-pinanone is metabolized faster than the trans isomer, primarily forming 2-hydroxy-cis-3-pinanone, a metabolite with reduced neurotoxicity and receptor binding inhibition, suggesting metabolic detoxification as a safety mechanism (Höld et al., 2002). Aqueous extract of *H. officinalis* at a dose of 100 mg/kg significantly delayed the onset of tonic-clonic seizures in PTZ-induced seizure models in rats, indicating potential anticonvulsant activity (Gholami et al., 2020). This effect was accompanied by a notable increase in hippocampal iNOS gene expression, suggesting that modulation of the nitric oxide pathway may play a role in the extract's seizure-modifying effects. The results of the animal antioxidant property experiment showed that the ethanol extract of *H. cuspidatus* can significantly enhance the activities of SOD, CAT, and GSH-px in mouse serum and reduce the content of MDA. In addition, the antioxygenation property of the ethanol extract of H. cuspidatus may be related to TPC and TFC (Zhao et al., 2020). *H. officinalis* demonstrated allelopathic activity through its aqueous extracts, with both stimulating and inhibitory effects on the root growth of watercress depending on plant part, age, and extract concentration. The strongest inhibitory effect was observed from flower extracts of second-year plants and from soil solutions in the third year of cultivation, particularly at a 1:10 concentration (Tkachova et al., 2022).

These diverse pharmacological and ecological activities underscore hyssop's potential for multi purpose applications but also highlight the importance of dose and formulation in determining its safety and efficacy.

Toxicity and Safety Profiles

Clinical research on *H. officinalis* has increased in the 21st century; however, most studies have been limited to animal models, highlighting the need for more robust human trials to validate these findings. A 2014 study suggested that hyssop may possess anticancer properties, and subsequent research in 2016 supported these effects in cervical cancer cells, though conducted in a lamb model. Additional animal studies have shown that hyssop may be beneficial in the treatment of gastric and peptic ulcers, and a 2016 study in rats indicated antiviral potential, particularly against herpes simplex virus, by delaying symptom onset. Hyssop is also recognized for its high polyphenol content, contributing to its strong antioxidant capacity and suggesting possible broader health benefits. Furthermore, animal models have demonstrated anticonvulsant activity, although caution is warranted, as hyssop has also been associated with seizure induction, making its neurological effects population- and dose-dependent. Interestingly, hyssop has also been explored as a natural curing agent for pork, offering a potential alternative to synthetic nitrates. Hyssop, widely used for treating respiratory and gastrointestinal conditions, is generally considered safe when used in extract or capsule form, with only mild side effects such as gastrointestinal discomfort or anxiety. While hyssop oil contains mildly toxic compounds like pinocamphone and may exacerbate seizure disorders—especially in children—there is no clinical evidence linking hyssop to liver toxicity. Despite its widespread use in traditional medicine and modern herbal supplements, no cases of hepatotoxicity have been reported, supporting its overall favorable safety profile when used appropriately ("Hyssop" 2012).

H. cuspidatus Boriss extract treatment significantly reduced cough frequency and decreased inflammatory cell infiltration around the bronchus, as observed in histological lung sections stained with H&E. Behavioral improvements in treated mice further supported the therapeutic benefit of *H. cuspidatus* Boriss extract in alleviating asthma symptoms (Ling-Fei et al., 2023). In the conducted study, *H. cuspidatus* Boriss. EO was administered orally to mice at doses up to 0.4 ml/kg for 7 days without any reported adverse effects on body weight or general behavior, suggesting good tolerability. Moreover, the anti-inflammatory effects observed at this dose indicate that *H. cuspidatus* Boriss. EO exerts pharmacological activity without evident toxicity. However, detailed acute or chronic toxicity evaluations, including organ histopathology and biochemical markers, have yet to be reported and are necessary to comprehensively assess the safety profile of *H. cuspidatus* Boriss. EO (Zhang et al., 2019).

Harvesting, Post-harvesting/ Preservation, Drying, and Processing Methodologies

The cultivation and post-harvest processing of *H. officinalis* significantly influences the yield and chemical composition of its EO. Optimal harvesting time is typically during the full flowering stage, when the concentration of major bioactive compounds—such as pinocamphone, isopinocamphone, and other monoterpenes—is highest (Kizil et al., 2016; Yousefzadeh and Naghdi Badi, 2017). Studies conducted under various climatic conditions, including semi-arid and Mediterranean zones, indicate that harvesting during the pre-flowering or post-flowering phases results in reduced EO yield and less favorable phytochemical profiles (Can et al., 2022; Yousefzadeh and Naghdi Badi, 2017).

For medicinal and EO production, only the aerial parts (leaves and flowering tops) are collected, as these contain the highest concentrations of volatile oils and phenolic compounds (Kizil et al., 2016). Manual or mechanical harvesting should ideally occur in the morning, under dry weather conditions, to preserve the integrity of the glandular trichomes, which contain EOs (Moro et al., 2011).

After harvesting, hyssop should be dried promptly under shade or using low-temperature drying (35–45 °C) to preserve its volatile compounds and prevent microbial contamination. Rapid drying at higher temperatures may result in the loss of thermo-labile constituents like thujone and pinocamphone (Kizil et al., 2016). Air-drying in well-ventilated conditions is the most traditional and widely accepted method, though industrial-scale drying may utilize forced-air systems to reduce moisture content to below 12%, ensuring safe storage and preventing mold growth (Saebi et al., 2021).

Processed hyssop raw material is usually stored in airtight, light-resistant containers under cool conditions to maintain oil stability and prevent oxidative degradation. EO extraction is commonly performed via steam distillation, and the yield varies depending on geographic origin, plant age, and harvest timing (Saebi et al., 2021).

Post-harvest preservation techniques, including vacuum packaging and storage in inert atmospheres, are employed to prevent oxidation of flavonoids and polyphenolic constituents. Chemical analysis of hyssop EO confirms that processing conditions directly influence the ratio of bioactive constituents and, thus, the herb's medicinal value.

Potential Food, Pharmaceutical, and Value-added Products

In modern applications, hyssop is valued for its potential health benefits, flavor enhancement in foods, and use in various value-added products. Hyssop has been studied for its potential therapeutic effects. Research indicates that its essential oil contains compounds like pinocamphone, isopinocamphone, and camphor, which may contribute to its pharmacological activities, including antioxidant, anti-inflammatory, and anti-asthmatic effects. However, it's important to note that the essential oil should be used cautiously, as it can cause convulsions and is contraindicated in children under seven years old, during pregnancy, and while breastfeeding. The leaves and flowers of hyssop are edible and have an intensely aromatic flavor, somewhat like a cross between sage and mint. They can be used raw in salads or as a flavoring in soups and stews. Hyssop is also utilized in the production of herbal teas and as a flavoring agent in certain liqueurs. Hyssop's essential oil is employed in perfumery and cosmetics for its fragrance. The oil's volatile compounds contribute to its aromatic profile, making it a valuable ingredient in various scented products. Beyond its basic uses, hyssop is incorporated into value-added products such as herbal teas, essential oils, and cosmetic formulations.

The versatility of hyssop allows for its inclusion in a range of products aimed at promoting health and well-being.

Conclusion

The multifaceted uses of hyssop species from their complex chemical profile and demonstrated biological activities. Their applications range from traditional remedies for respiratory and gastrointestinal ailments to promising candidates for natural antioxidants, antimicrobial agents, and bioinsecticides. Hyssop's essential oils and extracts have significant potential in pharmaceutical formulations, functional foods, and cosmetic products, offering a natural alternative to synthetic compounds. Despite their widespread traditional use and encouraging experimental results, standardized cultivation, harvesting, and processing protocols, alongside rigorous clinical validation, remain critical for ensuring efficacy, safety, and quality. Overall, hyssop represents a valuable botanical resource with considerable potential for expanding natural product development across multiple industries.

References

Ablizl, P., Cong, Y., Musa, M., Zhu, Y. and Kasimu, R. 2009. Chemical composition of the essential oil of *Hyssopus cuspidatus* from Xinjiang, China. *Chem. Nat. Compd.*, *45* (3): 445. https://doi.org/10.1007/s10600-009-9309-y.

Aćimović, Milica, Ana Varga, Mirjana Cvetković, Pezo Lato, Biljana Lončar, Maja Ignjatov, and Tijana Zeremski. 2021. "Chemical Composition of Hyssop cv. 'Domaci Ijubicasti' Essential Oil and Its Antimicrobial Activity." *Ratarstvo i Povrtarstvo*, 58 (1): 23–30. https://doi.org/10.5937/raptov58-31751.

Aihaiti, K., Li, J., Xu, N.N., Tang, D. and Aisa, H.A. 2023. Monoterpenoid Derivatives from *Hyssopus cuspidatus* Boriss. and Their Bioactivities. *Fitoterapia*, *165*: 105432. https://doi.org/10.1016/j.fitote.2023.105432.

Aihaiti, K., Li, J., Yaermaimaiti, S., Yin, A. and Aisa, H.A. 2023. A new macrocyclic spermidine alkaloid from the aerial part of *Hyssopus cuspidatus* Boriss. *Nat. Prod. Res.*, *37* (13): 2113–2119. https://doi.org/10.1080/14786419.2022.2027935.

Almasoudi, H.H., Jan, M.S., Nahari, M.H., Alhazmi, A.Y.M., Binshaya, A.S., Abdulaziz, O., Mahnashi, M.H., Ibrar, M., Zafar, R. and Sadiq, A. 2024. Phenolic phytochemistry, *in vitro*, *in silico*, *in vivo*, and mechanistic anti-inflammatory and antioxidant evaluations of Habenaria digitata. *Front. Pharmacol.*, *15*: 1346526. https://doi.org/10.3389/fphar.2024.1346526.

Amini, F., Jaladat, A.M., Daneshfard, B., et al. 2022. "Effect of Hyssop Syrup on Syndrome-Differentiated Mild to Moderate Asthma: A Pilot Clinical Trial." Planta Medica 88 (December): P-361B. https://doi.org/10.1055/s-0042-1759332.

Atazhanova, G., Ishmuratova, M., Levaya, Y., Smagulov, M. and Lakomkina, Y. 2024. The genus Hyssopus: Traditional use, phytochemicals and pharmacological properties. *Plants*, *13* (12): 1683. https://doi.org/10.3390/plants13121683.

Babovic, A., Djilas, S., Jadranin, M., Vajs, V., Ivanovic, J., Petrovic, S. and Zizovic, I. 2010. Supercritical carbon dioxide extraction of antioxidant fractions from selected Lamiaceae herbs and their antioxidant capacity. *Innov. Food Sci. Emerg. Technol.*, *11* (1): 98–107. https://doi.org/10.1016/j.ifset.2009.08.013.

Benea, A., Ciobanu, C., Ciobanu, N., Pompus, I. and Cojocaru-Toma, M. 2022. Polyphenolic content and antioxidant activity of *Hyssopus officinalis* L. from the Republic of Moldova. *Mold. Med. J.*, *65* (2): 41–46. https://doi.org/10.52418/moldovan-med-j.65-2.22.06.

Benedec, D., Oniga, I., Kozma-Imre, A., Hanganu, D., Ţarmure, V. and Bodoki, E. 2017. Determination of rosmarinic acid by HPTLC-image analysis in medicinal teas and their biological properties. *Farmacia*, *65* (4): 605–609.

Bernotienė, G. and Butkienė, R. 2010. Essential oils of Hyssopus officinalis L. cultivated in East Lithuania. *Chemija*, *21* (2–4): 135–138.

CABI digital library. Accessed September 18, 2025. https://www.cabidigitallibrary.org/doi/full/10.5555/20173127761.

Camele, I., De Feo, V., Altieri, L., Mancini, E., De Martino, L. and Rana, G.L. 2010. An attempt of postharvest orange fruit rot control using essential oils from Mediterranean plants. *J. Med. Food*, *13* (6): 1515–1523. https://doi.org/10.1089/jmf.2009.0285.

Can, M., Katar, N. and Katar, D. 2022. The influence of harvesting applied at different stages of plant development on *Hyssopus officinalis* L. production. *Acta Sci. Pol. Hortorum Cultus*, *21* (3): 113–120. https://doi.org/10.24326/asphc.2022.3.10.

Carević, T., Stojković, D. and Ivanov, M. 2023. Plant flavonoids as reservoirs of therapeutics against microbial virulence traits: A comprehensive review update. *Curr. Pharm. Des.*, *29* (12): 914–927. https://doi.org/10.2174/1381612829666230413085029.

Chagas, M.S.S., Behrens, M.D., Moragas-Tellis, C.J., Penedo, G.X.M., Silva, A.R. and Gonçalves-de-Albuquerque, C.F. 2022. Flavonols and flavones as potential anti-inflammatory, antioxidant, and antibacterial compounds. *Oxid. Med. Cell. Longev.*, *2022*: 9966750. https://doi.org/10.1155/2022/9966750.

Daneshfard, B., Amini, F., Jaladat, A. M., Momeni, B., Abdolahinia, A., Hosseinkhani, A. and Hosseini, L. 2024. Effect of Hyssop (*Hyssopus officinalis*) syrup on mild to moderate asthma: A randomized double-blind placebo-controlled trial. *Tanaffos*, *23* (2): 146–155.

Džamić, A. M., Soković, M. D., Novaković, M., Jadranin, M., Ristić, M. S., Tešević, V. and Marin, P. D. 2013. Composition, antifungal, and antioxidant properties of *Hyssopus officinalis* L. subsp. pilifer (Pant.) Murb. essential oil and deodorized extracts. *Ind. Crops Prod.*, *51*: 401–407. https://doi.org/10.1016/j.indcrop.2013.09.038.

Ebrahimzadeh, M.A., Nabavi, S.M., Nabavi, S.F., Bahramian, F. and Bekhradnia, A.R. 2010. Antioxidant and free radical scavenging activity of *H. officinalis* L. var. angustifolius, V. odorata, B. hyrcana, and C. speciosum. *Pak. J. Pharm. Sci.*, *23* (1): 29–34.

Fallahian, F., Mahdavi, N. and Haeri, M. 2019. The cytotoxic effect of Hyssop extract on breast cancer cell line. *Qom Univ. Med. Sci. J.*, *13* (7): 22–28. https://doi.org/10.29252/qums.13.7.22.

Fathiazad, F. 2011. A review on *Hyssopus officinalis* L.: Composition and biological activities. *Afr. J. Microbiol. Res.*, *5* (17). https://doi.org/10.5897/AJPP11.527.

Fathiazad, F., Mazandarani, M. and Hamedeyazdan, S. 2011. Phytochemical analysis and antioxidant activity of *Hyssopus officinalis* L. from Iran. *Adv. Pharm. Bull.*, *1* (2): 63–67. https://doi.org/10.5681/apb.2011.009.

Fleisher, A. and Fleisher, Z. 1988. Identification of biblical hyssop and origin of the traditional use of oregano-group herbs in the Mediterranean region. *Econ. Bot.*, *42*: 232–241. https://doi.org/10.1007/BF02858924.

Furukawa, M., Makino, M., Ohkoshi, E., Uchiyama, T. and Fujimoto, Y. 2011. Terpenoids and phenethyl glucosides from *Hyssopus cuspidatus* (Labiatae). *Phytochemistry*, *72* (17): 2244–2252. https://doi.org/10.1016/j.phytochem.2011.07.008.

Garg, S., Naqvi, A. A., Singh, A., Ram, G. and Kumar, S. 1999. Composition of essential oil from an annual crop of Hyssopus officinalis grown in Indian plains. *Flavour Fragr. J.*, *14* (3): 170–172. https://doi.org/10.1002/(SICI)1099-1026(199905/06)14:3.

Ghasempour, M., Hosseini, M., Soltani-Zangbar, M. S., Motavalli, R., Aghebati-Maleki, L., Dolati, S., Mehdizadeh, A., Yousefi, M. and Ahmadian Heris, J. 2022. The impact of Hyssop (Hyssopus officinalis) extract on activation of endosomal toll- like receptors and their downstream signaling pathways. *BMC Res. Notes*, *15* (1): 366. https://doi.org/10.1186/s13104-022-06253-3.

Gholami, M., Jafari, F., Baradaran, Z., Amri, J., Azhdari-Zarmehri, H. and Sadegh, M. 2020. Effects of aqueous extract of Hyssopus officinalis on seizures induced by pentylenetetrazole and hippocampus mRNA level of iNOS in rats. *Avicenna J. Phytomed.*, *10* (3): 213–221.

Glamoclija, J., Soković, M., Vukojevic, J., Milenković, I., Brkic, D. and Van Griensven, L. 2005. Antifungal activity of essential oil *Hyssopus officinalis* L. against micopathogen Mycogone perniciosa (Mang). *Zbornik Matice Srpske Za Prirodne Nauke*, *2005* (January). https://doi.org/10.2298/ZMSPN0519123G.

Gleason, H.A. 1991. *Manual of Vascular Plants of Northeastern United States and Adjacent Canada*. Bronx, N.Y.: New York Botanical Garden. http://archive.org/details/manualofvascular0000glea.

Gollapudi, S., Sharma, H. A., Aggarwal, S., Byers, L. D., Ensley, H.E. and Gupta, S. 1995. Isolation of a previously unidentified polysaccharide (MAR-10) from *Hyssopus officinalis* that exhibits strong activity against human immunodeficiency virus type 1. *Biochem. Biophys. Res. Commun.*, *210* (1): 145–151. https://doi.org/10.1006/bbrc.1995.1639.

Gonceariuc, M. and Balmuù, Z. 2013. Diversity of the essential oil content and chemical composition. Muzeul Olteniei Craiova. Oltenia. *Studii și comunicări. Științele Naturii*, *29* (1): 71–77.

Govaerts, R. 2023. *World Checklist of Vascular Plants* (*WCVP*) – *Version 12*. Board of Trustees of the Royal Botanic Gardens, Kew. https://doi.org/10.34885/JDH2-DR22.

Grebennikova, O., Paliy, A.E., Khlypenko, L.A. and Rabotyagov, V.D. 2017. Biologically active substances of Hyssopus officinalis L. *Orbital*, (1): 21–28.

Guerrini, A., Sacchetti, G., Echeverria Guevara, M.P., Paganetto, G., Grandini, A., Maresca, I., Menghini, L., Di Martino, L., Marengo, A. and Tacchini, M. 2021. Wild Italian Hyssopus officinalis subsp. aristatus (Godr.) Nyman: From morphological and phytochemical evidences to biological activities. *Plants*, *10* (4): 631. https://doi.org/10.3390/plants10040631.

Gupta, T., Kataria, R. and Sardana, S. 2022. A comprehensive review on current perspectives of flavonoids as antimicrobial agent. *Curr. Top. Med. Chem.*, *22* (6): 425–434. https://doi.org/10.2174/1568026622666220117104709.

Hartwell, J.L. 1971. Plants used against cancer: A survey. *Lloydia*, *34* (4): 386–425.

Hatipoğlu, G., Sökmen, M., Bektaş, E., Daferera, D., Sökmen, A., Demir, E. and Şahin, H. 2013. Automated and standard extraction of antioxidant phenolic compounds of Hyssopus officinalis L. ssp. angustifolius. *Ind. Crops Prod.*, *43*: 427–433. https://doi.org/10.1016/j.indcrop.2012.07.028.

Höld, K.M., Sirisoma, N.S., Sparks, S.E. and Casida, J.E. 2002. Metabolism and mode of action of cis- and trans-3-pinanones (the active ingredients of hyssop oil). *Xenobiotica*, *32* (4): 251–265. https://doi.org/10.1080/00498250110095745.

Huyut, Z., Beydemir, Ş. and Gülçin, İ. 2017. Antioxidant and antiradical properties of selected flavonoids and phenolic compounds. *Biochem. Res. Int.*, *2017*: 7616791. https://doi.org/10.1155/2017/7616791.

Hyssop. 2012. In: *LiverTox: Clinical and Research Information on Drug-Induced Liver Injury*. National Institute of Diabetes and Digestive and Kidney Diseases. http://www.ncbi.nlm.nih.gov/books/NBK548391/.

Hyssopus cuspidatus Boriss. 2002. In: *Flora of China* @ eFloras.org. Missouri Botanical Garden, St. Louis, MO & Harvard University Herbaria, Cambridge, MA. Accessed February 2018.

Hyssopus latilabiatus. In: *Flora of China* @ eFloras.org. Accessed September 4, 2025.

Hyssopus officinalis Linnaeus. 2005. In: *Flora of China* @ eFloras.org. Missouri Botanical Garden, St. Louis, MO & Harvard University Herbaria, Cambridge, MA. Accessed February 2018.

Jankovský, M. and Landa, T. 2002. Genus Hyssopus L.: Recent knowledge: A review. *Hortic. Sci.*, *29* (3): 119–123. https://doi.org/10.17221/4474-HORTSCI.

Jarvis, C.E. 2007. *Order Out of Chaos: Linnaean Plant Names and Their Types*. Linnean Society of London.

Judžentienė, A. 2016. Hyssop (Hyssopus officinalis L.) oils. In: V. R. Preedy (Ed.), *Essential Oils in Food Preservation, Flavor, and Safety*. Academic Press. https://doi.org/10.1016/B978-0-12-416641-7.00053-5.

Khaksar, S., Kiarostami, K. and Alinaghi, S. 2022. The effects of methanol extracts of *Hyssopus officinalis* on model of induced glioblastoma multiforme (GBM) in rats. *J. Mol. Neurosci., 72* (9): 2045–2066. https://doi.org/10.1007/s12031-022-02058-y.

Kiprovski, B. and Rat, M. 2019. Essential oil analysis of different hyssop genotypes from IFVCNS medicinal plant collection garden. *Ann. Agron., 43*(1): 38–45.

Kizil, S., Haşimi, N., Tolan, V., Kilinç, E. and Karataş, H. 2010. Chemical composition, antimicrobial, and antioxidant activities of hyssop (Hyssopus officinalis L.) essential oil. *Not. Bot. Horti Agrobo. Cluj-Napoca, 38* (3): 99–103. https://doi.org/10.15835/nbha3834788.

Kizil, S., Guler, V., Kirici, S. and Turk, M. 2016. Some agronomic characteristics and essential oil composition of hyssop (*Hyssopus officinalis* L.) under cultivation conditions. *Acta Sci. Pol. Hortorum Cultus, 15* (6): 193–207.

Kowalchik, C. and Hylton, W.H. 1987. *Rodale's Illustrated Encyclopedia of Gardening*. Emmaus: Rodale Press.

Kreis, W., Kaplan, M.H., Freeman, J., Sun, D.K. and Sarin, P.S. 1990. Inhibition of HIV replication by *Hyssopus officinalis* extracts. *Antiviral Res., 14*: 323–333.

Kumar, V., Kaur, N., Kaur, A. and Wadhwa, P. 2023. Phytochemistry and pharmacology of Indian traditional plant hyssop (*Hyssopus officinalis* L.): A review. *Nat. Prod. J., 13* (4): 11–41. https://doi.org/10.2174/2210315512666220811153919.

Kürkçüoğlu, M. 2016. Composition of the essential oil of the *Hyssopus officinalis* L. subsp. angustifolius (Bieb.) Arcangeli. *Nveo - Nat. Volatiles Essent. Oils J.*: 15–19.

Lakomkina, E.V., Akhmetova, S.B., Atazhanova, G.A. and Ishmuratova, M. Yu. 2023. Component composition and antimicrobial activity of essential oil samples of Hyssopus ambiguus (Trautv.) Iljin collected in the Karaganda region. *Pharm. Kazakhst., 5*: 374–379. https://doi.org/10.53511/pharmkaz.2023.46.10.049.

Lakomkina, E.V., Atazhanova, G.A., Akhmetova, S.B. and Zilfikarov, I.N. 2023. Development of composition and technology for obtaining antimicrobial composition based on mono- and sesquiterpenoids. *Pharm. Pharmacol., 11* (2): 114–126. https://doi.org/10.19163/2307-9266-2023-11-2-114-126.

Lakomkina, Y. V., Ishmuratova, M.Y. and Atazhanova, G.A. 2022. Morphometric study of Hyssopus ambiguus growing in the territory of Central Kazakhstan. *OnLine J. Biol. Sci., 22* (1): 112–117. https://doi.org/10.3844/ojbsci.2022.112.117.

Letessier, M.P., Svoboda, K.P. and Walters, D.R. 2001. Antifungal activity of the essential oil of hyssop (*Hyssopus officinalis*). *J. Phytopathol., 149* (12): 673–678. https://doi.org/10.1046/j.1439-0434.2001.00692.x.

Lewis, W.H., Lewis, W.H. and Elvin-Lewis, M.P.F. 1977. *Medical Botany: Plants Affecting Man's Health*. Wiley.

Ling-Fei, K., Xiao-Juan, R., Pan, Y., Tuo, Q., Xiao-Hui, Z., Yu-Tong, K., Bo, C., Wen-Ling, S., Tian-Le, G. and Cai, T. 2023. The influence of Hyssopus cuspidatus Boriss extract on lipid mediators metabolism network in asthmatic mice. *Front. Pharmacol., 14*: 1066643. https://doi.org/10.3389/fphar.2023.1066643.

Liu, X., Su, J., Wang, G., Zheng, L., Wang, G., Sun, Y., Bao, Y., Wang, S. and Huang, Y. 2021. Discovery of phenolic glycoside from Hyssopus cuspidatus attenuates LPS-induced inflammatory responses by inhibition of iNOS and COX-2 expression through suppression of NF-κB activation. *Int. J. Mol. Sci., 22* (22): 12128. https://doi.org/10.3390/ijms222212128.

Ma, X., Ma, X., Ma, Z., Wang, J., Sun, Z., Yu, W., Li, F. and Ding, J. 2014. Effect of Hyssopus officinalis L. on inhibiting airway inflammation and immune regulation in a chronic asthmatic mouse model. *Exp. Ther. Med.*, *8* (5): 1371–1374. https://doi.org/10.3892/etm.2014.1978.

Matsuura, H., Miyazaki, H., Asakawa, C., Amano, M., Yoshihara, T. and Mizutani, J. 2004. Isolation of alpha-glucosidase inhibitors from hyssop (*Hyssopus officinalis*). *Phytochemistry*, *65* (1): 91–97. https://doi.org/10.1016/j.phytochem.2003.10.009.

Michalak, M. 2023. Plant extracts as skin care and therapeutic agents. *Int. J. Mol. Sci.*, *24* (20): 15444. https://doi.org/10.3390/ijms242015444.

Michalczyk, M., Macura, R., Tesarowicz, I. and Banaś, J. 2012. Effect of adding essential oils of coriander (Coriandrum sativum L.) and hyssop (*Hyssopus officinalis* L.) on the shelf life of ground beef. *Meat Sci.*, *90* (3): 842–850. https://doi.org/10.1016/j.meatsci.2011.11.026.

Mićović, T., Topalović, D., Živković, L., Spremo-Potparević, B., Jakovljević, V., Matić, S., Popović, S., Baskić, D., Stešević, D., Samardžić, S., Stojanović, D. and Maksimović, Z. 2021. Antioxidant, antigenotoxic, and cytotoxic activity of essential oils and methanol extracts of *Hyssopus officinalis* L. subsp. aristatus (Godr.) Nyman (Lamiaceae). *Plants*, *10* (4): 711. https://doi.org/10.3390/plants10040711.

Mićović, T., Ušjak, D., Milenković, M., Samardžić, S. and Maksimović, Z. 2023. Antimicrobial activity of *Hyssopus officinalis* L. subsp. aristatus (Godr.) Nyman (Lamiaceae) essential oils from Montenegro and Serbia. *Lekovite Sirovine*, *43* (1). https://doi.org/10.61652/leksir2343e173M.

Mićović, T., Katanić Stanković, J.S., Bauer, R., Nöst, X., Marković, Z., Milenković, D., Jakovljević, V., Tomović, M., Bradić, J., Stešević, D., Stojanović, D. and Maksimović, Z. 2022. *In vitro*, *in vivo*, and *in silico* evaluation of the anti-inflammatory potential of *Hyssopus officinalis* L. subsp. aristatus (Godr.) Nyman (Lamiaceae). *J. Ethnopharmacol.*, *293*: 115201. https://doi.org/10.1016/j.jep.2022.115201.

Milica, A., Varga, A., Cvetković, M., Lato, P., Lončar, B., Ignjatov, M. and T. Zeremski. 2021. Chemical composition of *Hyssop* cv. 'Domaci Ijubicasti' essential oil and its antimicrobial activity. *Ratar. Povrt.*, *58*(1): 23–30. https://doi.org/10.5937/raptov58-31751.

Millet, Y., Jouglard, J., Steinmetz, M.D., Tognetti, P., Joanny, P. and Arditti, J. 1981. Toxicity of some essential plant oils. Clinical and experimental study. *Clinical Toxicology*, *18* (12): 1485–1498. https://doi.org/10.3109/15563658108990357.

Miyazaki, H., Matsuura, H., Yanagiya, C., Mizutani, J., Tsuji, M. and Ishihara, C. 2003. Inhibitory effects of hyssop (*Hyssopus officinalis*) extracts on intestinal alpha-glucosidase activity and postprandial hyperglycemia. *J. Nutr. Sci. Vitaminol.*, *49* (5): 346–349. https://doi.org/10.3177/jnsv.49.346.

Moro, A., Zalacain, A., Carmona, M. and Hurtado de Mendoza, J. 2011. Effects of agronomic practices on volatile composition of *Hyssopus officinalis* L. essential oils. *Molecules*, *16* (5): 4131–4139. https://doi.org/10.3390/molecules16054131.

Moulodi, F., Khezerlou, A., Zolfaghari, H., Mohamadzadeh, A. and Alimoradi, F. 2018. Chemical composition and antioxidant and antimicrobial properties of the essential oil of *Hyssopus officinalis* L. *J. Kermanshah Univ. Med. Sci.*, *22* (4). https://brieflands.com/articles/jkums-85256#abstract.

Muhammad, Akram., 2013. Diabetes Mellitus Type II: Treatment strategies and options: A review. J. *Diabetes Metab.*, *4* (9): 1–9. https://doi.org/10.4172/2155-6156.1000304.

Nistor, A.L., Pop, C.R., Mitrea, L., Cătunescu, G.M., Vârban, R., Lipşa, F.D., Rusu, C.C. and Rotar, A.M. 2025. Antimicrobial potential of *Hyssopus officinalis* L. and Agastache foeniculum (Pursh) Kuntze essential oils for food applications: A review of their chemical compositions and antimicrobial efficacy. *Applied Sciences*, *15* (9): 4772. https://doi.org/10.3390/app15094772.

Oalđe, M., Kolarević, S., Živković, J., Alimpić Aradski, A., Jovanović Marić, J., Kračun Kolarević, M., Đorđević, J., Marin, P.D., Šavikin, K., Vuković-Gačić, B. and Duletić-Laušević, S. 2021. A comprehensive assessment of the chemical composition, antioxidant, genoprotective, and antigenotoxic activities of Lamiaceae species using different experimental models *in vitro*. *Food and Function*, *12* (7): 3233–3245. https://doi.org/10.1039/d1fo00447f.

Ozer, Hakan, Fikrettin Sahin, Hamdullah Kilic and Medine Güllüce. 2004. Essential Oil Composition of *Hyssopus Officinalis* L. Subsp. Angustifolius (Bieb.) Arcangeli from Turkey. *Flavour and Fragrance Journal*, *20*: 42–44. https://doi.org/10.1002/ffj.1421.

Özlem, Bariş, Hakan Ozer, Atalay Sokmen, Ahmet Adiguzel, Hamdullah Kilic, Fikrettin Sahin and Gulluce. 2006. *In Vitro* Antimicrobial and Antioxidant Activities of the Essential Oils and Methanol Extracts of *Hyssopus Officinalis* L. Ssp. Angustifolius. *Italian Journal of Food Science*, *18*: 73–84.

Paun, Gabriela, Simona Carmen Litescu, Elena Neagu, Andreia Tache and Gabriel Lucian Radu. 2014. Evaluation of Geranium Spp., Helleborus Spp., and Hyssopus Spp. Polyphenolic Extracts Inhibitory Activity against Urease and α-Chymotrypsin. *Journal of Enzyme Inhibition and Medicinal Chemistry*, *29* (1): 28–34. https://doi.org/10.3109/14756366.2012.749399.

Pavela, Roman. 2005. Insecticidal Activity of Some Essential Oils against Larvae of Spodoptera Littoralis. *Fitoterapia*, *76* (7–8): 691–696. https://doi.org/10.1016/j.fitote.2005.06.001.

Rastkari, Noushin, Nasrin Samadi, Reza Ahmadkhaniha, Rozita Alemi and Leila Afarin. 2007. Chemical Composition and Biological Activities of *Hyssopus Officinalis* Cultivated in Iran. *NPAIJ*, *3*(2): 87–91.

Reader's Digest Association. 1979. *The "Reader's Digest" Complete Do-it-yourself Manual*. Reader's Digest Association.

Rezaei Savadkouhi, N., Ariaii, P. and Charmchian Langerodi, M. 2020. The Effect of Encapsulated Plant Extract of Hyssop (*Hyssopus Officinalis* L.) in Biopolymer Nanoemulsions of Lepidium perfoliatum and Orchis mascula on Controlling Oxidative Stability of Soybean Oil. *Food Science and Nutrition*, *8* (2): 1264–1271. https://doi.org/10.1002/fsn3.1415.

Saebi, Ali, Saeid Minaei, Ali Reza Mahdavian and Mohammad-Taghi Ebadi. 2021. Precision Harvesting of Medicinal Plants: Elements and Ash Content of Hyssop (*Hyssopus Officinalis* L.) as Affected by Harvest Height. *Biological Trace Element Research*, *199* (2): 753–762. https://doi.org/10.1007/s12011-020-02171-2.

Said-Al Ahl, H. 2015. Essential Oil Composition of *Hyssopus Officinalis* L. Cultivated in Egypt. *International Journal of Plant Research*, *1*(2): 49–53.

Saini, A. and Sharma, R. 2012. To Explore the Ulceroprotective and Antioxidant Potential of *Hyssopus Officinalis* in Ethanol-Induced Gastric Ulcers in Rats. *Journal of Pharmaceutical Research International*, 197–214. https://doi.org/10.9734/BJPR/2012/1930.

Salvatore, Giuseppe, Armando D'Andrea and Marcello Nicoletti. 1998. A Pinocamphone Poor Oil of *Hyssopus Officinalis* L. Var. Decumbens from France (Barton). *Journal of Essential Oil Research*, *10* (5): 563–567. https://doi.org/10.1080/10412905.1998.9700972.

Samany, Alavi Seyed Mohammad, Abdollah Ghasemi Pirbalouti and Fatemeh Malekpoor. 2022. Phytochemical and Morpho-Physiological Changes of Hyssop in Response to Chitosan-Spraying under Different Levels of Irrigation. *Industrial Crops and Products*, 176: 114330. https://doi.org/10.1016/j.indcrop.2021.114330.

Shanayda, M.I. 2015. Amino Acids Composition of Herb of Lamiaceae Family Representatives. *Farmatsevtychnyi Zhurnal*, *1*: 82–87. https://doi.org/10.32352/0367-3057.1.15.06.

Sharifi-Rad, J., Quispe, C., Kumar, M., Akram, M., Amin, M., Iqbal, M., Koirala, N., Sytar, O., Kregiel, D., Nicola, S., Ertani, A., Victoriano, M., Khosravi-Dehaghi, N., Martorell, M., Alshehri, M. M., Butnariu, M., Pentea, M., Rotariu, L.S., Calina, D., Cruz-Martins, N. and Cho, W.C. 2022. Hyssopus Essential Oil: An Update of Its Phytochemistry, Biological Activities, and Safety Profile. *Oxid. Med. Cell. Longev.*, *2022*: 8442734. https://doi.org/10.1155/2022/8442734.

Sharma, V., Sharma, D., Saini, M., Jain, A., Chaudhary, J., Kaur, N., Sahoo, S., Singh, S.K., Goyal, K., Rekha, A., Ali, H., Gupta, S., Hussain, M.S. and Gupta, G. 2025. Flavonoids as Antimicrobial Agents: A Comprehensive Review of Mechanisms and Therapeutic Potential. *Curr. Pharm.*

Biotechnol. Advance online publication. https://doi.org/10.2174/011389201038400225062816 3849.

Sharopov, F. S., Zhang, H., Wink, M. and W.N. Setzer. 2015. Aromatic medicinal plants from Tajikistan (Central Asia). *Medicines*, *2* (1): 28–46. https://doi.org/10.3390/medicines2010028.

Sheremeta, A.V., Ivlev, V.A., Murtazina, A.A., Polukhina, A.Y., Miridonova, L.A. and Vasiliev, V.G. 2023. Identification of diosmin in *Hyssopus officinalis* L. by ^{1}H NMR spectroscopy. In: *Proc. Conf. Modern Problems of Science and Education*: 233–235.

Shomirzoeva, O., Li, J., Numonov, S., Atolikshoeva, S. and Aisa, H.A. 2020. Chemical components of Hyssopus cuspidatus Boriss.: Isolation and identification, characterization by HPLC-DAD-ESI-HRMS/MS, antioxidant activity, and antimicrobial activity. *Nat. Prod. Res.*, *34* (4): 534–540. https://doi.org/10.1080/14786419.2018.1488710.

Simon, J.E., Chadwick, A.F. and Craker, L.E. 1984. *Herbs: An Indexed Bibliography, 1971–1980: The Scientific Literature on Selected Herbs, and Aromatic and Medicinal Plants of the Temperate Zone*. Archon Books.

Skrypnik, L., Feduraev, P., Styran, T., Golovin, A., Katserov, D., Nebreeva, S. and Maslennikov, P. 2022. Biomass, phenolic compounds, essential oil content, and antioxidant properties of hyssop (Hyssopus officinalis L.) grown in hydroponics as affected by treatment type and selenium concentration. *Horticulturae*, *8* (11): 1037. https://doi.org/10.3390/horticulturae8111037.

Skrzypek, Z. and Wysokińska, H. 2003. Sterols and triterpenes in cell culture of Hyssopus officinalis L. Z. *Naturforsch. C.*, *58* (5–6): 308–312. https://doi.org/10.1515/znc-2003-5-602.

Srivastava, A., Awasthi, K., Kumar, B., Misra, A. and Srivastava, S. 2018. Pharmacognostic and pharmacological evaluation of Hyssopus officinalis L. (Lamiaceae) collected from Kashmir Himalayas, India. *Pharmacogn. J.*, *10* (4): 690–693. https://doi.org/10.5530/pj.2018.4.114.

Suleimen, E.M., Tursynova, N.K., Ibataev, Zh. A., Iskakova, Zh. B. and Ishmuratova, M. Yu. 2015. Constituent composition and biological activity of essential oil from Hyssopus ambiguus. *Chem. Nat. Compd.*, *51* (6): 1186–1187. https://doi.org/10.1007/s10600-015-1527-x.

Tabatabaie, F., Keighobadi, A., Golestani, M., Shahmohammad, N., Voosoogh, M., Noori, M., Mokhtarian, K. and Maleki, F. 2014. Antileishmanial activity of *Hyssopus officinalis*, Tussilago farfara, Carum copticum extracts in comparison with Glucantime in Iran. *J. Med. Plants Stud.*, *2* (5): 12–18.

Tahir, M., Rahman, M. and Khushtar, M. 2019. Gastroprotective effect of Hyssopus officinalis L. leaves via reduction of oxidative stress in indomethacin-induced gastric ulcer in experimental rats. *Drug Chem. Toxicol.*, *45*: 1–10. https://doi.org/10.1080/01480545.2019.1685537.

Tkachova, Y., Fedorchuk, M. and Kovalenko, O. 2022. Allelopathic activity of plants *Hyssopus officinalis* L. *Ukr. Black Sea Reg. Agrar. Sci.*, *26* (4): 19–29. https://doi.org/10.56407/2313-092X/2022-26(4)-2.

Toppo, R. 1995. *Treatment of Diseases with Hyssopus officinalis extract. WO1995028168A1*. Filed April 12, 1995. Published October 26, 1995. https://patents.google.com/patent/WO1995028168A1/en.

Vallejo, M.C.G., Herraiz, J.G., Pérez-Alonso, M.J. and Velasco-Negueruela, A. 1995. Volatile oil of Hyssopus officinalis L. from Spain. *J. Essent. Oil Res.*, *7* (5): 567–568. https://doi.org/10.1080/10412905.1995.9698590.

Venditti, A., Bianco, A., Frezza, C., Conti, F., Maleci Bini, L., Giuliani, C., Bramucci, M., Quassinti, L., Damiano, S., Lupidi, G., Beghelli, D., Caterbi, S., Petrelli, D., Vitali, L. A., Papa, F., Caprioli, G. and Maggi, F. 2015. Essential oil composition, polar compounds, glandular trichomes, and biological activity of *Hyssopus officinalis* subsp. aristatus (Godr.) Nyman from central Italy. *Ind. Crops Prod.*, *77*: 353–363. https://doi.org/10.1016/j.indcrop.2015.09.002.

Vlase, L., Benedec, D., Hanganu, D., et al. 2014. "Evaluation of Antioxidant and Antimicrobial Activities and Phenolic Profile for Hyssopus Officinalis, Ocimum Basilicum and Teucrium Chamaedrys." *Molecules*, 19 (5): 5490–507. https://doi.org/10.3390/molecules19055490.

Wang, N. and X.-W. Yang. 2010. Two new flavonoid glycosides from the whole herbs of *Hyssopus officinalis*. *J. Asian Nat. Prod. Res.*, *12* (12): 1044–1050. https://doi.org/10.1080/10286020.2010.533120.

WFO. 2025. *Hyssopus officinalis L*. Published on the Internet. http://www.worldfloraonline.org/taxon/wfo-0000217205. Accessed on: 03 Sep 2025.

Yousefzadeh, S. and Naghdi Badi, H.A. 2017. Changes of essential oil, photosynthetic pigments, and morphological characteristics of hyssop (*Hyssopus officinalis* L.) at different harvesting time. *J. Med. Plants*, *16* (61): 79–88.

Zawislak, G. 2016. Essential oil composition of *Hyssopus officinalis* L. grown in Poland. *J. Essent. Oil Bear. Plants*, *19* (3): 699–705. https://doi.org/10.1080/0972060X.2014.935034.

Zhang, H., Hou, Y., Ehbal, T., Jiang, M. and Li, F. 2019. A comparative analysis of the anti-inflammatory effects of Hyssopus cuspidatus Boriss. essential oil and aspirin on chronic inflammation models in mice. *Int. J. Clin. Exp. Med.*, *12* (7): 8261–8270. http://www.ijcem.com/files/ijcem0090881.pdf.

Zhao, L., Ji, Z., Li, K., Wang, B., Zeng, Y. and Tian, S. 2020. HPLC-DAD analysis of Hyssopus cuspidatus Boriss extract and mensuration of its antioxygenation property. *BMC Complement. Med. Ther.*, *20* (1): 228. https://doi.org/10.1186/s12906-020-03016-0.

Zhou, X., Gong, H.-Y., Xu, T.-H. and Tian, S. 2010. Physicochemical evaluation and essential oil composition analysis of Hyssopus cuspidatus Boriss from Xinjiang, China. *Pharmacogn. Mag.*, *6* (24): 278–281. https://doi.org/10.4103/0973-1296.71790.

12 Nutritive Value, Bioactive Compounds, and Biological Activities of Curry Leaf (*Bergera koenigii* (L.)

Vijayalaxmi S. Dandin[1*], Jagdish Vishnu Dalavi[2], Vasundhara Vilas Khot[3], Nilesh Appaso Madhav[4], Joseph K.S.[5] and Praveen N.[5]

Introduction

The curry tree is popularly known for its culinary and medicinal uses and is one of the botanically and socioeconomically significant plants. This plant originated in India and is now widely cultivated all over tropical Asia and other parts of the world, and escaped into wilder areas. This plant is commonly referred to as "sweet neem" or "Salam India" and belongs to the family Rutaceae. It has various colloquial names in different regions of India, including Mahanimb or Girinimba (Sanskrit), Kari Patta (Hindi), Karibevu (Kannada), Kariveppalai (Tamil), Karivepaku (Telugu), Kadhi Patta or Kadhi Limb (Marathi), and Karapincha (Sinhalese). In addition to its essential role in conventional medical systems like Ayurveda, Siddha, and Unani, its pharmacological qualities are becoming more widely acknowledged (Bhandari, 2012). *B. koenigii* leaves have great cultural and traditional significance in Indian rituals. They stand for purity and good fortune. *B. koenigii* leaves are thought to repel negative energy and attract positive vibes when placed in houses and temples. Because curry leaves can aid in digestion, they are employed in pre-meal rituals in some societies as a blessing for health and energy. They are used in wedding rituals like the Mangal sutra ceremony, which represents the prosperity and well-being of the couple. The plant serves as a valuable source of essential micronutrients, including calcium, phosphorus, iron, dietary fibre, and vitamins

[1] Department of Biology, Alliance Pre-university College, Bengaluru 560099, India.
[2] Botanical Survey of India, Western Regional Centre, Pune, Maharashtra 411001, India.
[3] Department of Botany, Balwant College Vita, (Sangli), M.S., India.
[4] Department of Botany, HPT Arts and RYK Science College, Nashik, M.S., India.
[5] Department of Life Sciences, CHRIST (Deemed to be University), Bengaluru, India.
E-mail: , e-mail: vijayalaxmidandin@gmail.com[1]; , Email: jagdish.taxonomy@gmail.com[2]; khotvasundhara@gmail.com[3]; madhavnilesh1@gmail.com[4]; joseph.ks@christuniversity.in[5]; praveen.n@christuniversity.in[5]
* Corresponding author: vijayalaxmi.d@alliancepuc.edu.in

DOI: 10.1201/9781003687832-12

A, B-complex, C, and E. These nutritional qualities accentuate its importance as a functional food by contributing to metabolic regulation and cellular homeostasis. However, its therapeutic value is more profoundly attributed to carbazole alkaloids, namely, mahanimbine, girinimbine, koenimbine, and murrayazoline, along with a variety of flavonoids, terpenoids, and phenolic acids (Ma et al., 2016). Extensive pharmacological investigations have reported that these compounds display a wide range of biological effects, such as antidiabetic (Arulselvan et al., 2006a), hepatoprotective (Tembhurne and Sakarkar, 2010a), antioxidant (Ma et al., 2016), anticancer (Debnath et al., 2003), anti-inflammatory, antimicrobial (Nagappan et al., 2011a), and neuroprotective activities. In addition, *B. koenigii's* larvicidal and insect-repellent qualities show promise in targeted therapeutic contexts such as wound healing (Sastry and Moudgal, 2012), tissue regeneration, and vector control, in addition to its systemic health advantages. These effects of *B. koenigii* leaves are mediated by several mechanisms, such as cellular membrane stabilization, cytokine expression control, inhibition of important metabolic enzymes, and neutralization of reactive oxygen species (ROS). The *B. koenigii* nutritional value, phytochemical diversity, and potential for therapeutic use are presented in this chapter. Furthermore, the study summarizes novel uses of *B. koenigii* in nutraceuticals and phyto-therapeutics, highlighting its role in bridging traditional ethnomedicine with modern science for preventive and integrative health.

Taxonomic Background

Bergera koenigii L. in Mant. Pl. 2: 563. 1767; *Camunium koenigii* (L.) Kuntze in Revis. Gen. Pl. 1: 99. 1891; *Chalcas koenigii* (L.) Kurz in J. Asiat. Soc. Bengal, Pt. 2, Nat. Hist. 44: 132. 1875; *Murraya koenigii* (L.) Spreng. in Syst. Veg., ed. 16. 2: 315. 1825; *Bergera siamensis* (Craib) F.J. Mou in Eur. J. Taxon. 860: 168. 2023; *Chalcas siamensis* (Craib) Tanaka in Bull. Soc. Bot. France 75: 710. 1928; *Murraya foetidissima* Teijsm. and Binn. in Natuurk. Tijdschr. Ned. -Indië 27: 41. 1864; *Murraya siamensis* Craib in Bull. Misc. Inform. Kew 1926: 340. 1926; *Nimbo melioides* Dennst. in Schlüssel Hortus Malab.: 30. 1818.

Common names: Curry tree, Curry leaf tree, Curry bush, Curry leaves, Sweet neem.

Etymology: The generic name '*Bergera*' honors 'Christian von Bergera', a Swedish physician and botanist (Koenig, 1771), and the specific epithet '*koenigii*' honors the 'Johann Gerhard König', a well-known Botanist.

Taxonomical description: Perennial, much-branched shrub to medium-sized tree, grows to 3–6 metres high; sometimes grows as a clumpy shrub due to the formation of basal suckers. Stems erect, main trunk 10–20 cm thick, much-branched, woody, bark dark greenish brown to blackish, rough, exfoliating in longitudinal strips (Figure 1a and b). Leaves 15–30 cm long, pinnately compound, imparipinnate, alternate, spirally arranged, strongly aromatic when crushed; petiole terete, puberulous to glabrous (Figure 1c and d). Leaflets 2–4.5 × 1–2 cm, (9–) 11–21(–31) per leaf, alternate to subopposite, ovate to ovate-lanceolate or oblong-lanceolate, base oblique, apex acute to acuminate, margin entire to crenulate, glandular-punctate; surfaces glossy dark green adaxially, paler abaxially (Figure 1e and f), glabrous or minutely puberulous along veins (Figure 1h and i). Inflorescence 10–15 cm long, terminal, corymbose panicle, densely flowered (Figure 1j). Flowers bisexual, actinomorphic, 5-merous, white or creamy-white, to greenish, fragrant; pedicels 2–5 mm long (Figure 1k-n); calyx of 5 sepals, fused at base, sepals more or less ovate broadly triangular at apex, puberulous, distinctly gland dotted (Figure 1o); petals 5, polypetalous, petals 5–7 mm long, linear-oblong to elliptic, strongly recurved at anthesis, highly glandular, white above greenish below (Figure 1m – 1q); Stamens 10, free, in two whorls, didynamous, filaments subulate, 3–6 mm long, anthers oblong-ovoid, dorsifixed, dithecous, dehiscing longitudinally (Figure 1r – t); gynoecium of 1 carpel; ovary superior, ovoid, 2-locular, with 2 ovules per locule, style stout, persistent, stigma capitate, wet (Figure 1u and v). Fruit 1–1.6 × 1–1.2 cm, a sub-globose to ovoid berry, green ripening to glossy purplish-black, pulpy; seeds 1 or 2 per fruit, green, oblong to reniform, with sticky covering (Figure 1x, 1y, and 1z).

***Flowering*:** March–July, Fruiting: July–September (South and Central Asia)

***Native*:** Southern China to Tropical Asia and South West Asia Pacific.

***Distribution*:** Bangladesh, Cambodia, China Southeast, China South-Central, Hainan, East and West Himalaya, India, Indonesia, Laos, Maldives, Malaya, Myanmar, New Caledonia, Nepal, New Guinea, Pakistan, Philippines, Solomon Is., Queensland, Sri Lanka, Thailand, Taiwan, Vietnam, Vanuatu.

Traditional Uses

The aromatic leaves of *B. koenigii* are extensively used in Indian cuisine and traditional medicines. Fresh curry leaves hold a special place in South Indian cuisine, where they are often tempered in oil along with mustard seeds and asafoetida to prepare a flavourful 'tadka'. Typically, curry leaves, mustard seeds, and cumin are sautéed together in ghee or another heat-stable oil, releasing their rich aroma and taste. This seasoned mixture can either serve as the base for cooking additional ingredients or be drizzled over prepared dishes to enhance the flavour of curries, pickles, chutneys, soups, buttermilk preparations, and other traditional recipes. The leaves retain much of their aroma even when dried, although a mild bitterness can develop, making them suitable for spice powders. Traditionally, curry leaves have been employed to stimulate appetite, enhance digestion, and treat fever, diarrhoea, and gastrointestinal disturbances. Decoctions of the leaves, especially when combined with black pepper, are used to relieve nausea, morning sickness, and manage early-stage diabetes (Parrotta, 2001; Rahman and Gray, 2005). A wide range of non-food items, such as perfumes, soaps, cosmetics,

Fig. 1 *B. koenigii* morphology: a. Habit; b. stem magnified; c. & d. leaf; e. & f. leaflets lower & upper surface; g. leaflet margin; h. magnified image of gland; i. indumentum of rachis; j. inflorescence; k. bud; l. flower; m. bracts & bracteoles; n. & o. calyx outer and inner side; p. & q. front and back view of corolla; r. staminal column; s. staminal column dissected; t. dimorphic stamens; u. gynoecium; v. fertilized ovary; w., x. & y. different stages of fruits; z. seeds

aromatherapy products, massage oils, and air fresheners, are prepared from *B. koenigii* leaves. Topically, leaf paste is applied to cuts, burns, bruises, and skin infections. Additionally, the twigs are used in rural oral hygiene practices due to their antimicrobial activity attributed to α-pinene and α-copaene (Bhandari, 2012). *B. koenigii* leaves are used in several formulations of Ayurveda, Siddha, Unani, and Folk Medicinal systems to treat various disorders (Table 1). 7.5% curry leaf essential oil formulation increased follicle diameter and hair density in mice, outperforming 2% minoxidil in efficacy and safety (Gupta et al., 2011; Sharma et al., 2011; Rao et al., 2011). The leaves are known to exhibit antiemetic, anti-diarrheal, and purgative properties. Historical medicinal texts also reference the use of *B. koenigii* in treating renal disorders. Supporting this, a histopathological study showed that aqueous leaf extract of curry leaves alleviated renal pain symptoms in diabetic rats (Bhandari, 2012; Yankuzo et al., 2011a). However, some traditionally claimed benefits, such as those for piles, asthma, fever (cooling properties), and influenza, remain insufficiently validated by contemporary research. This highlights the need for more systematic investigations and controlled *in vivo* trials to scientifically substantiate these ethnobotanical applications and integrate *B. koenigii* into modern medicinal formulations.

In the market, *B. koenigii* leaf products such as rice powder blends, essential oils for nourishing the scalp, extract capsules for blood sugar regulation and liver health, herbal shampoos for hair growth, and dietary supplements fortified with curcumin are available today. Nanoparticles and nano-emulsions obtained by nano-processing of *B. koenigii* leaf material are used in a variety of applications beyond traditional ones. These include edible coatings to extend food shelf- life, oil stabilization to prevent rancidity, biosensors for rapid detection systems, targeted drug administration, functional fabrics, water purification solutions, and even sophisticated medical equipment. Collectively, these developments demonstrate the remarkable usefulness of *B. koenigii* leaves (Figure 2).

Nutritional Compositions

B. koenigii leaves are a nutritional powerhouse, contributing significantly to a balanced diet. The precise nutritional composition can vary depending on geographical location, soil type, climate, and processing methods (fresh vs. dried). The leaves demonstrate a robust nutritional profile supporting their traditional and medicinal use in South Asian diets. The moisture content in fresh leaves is remarkably high (694 mg/ g), which aids in enzymatic activity and palatability (Chathurika et al., 2021). The crude protein content (156 mg/g) indicates the potential of *B. koenigii* leaves as a supplementary protein source in vegan diets. According to Chathurika et al. (2021), the lipid-soluble phytochemicals, especially terpenoids and essential oils, are supported by the crude fat levels (110 mg/g), which add aroma to *B. koenigii* leaves. Carbohydrate content (2.3 g/100 g) is ideal for low-glycemic diets as it is relatively low. The rich crude fibre content (630 mg/g) is beneficial for gut health (Igara et al., 2016). Mineral content is

Fig. 2 Different applications of *Bergera koenigii* leaves.

Table 1 Traditional uses in Ayurveda, Siddha, Unani, and Folk Medicine.

Use	*Part Used*	*System*	*Traditional Use*	*References*
Antiemetic (nausea relief)	Leaf juice	South Indian folk medicine	to reduce nausea and vomiting, during pregnancy	Jain (1991)
Blood purification	Tender leaves	Ayurveda	Consumed to cleanse blood and balance doshas	Kirtikar & Basu (1935)
Anti-diarrheal	Leaves and roots	Siddha (Tamil Nadu)	Decoction or paste used to treat diarrhoea and dysentery	Ayyanar & Ignacimuthu (2005)
Digestive tonic	Fresh leaves	Ayurveda	Consumed as chutney or decoction to enhance digestion	Chopra et al. (1956)
Blood sugar regulation	Fresh leaves	Ayurveda and Siddha	Chewed raw or decocted for controlling hyperglycaemia	Arulselvan et al. (2006a)
Eye infection remedy	Leaf juice	Unani, rural traditions	Mild extract eye drop for soothing inflammation	Nadkarni (1976)
Fever remedy	Leaf + black pepper	Folk traditions	Decoction is administered to reduce fever	Singh et al. (2012)
Hair health support	Leaf paste/ oil	Indian folk practices	Applied externally to reduce dandruff and hair fall	Bhattacharya et al. (2011)

significant, with ash content indicating 275 mg/g of total minerals. Among these, magnesium (490.6 mg/g), calcium (197.5 mg/g), and zinc (0.4 mg/g) are predominant and are essential for enzymatic regulation, immune function, and bone health (Igara et al., 2016). The vital vitamins are β-carotene (6.04 mg/g) with antioxidant properties, along with B-complex vitamins such as thiamine (8.9 mg/g), riboflavin (0.9 mg/g), and niacin (27.3 mg/g). These vitamins are vital for nerve function and cell metabolism. The plant also possesses free amino acids like glycine, asparagine, serine, aspartic acid, threonine, alanine, proline, glutamic acid, tryptophan, leucine, tyrosine, α-aminobutyric acid, phenylalanine, isoleucine, lysine, arginine, and histidine. Though *B. koenigii* leaves possess comparatively lower ascorbic acid content (Vitamin C; 0.4 mg/g), it plays a significant role in antioxidant mechanisms and immune health. Nutritional value of *B. koenigii* leaves supports their traditional uses, suggesting they could be used in functional foods or supplements. Table 2 shows various nutritional components found in *B. koenigii* leaves.

Table 2 Nutritional composition of *B. koenigii* leaves.

Component	*Weight*	*Reference*
Moisture	694 mg/g	Chathurika et al. (2021)
Crude protein	156 mg/g	Chathurika et al. (2021)
Fat (crude fat)	110 mg/g	Chathurika et al. (2021)
Carbohydrate	23 mg/g	Chathurika et al. (2021)
Crude fibre	63 mg/g	Igara et al. (2016)
Ash content (mineral fraction)	275 mg/g	Chathurika et al. (2021)
Vitamin C (Ascorbic acid)	0.4 mg/g	Igara et al. (2016)
Thiamine (B_1)	8.9 mg/g	Igara et al. (2016)
Riboflavin (B_2)	0.9 mg/g	Igara et al. (2016)
Niacin (B_3)	27.3 mg/g	Igara et al. (2016)
Calcium (Ca)	197.5 mg/g	Igara et al. (2016)
Magnesium (Mg)	490.6 mg/g	Igara et al. (2016)
Manganese	400.0 mg/g	Igara et al. (2016)
Zinc (Zn)	0.4 mg/g	Igara et al. (2016)
β-Carotene (Vitamin A precursor)	6.04 mg/g	Igara et al. (2016)
Vitamin E (α-tocopherol)	0.03 mg/g	Igara et al. (2016)
Vitamin B_6 (Pyridoxine)	0.04–0.05 mg	Igara et al. (2016)

Bioactive Compounds

Alkaloids

B. koenigii foliage possesses main alkaloids in the form of carbazole alkaloids, which are biosynthesized from indole-based precursors. More than 100 carbazole alkaloids have been identified from different parts of *B. koenigii*, with mahanimbine and koenimbine being the predominant ones present in the leaves. These compounds have a tricyclic aromatic framework, consisting of a central pyrrole ring bordered by two benzene rings, and have anti-analgesic, anti-ulcerogenic, anti-obesity, and antitumour activities. Reported compounds include bicyclomahanimbine, 6,7-dimethoxy-1-hydroxy-3-methylcarbazole, euchrestine B, girinimbine, 7-hydroxymurrayazolinine, isomahanimbine, IMN, koenidine, koenigicine, koenine, mahanimbicine, mahanimbine, mahanine, mahaninebicine T, mahaninebine, 7-methoxymurrayacine, O-methylmahanine, O-methylmurrayamine A, mukoeic acid, mukonicine, murrayanine, murrayamine B & D, murrayamine A, murrayamine C, murrayamine C & J, murrayacine, murrayanol, murrayazoline, murrayazolinine, pyrayafoline D, bisisomahanine, 8,8′-biskoenigine, bismamanimboline, bismahanine, bismurrayafoline E, bispyrayafoline, mahabinine A, murrafoline I, 3,3′,5,5′,8-pentamethyl-3,3′-bis(4-methylpent-3-en-1-yl)-3,3′,11,11′-tetrahydro-10,10′-bipyrano[3,2-a]carbazole, and 8,10′-[3,3′,11,11′-tetrahydro-9,9′-dihydroxy-3,3′,5,5′,8′-tetramethyl-3,3′-bis(4-methyl-3-entenyl)]bipyrano[3,2-a] carbazole. This array of leaf-derived carbazole alkaloids highlights the plant's chemical diversity and underlines its therapeutic relevance (Chan and Wong, 2024). The first carbazole alkaloid from *B. koenigii*, murrayanine, was isolated from bark by Chakraborty et al. (2012). Subsequent discoveries included mahanine (Narasimhan et al., 1975), mahanimbine, and murrayanol (Reisch et al., 1994) from leaves (Ramsewak et al., 1999). Other notable findings were biskoenigine (Wang et al., 2008) and six alkaloids—bismahanimboline, murrastanine-A, murrastinine-A, murrastinine-B, murrastinine-C, and murrayatanine-A reported by Tan et al. (2015). Nalli et al. (2016) later identified four new alkaloids: murrayakonines-A and -B (stem) and murrayakonines-C and -D (leaves). Key compounds such as mahanimbine, girinimbine, koenimbine, and murrayafoline-A have been repetitively identified by using advanced analytical techniques like HPLC, LC–MS, and NMR in methanolic, chloroform, and ethyl acetate extracts (Natarajan et al., 2002). Among them, mahanimbine and girinimbine have demonstrated potent pharmacological effects (Nalli et al., 2016; Natarajan et al., 2002). Numerous early phytochemical analyses have revealed minor alkaloids, including pyridines, pyrrolidines, tropanes, pyrrolizidines, isoquinolines, indoles, and quinolines. According to Kumar et al. (2013), polar ethanol and aqueous extracts of *B. koenigii* leaves have been found to contain pyridine derivatives such as 3-hydroxypyridine and methylpyridine, while LC–QTOF-MS was used to identify tropane-like analogues, which could indicate either endogenous synthesis or microbial symbiosis (Singh et al., 2016). Using LC–MS/MS and NMR techniques, quinoline compounds such as koenigine have been identified and structurally verified in leaf and root tissues (Chakraborty et al., 2012). UPLC–MS analysis of hydrolysed samples revealed norboldine-like isoquinolines (Rao et al., 2018), while High-resolution mass spectrometry was used to confirm pyrrolizidine structures in methanol extracts. The alkaloid profile of *B. koenigii* emphasizes both its therapeutic value and the need for more investigation into lesser-known alkaloid subclasses and their biosynthetic sources. Figures 3a, b, and c show the chemical structures of carbazole alkaloids found in *B. koenigii* leaves.

Polyphenols

Polyphenols are important antioxidants present in plants with more than 8,000 phenolic structures currently known. Many polyphenolic groups, such as phenolic acids, flavonoids, stilbenes, lignans, tannins, and coumarins, are abundant in *B. koenigii* foliage. Several phenolic acids, like caffeic acid, ferulic acid, gallic acid, chlorogenic acid, vanillic acid, and *p*-coumaric acid, were found predominantly in methanolic and ethanolic leaf extracts when tested with advanced analytical techniques such as high-performance liquid chromatography and liquid chromatography coupled with tandem mass spectrometry.

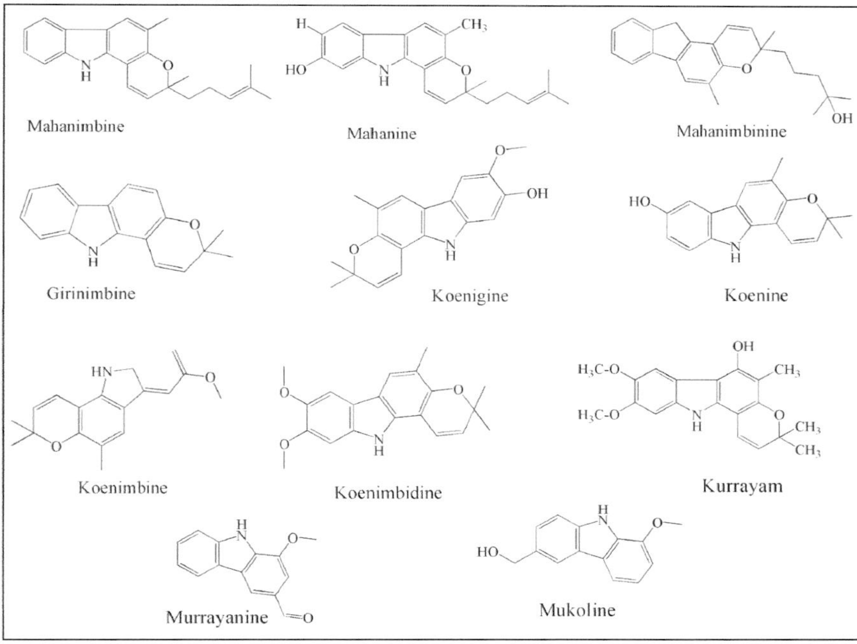

Fig. 3a Carbazole alkaloids found in *B. koenigii* leaves.

Apigenin, kaempferol, rutin, myricetin, naringenin, luteolin, and quercetin are among the significant flavonoids present in *B. koenigii* leaves and occur in both glycosidic and aglycone forms in significant amounts in the leaf tissue (Saini et al., 2020). 4-*O*-*β*-D-Rutinosyl-3-methoxyphenyl-1-propanone, 1-*O*-*β*-D-Rutinosyl-2(R)-ethyl-1-pentanol, 8-Phenylethyl-*O*-*β*-D-rutinoside are some other flavonoids found in leaves. Quercetin and kaempferol are very important ones. The spectrophotometric and colorimetric assays, such as the Folin–Ciocalteu and vanillin-HCl analyses, confirmed the presence of both hydrolysable and condensed tannin profiles of *B. koenigii* leaves (Patel et al., 2015). Tannins, such as tannic acid, increase the leaves' antibacterial strength and their distinctive astringency. Figure 4 depicts some of these compounds present in *B. koenigii* leaves.

Semi-polar chloroform and ethyl acetate solvent fractions revealed the presence of coumarins like murralonginol, murracarpin, murrangatin, isomurrangatin, murrayone, and scoparone (Rahman and Gray, 2005; Parmar et al., 2011). *B. koenigii* leaves contain a wide variety of polyphenols with different structures and functions, indicating their complex chemical composition. This understanding can help in standardizing formulations for developing effective therapeutic drugs and nutritional supplements.

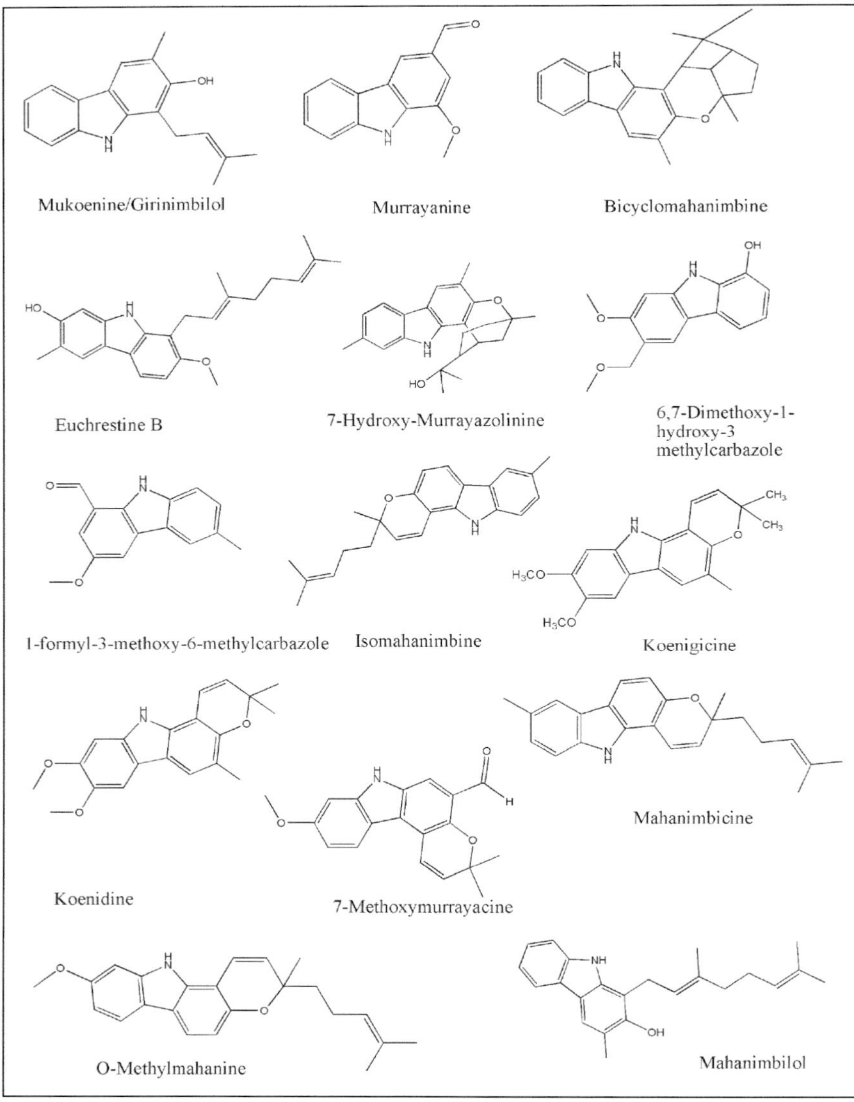

Fig. 3b Carbazole alkaloids found in *B. koenigii* leaves.

Terpenoids

B. koenigii leaves contain diverse terpenoids, including monoterpenes, sesterterpenes, sesquiterpenes, hemiterpenes, diterpenes, triterpenes, and tetraterpenes. Identified sesquiterpenes include β-elemene, β-bisabolene, β-caryophyllene, α-humulene, cadinene, germacrene D, α-amorphene, α-gurjunene, junipene, β-selinene, and

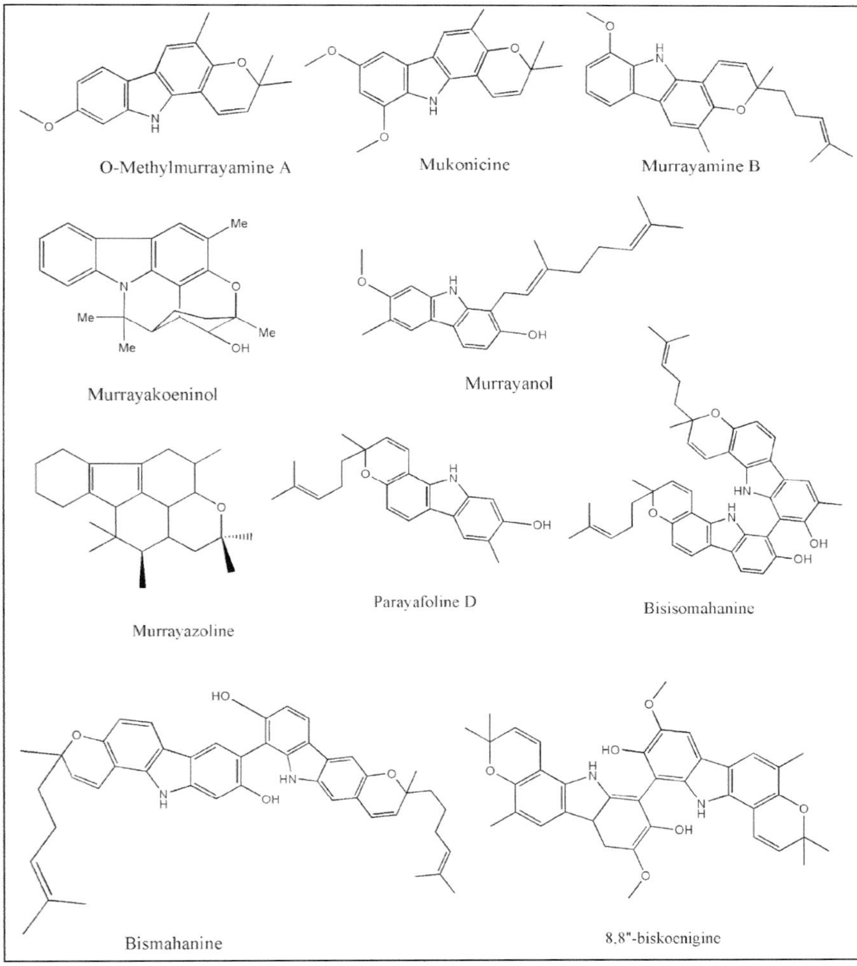

Fig. 3c Carbazole alkaloids found in *B. koenigii* leaves.

others (Yuviati et al., 2024). Several oxygenated sesquiterpenes such as elemol, viridiflorol, γ-eudesmol, B-eudesmol, and A-eudesmol have also been reported. GC-MS studies show monoterpenes and sesquiterpenes are major volatile oil components. The characteristic aroma is especially due to the blend of major terpenes and the synergistic presence of minor volatiles that amplify the overall sensory experience. Key monoterpenes include α-pinene, sabinene, β-pinene, α-terpinene, β-phellandrene, γ-terpinene, and terpinene-4-ol, and their abundance differs by extraction and leaf maturity. The predominant monoterpenes in *B. koenigii* leaves include α-pinene (19.0–19.7%), sabinene (31.8–44.8%), β-pinene (4.2–4.7%), α-terpinene (1.3–4.3%), β-phellandrene (6.5–7.9%), γ-terpinene (3.9–

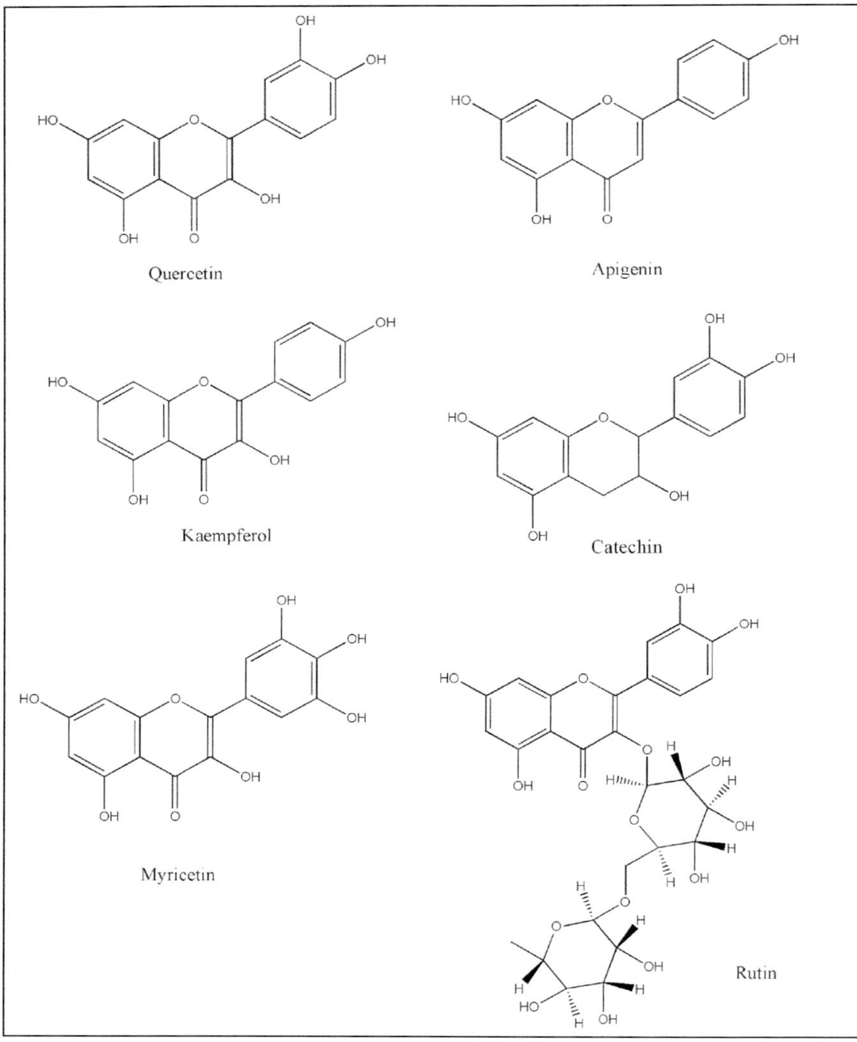

Fig. 4 Some important flavonoids of *B. koenigii* leaves.

7.1%), and terpinene-4-ol (5.2–9.9%). Monoterpenes like myrcene, *β*-pinene, limonene, *α*-terpinolene, *γ*-terpinene, [Z]- and [E]-*β*-ocimene, and allo-ocimene are prevalent, along with oxygenated forms such as 1,8-cineole, linalool, nerol, and carvone. Esters like 1-octen-3-yl acetate and 3-octanyl acetate have also been identified (Yuviati et al., 2024). Hemiterpenes, like isoprenes, are found in trace amounts. Diterpenes, including phytol, occurred in polar extracts. Evidence for the presence of sester-terpenes, though limited, has emerged through interpretation from nonpolar extract fractions of mass spectral fragmentation data (Ramesh and Raghu, 2019). Additionally, unique sodium-containing derivatives, including (3R,5S,6R)-3,5,6-trihydroxy-1,1,5-trimethylcyclohexyl-8-butyn-9-one (MW = 263.1254 g/mol) and (8E,9R)-ethyl-7-(3S,5R,6S)-3,6-dihydroxy-1,1,5-trimethylcyclohexyl-9-hydroxybut-8-enoate (MW = 309.1672 g/mol), along with (3R)-3-*O*-*β*-*D*-glucoside-6′-*O*-*D*-apiose-*β*-ionone (MW = 525.2306 g/mol), expand the chemical diversity of *B. koenigii* terpenoid profile. Furthermore, methanolic and hexane extracts of *B. koenigii* leaves have been identified as triterpenoids like β-amyrin, friedelin, and lupeol (Balakrishnan et al., 2012). Lupeol is widely recognized for its anti-diabetic, hepatoprotective, and anti-inflammatory actions (Sundararajan et al., 2006). In fresh leaves subjected to light-induced oxidative stress, tetraterpenes like *β*-carotene and other carotenoids have been quantified using HPLC and spectrophotometric methods (Ravindran et al., 2014). Linalool, elemol, geranyl acetate, myrcene, allo-ocimene, *α*-terpinene, (E)-*β*-Ocimene, *β*-caryophyllene, 3-carene, limonene, and *α*-pinene are the compounds that give leaves their scent. The majority of these are monoterpenes and oxygenated monoterpenes, which are what give food and fragrance their powerful, crisp, and delicious perfume. Although hemiterpenes and sesterterpenes are present in lower concentrations, their possible biological effects need further exploration. Figure 5 depicts a few important terpenoids of *B. koenigii* leaves.

Sulphur-containing Compounds

Research on *B. koenigii* leaves has predominantly focused on alkaloids and flavonoids, while sulphur-containing compounds are less studied. Steinhaus et al. (2015) studied the volatile fraction of curry leaves, and they found several sulphur-containing odourants, such as 1-phenylethanethiol, 3-(methylsulfanyl) propanal, which contribute strongly to the characteristic sulphury/burnt notes. So, the sulphur volatiles are a smaller but significant portion for odour profile. Thorough metabolomic and transcriptome investigations are necessary to clarify their structural variety, biosynthetic origins, and pharmacological importance of sulphur-containing metabolites found in *B. koenigii* leaves.

Phenylpropanoids and Alkanes

Bergera koenigii leaves possess various complex phenylpropanoids like murrayanol, murrayanagetin, and various methoxyphenyl derivatives like [7'E,8S]-9'-Hydroxy-7'-propen-3',5'-dimethoxyphenyl glucopyranosides and citrusin B, which exhibit potential therapeutic benefits. The alkanes include compounds such as [3R,5S,6E,8S,10E]-3,7,11-Trimethyl-1,6,10-dodecatriene-triol, [3S,4E,6E,10R]-2,10-Dihydroxy-2-methylethyl-sencolaniinic glucopyranoside and oxylipin (Ma et al. 2016; Figures 6 and 7). The leaves also possess a few other compounds like Imperatorin, Iso menthone, Nicotinic acid, Geranyl acetate, Lavandulyl acetate, Tocopherol, Z-Jasmone, Myrtenyl acetate, Heraclenin, Lutein, Linalyl acetate, Neryl acetate, and Menthol (Figure 8).

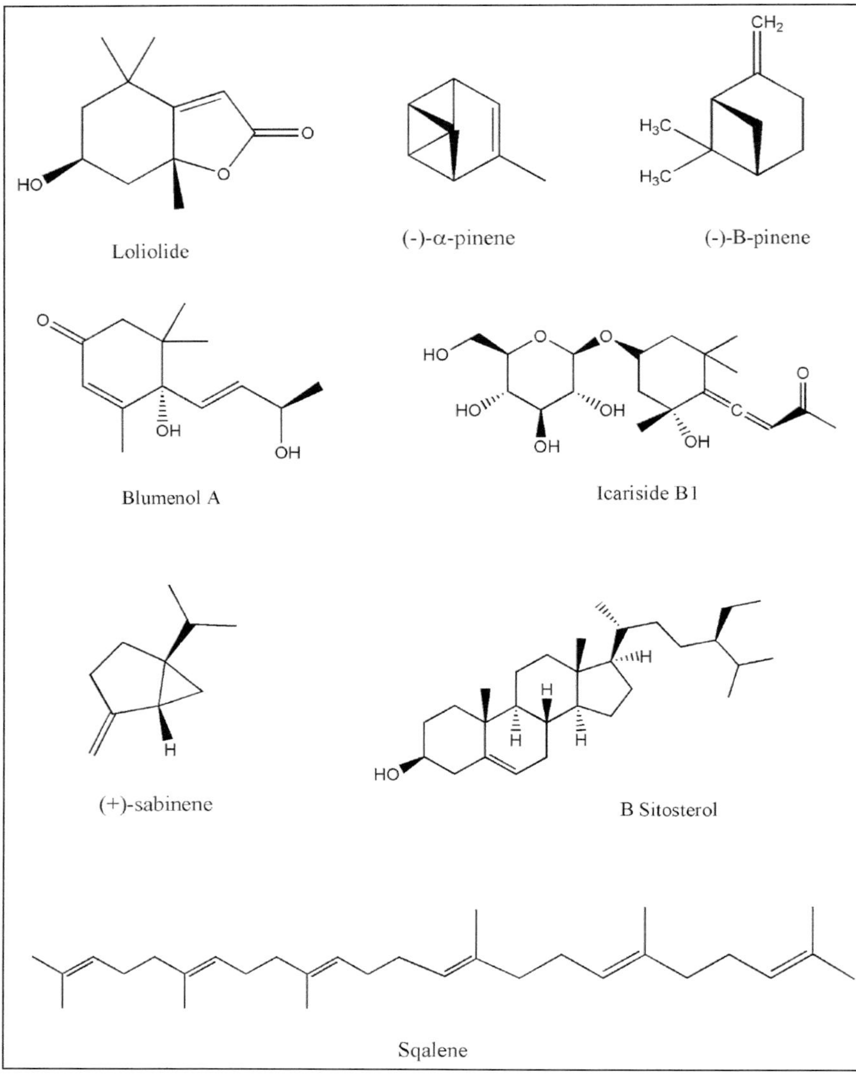

Fig. 5 Terpenoids found in leaves of *B. koenigii.*

Biological Activities

Antioxidant Properties

In vitro and *in vivo* antioxidant activity of *B. koenigii* leaves is largely attributed to their rich content of carbazole alkaloids, polyphenols, and flavonoids. Carbazole alkaloids such as mahanimbine, koenimbine, girinimbine, isolongifolene, koenoline, isomahanine, and *O*-methylmurrayamine have demonstrated potent

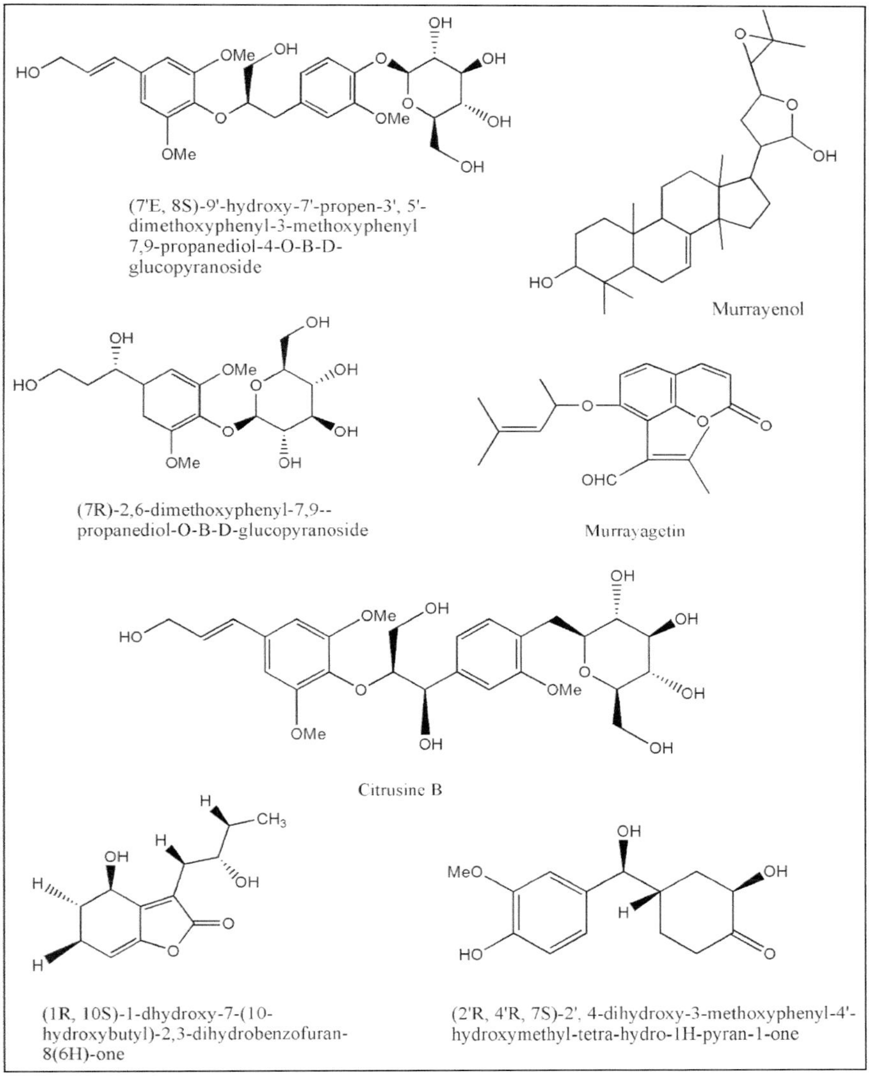

Fig. 6 Various other bioactive compounds present in *B. koenigii* leaves.

redox-modulating properties, contributing to the neutralization of reactive oxygen species and protection against lipid peroxidation and DNA damage (Gill et al., 2013; Rehana et al., 2017).

Polyphenols, as flavonoids, show antioxidant effects with respect to the scavenging of the different oxidizing species, superoxide anion, hydroxyl radical, and peroxyl radical (Harbone and Williams, 2000). *B. koenigii* leaf extracts have been effective in several antioxidant assays in different formulations. In an ethanol-

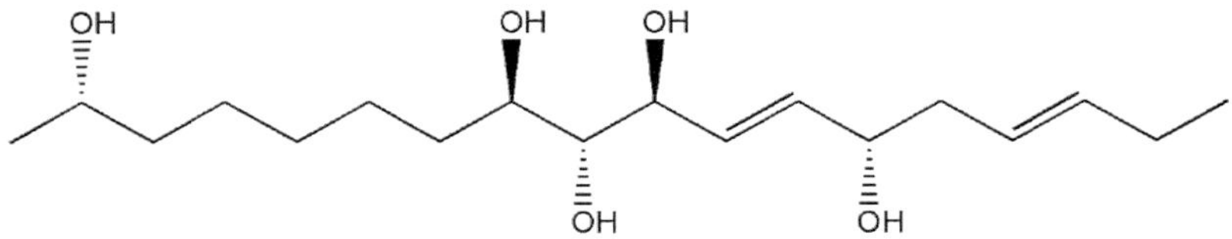

(3E, 6S, 7E, 9R, 10S, 11S, 17R)-octadeca-3,7-diene-6,9,10,11,17-pentaol

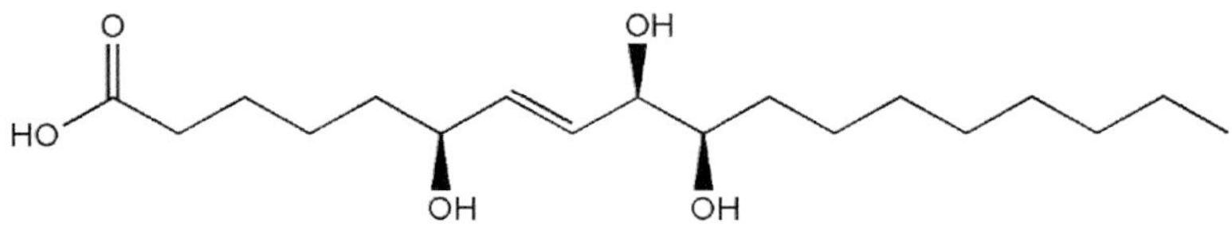

(6R, 7E, 9S, 10R)-6, 9, 10-trihydroxy-7-octadecenoic acid

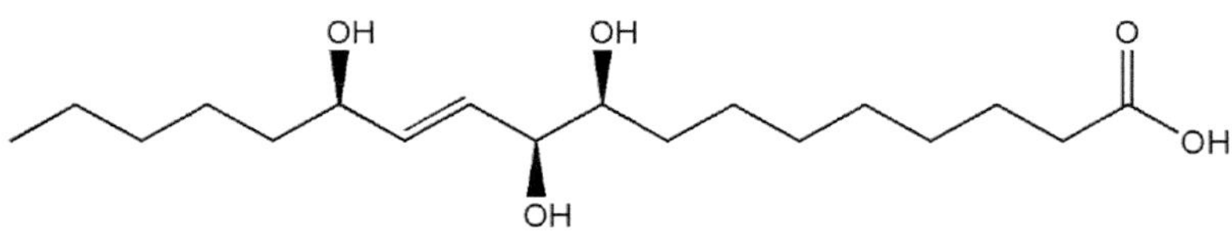

(9S, 10R, 11E, 13S)-9, 10, 13-trihydroxyoctadec-11-enoic acid

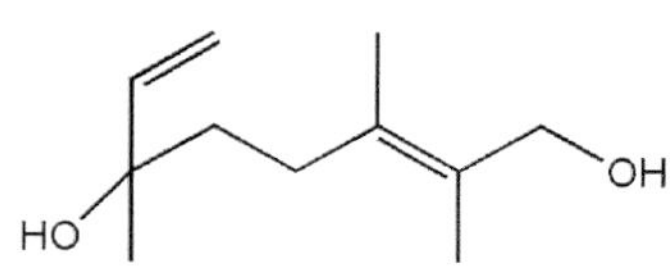

(2E, 6R)-2, 6 -dimethyl-2,7-octadiene-1,6-diol

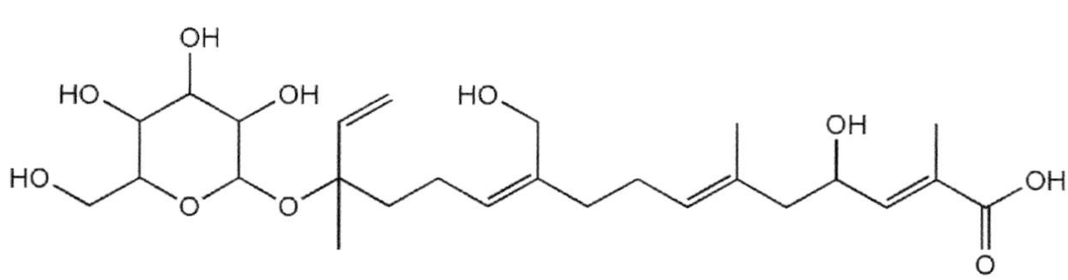

Capsianoside V

Fig. 7 Phenylpropanoids present in *B. koenigii* leaves.

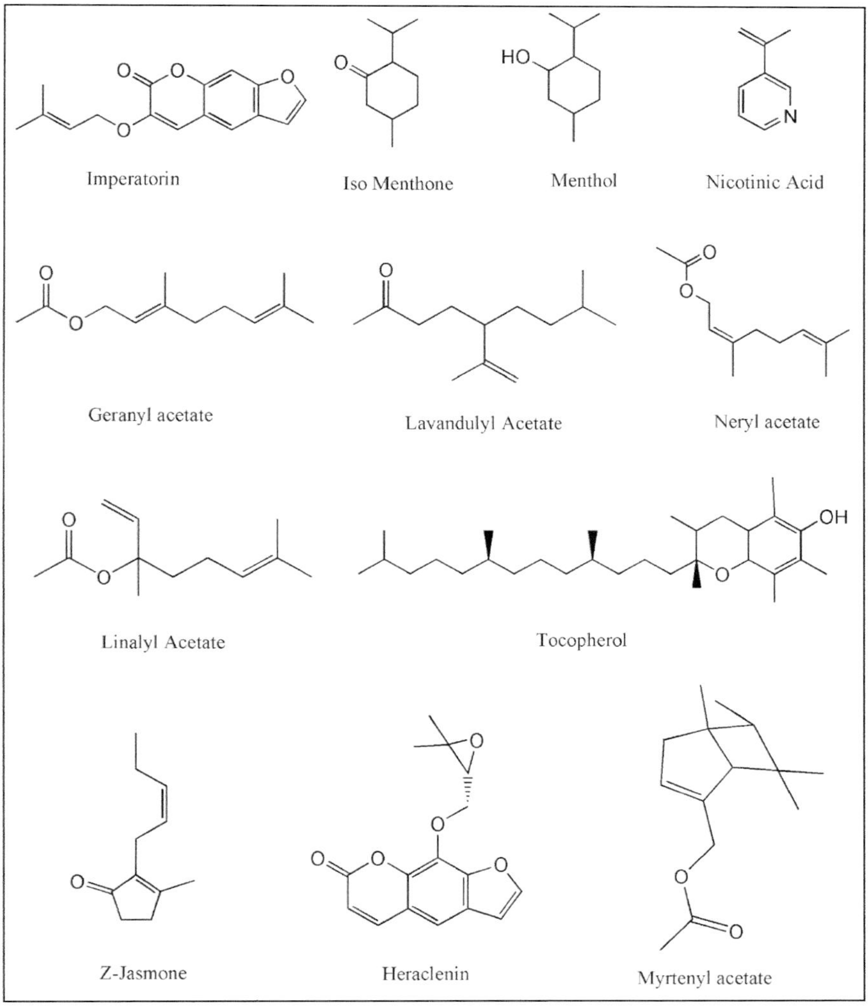

Fig. 8 A few important alkanes present in *B. koenigii* leaves.

induced hepatotoxicity *in vitro* model, carbazole alkaloid-containing aqueous extracts displayed antioxidant potential in TBARS, catalase (CAT), superoxide dismutase (SOD), and glutathione (GSH) assays, with an IC_{50} value of 120 μg/mL (Sathaye et al., 2011). Ethanol extracts of the *B. koenigii* leaves showed potent free radical scavenging activity with IC_{50} values ranging from 21.4 to 49.5 μg/mL in the DPPH assay (Ma et al., 2016). Furthermore, in DPPH (IC_{50} = 36.46 μg/mL), ABTS (IC_{50} = 11.55 μg/mL), superoxide (IC_{50} = 11.47 μg/mL), and H_2O_2 (IC_{50} = 54.06 μg/mL) radical scavenging assays, zinc oxide nanoparticle-synthesized aqueous leaf extracts of *B. koenigii* exhibited significantly enhanced antioxidant activity

(Rehana et al., 2017). When paired with formulations based on nanotechnology, the significant antioxidant capability of *B. koenigii*'s leaf extracts will show only an increase. *B. koenigii* leaf extract exhibits potent antioxidant activity and also mitigates cadmium-induced oxidative stress, cardiac damage, and functional alterations in rats (Khan et al., 1997), and shows heavy metal chelating activity, significantly lowering tissue cadmium levels when administered before exposure (Mitra et al., 2013). Among the various extracts, *B. koenigii* oleoresin exhibited a maximum activity of 83.2% at 100 ppm (Rao et al., 2007).

In tests like DPPH, FRAP, CUPRAC, and ABTS, methanolic and ethanolic extracts of the leaves have demonstrated a strong ability to scavenge free radicals, with efficacy on par with or better than that of common antioxidants like ascorbic acid and butylated hydroxytoluene (Rao et al., 2011; Zahin et al., 2010). The ethanolic extract, for example, showed over 80% suppression of DPPH radicals, while the benzene fractions showed 88.3% inhibition at 100 μg/mL, which is comparable to the 93.1% observed with ascorbic acid (Zahin et al., 2010). *B. koenigii* leaf supplementation increased the activity of endogenous antioxidant enzymes such as glutathione peroxidase (GPx), catalase (CAT), and superoxide dismutase (SOD), indicating systemic oxidative stress reduction (Khan et al., 1997).

Furthermore, the cardioprotective potential of *B. koenigii* was evidenced in cadmium-exposed rats, where pre-treatment with leaf extracts reduced oxidative damage and cadmium accumulation in cardiac tissue (Khan et al., 2012; Mitra et al., 2012). Overall, *B. koenigii* emerges as a rich source of natural antioxidants with significant potential in mitigating oxidative stress, protecting organs, and preserving mitochondrial health (Table 3).

Table 3 Antioxidant activities of *B. koenigii* leaves.

Compound Class	*Representative Compounds*	*Role in Antioxidant Activity*	*Reference*
Flavonoids	Quercetin, Kaempferol, Rutin	Scavenge reactive oxygen species (ROS); metal-ion chelation	Balakrishnan et al. (2020); Syam et al. (2011)
Phenolic acids	Gallic acid, Caffeic acid, Ferulic acid	Inhibit lipid peroxidation and reduce oxidative damage	Ghasemzadeh & Jaafar (2013); Syam et al. (2011)
Carbazole alkaloids	Mahanine, Mahanimbine, Koenimbine	Protect DNA; prevent lipid peroxidation via radical scavenging	Nagappan et al. (2011b); Debnath et al. (2003a)
Vitamins	Ascorbic acid (Vitamin C), β-Carotene	Contribute to radical quenching; work synergistically with flavonoids	Syam et al. (2011)

Contd...

Table 3 *Contd...*

Compound Class	*Representative Compounds*	*Role in Antioxidant Activity*	*Reference*
Essential oils	Linalool, β-Caryophyllene	Neutralize free radicals and inhibit oxidative stress	Nagappan et al. (2011b)
Glutathione (GSH)	—	Maintains redox homeostasis; detoxifies reactive intermediates	Syam et al. (2011)

Antimicrobial Activities

The essential oil of *B. koenigii* leaves exhibits significant antibacterial effectiveness, especially inhibiting species such as *Proteus mirabilis, Staphylococcus aureus, Escherichia coli, Pseudomonas aeruginosa, Klebsiella pneumoniae,* and *Salmonella enterica* (Ramamurthy and Prasanth, 2018). Pyranocarbazoles isolated from *B. koenigii* leaves display selective antibacterial action against S. *aureus* and *K. pneumoniae*. Several pathogenic fungi and a variety of Gram-positive and Gram-negative bacteria have been significantly inhibited by essential oils and solvent-based extracts of *B. koenigii* leaves. Important bioactive components like spathulenol, cinnamaldehyde, and linalool were responsible for the antibacterial property and for 80% suppression of *P. aeruginosa* biofilm formation. *Trichoderma viride, Penicillium species, Aspergillus species,* and *Alternaria alternata* were among the phytopathogens that were suppressed by the essential oil and crude extracts of *B. koenigii* leaves (Tripathi et al., 2018). *B. koenigii* extracts contain terpenoids and carbazole alkaloids, such as mahanimbine and koenimbine, which act on fungal infections by *Aspergillus niger, Candida albicans,* and *Fusarium oxysporum* (Bhavaniramya et al., 2019). Furthermore, the ethanolic extract of the leaves effectively inhibited the growth of *Trichophyton mentagrophytes* and *Microsporum gypseum* (Jayaprakash and Ebenezer, 2012).

The antifungal activity of *B. koenigii* has been attributed to phytochemicals such as alkaloids, terpenoids, flavonoids, phenolics, tannins, saponins, and essential oils (Goutam et al., 1974). Essential oil and ethanolic leaf extracts exhibit activity against pathogens, including *Penicillium notatum*, *Aspergillus flavus*, *A. niger*, *Fusarium moniliforme*, *Mucor mucedo*, and *P. funiculosum*, isolated from infected saplings and spoiled foods. Bioactive compounds such as girinimbine, murrayanine, marmesin-1′-O-β-D-galactopyranoside, mahanine, murrayacine, mukoeic acid, murrayazolinine, girinimbilol, pyrafoline-D, and murrafoline-I—especially girinimbine, murrayanine, and marmesin-1′-O-β-D-galactopyranoside—show notable antifungal effects. Ethanol extracts from *B. koenigii* have shown anti-mycobacterial action in concert with rifampicin to combat strains of *Mycobacterium smegmatis* and *Mycobacterium bovis* BCG. These extracts have an IC_{50} of about 400 μg/mL, which suggests that they may be used as an adjuvant therapy for tuberculosis. Leaf extracts, including ethanol and methanol, have strong antibacterial qualities and work well against *Enterococcus faecalis, Bacillus subtilis*, and other bacteria. The successful manufacture of silver nanoparticles (AgNPs) utilizing *B. koenigii* leaf extracts is highlighted by recent developments. Moreover, methanolic leaf extracts reportedly inhibited Herpes Simplex Virus-1 (HSV-1) replication (Satyavani et al., 2011). These antimicrobial activities are thought to be caused via disruption of microbial cell membranes, suppression of enzyme function, and interference with the production of nucleic acids. In the context of rising resistance to conventional antibiotics, *B. koenigii* can be placed as a viable source for the development of antimicrobial agents.

Anticancer Activities

The key carbazole alkaloids, such as mahanine, mahanimbine, girinimbine, and koenimbine, present in *B. koenigii* leaves are responsible for cytotoxic, pro-apoptotic, and antiproliferative properties against a broad spectrum of human cancer cell lines. The carbazole alkaloids have shown their effect against leukemia, breast, colon, and liver cancer models. These bioactive compounds induce apoptosis, disrupt mitochondrial membrane potential, and inhibit NF-κB activation, which is a vital transcription factor involved in the survival and proliferation of cancer cells (Debnath et al., 2003; Nagappan et al., 2011b). Mahanine, which is one of the carbazole alkaloids, exhibits broad-spectrum activity in lung cancer cells with reported IC_{50} values of 10–12.5 μM and increases the efficiency of cisplatin in oral squamous carcinoma many-fold (Kandimalla et al., 2022) and also exhibits cytotoxicity against leukemic, lung, colon, pancreatic, oral, breast, and bladder cancer cells, with IC_{50} values ranging from 3.5–42.6 μM (Bhattacharya et al., 2010; Hobani, 2022). Another alkaloid, mahanimbine, displays strong effects against pancreatic (CAPAN, SW119), breast (MCF-7), and bladder (Hs172.T) cancer cells, with IC_{50} values as low as 3.5 μM. Mahanimbine has been associated with apoptotic induction in cancer cells. Likewise. Girinimbine and koenimbine cause cell cycle arrest, apoptosis, and suppress inflammatory mediators like TNF-α and IL-1β by affecting many oncogenic routes, including PI3K/AKT/mTOR and Wnt/β-catenin, via mitochondrial dysfunction and caspase activation. Koenimbine has also been shown to suppress nuclear factor-kappa B (NF-κB) activity and induce cytochrome c release, further promoting intrinsic apoptotic pathways (Ahmadipour et al., 2015). Additional alkaloids such as murrayazoline and *O*-methylmurrayamine have demonstrated the ability to suppress the Akt/mTOR axis in colon cancer models, while pyrayafoline-D and murrafoline-I trigger apoptosis in HL-60 leukemia cells via mitochondrial membrane disruption and caspase cascade activation. *B. koenigii* leaf chemo preventive potential is proven through multi targeted mechanisms combined with the antioxidant capacity by contributing to the oxidative DNA damage, inhibition of angiogenesis, and metastasis. Leaf extracts prepared in aqueous methanol have been reported to induce dose-dependent cytotoxicity, with IC_{50} values ranging from 4.8 to 17.5 μg/mL across HeLa, CaCO2, HepG2, and LNCaP cell lines.

The flavonoid constituents of *B. koenigii* leaves, including quercetin, kaempferol, and rutin, are also effective phytocompounds against various cancer types. These compounds work to inhibit cancer progression in breast, prostate, and lung cancers by modifying antioxidant defences and inhibiting the PI3K/Akt and MAPK signalling pathways (Syam et al., 2014; Balakrishnan et al., 2020). Moreover, the chemoprotective effects of phenolic acids like gallic acid and ferulic acid are proven in colon and gastric cancers. In breast and skin cancer models, essential oils such as linalool and caryophyllene have also exhibited antitumor activities. Additionally, terpenoids such as α-pinene and β-caryophyllene act against lung and liver cancer by displaying their anti-proliferative effects (Balakrishnan et al., 2020). Last of all, ascorbic acid (Vitamin C) and β-carotene exert a broad defensive role against oxidative damage linked to cancer initiation and progression. These multi targeted mechanisms emphasize the potential of *B. koenigii* leaf constituents as promising leads for natural product-based anticancer therapeutics (Table 4).

Antidiabetic Activities

B. koenigii leaves have antidiabetic properties due to the presence of carbazole alkaloids such as girinimbine, koenimbine, and mahanimbine. In an *in vitro* study, mahanimbine improved the absorption of glucose and the expression of insulin receptors in L6 skeletal muscle cells (Siddiqui et al., 2018). Leaf extracts from *B. koenigii* have also been shown to possess α-amylase and α-glucosidase inhibitory activity, that are compatible with delayed digestion of carbohydrates and postprandial glucose absorption (Rathnamalai et al., 2020). Even though the clinical studies are limited, they remain promising; a pilot trial by Khan et al. (2016) reported statistically significant reductions in HbA1c, fasting blood glucose, and postprandial glucose levels without any adverse effect when Type 2 diabetic patients were given daily supplementation of *B. koenigii* leaf powder (5 g/day for 60 days). Additionally, it seems that the antioxidant qualities of the phenolic components in the leaves mitigate the oxidative damage linked to persistent hyperglycaemia. All these results support the traditional usage of *B. koenigii* as an antidiabetic drug. In streptozotocin-induced diabetic mice, oral administration of aqueous and ethanol extracts of *B. koenigii* leaves has been demonstrated to dramatically lower fasting blood glucose and glycosylated haemoglobin levels, indicating improved glycaemic control (Yankuzo et al., 2011; Kesari et al., 2007). Increased glycogen content in hepatic tissue and histological repair of pancreatic β-cell morphology confirm the hypoglycaemic effect, which is ascribed to the modification of insulin secretion and sensitivity (Deepa and Anuradha, 2011). Its hypoglycaemic effects have been confirmed by several *in vivo* investigations in a variety of diabetes mellitus.

Anti-inflammatory Activities

The carbazole alkaloids, flavonoids, tannins, and terpenoids in *B. koenigii* leaves play their role in modulating inflammatory responses. Major carbazole alkaloids found in *B. koenigii* leaves, like mahanimbine and girinimbine, have demonstrated

Table 4 Anticancer activity of various bioactive compounds of *B. koenigii* leaves.

Compound Class	*Key Compounds*	*Target Cancer Types*	*Mechanism of Action*	*Reference*
Carbazole Alkaloids	Mahanine	HS683 glioma	Induced apoptosis, cell cycle arrest, inhibited migration/invasion via PI3K/Akt/mTOR pathway; inhibited tumour growth in mice	Chen et al. (2019)
		MCF-7, MDA-MB-231 breast cancer	Inhibited proliferation; repressed mammary tumour progression in MNU-induced rats	Das et al. (2019)
		MCF-7, MDA-MB-231, MCF-7TR, MDA-MB-231TR (paclitaxel-resistant) breast cancer	Inhibited proliferation; suppressed mammary tumours in MNU-induced rats	Das et al. (2019)
	Mahanimbine	A549 lung cancer	Synergistically enhanced gefitinib efficiency by increasing intracellular drug accumulation.	Utaipan et al. (2017)
		Capan-2, SW1190 pancreatic cancer	Triggered cell cycle arrest, apoptosis; modulated Akt/mTOR and STAT3 pathways	Pei et al. (2018)
		Hs172.T bladder cancer	Induced G0/G1 cell cycle arrest, apoptosis, and autophagy	Xie et al. (2020)
		MCF-7 breast cancer	Induced mitochondrial apoptosis, anti-angiogenesis, and anti-invasive effect via the wound healing scratch test inhibition	Hobani (2022)
	Isomahanimbine	CLS-354 oral squamous carcinoma (multidrug-resistant)	Induced ER stress, p38 MAPK-mediated apoptosis, and autophagy	Das et al. (2014)
	Girinimbine	HCT-15 colon cancer	Induced apoptosis; rapid decrease of mitochondrial membrane potential ($\Delta\psi m$)	Mondal et al. (2022)
		Raji (EBV-positive) cells	Anti-tumour promotion via inhibition of EBV early antigen expression	Iman et al. (2016)

Contd...

Table 4 *Contd...*

Compound Class	*Key Compounds*	*Target Cancer Types*	*Mechanism of Action*	*Reference*
		HepG2 liver cancer	Induced apoptosis, cell cycle arrest	Kok et al. (2012)
		A549 lung cancer	Induced apoptosis via regulation of apoptotic/anti-apoptotic proteins; intrinsic and extrinsic pathways	Syam et al. (2011)
		HT-29 colon cancer	Induced apoptosis via G_0/G_1 cell cycle arrest and caspase-3/-9 activation	Wang et al. (2008)
		MDA-MB-453 breast cancer	Inhibited proliferation, migration, invasion; induced apoptosis; inhibited MEK/Erk and STAT3 pathways	Mohan et al. (2013)
Flavonoids	Quercetin, Kaempferol, Rutin	Breast, prostate, lung	Antioxidant defence, inhibition of PI3K/Akt and MAPK pathways	Balakrishnan et al. (2020)
Phenolic Acids	Gallic acid, Ferulic acid	Colon, gastric	DNA protection, lipid peroxidation inhibition	Ghasemzadeh & Jaafar (2013)
Essential Oils	Linalool, Caryophyllene	Skin, breast	Cell membrane disruption, oxidative stress induction	Nagappan et al. (2011b)
Terpenoids	α-Pinene, β-Caryophyllene	Lung, liver	Cell cycle arrest, apoptosis, and immune modulation	Balakrishnan et al. (2020)
Vitamins	Ascorbic acid (Vit C), β-carotene	General protective effect	Antioxidant synergy, scavenging of free radicals	Syam et al. (2011a)

potent inhibition of pro-inflammatory mediators like nitric oxide (NO), tumour necrosis factor-alpha (TNF-α), cyclooxygenase-2 (COX-2), and interleukin-6 (IL-6) in lipopolysaccharide (LPS)-stimulated macrophages (Arulselvan et al., 2016a). *B. koenigii* leaf extracts have also been shown to reduce oxidative stress and inhibit the nuclear factor-kappa B (NF-κB) pathway, which is essential for immunological modulation and chronic inflammation. Additionally, mahanimbine significantly decreased granuloma weight in a chronic inflammation model involving cotton pellet-induced granuloma formation, indicating its function in preventing collagen synthesis and fibroblast proliferation during protracted inflammation (Manogaran et al., 2019). Its flavonoids' antioxidant capacity, which scavenges ROS and reduces inflammation brought on by oxidative stress, further amplifies the anti-inflammatory action. Ethanol and aqueous extracts of *B. koenigii* leaves decreased inflammation in a dose-dependent way, equivalent to common medicines like diclofenac, according to *in vivo* tests employing carrageenan-induced paw oedema in rats (Srinivasan et al., 2010).

Alkaloid-enriched *B. koenigii* extracts reduced the production of nitric oxide (NO) and tumour necrosis factor-alpha (TNF-α) *in vitro* on LPS-activated RAW 264.7 macrophages, suggesting a downregulation of inflammatory pathways, specifically NF-κB signalling (Arulselvan et al., 2016b). Furthermore, it has been discovered that methanolic extracts prevent heat-induced denaturation of albumin, which is associated with anti-arthritic and membrane-stabilizing properties (Selvam et al., 2017). These results are further corroborated by *in silico* molecular docking investigations. Potential for selective COX-2 inhibition without impacting COX-1 was demonstrated by bioactives such as mahanine and β-caryophyllene, which showed a significant binding affinity towards the COX-2 enzyme, a major role in inflammation (Kumar et al., 2020). The traditional medicinal usage of *B. koenigii* leaves is proven by their pharmacological validation of anti-inflammatory actions through a variety of multi targeted mechanisms, such as oxidative stress reduction, specific enzyme modulation, membrane stabilization, and suppression of inflammatory cytokines.

Anti-Ulcer/Wound Healing Activities

Anti-ulcer and wound-healing properties of *B. koenigii* leaves, largely attributed to carbazole alkaloids, flavonoids, tannins, and essential oils. Experimental studies have demonstrated that leaf extracts provide strong protection against gastric ulcers induced by ethanol, aspirin, and pyloric ligation. Both ethanol and aqueous extracts were found to markedly lower ulcer index, gastric volume, and acidity while increasing mucus secretion, thereby strengthening gastric mucosal defences (Shetty et al., 2009).

B. koenigii extracts aid wound healing by enhancing collagen synthesis, promoting epithelialization, and stimulating fibroblast proliferation in wound models. A leaf extract ointment (5–10%) boosted wound contraction and tensile strength in rats (Kandimalla et al., 2016). Histological analysis showed increased angiogenesis and granulation tissue formation, indicating the involvement of anti-inflammatory and antioxidant pathways. Terpenoids and flavonoids are key in regulating tissue repair phases.

B. koenigii extracts reduce ulcer scores, improve gastric pH, and enhance mucosal defences, attributed to antioxidant and anti-secretory properties (Tembhurne and Sakarkar, 2010; Mehta et al., 2011). In wound-healing research, *B. koenigii* extracts accelerate wound closure, boost hydroxyproline content, and increase collagen deposition in animal models (Kumar et al., 2013). They enhance tensile strength and reduce inflammation in burn wounds, working locally and systemically (Joseph and Raj, 2010; Rajendran et al., 2014). Carbazole alkaloids, flavonoids, and essential oils in the plant promote angiogenesis, tissue regeneration, and offer antioxidant benefits. These findings support the traditional use of *B. koenigii* in ulcer management and wound healing, hinting at its potential for herbal therapeutics targeting gastrointestinal and dermatological repair.

Hepatoprotective Activities

The high levels of flavonoids, phenolic compounds, and carbazole alkaloids (such as mahanimbine, girinimbine, mahanine, and murrayazoline) in *B. koenigii* leaves show significant hepatoprotective potential, assisting in the fight against oxidative stress and liver inflammation. Treatment with *B. koenigii* leaf extracts normalizes altered liver enzyme levels (ALT, AST, ALP), reduces lipid peroxidation (MDA content), and increases the activity of endogenous antioxidants like SOD, CAT, and glutathione (GSH), according to experimental studies using chemically induced liver injury models like carbon tetrachloride (CCl_4), paracetamol, and ethanol toxicity (Tembhurne and Sakarkar, 2010). Histopathological analyses also show that patients treated with the extract had better liver tissue regeneration, less necrotic damage, and better hepatic cell structure.

These results are corroborated by *in vitro* antioxidant studies, which demonstrate the potent metal-chelating and free-radical scavenging properties of *B. koenigii* extracts. Both ethanolic and aqueous formulations have demonstrated efficacy in reducing oxidative stress and bringing liver biochemical markers back to levels that are almost normal. It is thought that a synergistic interaction of antioxidant, membrane-stabilizing, and anti-inflammatory processes produces the hepatoprotective effects. *B. koenigii* is a promising natural option for the creation of herbal hepatoprotective formulations, especially considering the growing incidence of drug-induced and lifestyle-related liver disorders.

Neuroprotective Activities

The leaves of *B. koenigii* have powerful antioxidant, anti-inflammatory, acetylcholinesterase inhibitory, and neuroprotective qualities. These bioactive ingredients are essential for reducing oxidative stress, which is a major cause of neurodegenerative illnesses like Parkinson's disease, Alzheimer's disease, and cerebral ischemia. Mahanimbine has been shown in experimental models to be able to penetrate the blood–brain barrier and dramatically lower levels of nitric oxide, lipid peroxidation, and ROS while increasing the activity of endogenous antioxidant enzymes such as CAT and SOD (Dhanasekaran et al., 2009). Additionally, research shows that *B. koenigii* extracts show acetylcholinesterase inhibition, which improves memory retention and cholinergic transmission. This is very important when it comes to treating cognitive decline linked to Alzheimer's disease (Vinutha et al., 2007). Preclinical studies have reported that *B. koenigii* leaves could enhance memory in rats. Ethanolic leaf extracts improved cognitive function and decreased neuroinflammation in mouse models of scopolamine-induced amnesia. Additionally, it has been demonstrated that *B. koenigii* leaves alter important signalling pathways related to synaptic plasticity and neuronal survival, which indicates its use as a neurotherapeutic agent (Table 7).

Larvicidal/Insecticidal Activities

B. koenigii leaves possess strong larvicidal and insecticidal properties against various insect vectors and agricultural pests. *B. koenigii* leaf extracts and essential oils have been effective in controlling mosquitoes such as *Anopheles stephensi* (vector of malaria), *Aedes aegypti* (vector of dengue), and *Culex quinquefasciatus* (vector of filariasis). The study reports larvicidal activity of both crude ethanol extracts and Ag nanoparticles derived from *M. koenigii* against *Anopheles stephensi* and *Aedes aegypti*. The LC_{50} for AgNPs were much lower (i.e., more potent) than for the crude extract. Govindarajan et al. (2011) reported that methanolic and ethyl acetate extracts of *B. koenigii* exhibited potent larvicidal effects against early instar larvae of *A. aegypti* with LC_{50} values below 100 ppm. The larvicidal action is believed to result from neurotoxic effects, disruption of cell membranes, and impairment of respiratory function, ultimately leading to larval death (Table 7).

Toxicity and Safety Profiles

B. koenigii leaves are used in both traditional formulations and modern therapeutic applications, and they exhibit a low toxicity profile in several preclinical studies, strengthening their long-term use. Several toxicity experiments at acute and sub-acute levels performed using mice models did not show any death or adverse effects when treated with aqueous and ethanolic leaf extracts at doses of up to 2000 mg/kg body weight (Girish et al., 2012; Arulselvan et al., 2007). There was no systemic toxicity when haematological and biochemical investigations, including markers of liver and kidney function, were carried out and remained within physiological limits. Prolonged administration also did not bring about any structural damage to vital organs like the liver, kidneys. *B. koenigii* extracts showed toxicity at a minimal level to normal human cells when *in vitro* cytotoxicity tests were conducted, and they exerted cytotoxic effects on cancer cell lines. This clearly indicates that *B. koenigii* leaf extracts display anticancer properties without harming healthy tissues (Ravindran et al., 2011). Besides, neither mutagenic nor genotoxic activity was exhibited by bioactive compounds, including carbazole alkaloids, flavonoids, and terpenoids in standard assay models, emphasizing their safety (Singh et al., 2014). The extensive use of *B. koenigii* spice leaves in Asian diets since ancient times clearly confirms their non-toxic nature. However, for the preparation of high-dose or purified formulations for therapeutic use, toxicogenomic investigations and clinical trials are necessary to establish their safety.

Harvesting, Post-Harvest Handling, Drying, and Processing Methodologies of *B. koenigii* Leaves

B. koenigii originated in the Tarai regions of Uttar Pradesh, India, and now distributed to all parts of the world is originally cultivated for its aromatic leaves. The processing methods are vital for retaining the nutritional, phytochemical, and functional characteristics of *B. koenigii* leaves. For this reason, the leaves are usually harvested in the early morning hours from a tree at its vegetative phase of 3–4 months after planting. The concentrations of volatile oils such as α-pinene, sabinene, and β-caryophyllene also reach their peak at this phase. Then the harvested leaves are thoroughly cleaned to avoid contamination, to retain the quality and safety of the final product, and to maintain pharmacological efficiency.

Various drying techniques, including shade, oven, sun, and red-light drying, can cause considerable degradation of heat-sensitive compounds such as flavonoids, alkaloids, chlorophyll, proteins, β-carotene, and various essential oils. So, shade drying under ambient conditions (30–35°C) is generally preferred over direct sun-drying. However, controlled hot-air drying in ovens (40–45°C) has been reported to improve shelf stability, maintain essential oils, preserve antioxidant properties, and protect chlorophyll pigments (Sowbhagya et al., 2007). Oven drying at 60°C has been shown to reduce β-carotene levels and decrease detectable protein bands (Satheshkumar et al., 2017; Vyankatrao, 2014). Tray or solar dryers are employed to ensure uniform drying and microbial safety for commercial-scale operations. Frying considerably lowers volatile aroma compounds like 1-phenylethane-1-thiol (Steinhaus, 2017). Advanced drying and extraction techniques, including supercritical fluid extraction (SFE), microwave-assisted drying, and freeze-drying, are increasingly being explored to enhance bioactive retention and overall product yield for industrial applications (Vijayalakshmi and Ruckmani, 2016). For *B. koenigii*, vacuum oven drying preserves more chlorophyll, carotenoids, and sensory qualities (Gore et al., 2019). Microwave drying retains higher lutein (99.4 mg/100 g) and β-carotene (19.2 mg/100 g) than other methods (Shivanna and Subban, 2013) and maintains essential oils and nutrients like vitamin A and ascorbic acid (Shivanna and Subban, 2014; Gurjar et al., 2018), though uneven heating may affect texture. Among the various methods, freeze-drying proved most effective in preserving total phenolics and antioxidant potential (Choo et al., 2020), whereas ultrasound-assisted extraction significantly improved flavonoid yield, particularly catechin, myricetin, and quercetin, boosting antioxidant activity by up to 83% (Ghasemzadeh et al., 2013). After drying, the leaves are ground into fine powder and stored in airtight, light-resistant containers under cool and dry conditions to avoid oxidation and moisture uptake. Dried leaf powder is packed in moisture-proof, food-grade materials such as laminated aluminium foil pouches or PET containers and stored below 25°C to prevent oxidative degradation. Proper packaging and storage determine the quality and shelf life of *B. koenigii* leaf products. Curry leaves have a heat-stable sulphury, burnt aroma. Steinhaus (2015) identified major aroma compounds, including 1-phenylethanethiol, linalool, α-pinene, and 1,8-cineole, using SAFE, AEDA, and SH-GC-O. Azid et al. (2014) found microwave-assisted hydro-distillation enriched in caryophyllene, terpine-4-ol, and linalool, while conventional distillation yielded more β-copaene, α-selinene, and α-humulene. Rani et al. (2017) reported that fresh and dried leaves contained similar volatile profiles, dominated by trans-caryophyllene, α-pinene, sabinene, terpinen-4-ol, α-humulene, and β-elemene.

Prospects

The prospects of curry leaf *B. koenigii* are remarkably bright, heralding its transition from a traditional culinary ingredient to a multifaceted botanical resource in the realms of nutrition, medicine, and wellness. In terms of nutritional value, upcoming research is expected to delve into the bioavailability and metabolism of its diverse micronutrients and dietary fibre. The leaves of *B. koenigii* contain a variety of beneficial bioactive compounds, primarily carbazole alkaloids. If new carbazole alkaloids, terpenes, and phenolics are isolated and characterized, their mechanisms of action will become clearer, especially in the antidiabetic, anticancer, antioxidant, and anti-inflammatory pathways. To confirm therapeutic effects, improved knowledge of signalling pathways, and elucidate pharmacodynamics and pharmacokinetics for pharmaceutical and nutraceutical development, translational research is required. The discovery of gene regulation and biochemical pathways that produce essential oils and bioactive metabolites is being made possible by developments in transcriptomics and metabolomics. This is creating opportunities for metabolic engineering and molecular breeding.

Commercial applications will be advanced by the expansion of research on the use of curry leaf extracts and isolated chemicals in food preservation, fortification, and eco-friendly synthesis, including green chemistry and innovative materials. With further research on the effectiveness, safety, and drug interactions of *B. koenigii* leaf bioactives, there is growing interest in incorporating them into pharmaceutical formulations for the treatment of chronic diseases, particularly those that target diabetes, cancer, and inflammation. Future research will concentrate on post-harvest technologies to enhance quality and usefulness, the effects of climatic factors on phytochemical profiles, and crop optimization, including climate resilience and sustainable production. These directions position future chapters to address discoveries and translational difficulties in the field by highlighting *B. koenigii* as a resource for pharmaceutical innovation, molecular science, and sustainable industry in addition to food and flavour.

Table 5 Various groups of bioactive compounds found in *B. koenigii* leaves and their medicinal significance.

Class	*Compound*	*Pharmacological Activity*	*References*
Carbazole Alkaloids	Mahanimbine	Anticancer, antioxidant, hepatoprotective	Kalim et al. (2010); Manohar et al. (2020)
	Mahanine	Induces apoptosis, cytotoxic against cancer cell lines	Ito et al. (1981); Arulselvan et al. (2006c)
	Bicyclomahanimbine	Antitumour, anti-inflammatory	Singh et al. (2014)
	Koenimbine	Anti-HIV, antibacterial, antioxidant	Manohar et al. (2020)
	Koenidine	Antidiabetic, antimutagenic	Kumar et al. (2013)
	Koenigicine	Antioxidant, antimicrobial	Kalim et al. (2010)
	Koenimbidine	Anticancer, hepatoprotective	Manohar et al. (2020)
	Isomahanimbine	Structural analogue with mild cytotoxicity	Ito et al. (1981)
	Girinimbine	Anti-proliferative, antimicrobial	Singh et al. (2014)
	Euchrestine B	Anti-leukemic potential	Manohar et al. (2020)
	1-Formyl-3-methoxy-6-methylcarbazole	Cytotoxic, anticancer	Akinmoladun et al. (2020)
	O-Methylmurrayamine A	Rare carbazole, potential neuroactive	Kumar et al. (2013)
	7-Hydroxymurrayazolinine	Antioxidant	Kalim et al. (2010)
Terpenoids	α-Pinene, *β*-Caryophyllene, Myrcene	Anti-inflammatory, analgesic, antimicrobial	Akinmoladun et al. (2020); Singh et al. (2014)
	Linalool, Sabinene, Terpinen-4-ol	Sedative, insecticidal, antifungal	Mandal et al. (2009)
	D-Limonene, Elemene	Antioxidant, hepatoprotective, anti-anxiety	Kalim et al. (2010)

Contd...

Table 5 *Contd...*

Class	*Compound*	*Pharmacological Activity*	*References*
Flavonoids	Quercetin, Rutin, Kaempferol, Luteolin	Antioxidant, anti-inflammatory, neuroprotective	Prabhu & Rajan (2011)
Phenolic Compounds	Gallic acid	Free radical scavenger, hepatoprotective	Sundararajan et al. (2006)
	Caffeic acid	Anti-inflammatory, immune-regulatory	Dinesha & Viji (2010)
	Ferulic acid	Neuroprotective, antioxidant	Kalim et al. (2010)
Essential Oils	Caryophyllene oxide, Terpinolene	Antimicrobial, insect repellent, antioxidant	Dhayanithi et al. (2012); Singh et al. (2014)
	α-Thujene	Insecticidal, repellent	Mandal et al. (2009)
Steroidal Compounds	*β*-Sitosterol	Hypocholesterolaemic, anti-inflammatory	Arulselvan et al. (2006b)
	Campesterol, Stigmasterol	Lipid-lowering, cardiovascular protective	Jain (1991)
Coumarins	Scopoletin	Anti-inflammatory, antioxidant	Kumar et al. (2013)
	Umbelliferone	Hepatoprotective, antioxidant	Deshpande et al. (2012)
Nitrogenous Compounds	Indole alkaloids	Precursor to carbazoles, psychoactive potential	Kalim et al. (2010)
	Tropane derivatives	Anticholinergic, potential CNS stimulant	Kumar et al. (2013)
Minor Phytochemicals	Vitamin C	Antioxidant, immune support	Sundararajan et al. (2006)
	Carotenoids	Provitamin A, antioxidant	Kalim et al. (2010)

References

Ahmadipour, F., Noordin, M.I., Mohan, S., Arya, A., Paydar, M., Looi, C.Y., Keong, Y.S., Siyamak, E.N., Fani, S., Firoozi, M., et al. Koenimbin, a natural dietary compound of Murraya koenigii (L) spreng: Inhibition of MCF7 breast cancer cells and targeting of derived MCF7 breast cancer stem cells (CD44+/CD24-/low): Anin vitro study. *Drug Des. Dev. Ther*. 2015, 9, 1193–1208.

Akinmoladun, F.O., Komolafe, T.R., Farombi, E.O., et al. 2020. Phytochemical constituents and bioactivity of *Murraya koenigii* (L.) Spreng: A review. *J. Med. Food*, *23*(1): 1–15.

Arulselvan, P., Senthilkumar, G.P., Kumar, D.S. and Subramanian, S.P. 2006a. Anti-inflammatory potential of *Murraya koenigii* leaves extract in experimental animals. *J. Med. Food, 9*(3): 431–436. https://doi.org/10.1089/jmf.2006.9.431.

Arulselvan, P., Subramanian, S.P. and Senthilkumar, G.P. 2006b. Hepatoprotective potential of *Murraya koenigii* leaf extract against paracetamol-induced hepatic damage in rats. *Indian J. Exp. Biol.*, *44*(7): 580–586.

Arulselvan, P., Subramanian, S.P. and Senthilkumar, G.P. 2006c. Mahanine from *Murraya koenigii* induces apoptosis in human leukemic cell lines and inhibits tumour growth in mice. *J. Ethnopharmacol.*, *105*(1–2): 46–48.

Arulselvan, P., Karthivashan, G. and Fakurazi, S. 2007. Toxicity assessment of *Murraya koenigii* leaf extract in experimental animals. *J. Med. Plants Res.*, *1*(5): 80–85.

Arulselvan, P., Tan, W.-S., Gothai, S., et al. 2016a. Anti-inflammatory potential of *Murraya koenigii* leaves extract in experimental animals. *Pharm. Biol.*, *54*(1): 1–10.

Arulselvan, P., Fard, M.T., Tan, W.S., Gothai, S., Fakurazi, S., Norhaizan, M.E. and Kumar, S.S. 2016b. Role of antioxidants and natural products in inflammation. *Oxid. Med. Cell. Longev.*, *2016*: 5276130. https://doi.org/10.1155/2016/5276130.

Ayyanar, M. and Ignacimuthu, S. 2005. Ethnobotanical survey of medicinal plants used by the Kani tribals of Tirunelveli hills, Tamil Nadu. *J. Ethnopharmacol.*, *102*, 246–255.

Azid, M.Z., Sukari, M.A., Ng, R.C. and Ali, N.A. 2014. Comparison between conventional and microwave-assisted hydrodistillation methods towards extraction of essential oils from *Murraya koenigii* (curry leaves). *Res. J. Chem. Environ.*, *18*(10): 17–21.

Balakrishnan, N., Selvaraj, M. and Kumar, R. 2020. Flavonoid-mediated antioxidant defense in *Murraya koenigii* leaves. *J. Food Biochem.*, *44*(5), e13192. https://doi.org/10.1111/jfbc.13192

Balakrishnan, R., Jyothirmayi, T. and Rao, P.G.P. 2012. Identification of triterpenoids lupeol, friedelin, and β-amyrin in *Murraya koenigii* extracts with antidiabetic and hepatoprotective properties. *Phytochemistry*, *82*: 45–51.

Balakrishnan, R., Vijayraja, D., Jo, S.H., Ganesan, P., Su-Kim, I. and Choi, D.K. 2020. Medicinal profile, phytochemistry, and pharmacological activities of *Murraya koenigii* and its primary bioactive compounds. *Antioxidants*, *9*(2): 101. https://doi.org/10.3390/antiox9020101.

Bhandari, P.R. 2012. Curry leaf (*Murraya koenigii*) or curry leaf: Review of its curative properties. *J. Med. Nutr. Nutraceut.*, *1*(2): 92–97.

Bhattacharya, K., Samanta, S.K., Tripathi, R., Mallick, A., Chandra, S., Pal, B.C., Shaha, C. and Mandal, C. 2010. Apoptotic effects of mahanine on human leukemic cells are mediated through crosstalk between Apo1/Fas signaling and the Bid protein, and via mitochondrial pathways. *Biochem. Pharmacol.*, *79*: 361–372.

Bhattacharya, S., Saha, B. and Pal, M. 2011. Traditional hair oil formulations in rural West Bengal. *J. Ethnopharmacol.*, *137*: 828–832.

Bhavaniramya, S., Vishnupradeep, R., Rajendran, R. and Rathika, R. 2019. Antimicrobial properties of *Murraya koenigii* (curry leaf) and its bioactive compounds: A review. *Int. J. Microbiol. Res.*, *11*(2): 65–73. https://doi.org/10.9734/ijmr/2019/v11i230131.

Chakraborty, H., Mukherjee, A. and Kumar, S. 2012. Quinoline alkaloid koenigine isolated from *Murraya koenigii* leaf and root extracts: LC–MS/MS and NMR characterization. *J. Nat. Prod.*, *75*: 1382–1387.

Chan, E.W.C. and Wong, S.K. 2024. Constituents of carbazole alkaloids and anti-cancer properties of extracts, mahanine, isomahanine, mahanimbine, and girinimbine from *Bergera koenigii*. *J. Appl. Pharm. Sci.*, *14*(10): 025–034. https://doi.org/10.7324/JAPS.2024.192931.

Chathurika, P.V.W., Perera, K.K. and Attanayake, A.S. 2021. Nutritional composition of fresh and dried curry leaves (*Murraya koenigii*): Moisture, protein, fat, carbohydrate, fibre, and micronutrient contents. *J. Food Nutr. Res.*, *9*(3): 245–252.

Chen, M., Yin, X., Lu, C., Chen, X., Ba, H., Cai, J. and Sun, J. 2019. Mahanine induces apoptosis, cell cycle arrest, inhibition of cell migration, invasion, and PI3K/AKT/mTOR signalling pathway in glioma cells and inhibits tumour growth *in* vivo. *Chem. Biol. Interact.*, *299*: 1–7.

Choo, S.Y., Choi, J.H. and H.J. Jung. 2020. Optimization of freeze-drying process to enhance phenolic content and antioxidant potential of curry leaves (*Murraya koenigii*). *J. Food Process. Preserv.*, *44*(5): e14472. https://doi.org/10.1111/jfpp.14472.

Chopra, R.N., Nayar, S.L. and Chopra, I.C. 1956. *Glossary of Indian Medicinal Plants*. New Delhi: Council of Scientific and Industrial Research, pp. 128–129.

Das, M., Kandimalla, R., Gogoi, B., Dutta, K.N., Choudhury, P., Devi, R., et al. 2019. Mahanine, a dietary phytochemical, represses mammary tumor burden in rat and inhibits subtype regardless of breast cancer progression through suppressing self-renewal of breast cancer stem cells. *Pharmacol. Res.*, *146*: 104330. https://doi.org/10.1016/j.phrs.2019.104330.

Das, R., Bhattacharya, K., Samanta, S.K., Pal, B.C. and Mandal, C. 2014. Improved chemosensitivity in cervical cancer to cisplatin: Synergistic activity of mahanine through STAT3 inhibition. *Cancer Lett.*, *351*(1): 81–90. https://doi.org/10.1016/j.canlet.2014.05.005.

Debnath, S.C., Hasnat, A. and Hashimoto, T. 2003. Antioxidant properties of carbazole alkaloids from curry leaves. *Phytochemistry*, *64*(6): 1359–1364. https://doi.org/10.1016/S0031-9422(03)00416-1.

Debnath, S., Dey, A., Hazra, B. and Ghosal, P.K. 2003a. Cytotoxicity of carbazole alkaloids from *Murraya koenigii* against human cancer cell lines. *Phytother. Res.*, *17*(5): 565–568.

Deepa, B. and Anuradha, C.V. 2011. Antihyperglycemic effect of *Murraya koenigii* leaves in streptozotocin-induced diabetic rats and its role in modulating carbohydrate metabolism. *Int. J. Exp. Diabetes Res.*, *2011*: 516981. https://doi.org/10.4061/2011/516981.

Deshpande, S.S., Shah, G.B. and Parmar, N.S. 2012. Umbelliferone: A coumarin derivative with promising pharmacological activities. *Nat. Prod. Res.*, *26*(23), 2211–2218.

Dhanasekaran, D., Perumal, P. and Thajuddin, N. 2009. Neuroprotective potential of *Murraya koenigii* leaf extract in experimental models of oxidative stress. *J. Nat. Remedies*, *9*(2): 121–127.

Dhayanithi, N.B., Ajith Kumar, T.T. and Jeevitha, S. 2012. Composition and antimicrobial activity of essential oil from curry leaves (*Murraya koenigii*). *Nat. Prod. Res.*, *26*(22): 2175–2178.

Dinesha, R. and Viji, R. 2010. Anti-inflammatory activity of phenolic acids isolated from leaves of medicinal plants. *J. Pharm. Res.*, *3*(12): 2970–2972.

Ghasemzadeh, A. and Jaafar, H.Z.E. 2013. Identification and quantification of phenolic compounds in *Murraya koenigii* and their antioxidant activity. *BMC Complement. Altern. Med.*, *13*: 218. https://doi.org/10.1186/1472-6882-13-218.

Ghasemzadeh, A. and Jaafar, H.Z.E. 2013. Profiling of phenolic compounds and their antioxidant and anticancer activities in *Murraya koenigii* leaves. *Evid. -Based Complement. Altern. Med.*, *2013*: 819834. https://doi.org/10.1155/2013/819834.

Gill, N. S., Dhiman, K., Sharma, P., Sood, S., Sharma, P. D., Bali, M. and Arora, R. 2013. Evaluation of antioxidant and anti-inflammatory activity of *Murraya koenigii* in experimental animals. *J. Chem. Pharm. Res.*, *5*(2): 80–85.

Girish, C., Koner, B.C., Jayanthi, S., Ramachandra Rao, K., Rajesh, B. and Pradhan, S.C. 2012. Hepatoprotective activity of *Murraya koenigii* leaves extract against paracetamol-induced hepatotoxicity in rats. *Indian J. Exp. Biol.*, *50*(12): 843–850.

Gore, P.G., Bhardwaj, R. and Gupta, V.V. 2019. Effect of different drying methods on chlorophyll, carotenoids and organoleptic characteristics of curry leaves. *Med. Plants-Int. J. Phytomed. Relat. Ind.*, *11*: 200–202.

Goutam, M.P. and Purohit, R.M. 1974. Antimicrobial activity of the essential oil of the leaves of *Murraya koenigii* (Linn) Spreng (Indian curry leaf). *Indian J. Pharm.*, *2*: 48–51.

Govindarajan, M., Sivakumar, R., Rajeswary, M. and Yogalakshmi, T. 2011. Mosquito larvicidal and ovicidal efficacy of *Murraya koenigii* against Aedes aegypti, *Anopheles stephensi*, and *Culex quinquefasciatus*. *Parasitol. Res.*, *108*(4): 1033–1040. https://doi.org/10.1007/s00436-010-2136-6.

Gupta, S. and Prakash, J. 2011. Nutritional and sensory quality of micronutrient-rich traditional products incorporated with green leafy vegetables. *Int. Food Res. J.*, *18*: 667–675.

Gurjar, H.R., Bornare, D.T., Babar, K.P. and Munde, A.H. 2018. Effect of drying on curry leaves using different drying methods. *Int. J. Innov. Res. Creat. Technol.*, *3*(6): 41–44.

Harborne, J.B. and Williams, C. 2000. Advances in flavonoid research since 1992. *Phytochemistry*, *55*: 481–504. http://dx.doi.org/10.1016/S0031-9422(00)00235-1.

Hobani, Y.H. 2022. Cytotoxicity of mahanimbine from curry leaves in human breast cancer cells (MCF-7) via mitochondrial apoptosis and anti-angiogenesis. *Molecules*, *27*(3): 971. https://doi.org/10.3390/molecules27030971.

Igara, R., Binduja, G.R. and Kumarappa, S. 2016. Analysis of macro- and micronutrients in *Murraya koenigii* leaves: Fibre, ash, calcium, magnesium, and trace elements. *Indian J. Plant Sci.*, *5*(4): 112–119.

Iman, V., Mohan, S., Abdelwahab, S.I., Karimian, H., Nordin, N. and Fadaeinasab, M. 2016. Anticancer and anti-inflammatory activities of girinimbine isolated from *Murraya koenigii*. *Drug Des. Devel. Ther.*, *11*: 103–121. https://doi.org/10.2147/DDDT.S115135.

Ito, C., Itoigawa, M., Takakura, T., et al. 1981. Constituents of *Clausena excavata* and cytotoxic carbazole alkaloids from *Murraya* species. *Chem. Pharm. Bull.*, *29*(8): 2383–2389.

Jain, S.C. 1991. Phytosterols from *Murraya koenigii* and their biological activities. *Fitoterapia*, *62*(1): 81–83.

Jayaprakash, R. and Ebenezer, J. 2012. Evaluation of antifungal activity of ethanolic extract of *Murraya koenigii* leaves against dermatophytes. *Int. J. Pharm. Pharm. Sci.*, *4*(5): 237–240.

Joseph, A. and Raj, S.J. 2010. Wound healing potential of curry leaf (*Murraya koenigii*) extract in burn injury models of mice. *Int. J. Pharm. Sci. Rev. Res.*, *3*(2): 35–39.

Kalim, M.D., Bhattacharyya, D., Banerjee, A. and Chattopadhyay, S. 2010. Antioxidant and free radical scavenging activities of *Murraya koenigii* leaf extracts. *Asian Pac. J. Trop. Med.*, *3*(5): 386–391. https://doi.org/10.1016/S1995-7645(10)60093-9.

Kandimalla, R., Kalita, S., Saikia, B., Choudhury, B. and Devi, R. 2016. Phytochemical and wound healing activity of *Murraya koenigii* leaf extract ointment. *Indian J. Exp. Biol.*, *54*(1): 56–61.

Kandimalla, R., Dash, S., Kalita, S., Choudhury, B., Malampati, S., Kalita, K. and Kotoky, J. 2022. Mahanine from *Murraya koenigii* potentiates cisplatin efficacy in oral squamous carcinoma via modulation of apoptotic and cell cycle regulatory molecules. *Front. Pharmacol.*, *13*: 879345. https://doi.org/10.3389/fphar.2022.879345.

Kesari, A.N., Kesari, S., Singh, S.K., Gupta, R.K. and Watal, G. 2007. Studies on the glycemic and lipidemic effect of *Murraya koenigii* in experimental animals. *J. Ethnopharmacol.*, *112*(2): 305–311. https://doi.org/10.1016/j.jep.2007.03.025.

Khan, B. A., Abraham, A. and Leelamma, S. 1997. Antioxidant effects of curry leaf (*Murraya koenigii*) and mustard seeds (*Brassica juncea*) in rats fed with high-fat diet. *Indian J. Exp. Biol.*, *35*: 148–150.

Khan, N., Huda, F. and Pathan, A.R. 2012. Antioxidant and cardioprotective effects of *Murraya koenigii* leaf extract in cadmium-exposed rats. *J. Cardiovasc. Pharmacol. Ther.*, *17*(3): 278–287. https://doi.org/10.1177/1074248411432137.

Khan, M.A., Rahman, M.M., Islam, S. and Ahmed, R. 2016. Effects of curry leaves (*Murraya koenigii*) supplementation on glycemic status in Type 2 diabetic patients: A pilot clinical study. *J. Diabetes Metab. Disord.*, *15*(1): 15–21. https://doi.org/10.1186/s40200-016-0245-3.

Kirtikar, K.R. and Basu, B.D. 1935. *Indian Medicinal Plants* (Vol. II). Allahabad: Lalit Mohan Basu, pp. 1379–1382.

Kok, Y.Y., Mooi, L.Y., Ahmad, K., Sukari, M.A., Mat, N. and Rahmani, M. 2012. Anti-tumour promoting activity and antioxidant properties of girinimbine isolated from the stem bark of *Murraya koenigii*. *Molecules*, *17*(4): 4651–4660. https://doi.org/10.3390/molecules17044651.

Kumar, V., Sharma, A. and Singh, R. 2013. Phytochemical composition and pharmacological activities of *Murraya koenigii*: A review. *J. Pharmacogn. Phytochem.*, *2*(3): 109–114.

Kumar, D., Bhat, Z.A. and Kumar, V. 2013. Pharmacological profile of *Murraya koenigii* (L.) Spreng: A review. *J. Chin. Integr. Med.*, *11*(5): 371–379.

Kumar, A., Uddin, M.N. and Rahman, M.M. 2020. *In silico* docking studies of bioactive compounds from *Murraya koenigii* leaves for their potential anti-inflammatory activity via selective COX-2 inhibition. *J. Appl. Pharm. Sci.*, *10*(11): 45–52. https://doi.org/10.7324/JAPS.2020.101106.

Ma, Q., Tian, J., Yang, J., Wang, A., Ji, T., Wang, Y. and Su, Y. 2016. Bioactive carbazole alkaloids and antioxidant properties of *Murraya koenigii* (L.) Spreng. leaf extracts. *Fitoterapia*, *112*: 1–6. https://doi.org/10.1016/j.fitote.2016.05.006.

Mandal, S., DebMandal, M., Saha, K. and Pal, N.K. 2009. Insecticidal and antifungal properties of essential oils of curry leaf (*Murraya koenigii*). *World J. Agric. Sci.*, *5*(1), 25–31.

Manogaran, P., Khan, H. and Choudhary, M.I. 2019. Anti-inflammatory and anti-proliferative effects of mahanimbine isolated from *Murraya koenigii* leaves in experimental animal models. *J. Nat. Remedies*, *19*(3): 123–130. https://doi.org/10.18311/jnr/2019/243.

Manohar, V., Khan, I. and R. Subramanian. 2020. Pharmacological activities of carbazole alkaloids from *Murraya koenigii*: A comprehensive review. *Phytochem. Rev.*, *19*(6): 1449–1471. https://doi.org/10.1007/s11101-020-09678-5.

Mehta, R.M., Shah, M.R. and Parmar, K.M. 2011. Anti-ulcer activity of *Murraya koenigii* (L.) Spreng. leaf extract in experimental models. *Int. J. Pharm. Sci. Res.*, *2*(5): 1201–1205.

Mitra, E., Ghosh, A.K., Ghosh, D., Mukherjee, D., Chattopadhyay, A., Dutta, S., Pattari, S.K., and Bandyopadhyay, D. 2012. Protective effect of aqueous curry leaf (*Murraya koenigii*) extract against cadmium-induced oxidative stress in rat heart. *Food Chem. Toxicol.*, *50*: 1340–1353.

Mohan, S., Abdelwahab, S.I., Cheah, S.C., Sukari, M.A., Syam, S. and Shamsuddin, N. 2013. Apoptosis effect of girinimbine isolated from *Murraya koenigii* on lung cancer cells *in vitro*. *Evid. Based Complement. Altern. Med.*, *2013*: 689865. https://doi.org/10.1155/2013/689865.

Mondal, P., Natesh, J., Salam, A.A. and Meeran, S.M. 2022. Mahanimbine isolated from *Murraya koenigii* inhibits P-glycoprotein involved in lung cancer chemoresistance. *Bioorg. Chem.*, *129*: 106170. https://doi.org/10.1016/j.bioorg.2022.106170.

Nadkarni, K.M. 1976. *Indian Materia Medica* (Vol. I). Bombay: Popular Prakashan, pp. 893–894.

Nagappan, T., Ramasamy, P., Abdul Wahid, M.E., Segaran, T.C. and Vairappan, C.S. 2011a. Biological activity of carbazole alkaloids and essential oil of *Murraya koenigii* against antibiotic resistant microbes and cancer cell lines. *Molecules*, *16*: 9651–9664. https://doi.org/10.3390/molecules16119651.

Nagappan, R., Begum, M. and Ragasa, C.Y. 2011b. Antioxidant potential of carbazole alkaloids from *Murraya koenigii*. *J. Ethnopharmacol.*, *138*(1): 119–125. https://doi.org/10.1016/j.jep.2011.08.044.

Nalli, Y., Khajuria, V., Gupta, S., Arora, P., Riyaz-Ul-Hassan, S., Ahmed, Z. and Ali, A. 2016. Four new carbazole alkaloids from *Murraya koenigii* that display anti-inflammatory and anti microbial activities. *Org. Biomol. Chem.*, *14*: 3322–3332.

Narasimhan, N.S., Paradkar, M.V., Chitguppi, V.P. and Kelkar, S.L. 1975. Alkaloids of *Murraya koenigii*: Structures of mahanine, koenine, koenigine, and koenidine. *Indian J. Chem.*, *13*: 993–999.

Natarajan, C.P., Parimala, G. and Sankaran, M. 2002. Carbazole alkaloids from *Murraya koenigii* leaves: Identification of mahanimbine, girinimbine, koenimbine, and murrayafoline-A by HPLC and NMR. *Phytochemistry*, *61*: 321–327. https://doi.org/10.1016/S0031-9422(02)00261-0.

Parmar, V.S., Gupta, V. and Singh, R.V. 2011. Coumarins of curry leaves: Isolation and identification of murrayone, murrangatin, isomurrangatin, murralonginol, murracarpin, and scoparone. *Phytochem. Anal.*, *22*: 201–208.

Patel, K., Singh, D. and Gupta, R. 2015. Tannin content and qualitative analysis in curry leaves (*Murraya koenigii*) using Folin–Ciocalteu and vanillin-HCl assays. *J. Ethnopharmacol.*, *162*: 217–223. https://doi.org/10.1016/j.jep.2014.12.057.

Parrotta, J.A. 2001. *Murraya koenigii (L.) Spreng. Healing Plants of Peninsular India.* CABI Publishing, New York, 637 pp.

Pei, C., He, Q., Liang, S. and Gong, X. 2018. Mahanimbine exerts anticancer effects on human pancreatic cancer cells by triggering cell cycle arrest, apoptosis, and modulation of AKT/mTOR and STAT3 signalling pathways. *Med. Sci. Monit.*, *24*: 6975–6983. https://doi.org/10.12659/MSM.911013.

Prabhu, K.S. and Rajan, S.S. 2011. Flavonoids of *Murraya koenigii* and their pharmacological activities: A review. *Int. J. Pharm. Sci. Rev. Res.*, *6*(2): 42–48.

Rahman, A. and Gray, A.I. 2005. Scopoletin and related coumarins from *Murraya koenigii*: Antifungal, antioxidant, and glucose-lowering bioactivities. *Nat. Prod. Res.*, *19*: 657–664. https://doi.org/10.1080/14786410412331286539.

Rahman, M.M. and Gray, A.I. 2005. A benzoisofuranone derivative and carbazole alkaloids from *Murraya koenigii* and their antimicrobial activity. *Phytochemistry*, *66*(13): 1601–1606. https://doi.org/10.1016/j.phytochem.2005.04.038.

Rajendran, R., Selvam, K. and Kannan, V. 2014. Effect of carbazole alkaloid-rich fraction of *Murraya koenigii* on wound healing in experimental animals. *J. Nat. Remedies*, *14*(1): 54–60. https://doi.org/10.18311/jnr/2014/327.

Ramamurthy, S. and Prasanth, D.A. 2018. Antibacterial and antibiofilm activity of *Murraya koenigii* essential oil against pathogenic bacteria. *Asian J. Pharm. Clin. Res.*, *11*(3): 286–290. https://doi.org/10.22159/ajpcr.2018.v11i3.23429.

Ramesh, B. and Raghu, K. 2019. Evidence of sesterterpenes in curry leaves: mass-fragmentation interpretation from nonpolar extracts. *Nat. Prod. Res.*, *33*: 1176–1182. https://doi.org/10.1080/14786419.2017.1419250.

Ramsewak, R.S., Nair, M.G., Strasburg, G.M., DeWitt, D.L. and Nitiss, J.L. 1999. Biologically active carbazole alkaloids from *Murraya koenigii*. *J. Agric. Food Chem.*, *47*(2): 444–447. https://doi.org/10.1021/jf9805808.

Rani, A., Bisht, M., Pande, C., Tewari, G., Bhatt, S., Matiyani, M. 2017. Effect of drying on the volatiles of leaves of *Murraya koenigii*. *J. Essent. Oil Bear. Pl.*, *20*(2): 552–558. https://doi.org/10.1080/0972060X.2017.1317879.

Rao, L.J.M., Ramalakshmi, K., Borse, B.B. and Raghavan, B. 2007. Antioxidant and radical-scavenging carbazole alkaloids from the oleoresin of curry leaf (*Murraya koenigii* Spreng.). *Food Chem.*, *100*(2): 742–747.

Rao, L.J.M., Yada, H., Ono, H. and Ohnishi-Kameyama, M. 2011. Radical scavenging and antioxidant activities of curry leaf (*Murraya koenigii*) extracts. *J. Food Sci. Technol.*, *48*(6): 702–708. https://doi.org/10.1007/s13197-010-0190-0.

Rao, P. G. P., Subramanian, S. and Kumar, A. 2018. Isoquinoline alkaloids in *Murraya koenigii* revealed by UPLC-MS: Identification of norboldine-like structures. *J. Chromatogr. B*, *1090*: 35–42.

Rathnamalai, A., Sethupathy, S., Sarathchandiran, I. 2020. Inhibitory effect of *Murraya koenigii* leaves on α-amylase and α-glucosidase activity: An *in vitro* approach. *Asian J. Pharm. Clin. Res.*, *13*(5): 87–91. https://doi.org/10.22159/ajpcr.2020.v13i5.37334

Ravindran, P. N., Nirmal Babu, K. and Sivakumar, V. 2014. β-Carotene and other carotenoids in fresh curry leaves quantified by HPLC under light-induced stress: Antioxidant and ocular benefits. *J. Food Sci. Technol.*, *51*: 1683–1689.

Ravindran, P.N., Nirmal Babu, K. and Shylaja, V. 2011. Safety evaluation of curry leaf (*Murraya koenigii*) extracts in normal cell lines. *Food Chem. Toxicol.*, *49*(9): 2247–2252.

Rehana, D., Mahendiran, D., Kumar, R.S. and Rahiman, A.K. 2017. Evaluation of antioxidant, antibacterial and cytotoxic effects of green synthesized silver nanoparticles using *Murraya koenigii* leaf extract. *J. Photochem. Photobiol. B Biol.*, *174*: 133–143. https://doi.org/10.1016/j.jphotobiol.2017.07.025.

Reisch, J., Adebajo, A.C., Kumar, V. and Aladesanmi, A.J. 1994. Two carbazole alkaloids from *Murraya koenigii*. *Phytochemistry*, *36*(4): 1073–1076.

Saini, S., Sharma, A., Kumar, P. 2020. Flavonoid profiling of *Murraya koenigii* leaves: Identification and quantification of quercetin, kaempferol, rutin, myricetin, naringenin, luteolin, and apigenin in glycosidic and aglycone forms. *Phytochem. Lett.*, *35*: 75–82.

Sastry, A. and Moudgal, R. 2012. Wound healing and tissue regeneration potential of *Murraya koenigii*: A botanical review. *J. Ethnopharmacol.*, *143*(2): 519–523.

Sathaye, S., Bagul, M., Gupta, S., Kaur, H., Redkar, R. and Rajendran, R. 2011. Hepatoprotective potential of *Murraya koenigii* leaf extracts: *In vitro* and *in vivo* evaluation. *Pharm. Biol.*, *49*(2): 210–217. https://doi.org/10.3109/13880209.2010.504498.

Satheshkumar, R., Ramasamy, K. and Shanmugam, S. 2017. Effect of oven drying on nutritional quality of curry leaves (*Murraya koenigii*). *Int. J. Food Prop.*, *20*(1): 225–233.

Satyavani, K., Gurudeeban, S., Ramanathan, T. and Balasubramanian, T. 2011. Anti-HSV activity of *Murraya koenigii* leaf extracts against Herpes Simplex Virus type-1. *Int. J. Phytomed.*, *3*(1): 22–26.

Selvam, K., Balakrishnan, N. and Rajendran, R.R. 2017. Anti-arthritic activity of *Murraya koenigii* leaf extract: An *in vitro* and *in vivo* evaluation. *Int. J. Pharm. Sci. Rev. Res.*, *45*(1): 92–96.

Sharma, P., Vidyasagar, G., Bhandari, A., Singh, S., Ghule, S., Agrawal, A., Goyal, S. and Panwar, M.S. 2011. Antiulcer activity of leaves extract of *Murraya koenigii* in experimentally induced ulcer in rats. *Pharmacologyonline*, *2*: 818–824.

Shetty, R., Joseph, S. and Jena, P. 2009. Anti-ulcer activity of curry leaf (*Murraya koenigii*) extract in pyloric ligation and aspirin-induced ulcer models. *Indian J. Pharmacol.*, *41*(2): 65–70.

Shivanna, V.B. and Subban, N. 2014. Effect of various drying methods on flavor characteristics and physicochemical properties of dried curry leaves (*Murraya koenigii* L. Spreng.). *Dry. Technol.*, *32*: 882–890.

Shivanna, V.B. and Subban, N. 2013. Carotenoids retention in processed curry leaves (*Murraya koenigii* L. Spreng.). *Int. J. Food Sci. Nutr.*, *64*: 58–62.

Siddiqui, A., Khan, M.R., Shah, N.A. et al., 2018. Mahanimbine enhances glucose uptake through insulin receptor expression in L6 skeletal muscle cells. *J. Ethnopharmacol.*, *215*: 12–20. https://doi.org/10.1016/j.jep.2017.12.041.

Singh, G., Maurya, S. and Catalán, C.A. 2014. Genotoxicity and mutagenicity studies of curry leaf (*Murraya koenigii*) phytoconstituents. *Toxicol. Rep.*, *1*: 411–419. https://doi.org/10.1016/j.toxrep.2014.06.006.

Singh, R., Kumar, S. and Tewari, P. 2012. Folk medicinal plants used in fevers of North India. *Ethnobot. Leafl.*, *16*: 1–6.

Singh, R., Kumar, V. and Sharma, A.A., 2014. Phytochemical constituents and biological activities of *Murraya koenigii*: A review. *J. Pharm. Res.*, *8*(12): 1743–1752.

Singh, R., Rao, S., Patel, G., 2016. Detection of tropane and pyridine alkaloid analogues in *Murraya koenigii* via LC–QTOF-MS profiling. *J. Ethnopharmacol.*, *183*, 50–56.

Sowbhagya, H.B., Suma, P.F., Mahadevamma, S. and Tharanathan, R.N. 2007. Spice extracts as natural antioxidants in meat products. *J. Food Lipids*, *14*(1): 106–118.

Srinivasan, K., Sambaiah, K. and Chandrasekhara, N. 2010. Beneficial influence of curry leaf (*Murraya koenigii*) on serum and liver lipid profile in rats fed with high fat diet. *Plant Foods Hum. Nutr.*, *65*(1): 32–38. https://doi.org/10.1007/s11130-009-0142-5.

Steinhaus, P., 2017. Impact of frying on the volatile profile of curry leaves (*Murraya koenigii*). *Eur. Food Res. Technol.*, *243*(3): 449–458. https://doi.org/10.1007/s00217-016-2752-1.

Steinhaus, M., 2015. Characterization of the major odor-active compounds in the leaves of the curry tree *Bergera koenigii* L. by aroma extract dilution analysis. *J. Agric. Food Chem.*, *63*: 4060–4067.

Sundararajan, R., Koduru, R. and Bandyopadhyay, S. 2006. Antioxidant potential of *Murraya koenigii* leaves: A natural source of vitamin C and polyphenols. *J. Med. Food*, *9*(1): 68–74. https://doi.org/10.1089/jmf.2006.9.68.

*Suthar, A.C., Banavalikar, M.M. and Biyani, M.K. 2014. Anti-inflammatory activity of mahanimbine, a carbazole alkaloid isolated from *Murraya koenigii* leaves. *J. Nat. Remedies*, *14*(1): 22–27. https://doi.org/10.18311/jnr/2014/243.

Syam, D.S., Kuttan, G. and Kuttan, R. 2011. Antioxidant and anti-tumour activity of flavonoids in *Murraya koenigii*. *Indian J. Exp. Biol.*, *49*(4): 282–288.

Syam, S., Abdul, A.B., Sukari, M.A., Mohan, S., Abdelwahab, S.I. and Wah, T.S. 2011. The growth suppressing effects of girinimbine on HepG2 involve induction of apoptosis and cell cycle arrest. *Molecules*, *16*(8), 7155–7170. https://doi.org/10.3390/molecules16087155

Tan, S.-P., Ali, A.M., Nafiah, M.A., Awang, K. and Ahmad, K. 2015. Isolation and cytotoxic investigation of new carbazole alkaloids from *Murraya koenigii* (Linn.) Spreng. *Tetrahedron*, *71*: 3946–3953.

Tembhurne, S.V. and Sakarkar, D.M. 2012. Anti-obesity and hypoglycemic effect of ethanolic extract of *Murraya koenigii* (L.) leaves in high fatty diet rats. *Asian Pac. J. Trop. Dis.*, *2*(Suppl. 1): 166–168.

Tembhurne, S.V. and Sakarkar, D.M. 2010. Anti-ulcer and hepatoprotective activity of *Murraya koenigii* leaves extract in rats. *Res. J. Pharmacol. Pharmacodyn.*, *2*(1): 46–50.

Tripathi, A., Upadhyay, A.K. and Singh, R. 2018. Antifungal activity of curry leaf (*Murraya koenigii*) essential oil and its constituents. *J. Essent. Oil Res.*, *30*(4): 290–296. https://doi.org/10.1080/10412905.2018.1453786.

Utaipan, T., Athipornchai, A., Suksamrarn, A., Chunsrivirot, S. and Chunglok, W. 2017. Isomahanine induces endoplasmic reticulum stress and simultaneously triggers p38 MAPK-mediated apoptosis and autophagy in multidrug-resistant human oral squamous cell carcinoma cells. *Oncol. Res.*, *37*(2): 1243–1252. https://doi.org/10.3892/or.2017.5352.

Vijayalakshmi, D. and Ruckmani, K. 2016. Influence of drying techniques on nutritional quality and antioxidant potential of curry leaves (*Murraya koenigii*). *Int. J. Food Sci. Nutr.*, *67*(8): 970–977.

Vinutha, B., Prashanth, D., Salma, K., Sreeja, S.L., Pratiti, D., Padmaja, R., Radhika, S., Amit, A., Venkateshwarlu, K. and Deepak, M. 2007. Screening of selected Indian medicinal plants for acetylcholinesterase inhibitory activity. *J. Ethnopharmacol.*, *109*(2): 359–363. https://doi.org/10.1016/j.jep.2006.06.014.

Vyankatrao, N., 2014. Effect of different drying methods on nutrient composition of curry leaves (*Murraya koenigii*). *Asian J. Dairy Food Res.*, *33*(4): 283–288.

Wang, S.L., Cai, B., Cui, C.B., Yan, S.Y. and Wu, C.F. 2008. Study on induction of apoptosis by girinimbine in HCT-15 cell *in vitro*. *Chin. J. Pharm. Anal.*, *28*(2): 176–181.

Xie, H., Zhang, T., Yang, N., Li, Z. and Liu, Y. 2020. Anticancer effects of mahanimbine alkaloid on the human bladder cancer cells are due to the induction of G0/G1 cell cycle arrest, apoptosis, and autophagy. *J. Buon.*, *25*: 1166–1171.

Yankuzo, C., Bhandari, P. and Sarkar, A. 2011. Histopathological study of aqueous leaf extract of *Murraya koenigii* in diabetic-induced renal injury in rats. *J. Ethnopharmacol.*, *134*: 286–292.

Yankuzo, H., Ahmed, A., Ayo, J.O., Muhammad, B.Y. and Zezi, A.U. 2011. Hypoglycemic and hypolipidemic effects of *Murraya koenigii* leaf extract in alloxan-induced diabetic rats. *Asian J. Med. Sci., 3*(1): 24–28.

Yuvianti, D.F., Arief, N. and Nanang, F. 2024. *Murraya koenigii* L. Spreng.: An updated review of chemical composition, pharmacological effects, and toxicity studies. *J. Appl. Pharm. Sci., 14*(6): 11–27.

Zahin, M., Aqil, F. and Ahmad, I. 2010. Broad spectrum antimutagenic activity of antioxidant active fraction of *Murraya koenigii* (L.) Spreng. *Toxicol. In Vitro*, *24*(2): 515–522. https://doi.org/10.1016/j.tiv.2009.09.009.

13 Nutritional Value, Phytochemicals, and Biological Activities of Spindle Pod (*Cleome monophylla* L.)

Adeyemi O. Aremu[1,2*], Ahmed Olatunde[3],
Keamogela I. Nko[1] and Makhotso Lekhooa[4]

Introduction

Cleome monophylla L. (commonly known as spindle pod or spider flower) is a fast-growing annual herbaceous plant belonging to the genus *Cleome*. It is widely distributed across tropical and subtropical regions of Africa and Asia (Elffers et al., 1964). Amongst different cultural groups in Africa and Asia, it is traditionally valued as a leafy vegetable and medicinal plant (Maseko et al., 2017; Vijayashalini et al., 2023). Historically, members of the genus *Cleome* have been used for diverse health conditions such as fever, cough, snake bites, nephritis, diarrhea, bronchitis, malaria-related fevers, liver disorders, and skin diseases (Nnadozie et al., 2023; Moyo and Aremu, 2022). Spindle pod is gaining increased scientific attention due to its rich nutritional profile and phytochemicals (Khuntia et al., 2022). The increasing interest in spindle pod underscores its potential for improving food security, health, and income generation, especially in rural communities of developing countries. The leaves are known to be a good source of essential nutrients such as vitamins (notably vitamin C and provitamin A), minerals (including calcium, iron, and phosphorus), proteins, and dietary fiber (Odhav et al., 2007; Uusiku et al., 2010). These attributes position locally sourced leafy vegetables as a valuable component in combating micronutrient deficiencies, particularly in limited-resource settings often common in rural areas (Nko et al., 2024).

In addition to its nutritional benefits, *Cleome monophylla* is a reservoir of compounds, especially the terpenes (Khuntia et al., 2022). As a result of this rich pool of phytochemicals, spindle pod exerts diverse biological activities that

1 Indigenous Knowledge Systems Centre, Faculty of Natural and Agricultural Sciences, North-West University, Private Bag X2046, Mmabatho 2790, South Africa.
2 School of Life Sciences, College of Agriculture, Engineering and Science, University of KwaZulu-Natal, Private Bag X54001, Durban 4000, KwaZulu-Natal, South Africa.
3 Department of Medical Biochemistry, Abubakar Tafawa Balewa University, Bauchi 740272, Nigeria.
4 DSI/NWU Preclinical Drug Development Platform, Faculty of Health Sciences, North-West University, Private Bag X6001, Potchefstroom 2521, South Africa.
* Corresponding author: email: Oladapo.Aremu@nwu.ac.za

DOI: 10.1201/9781003687832-13

include antioxidant, antimicrobial, anti-inflammatory, and antidiabetic properties, mainly based on *in vitro* test systems (Vijayashalini et al., 2023). Additional empirical evidence on the nutritional attributes and pharmacological potential remains essential to support its extensive and continuous uses as food and medicine (Mabhaudhi et al., 2016; Yeung et al., 2018; Akinola et al., 2020; Nxusani et al., 2023). Furthermore, the widespread distribution and utilization of the spindle pod across different cultural groups warrants further scientific investigation for the development of functional foods, nutraceuticals, and/or plant-based therapeutic agents.

Although an overview on spindle pod exists with emphasis on its taxonomy and morphological aspects (Vijayashalini et al., 2023), a critical and comprehensive appraisal is essential to concisely identify the research gaps as a means of pursuing the bioeconomy potential for *Cleome monophylla*. This chapter explores the nutritional value, phytochemical composition, and biological activities of *Cleome monophylla*, emphasizing its potential relevance in promoting health and well-being, as well as its role in sustainable food systems, traditional health practices, alternative and complementary medicines.

Taxonomic Background, Distribution, and Conservation Status of *Cleome monophylla*

Cleome monophylla L. (family: Cleomaceae) is one of the lesser-known members of the approximately 200 species in the genus *Cleome*. It is unique in being the only single-leaved species in the genus *Cleome,* which is generally characterized by compound leaves (Vijayashalini et al., 2023). Macroscopically, the spindle pod is an annual plant that grows up to 1 m in height, herbaceous, and possesses glandular hairs (Ruffo et al., 2002). To complement the simple and alternate spirally arranged leaves are dull purple to pink flowers, four tiny sepals, and six stamens. The fruit is a siliqua ranging from 3–8 cm long, linear, beaked, strongly ribbed, glandular, and hairy.

Cleome monophylla has several common names, including single-leaved cleome, spindle pod, and spider flower (Table 1). It is also known by diverse vernacular names across the world, including Africa. In India, some of the names are "Elluku sakkalathi", "Kaatu kadugu", and "Naai kadugu" (Vijayashalini et al., 2023) and ' Nyakamage' in Tanzania (Ruffo et al., 2002). It is known as Enkelblaarcleome (Afrikaans), Mujakari (Ndebele), Isiwisa Esiluhlaza (IsiZulu), Matlepelo, Munyenyae, and Musa Pelo (Sesotho sa Leboa, Setswana) among the different ethnic groups in South Africa (Akula and Odhav, 2008; von Staden, 2020). Generally, the existence of vernacular names for plants is an indication of a strong and long-term relationship from an ethnobotanical perspective (Singh, 2008; Chakale et al., 2025).

Cleome monophylla is widespread across northern, central, and eastern South Africa. Furthermore, it occurs across Africa (e.g., Angola, Ghana, Ethiopia, Kenya, and Tanzania), extending to the Middle East and Asia as far as India (Figure 1).

Table 1 Taxonomical aspects and ethnobotanical value of *Cleome monophylla*.

Feature	*Details*	*Reference*
Botanical name	*Cleome monophylla* L.	http://www.theplantlist.org/tpl1.1/record/kew-4776111._
Family	Cleomaceae	https://powo.science.kew.org/taxon/urn:lsid:ipni.org:names:147187-1
Synonym	*Cleome monophylla* var. *madagascariensis* Hadj-Moust. *Cleome monophylla* var. *cordata* (Burch. ex DC.) Sond. *Cleome aculeata* L. *Cleome indica* Lam. *Cleome pentaphylla* L. *Sieruela monophylla* (L.) Roalson & J.C. Hall	https://wfoplantlist.org/taxon/wfo-0000611439-2025-06._
Common names	Spindle pod, spider flower, single-leaved cleome (English) Elluku sakkalathi (Tamil - India), Hurhura (Beng), Velaichedi (Irula). Menasinakare (Kannada), Cheruvela (Malayalam - India) Nyakamage (Hehe - Tanzania)	Vijayashalini et al. (2023) Ruffo et al. (2002).
Local names in South Africa	Enkelblaarcleome, Rusperbossie (Afrikaans), Mujakari (Ndebele), Isiwisa Esiluhlaza (IsiZulu), Matlepelo, Munyenyae, Musa Pelo (Sesotho sa Leboa, Setswana)	Akula & Odhav (2008); von Staden (2020)
Uses	Paste made from the whole plant is applied externally to reduce swellings, and crushed leaves are used to relieve headaches.	Lewis (1986)
	An unspecified part is treating bile enlargement due to disease or infection.	Kumari & Kumar (2001)
	Violet-flowered form of the plant, used in the Ayurvedic medicine ‘ hurhur’ applied to ulcers, boils, and wounds to prevent pus formation.	Lewis, (1986)
	Roots are chewed to treat coughs, and a pounded root paste is applied to the lips to revive individuals who have fainted.	Chopra et al. (1986)

Contd...

Table 1 *Contd...*

Feature	*Details*	*Reference*
	Finely ground leaves are also traditionally used to remove foreign particles from the eye.	Ruffo et al. (2002; Lewis (1986)
	Roots used for spiritual protection are placed at the entrance of a home to ward off evil spirits, tied around the arm or hip to repel negative energies, and placed under pillows to promote sleep.	Vijayashalini et al. (2023)
	The whole plant is used as a traditional remedy for various conditions, including liver and heart diseases, bronchitis, flatulence, colic, indigestion, constipation, and cough.	Vijayashalini et al. (2023)
	As a pot herb vegetable, and as an anthelmintic in southern and Eastern Africa.	Watt & Breyer-Brandwijk (1962)
Other non-medicinal uses	Leaves, young shoots, and flowering stems are cooked and consumed much like spinach. Leaves are chopped and cooked alone, or with beans and peas. Given that they are slightly toxic, it is advisable to eat them a day after cooking to allow for the toxic enzymes to break down.	Facciola (1990); Yesodharan & Sujana (2007) Ruffo et al. (2002)
Conservation status	Least concern in South Africa Common in Tanzania	Raimondo (2009) Ruffo et al. (2002)
Distribution in Africa	Africa (native): Botswana, Lesotho, Malawi, Mozambique, Namibia, Niger, Nigeria, Mali, Chad, Angola, Benin, Guinea, Ghana, Tanzania, Burkina, Burundi, Cameroon, Congo, DR Congo, Madagascar, Mauritania Africa (exotic): Kenya, Ethiopia, Sudan-South, Sudan, Rwanda, Senegal, Somalia	https://powo.science.kew.org/taxon/urn:lsid:ipni.org:names:147187-1#source-KBD
Occurrence in South African provinces	Eastern Cape, Free State, Gauteng, KwaZulu-Natal, Limpopo, Mpumalanga, Northern Cape, North West	von Staden (2020)
Habitats	It occurs in sandy soil in open woodland and is often a weed of disturbed places.	Raimondo (2009)

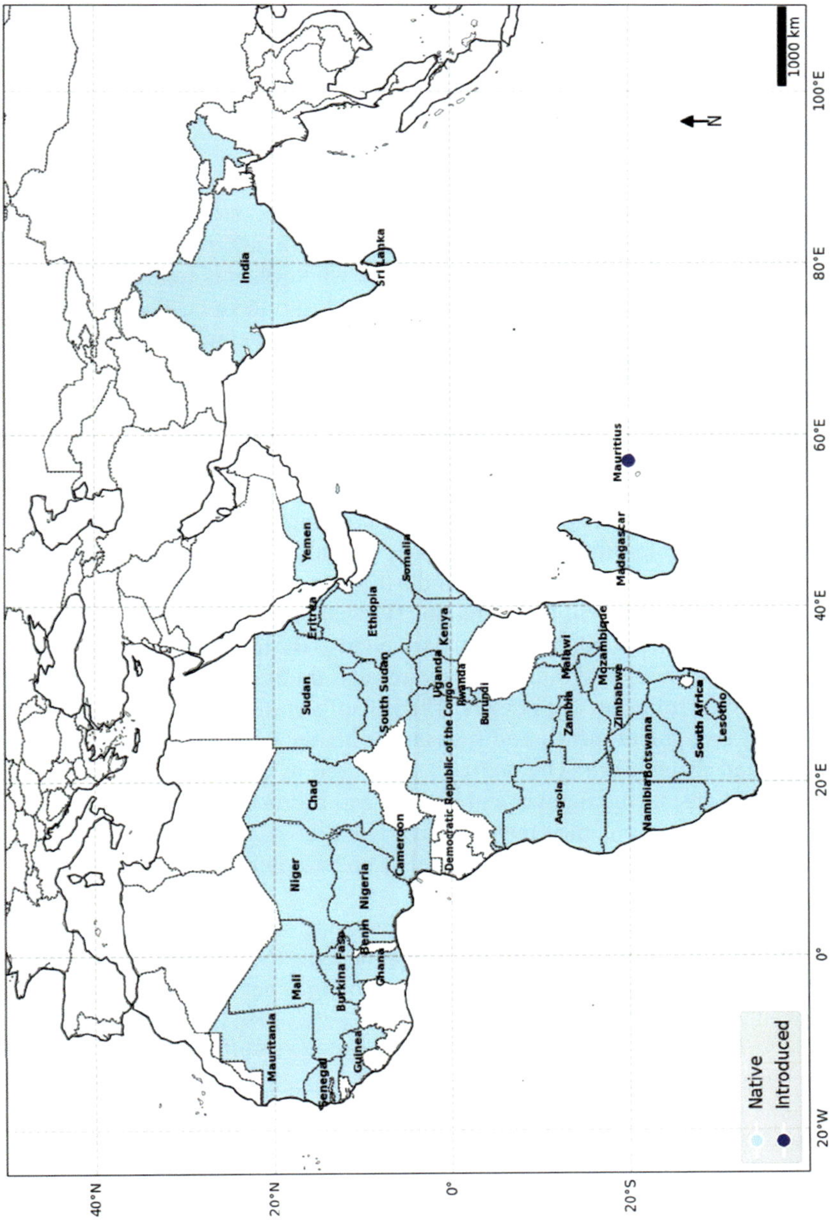

Fig. 1 Global distribution of *Cleome monophylla*. https://powo.science.kew.org/taxon/urn:lsid:ipni.org:names:147187-1.

It is also considered a native species in most of the countries where it is currently reported. In terms of its conservation status, *Cleome monophylla* is generally widespread and considered a weedy species without the danger (low risk) of extinction. It is classified as least concern, which denotes plant species that are widespread and abundant in South Africa.

Uses of *Cleome monophylla*

Cleome monophylla is a versatile plant that holds great value among various cultures in Africa and Asia (Table 1). Different parts (e.g., leaves, stem, and flowers) of the spindle pod are utilised for diverse purposes, which medicinal, nutritional, and spiritual uses. For instance, a paste made from the whole plant is applied externally to reduce swellings, while crushed leaves are used to relieve headaches (Lewis, 1986). It is also considered effective in treating bile enlargement due to infection (Kumari and Kumar, 2001). A decoction of the plant is traditionally used to manage conditions such as menstrual disorders, ovarian cysts (polycystic ovary syndrome), rhinorrhoea (runny nose), earaches, indigestion, ulcers, excessive body heat, and boils. Furthermore, some communities administer a mixture of crushed leaves and goat milk to enhance lactation in breastfeeding mothers, while the leaf paste is applied externally to reduce inflammation (Chopra et al., 1986; Ruffo et al., 2002).

In India, the violet-flowered form of the plant, often used with *Gynandropsis pentaphylla*, is an ingredient in the Ayurvedic medicine ‘hurhur’, applied to ulcers, boils, and wounds to prevent pus formation (Lewis, 1986). The roots are chewed to treat coughs, and a pounded root paste is applied to the lips to revive individuals who have fainted. Finely ground leaves are also traditionally used to remove foreign particles from the eye (Lewis, 1986; Ruffo et al., 2002). The Malayali tribe (India) uses a leaf infusion to manage dementia, neurasthenia, respiratory problems, and rheumatism. The entire plant is used as a traditional remedy for various conditions, including liver and heart diseases, bronchitis, flatulence, colic, indigestion, constipation, and cough (Vijayashalini et al., 2023). Additionally, a hot root infusion is used to revive individuals who have fainted and to treat scurvy and rheumatism.

Across different ethnic groups in Africa, members of the *Cleome* genus are important leafy vegetables for their nutritional benefits (Moyo and Aremu, 2022; Nko et al., 2024). For spindle pod, the leaves, young shoots, and flowering stem are cooked and consumed much like spinach (Lewis, 1986; Ruffo et al., 2002; Yesodharan and Sujana, 2007), either singly or combined with beans and peas, and sometimes prepared as a relish (Lewis, 1986). Due to slight toxicity, it is advised to consume the cooked leaves a day later, allowing time for toxic compounds to break down (Ruffo et al., 2002). Among the Soliga tribe in India, the root of *Cleome monophylla* is used for spiritual protection by placing it at the entrance of the home to ward off evil spirits. It may also be tied around the arm or hip to repel negative energies, and placed under pillows to promote sleep (Vijayashalini et al., 2023).

Nutritional Composition of *Cleome monophylla*

Generally, *Cleome* species have a high level of fiber and low concentration of fat; the seed oil possesses high nutrient compositions (Ekpong, 2009). Plants from the genus *Cleome* also contain a remarkable quantity of glutamate, aspartate, arginine, tyrosine, lysine, and histidine (Khuntia et al., 2022). Even though specific proximate composition data (such as moisture, protein, fat, fiber, ash, and carbohydrate content) for *Cleome monophylla* are not readily available in the literature, studies on related species within the *Cleome* genus suggest that these plants can be sources of nutrients and bioactive compounds. For instance, other *Cleome* species have been studied for their nutritional compositions, indicating their potential health benefits (Akinsola and Oluwafemi, 2022; Abdulaziz, 2023; Vijayashalini et al., 2023). For instance, *Cleome gynandra* contains elements such as selenium, copper, iron, manganese, zinc, and cobalt that are of nutritional importance (Wakhisi et al., 2020). The plant also contains beta carotene, and tocopherol, L-ascorbic acid as essential nutrients (Sivakumar et al., 2018; Wakhisi et al., 2020). Furthermore, *Cleome rutidosperma* was reported to contain crude protein, fiber, fat, and carbohydrate. Other elements such as calcium, potassium, phosphorus, sodium, manganese, magnesium, and iron were also reported in this plant (Akinsola and Oluwafemi, 2022). Given the limited data on *Cleome monophylla*, further scientific studies are necessary to determine its nutritional composition accurately and comprehensively.

Phytochemicals in *Cleome monophylla*

For different *Cleome* species, numerous compounds have been identified and quantified using techniques such as spectrophotometry, gas chromatography–mass spectrometry (GC-MS), liquid chromatography–mass spectrometry (LC-MS), and proton nuclear magnetic resonance (^{1}H NMR) spectroscopy (Upadhyay, 2015; Moyo and Aremu, 2022). For instance, the seeds of the *Cleome* genus are rich in fatty acids and contain unique compounds such as cleomiscosin A, cleomiscosin B, and other coumarinolignoids (Chand et al., 2022). *Cleome monophylla* has been studied for its phytochemical constituents, revealing many bioactive compounds that may contribute to its therapeutic potential (Figure 2). Extensive qualitative screenings have explored *Cleome monophylla* for chemicals such as cardiac glycosides, saponin glycosides, steroids, triterpenes, tannins, alkaloids, flavonoids, and steroidal glycosides (Uwaiya et al., 2019).

Polyphenols

Most of the polyphenols are studied in other species of *Cleome* other than *Cleome monophylla*. These include phenols, flavonoids, and tannins (Khuntia et al., 2022). These polyphenols are likely to be present in *Cleome monophylla*. Using different qualitative tests, the absence of tannins and flavonoids in the chloroform and hexane extracts of the whole plant of *Cleome monophylla* was indicated (Uwaiya

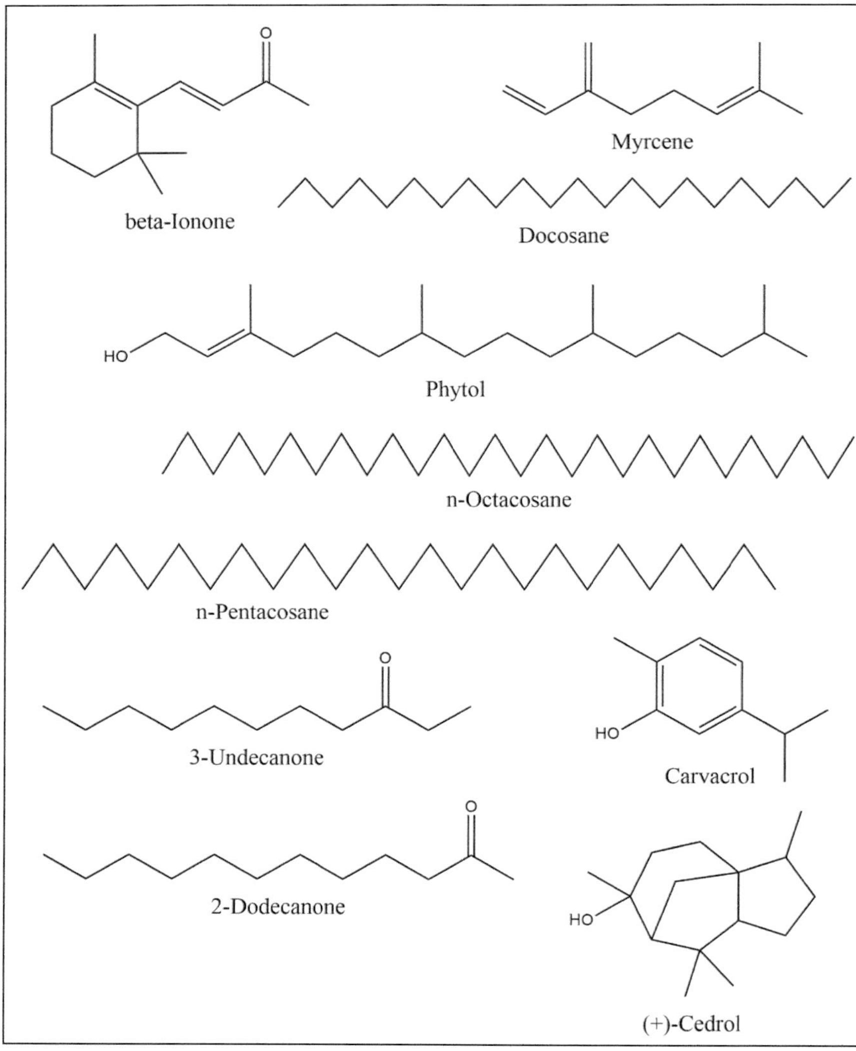

Fig. 2 Overview of the chemical structure for some essential oils in *Cleome monophylla*.

et al., 2019). Using the Folin-Ciocalteu method, the total phenolic compounds in the water and methanol extracts of *Cleome monophylla* were 12.5 and 32.6 mg/g, respectively (Akula and Odhav, 2008). Nevertheless, more studies are required to establish the detailed polyphenol content of *Cleome monophylla,* given the availability of diverse and sophisticated techniques with a high degree of sensitivity and reproducibility (Heinrich et al., 2022).

Terpenoids and Essential Oils

Phytochemical screening of *Cleome monophylla* has confirmed the presence of terpenoids in different parts of the plant (leaves, stem, and roots). Based on qualitative screening of n-hexane and chloroform extracts of the whole plant part of *Cleome monophylla*, the presence of triterpenes was confirmed by Uwaiya et al. (2019). The presence of triterpenoids and terpenoids has been reported in several members of the genus *Cleome,* including *Cleome gynandra*, *Cleome rutidosperma,* and *Cleome monophylla*. A study showed that identified dammarane-type triterpenoids, including cleogynol, were found in *Cleome gynandra,* while terpenoids were present in the methanol extract of *Cleome rutidosperma* (Akinsola and Oluwafemi, 2022).

Furthermore, *Cleome monophylla* contains essential oils, particularly those responsible for its characteristic aroma, and may play a role in its traditional medicinal uses. It contains β-ionone, n-docosane, myrcene, terpinolene, phytol, n-octacosane, n-pentacosane, (α + β)-humulene, (-)-cubebene, 3-undecanone, carvacrol, 2-dodecanone, l-α-terpineol, and (+)-cedrol (Ndungu et al., 1995).

Although detailed GC-MS analysis specific to *Cleome monophylla* essential oils is limited, related species such as *Cleome viscosa and Cleome gynandra* are known to have essential oils rich in monoterpenes and sesquiterpenes, including limonene, pinene, and caryophyllene. This suggests a similar profile may be expected in *Cleome monophylla*. However, these data on the comprehensive phytochemicals in *Cleome monophylla* are an essential requirement as a means of exploring its potential.

Alkaloids

Phytochemical investigations into *Cleome monophylla* have yielded divergent views regarding the presence of alkaloids. Some studies have reported their absence, while others suggest their presence, indicating potential variability due to factors such as plant parts analyzed, extraction method, or environmental factors. Phytochemical screening on n-hexane and chloroform extracts of the whole plant parts of *Cleome monophylla,* but alkaloids were not detected in either extract (Uwaiya et al., 2019). The authors noted that the absence of alkaloids might be attributed to the specific solvents used or the plant parts analyzed.

Sulfur-containing Compounds

Plants within the *Cleome* genus contain various chemical classes, including glucosinolates, which are sulfur- and nitrogen-containing compounds derived from glucose and amino acids. These compounds are known for their role in plant defense mechanisms and potential health benefits (Mano et al., 2024). Further studies on related species, such as *Cleome gynandra,* have identified the presence of thiol compounds, which are sulfur-containing compounds known for their antioxidant property. These findings suggest that sulfur-containing compounds are characteristic of the *Cleome* genus (Wakhisi et al., 2020). While direct evidence of sulfur-containing compounds in *Cleome monophylla* is currently unknown, the presence of such compounds in related species and the *Cleome* genus suggests that *Cleome monophylla* may also contain these bioactive constituents. Further phytochemical investigations, including targeted analyses with GC-MS, are recommended to elucidate the sulfur-containing compound profile of *Cleome monophylla* (Singh et al., 2018).

Biological Activities of *Cleome monophylla*

Cleome monophylla has been screened for a variety of biological activities such as antioxidant, anti-inflammatory, antimicrobial, anticancer, and antidiabetic actions (Table 2).

Antioxidant Effects

Cleome monophylla has considerable data reported for its antioxidant activity. Although, it is less extensively studied compared to some other *Cleome* species, the limited evidence from biological screenings suggests that *Cleome monophylla* possesses notable antioxidant activity, primarily due to its bioactive phytochemicals (Vijayashalini et al., 2023). Based on an *in vitro* assay (DPPH radicals), ethanol extracts of *Cleome monophylla* showed dose-dependent activity with an IC_{50} value of 81.42 μg/mL, compared to 42.67 μg/mL for ascorbic acid. Also, the extract exhibited an IC_{50} value of 49.88 μg/mL, while ascorbic acid had an IC_{50} value of 24.93 μg/mL against superoxide radicals. Using nitric oxide radicals, the extract demonstrated an IC_{50} value of 49.94 μg/mL, which was similar to the effect of the positive control (ascorbic acid) with an IC_{50} of 42.44 μg/mL. These results suggest that *Cleome monophylla* possesses significant antioxidant activity, likely due to its phytochemical constituents such as phenolics and flavonoids (Vijayashalini et al., 2023). In another study, both aqueous and methanol extracts of *Cleome monophylla* conferred antioxidant activities with DPPH scavenging values of 36.4 and 84.7%, respectively.

Anti-inflammatory Effects

In addition to its antioxidant activity, the plant exerted anti-inflammatory activity by inhibiting 5-lipoxygenase with an IC_{50} value of 32 μg/mL. This effect by *Cleome monophylla* was linked to the presence of total phenolic compounds in the plant extract (Akula and Odhav, 2008). However, both *in vitro* and *in vivo* studies are recommended to further establish the anti-inflammatory efficacy of *Cleome monophylla*.

Antimicrobial Effects

Cleome monophylla extracts exhibit varying antimicrobial effects against more than 10 microbial strains (Table 2). In terms of antibacterial screening, hexane and chloroform extracts were tested against various bacterial strains (e.g.,

Table 2 Examples of the biological screening of *Cleome monophylla* L based on different *in vitro* test systems.

Biological Activity	*Plant Part*	*Extract Type*	*Assay*	*Response(s)*	*Positive Control*	*Reference*
Antibacterial	Whole plants	Hexane, chloroform	Agar well-diffusion method against seven bacterial strains [*Staphylococcus aureus*, Methicillin-Resistant *Staphylococcus aureus* (MRSA), *Escherichia coli*, *Helicobacter pylori*, *Proteus mirabilis*, *Salmonella typhi*, Vancomycin-Resistant *Enterococci* (VRE)]	Hexane extracts inhibited MRSA, *Staphylococcus aureus*, *Escherichia coli*, and *Salmonella typhi*. Chloroform extracts inhibited MRSA, *Escherichia coli*, *Helicobacter pylori*, and *Salmonella typhi*	Ciprofloxacin inhibited *Staphylococcus aureus*, *Escherichia coli*, *Helicobacter pylori*, *Salmonella typhi*, Vancomycin-Resistant *Enterococci* (VRE)]	Uwaiya et al. (2019)
Antibacterial	Whole plants	Hexane, chloroform	Zones of Inhibition (mm) against seven bacterial strains *Staphylococcus aureus*, Methicillin-Resistant *Staphylococcus aureus* (MRSA), *Escherichia coli*, *Helicobacter pylori*, *Proteus mirabilis*, *Salmonella typhi*, Vancomycin-Resistant *Enterococci* (VRE)]	Hexane extracts had 26 mm (MRSA), 25 mm (*Staphylococcus aureus*), 23 mm (*Escherichia coli* and *Salmonella typhi*); chloroform extracts: 21 mm (MRSA), 23 mm (*Escherichia coli*), 22 mm (*Helicobacter pylori*)	Ciprofloxacin: 32 mm (VRE), 34 mm (*Staphylococcus aureus*), 38 mm (*Escherichia coli*), 30 mm (H. pylori), 41 mm (S. typhi)	Uwaiya et al. (2019)

Contd...

Table 2 *Contd...*

Biological Activity	*Plant Part*	*Extract Type*	*Assay*	*Response(s)*	*Positive Control*	*Reference*
Antidiabetic	Leaves, roots, fruit, and flowers, stem	Water	alpha-amylase inhibitory assay	Leaves = 13.4, 35.7, and 68.9% at 1, 3, and 5 mg/mL Roots = 20.7, 40.8, and 98% at 1, 3, and 5 mg/mL Fruit and flowers = 8.7, 80, and 51.1% at 1, 3, and 5 mg/mL Stem = 33.2, 79.7, and 34.4% at 1, 3, and 5 mg/mL	Acarbose (1 mg/mL) = 99.21%	Odhav et al. (2010)
Antifungal	Whole plants	Hexane, chloroform	Agar well-diffusion method against three fungal strains [*Candida albicans*, *Candida krusei,* and *Candida stellatoidea*]	Hexane extracts inhibited *Candida stellatoidea*; Chloroform extracts inhibited *Candida stellatoidea*	Fluconazole inhibited the three fungal strains	Uwaiya et al. (2019)
Antifungal	Whole plants	Hexane, chloroform	Zone of inhibition (mm) against three fungal strains [*Candida albicans*, *Candida krusei,* and *Candida stellatoidea*]	Hexane extract 25 mm (*Candida stellatoidea*); chloroform extract: 23 mm (*Candida stellatoidea*)	Fluconazole: 35 mm (*Candida albicans*, 31 mm (*Candida krusei*), 34 mm (*Candida stellatoidea*)	Uwaiya et al. (2019)

Contd...

Table 2 *Contd...*

Biological Activity	*Plant Part*	*Extract Type*	*Assay*	*Response(s)*	*Positive Control*	*Reference*
Anti-HIV and cytotoxicity	Leaves, fruit, and flowers, roots	80% methanol, water	Anti-HIV assay using MT-4 cell line infected with HIV to determine the 50% effective inhibitory concentration (EC_{50}), 50% cellular cytotoxicity concentration (CC_{50}), and selectivity index, SI (CC_{50}/EC_{50})	Leaves: not active (EC_{50}), 333 μg/mL (CC_{50}) for water extract; 26 μg/mL (EC_{50}), 278 μg/mL, 10.87 (SI) for 80% methanol extract Fruit and flowers: not active (EC_{50}), 62 μg/mL (CC_{50}) for water extract; not active (EC_{50}), 323 μg/mL (CC_{50}) for 80% methanol extract Roots: not active (EC_{50}), 109 μg/mL (CC_{50}) for water extract, not active (EC_{50}), 500 μg/mL (CC_{50}) for 80% methanol extract	Lopinavir 0.00102 μg/mL (EC_{50}), 32.75 μg/mL (CC_{50}), 32239.64 (SI)	Hurinanthan (2013)
Anti-inflammatory	Leaves	Methanol	5-Lipoxygenase inhibition	IC_{50}: 32.0 μg/mL	Not indicated	Akula & Odhav (2008)
Antioxidant	Leaves	Methanol, water	DPPH scavenging test	Inhibition = 84.7% (methanol), 36.4% (water)	Not indicated	Akula & Odhav (2008)
Antioxidant	Not indicated	Ethanol	DPPH scavenging test	IC_{50}: 81.42 μg/mL	Ascorbic acid IC_{50}: 42.67 μg/mL	Vijayashalini et al. (2023)

Contd...

Table 2 *Contd...*

Biological Activity	*Plant Part*	*Extract Type*	*Assay*	*Response(s)*	*Positive Control*	*Reference*
Antioxidant	Not indicated	Ethanol	Superoxide radical scavenging test	IC_{50}: 49.88 μg/mL	Ascorbic acid IC_{50}: 24.93 μg/mL	Vijayashalini et al. (2023)
Antioxidant	Not indicated	Ethanol	Nitric oxide radical scavenging test	IC_{50}: 49.94 μg/mL	Ascorbic acid IC_{50}: 42.44 μg/mL	Vijayashalini et al. (2023)
Anti-parasitic (ectoparasites)	Aerial part	Essential oils were hydrodistilled using a Clevenger-type apparatus with a yield (0.5%)	Tick climbing repellency bioassay using livestock (brown ear) tick, *Rhipicephalus appendiculatus*	89.9% inhibition at 0.1 μL	N, N-diethyl toluamide (DEET) = 84% inhibition at 0.1 μL	Ndungu et al. (1995).
Anti-parasitic (ectoparasites)	Aerial part	Essential oils were hydrodistilled using a Clevenger-type apparatus with a yield (0.5%)	Y-tube olfactometer bioassay using the maize weevil, *Sitophillus zeamais*	83.4% inhibition at 0.1 μL	N, N-diethyl toluamide (DEET) = 52.2% inhibition at 0.1 μL	Ndungu et al. (1995).
Insecticidal	Seeds	50% methanol	An insecticidal test was carried out under laboratory conditions against the 3rd instar larvae of *Spilarctia obliqua*	56.49% mortality against the 3rd instar larvae of *Spilarctia obliqua*	Not indicated	Srivastav & Chandel (2021)

Notes: DPPH - 2, 2-diphenyl - 1 - picryl hydrazyl radical; HIV -Human immunodeficiency virus.

Staphylococcus aureus, *Escherichia coli, Pseudomonas aeruginosa,* and *Klebsiella pneumoniae*) using the agar well diffusion technique (Uwaiya et al., 2019). The findings revealed that these extracts exhibited inhibitory effects against the tested bacterial strains, suggesting potential antibacterial activity. Particularly, the hexane extract had a higher antibacterial effect than chloroform, against methicillin-resistant *Staphylococcus aureus* (MRSA – 26 mm) and *Staphylococcus aureus* (25 mm). On the other hand, chloroform extract exerted a higher antibacterial effect against *Salmonella typhi* with a zone of inhibition of 24 mm (Uwaiya et al., 2019). The antifungal effect of *Cleome monophylla* whole plant has been explored against three fungal strains, namely *Candida albicans*, *Candida krusei,* and *Candida stellatoidea*. The antimicrobial activities of related species within the genus *Cleome,* including *Cleome gynandra* and *Cleome viscosa* (Khuntia et al., 2022). Given the limited empirical data on the antimicrobial effect of *Cleome monophylla*, further studies are warranted to comprehensively understand the antimicrobial potential of the plant, including investigations into the effects of different extraction methods, bioactive compounds, and mechanisms of action.

Antidiabetic Effects

Medicinal plants, including traditional vegetables, are considered a remedy for the management of diabetes in Africa (Asong et al., 2024). Along this line, Odhav et al. (2010) investigated the inhibitory activity of the leaves, flowers, stem, and fruit of *Cleome monophylla* against the alpha-amylase, an enzyme that regulates blood glucose levels. Only the highest concentration of *Cleome monophylla* stem extract showed significant inhibition of the alpha-amylase, providing preliminary evidence on the anti-diabetic potential *in vitro* that needs further evaluation through *in vivo* models of diabetes (Odhav et al., 2010). The presence of bioactive compounds and the activities observed in related species, including *Cleome aspera* (Goverdhan and Murthy, 2009), suggest potential antidiabetic properties. Further research, including *in vitro* and *in vivo* studies, is essential to explore and validate the antidiabetic potential of *Cleome monophylla*.

Anti-human Immunodeficiency Virus (HIV) Effects

Based on the overwhelming prevalence and incidence rates of HIV, especially in developing countries, the need to explore the wealth of medicinal plants to develop new and affordable HIV treatments cannot be overemphasized (Hurinanthan, 2013; Nakibuuka and Mugabi, 2022). Particularly, the untapped potential associated with nutri-medicinal plants for managing HIV/AIDS-related illnesses, especially in cases where the ethnobotanical evidence of their uses among local communities exists, remains pertinent (Nakibuuka and Mugabi, 2022). This is strongly supported by the benefits of treating the opportunistic infections and boosting the immune system as a key aspect of HIV/AIDS management. The anti-HIV effect of different parts of *Cleome monophylla* extracts was assessed against MT-4 cell lines (Hurinanthan, 2013). Among the tested plant parts (leaves, roots, fruits, and flowers), 80% methanol extract of the leaves had the most promising response at

26 μg/mL. Although limited data exist to explicitly support the anti-HIV value of *Cleome monophylla*, its diverse applications in folk medicine remain a motivation to further assess the anti-HIV effect in different test systems.

Biocontrol Effects

In recent times, increasing global effort has been geared toward screening plant species for bioactivity against pest species affecting different crops (Aremu et al., 2024). Exploring plant natural products offers new avenues for pest control in a sustainable manner, which aligns with the global shift towards green-based approaches in agriculture. Ndungu et al. (1995) investigated the biocontrol potential of *Cleome monophylla* against livestock tick (*Rhipicephalus appendiculatus*) and the maize weevil (*Sitophillus zeamais*). Findings from the study indicated that the oil from *Cleome monophylla* exerted repellent activity and could be explored for the control of tick and maize weevil actions. In addition, 14 compounds were identified in the oil of *Cleome monophylla* with terpenolene (14%) as the dominant compound and followed by 1-α-terpeneol (10%). Even though the compounds displayed repellent activity, 2-dodecanone and 1-α-terpeneol had the highest repellency activity against *Rhipicephalus appendiculatus* and *Sitophillus zeamais* (Ndungu et al., 1995). In terms of the insecticidal effect, *Cleome monophylla* extract demonstrated 56.49% mortality against third instar larvae of Bihar hairy caterpillars (*Spilarctia obliqua*), which is known to cause stunted growth and results in deterioration of yield in crops (Srivastav and Chandel, 2021).

Other Potential Biological Effects

Given the diverse medicinal uses of *Cleome monophylla* in folk medicine (Table 1), the existing biological screening is grossly limited (Table 2). According to Moteetee and Van Wyk (2007), *Cleome monophylla* is one of the plants categorized as "*Musa-pelo* ", a traditional concoction administered to reduce bereavement-related stress used by the Basotho ethnic group. The purported ethnobotanical effects are sedative and adaptogenic, illustrating potential for further validation on neuro modulation using scientific methods (Moteetee and Van Wyk, 2007). It is also pertinent to assess the anthelmintic effect, given the use of *Cleome monophylla* as an anthelmintic in Southern and Eastern Africa (Watt and Breyer-Brandwijk, 1962). Research on related species, such as *Cleome gynandra*, *Cleome viscosa,* and *Cleome felina*, suggests potential anticancer properties (Bose et al., 2011; Joseph et al., 2014), which justify the need for further investigation into the anticancer efficacy of *Cleome monophylla*. Currently, there are no specific studies detailing the anticancer activities of *Cleome monophylla*.

Toxicity and Safety Profiles of *Cleome monophylla*

The toxicity and safety profile of *Cleome monophylla* remains underexplored, with limited direct studies available. However, ethnobotanical records and research on related species provide some insights. The plant is traditionally utilised in various regions for medicinal purposes (Table 1). However, certain sources of advice caution against its consumption. For instance, it is recommended to consume the leaves a day after cooking to allow slightly toxic enzymes to break down, indicating the presence of mild toxins that degrade over time (Ruffo et al., 2002). In the absence of specific toxicity studies on *Cleome monophylla*, research on related species offers valuable context on the potential safety of the plant. For instance, it was revealed that the ethanol extract of *Cleome viscosa* leaves had an LD_{50} value greater than 5000 mg/kg in rats. Also, no alterations were recorded in body weight and relative organ weights of the experimental animals. No significant changes were observed in hematological indices or biochemical parameters, suggesting a favorable non-toxic profile (Adanlemegbe et al., 2023). In another study, the aerial parts of *Cleome rutidosperma* indicated no toxicity at doses up to 2000 mg/kg in rats, with no adverse effects on vital organs and hematological parameters (Ghosh et al., 2019). While these findings are promising, it is important to note that safety profiles can vary between species, and further research is required to establish the safety profile of *Cleome monophylla*.

Cleome species are known to contain various bioactive compounds, including flavonoids, terpenoids, and alkaloids. While these compounds contribute to therapeutic effects, some may also pose toxicity risks at certain concentrations. The specific phytochemical composition of *Cleome monophylla* has not been extensively characterized, underscoring the need for detailed studies. Given the limited direct research on *Cleome monophylla*, its safety profile remains inadequately defined. Traditional practices suggest that proper preparation can mitigate potential mild toxicity. However, comprehensive toxicological studies are essential to establish safe usage guidelines. Until such data are available, caution is advised when considering *Cleome monophylla* for medicinal or dietary purposes (Lewis, 1986).

Harvesting, Preservation, Drying, and Processing for *Cleome monophylla*

Cleome monophylla leaves are typically harvested when the plant is young and before flowering, usually 3–5 weeks after planting, as this is when the leaves are most tender and nutritious (Vijayashalini et al., 2023). Harvesting is done manually by plucking individual leaves or cutting leafy shoots using knives or hands, and in some regions, multiple harvests are possible from a single planting if the plant is pruned carefully (Van Rensburg et al., 2004). After harvesting, the leaves are sorted to remove any damaged or diseased parts and thoroughly washed to eliminate dirt and insects (Vijayashalini et al., 2023). Drying is commonly done through sun drying, where leaves are spread on clean surfaces for several days, or shade drying, which is preferred for better nutrient retention (Abugre et al., 2011).

Potential Food, Pharmaceutical, and Value-added Products from *Cleome monophylla*

Cleome monophylla is commonly used as a leafy vegetable, with its young leaves cooked and consumed as much as applicable with spinach (Facciola, 1990; Yesodharan and Sujana, 2007). Leafy vegetables generally play a significant role in traditional African diets and are often mixed with groundnuts or other ingredients to enhance both taste and nutritional value (Mabhaudhi et al., 2016; Akinola et al., 2020; Nko et al., 2024). The leaves can also be sun-dried and ground into powder for year-round use, typically added to soups, stews, and porridges, and are rich in essential nutrients such as iron, calcium, vitamins A and C (Vijayashalini et al., 2023).

In addition to its nutritional value, *Cleome monophylla* holds notable pharmaceutical potential due to its high content of polyphenols and flavonoids, which possess antioxidant and anti-inflammatory properties, making the plant a promising candidate for supplements aimed at reducing oxidative stress and inflammation (Chand et al., 2022; Nnadozie et al., 2023). It has also been traditionally used to treat wounds and infections, suggesting antimicrobial activity and potential for development into topical creams or plant-based antiseptics (Vijayashalini et al., 2023). Furthermore, the plant presents a wide range of opportunities for value-added products, especially in the functional food and personal care sectors. Dried leaves can be blended into herbal teas for digestion and detoxification benefits (Vijayashalini et al., 2023), while the powdered form can be incorporated into baked goods such as biscuits and flatbreads to enhance nutritional content. Its traditional use for treating skin ailments also supports the development of skincare products such as creams or soaps for managing acne, minor wounds, and other dermatological conditions (Singh et al., 2018). In addition, the biocontrol potential is another promising avenue to expand the value-added product from *Cleome monophylla*.

Conclusion

Cleome monophylla is a vegetative plant from the *Cleome* genus, and it is mainly found in African and Asia countries. Ethnobotanical reports from both regions support its cultural relevance and wide acceptance, making it an ideal candidate for further development towards value-added products. Moreover, preliminary studies further support increasing interest in *Cleome monophylla* due to its nutritional composition, consisting of carbohydrates, amino acids, vitamins, minerals, and fibers. *Cleome monophylla* has considerable ethnobotanical applications, which support some of the documented pharmacological uses. Current data have illustrated the biological activities of the *Cleome* genus, with *Cleome gynandra*, *Cleome viscosa* being the most studied. The existing data is a positive indicator of the potential of the genus, thereby necessitating a closer look into *Cleome monophylla*. However, many other biological activities of the plant are still speculative, without scientific validation, as well as its lack of toxicity and safety profile data. Therefore, there is a need to further explore the extensive chemical profiling of *Cleome monophylla* coupled to computational analysis for predictive studies. Furthermore, *in vitro* and *in vivo* studies are necessary for biological activity, safety profiling, and pharmacological effect to understand its therapeutic effects for both humans and animals. In addition, the experimental applications' profile of *Cleome monophylla* to further establish its use as a functional food to curb various nutritional deficiencies and diseases, and in personal care for skin-related benefits, is recommended. Similarly, it is paramount to explore its propagation possibilities and market research to predict its economic potential of value-added products and potential market access.

Acknowledgements

AOA appreciates the financial support from the National Research Foundation (NRF, Grant no: SRUG2204224395, 150734) of South Africa. AOA was supported by the South African Research Chairs Initiative of the Department of Science, Technology, and Innovation (DSTI), the National Research Foundation (NRF) of South Africa (Grant No: RCHDI2411105279212). We appreciate additional funding from the North-West University (NWU) and the University of KwaZulu-Natal (UKZN), South Africa. We thank Mr Temitope Omogbene for assisting with the map used in this chapter.

The opinion, finding, conclusion, or recommendation expressed in this material is that of the authors; NRF, NWU, UKZN, and Abubakar Tafawa Balewa University (Bauchi, Nigeria) do not accept any liability in this regard.

References

Adanlemegbe, K.M.F., Evenamede, K.S., Idoh, K., Kpegba, K. 2023. Toxicological study of the ethanolic extract of *Cleome viscosa* leaves in Wistar rats. *International Journal of Biological and Chemical Sciences*, 17:1929–1938.

Akinola R, Pereira LM, Mabhaudhi T, de Bruin F-M, Rusch L. 2020 A review of indigenous food crops in Africa and the implications for more sustainable and healthy food systems. *Sustainability* 12(8). 3493.

Akinsola, A. and Oluwafemi, A. 2022. Phytochemical analysis, proximate composition, and mineral profile of *Cleome rutidosperma* found in Ado-Ekiti, Nigeria. *Int. J. Adv. Biochem. Res., 6* (2): 94–100.

Akula, U. and Odhav, B. 2008. *In vitro* 5-lipoxygenase inhibition of polyphenolic antioxidants from undomesticated plants of South Africa. *J. Med. Plants Res., 2* (9): 207–212.

Aremu, A.O., Omogbene, T.O., Fadiji, T., Lawal, I.O., Opara, U.L. and Fawole, O.A. 2024. Plants as an alternative to the use of chemicals for crop protection against biotic threats: Trends and future perspectives. *Eur. J. Plant Pathol., 170* (4): 711–766. https://doi.org/10.1007/s10658-024-02924-y.

Asong, J.A., Ndhlovu, P.T., Olatunde, A. and Aremu, A.O. 2024. Uses of African plants and associated indigenous knowledge for the management of diabetes mellitus. *Diabetology, 5* (5): 476–492.

Bose, U., Bala, V., Ghosh, T., Gunasekaran, K. and Rahman, A. 2011. Antinociceptive, cytotoxic, and antibacterial activities of *Cleome viscosa* leaves. *Rev. bras. farmacogn., 21*: 165–169.

Chakale, M.V., Mooki, O., Nko, K.I., Sedupane, G. and Aremu, A.O. 2025. Batswana ethnobotany: An appraisal of the uses of botanicals for diverse purposes among local communities in Botswana and South Africa. *South Afr. J. Bot., 184*: 1010–1050. https://doi.org/10.1016/j.sajb.2025.06.040.

Chand, J., Panda, S., Jain, S., Murty, U., Das, A., Kumar, G. and Naidu, V. 2022. Phytochemistry and polypharmacology of *Cleome* species: A comprehensive ethnopharmacological review of the medicinal plants. *J. Ethnopharmacol., 282*: 114600. https://doi.org/10.1016/j.jep.2021.114600.

Chopra, R., Nayar, S. and Chopra, I. 1986. Glossary of Indian medicinal plants (including the supplement). *Council Sci. Ind. Res., New Delhi, 89*: 77–100.

Ekpong, B. 2009. Effects of seed maturity, seed storage and pre-germination treatments on seed germination of cleome (*Cleome gynandra* L.). *Sci. Hortic., 119* (3): 236–240.

Elffers, J., Graham, G. and DeWolf, G. 1964. Capparidaceae. In: Hubbard, C.E. and Milne-Redhead, M.A. (Eds.), *Flora of Tropical East Africa*. London, UK.

Facciola, S. 1990. *Cornucopia: A Source Book of Edible Plants*. Vista, United States: Kampong publications.

Ghosh, P., Chatterjee, S., Das, P., Karmakar, S. and Mahapatra, S. 2019. Natural habitat, phytochemistry and pharmacological properties of a medicinal weed—*Cleome rutidosperma* DC.(Cleomaceae): A comprehensive review. *Int. J. Pharmaceut. Sci. Res., 10* (4): 1605–1612.

Goverdhan, P. and Murthy, E. 2009. Antidiabetic effect of *Cleome aspera* in Type 2 diabetic rats. *J. Pharm. Res., 2* (6): 1072–1074. http://jpronline.info/article/view/389/341.

Heinrich, M., Jalil, B., Abdel-Tawab, M., Echeverria, J., Kulić, Ž., McGaw, L.J., Pezzuto, J.M., Potterat, O. and Wang, J.-B. 2022. Best practice in the chemical characterisation of extracts used in pharmacological and toxicological research—The ConPhyMP—Guidelines 12. *Front. Pharmacol., 13*: 953205. https://doi.org/10.3389/fphar.2022.953205.

Hurinanthan, V. 2013. *Anti-HIV Activity of Selected South African Medicinal Plants*. Degree of Doctorate of Technology (Biotechnology), Department of Biotechnology and Food Technology, Durban University of Technology.

Joseph, M., Vincent, A. and Charles, A. 2014. The anticancer activity of ethanolic extract of *Cleome felina* linn. *J. Pharm. Res., 8* (9): 1223–1225.

Khuntia, A., Martorell, M., Ilango, K., Bungau, G., Radu, A., Behl, T. and Sharifi-Rad, J. 2022. Theoretical evaluation of *Cleome* species bioactive compounds and therapeutic potential: A literature review. *Biomed. Pharmacother., 151*: 113161.

Kumari, B. and Kumar, S. 2001. A checklist of some leafy vegetables used by tribals in and around Ranchi, Jharkhand. *Zoos J., 16* (3): 442–444.

Lewis, W. 1986. *The Useful Plants of West Tropical Africa*. University Press of Virginia, Charlottesville: Springer.

Mabhaudhi, T., O'Reilly, P., Walker, S. and Mwale S. 2016. Opportunities for underutilised crops in Southern Africa's post- 2015 Development Agenda. *Sustainability, 8* (4): 302. https://doi.org/10.3390/su8040302.

Mano, J., Dicko, C., Ouédraogo, J. C. W. and Bonzi-Coulibaly, Y. L. (2024). Glucosinolate profiling in Cleome gynandra L. aerial parts based on two extraction methods. *Canadian Journal of Chemistry*, 103(1), 27–36. https://doi.org/10.1139/cjc-2024-0118.

Maseko, I., Mabhaudhi T., Tesfay S., Araya H., Fezzehazion, M. and Plooy, C. 2017. African leafy vegetables: A review of status, production, and utilization in South Africa. *Sustainability, 10* (1): 16.

Moteetee, A. and Van Wyk, B. 2007. The concept of*Musa-pelo* and the medicinal use of shrubby legumes (Fabaceae) in Lesotho. *Bothalia, 37* (1): 75–77.

Moyo, M. and Aremu, A.O. 2022. Nutritional, phytochemical, and diverse health-promoting qualities of *Cleome gynandra. Crit. Rev. Food Sci. Nutri., 62* (13): 3535–3552. https://doi.org/10.1080/10408398.2020.1867055.

Nakibuuka, M.M. and Mugabi, R. 2022. Ethnobotanical study of indigenous nutri-medicinal plants used for the management of HIV/AIDS opportunistic ailments among the local communities of central Uganda. *Sci. Afr., 16*: e01245. https://doi.org/10.1016/j.sciaf.2022.e01245.

Ndungu, M., Lwande, W., Hassanali, A., Moreka, L. and Chhabra, S. 1995. *Cleome monophylla* essential oil and its constituents as tick (*Rhipicephalus appendiculatus*) and maize weevil (*Sitophilus zeamais*) repellents. *Entomol. Exp. Appl., 76* (3): 217–222.

Nko, K.I., Mpolokeng, T.G., Mokgau, K., Asong, J.A., Omotayo, A.O. and Aremu, A.O. 2024. Ethnobotanical, biological, and phytochemical qualities of locally sourced leafy vegetables for food security, good health, and general well-being in South Africa: A review. *South Afr. J. Bot., 172*: 52–68.

Nnadozie, N., Adeniran, O. and Okhale, S. 2023. Medicinal uses, phytochemistry, and pharmacological activities of *Cleome* species (Cleomaceae): A review. *Afr. J. Biol. Chem. Phys. Sci., 2* (1): 1–10.

Nxusani, Z.N., Zuma, and Mbhenyane, X.G. 2023. A systematic review of indigenous food plant usage in Southern Africa. *Sustainability, 15* (11): 8799.

Odhav, B., Beekrum, S., Akula, U. and Baijnath, H. 2007. Preliminary assessment of nutritional value of traditional leafy vegetables in KwaZulu-Natal, South Africa. *J. Food Composit. Anal., 20* (5): 430–435.

Odhav, B., Kandasamy, T., Khumalo, N. and Baijnath, H. 2010. Screening of African traditional vegetables for their alpha-amylase inhibitory effect. *J. Med. Plants Res., 4* (14):1502–1507. https://doi.org/10.5897/JMPR09.090.

Raimondo, D. 2009. *Red List of South African Plants 2009*. South African National Biodiversity Institute.

Ruffo, C., Birnie, A. and Tengnäs, B. 2002. *Edible Wild Plants of Tanzania, RELMA Technical Handbook Series*. Nairobi, Kenya: Regional Land Management Unit (RELMA), Swedish International Development Cooperation Agency (Sida).

Singh, H. 2008. Importance of local names of some useful plants in ethnobotanical study. *Indian J. Trad. Knowled., 7* (2): 365–370.

Singh, H., Mishra, A. and Mishra, A. 2018. The chemistry and pharmacology of *Cleome* genus: A review. *Biomed. Pharmacother., 101*: 37–48.

Sivakumar, D., Chen, L. and Sultanbawa, Y. 2018. A comprehensive review on beneficial dietary phytochemicals in common traditional Southern African leafy vegetables. *Food Sci. Nutri., 6* (4): 714–727.

Srivastav, A.K. and Chandel, B.S. 2021. Laboratory assessment of herbal formulation as biorational natural resources for the management of Bihar hairy caterpillar, *Spilarctia obliqua* Walker (Arctiidae: Lepidoptera) on cabbage, *Brassica oleracea* Var. *capitata*. *J. Entomol. Zool. Stud., 9*: 474–478.

Upadhyay, R. 2015. *Cleome viscosa* Linn: A natural source of pharmaceuticals and pesticides. *Int. J. Green Pharm., 9* (2): 71–85.

Uusiku, N., Oelofse, A., Duodu, K., Bester, M. and Faber, M. 2010. Nutritional value of leafy vegetables of sub-Saharan Africa and their potential contribution to human health: A review. *J. Food Composit. Anal., 23* (6): 499–509.

Uwaiya, E., Ndukwe, G., Ibrahim, H. and Aliyu, A. 2019. Phytochemical constituents and antimicrobial activities of the crude extracts of *Cleome monophylla* (Linn). *Nig. Res. J. Chem. Sci., 7* (2): 123–132.

Van Rensburg, W., Venter S., Netshiluvhi T., Van Den Heever, E., Vorster, H., De Ronde, J. and Bornman, C. 2004. Role of indigenous leafy vegetables in combating hunger and malnutrition. *South Afr. J. Bot., 70* (1): 52–59.

Vijayashalini, P., Abirami, P. and Menaga, S. 2023. *In vitro* antioxidant activities of *Cleome monophylla* L. A sporadic medicinal herb. *Int. J. Bot. Stud., 8* (1): 53–58.

Vijayashalini, P., Abirami, P., Sajitha, B. and Menaga, S. 2023. *Cleome monophylla* L.(Spindle pod): An overview. *Int. J. Acad. Res. Dev., 8* (1): 22–25.

von Staden, L. 2020. *Cleome monophylla* L. *National Assessment: Red List of South African Plants version*. https://redlist.sanbi.org/species.php?species=1132-30. Accessed on 2025/07/06.

Wakhisi, C., Michael, G. and Mwangi, E. 2020. Mineral and phytochemical composition of *Cleome gynandra* methanolic extract. *Adv. J. Grad. Res., 8* (1): 18–26.

Watt, J.M. and Breyer-Brandwijk, M.G. 1962. *The Medicinal and Poisonous Plants of Southern and Eastern Africa* (2nd Edn.). London, UK: Livingstone.

Yesodharan, K. and Sujana, K. 2007. Wild edible plants traditionally used by the tribes in the Parambikulam Wildlife Sanctuary, Kerala, India. *Nat. Prod. Rad., 6* (1): 74–80.

Yeung, A.W.K., Heinrich, M. and Atanasov, A.G. 2018. Ethnopharmacology: A bibliometric analysis of a field of research meandering between medicine and food science? *Front. Pharmacol., 9*: 215. https://doi.org/10.3389/fphar.2018.00215.

14 Nutritional Value, Bioactive Compounds, and Biological Activities of Petroselinum *crispum*

Sowmya Viswanathan[1], Jameel M. Al-Khayri[2*] and Praveen Nagella[1*]

Introduction

Petroselinum crispum (parsley), belongs to the Apiaceae family. It is not only used as a leafy vegetable in cuisine, but also valued for its rich source of bioactive compounds with a wide range of medicinal effects. Its therapeutic potential is characterised by the presence of both fixed and essential oils (Piras et al., 2022). The name *Petroselinum* is derived from the Latin words ' petros', which means rock, and ' selinon', which means celery. Traditionally, parsley was used to treat digestive issues, urolithiasis, and as a diuretic and liver and stomach tonic. Several biological actions, such as hepatoprotective, nephroprotective, cardioprotective, anti-inflammatory, immunomodulatory, and neuroprotective properties, were confirmed by modern scientific approaches (Bahramsoltani et al., 2024). Parsley is native to Europe and Western Asia, and it is now widely cultivated all over the globe for its benefits. Its nutritional importance is emphasized by the wealth of minerals, including calcium, zinc, iron, and phosphorus, along with vitamins such as A, B, C, and K, and omega-3 fatty acids. The plant's significant antioxidant effect is also attributed to a variety of bioactive compounds, like flavonoids, carotenoids, coumarins, tocopherol, and essential oils (Chauhan and Aishwarya, 2018). Because of their biological effects, including cellular proliferation and free radical suppression, bioactive compounds, particularly phenolic compounds, are major providers of health benefits (Del Rio et al., 2020). Its phytochemical profiles, like essential oils, contain substances such as myristicin, apiol, alpha-pinene, beta-pinene, and flavonoids like apigenin, kaempferol, and diosmetin, which contribute to its aroma, flavour, and medicinal qualities (Sarwar et al., 2016). The phytochemicals present in parsley are extracted from the seed, petioles, roots, and leaves using the bioassay method of separation. While essential oils and other secondary metabolites are obtained using protocols like hydro-distillation and

[1] Department of Life Sciences, School of Sciences, Christ University, Bengaluru 560029, India.
[2] Department of Agriculture Biotechnology, College of Agriculture and Food Sciences, King Faisal University, Al-Ahsa 31982, Saudi Arabia.
* Corresponding authors: jkhayri@kfu.edu.sa[2]; praveen.n@christuniversity.in[1]

DOI: 10.1201/9781003687832-14

simultaneous distillation extraction (SDE) and are analysed by the methods like gas chromatography (GC), gas chromatography mass spectrometry (GC-MS), ultraviolet visible spectroscopy (UV-VIS), high performance liquid chromatography (HPLC) or by nuclear magnetic resonance (NMR) (Boutsika et al., 2021). The roots of parsley are used as potent diuretics and to treat carminatives, especially in Turkey. In addition, it is also used to treat myalgia, functional amenorrhea, cystitis, and dysamenorrhea. Besides these functions, parsley is also efficient to treat any menstrual disorders, and is used externally to combat head lice (Agyare et al., 2017). Though there are many bioactive compounds and the activities of which are identified against certain treatments, the bioactive compounds involving allelopathy are still unknown (Sbai et al., 2016).

Taxonomy

The Apiaceae family has 275 genera and 2,850 species growing worldwide. It is the most prevalent of the seven families in the Umbelliferae order. In the Republic of Serbia, this family is represented by 53 genera and 138 species. *P. crispum* is a vibrant green, biennial, hairless, glabrous tripennate plant with a height ranging from 60 to 100 cm. Its leaves are oval in shape, the tips of the petals are curled, and their style thickening is well developed. The roots of the plant might be tuberous, vertical, or thin or thick fusiform. Because of the plant's distinctive shield-like inflorescences, it is commonly referred to as the umbelliferae family. In the 1st year of growth, it develops a hollow, cylindrical, corrugated taproot with a tripennate rosette. The flat-topped umbels with yellow to yellowish-green blooms are observed in the 2nd year. The stems, which are striated, cylindrical in shape, can achieve a maximum length of 50–80 cm. The leaves are triangular-ovate in shape, either they can be curled or straight, and they measure 3–10 × 2–7 cm. Flower openings can be seen in August and September. The pear brown-shaped parsley seeds consist of mericaps which measure 2.5–3 × 0.5 mm and develop into slightly arcuate at maturity (Punosevac et al., 2021; Subas et al., 2024).

Distribution

Parsley is said to have originated naturally in the western Mediterranean region with different temperate zones. Historical evidence showed that it was a widely grown crop in Rome and Greece, giving its long-standing cultural and dietary significance. Currently, the crop is well cultivated in tropical regions like East and West Africa. The presence of aromatic leaves in the plant is one important aspect that these plants are grown in Mediterranean nations, North America, and Europe (Agyare et al., 2017). In India, parsley is well known for its medicinal and nutritional aspects, and the plant is mainly cultivated in the cooler regions (Mirmohammadmakki et al., 2020). Due to its ability to flourish in well-drained soils and temperatures, parsley is widely cultivated in nations like Algeria, Tunisia, and Italy, as shown in Figure 1. Because of its versatility, it has been introduced and naturalized in several locations like Europe, Asia, and North Africa (Taheri et al., 2015).

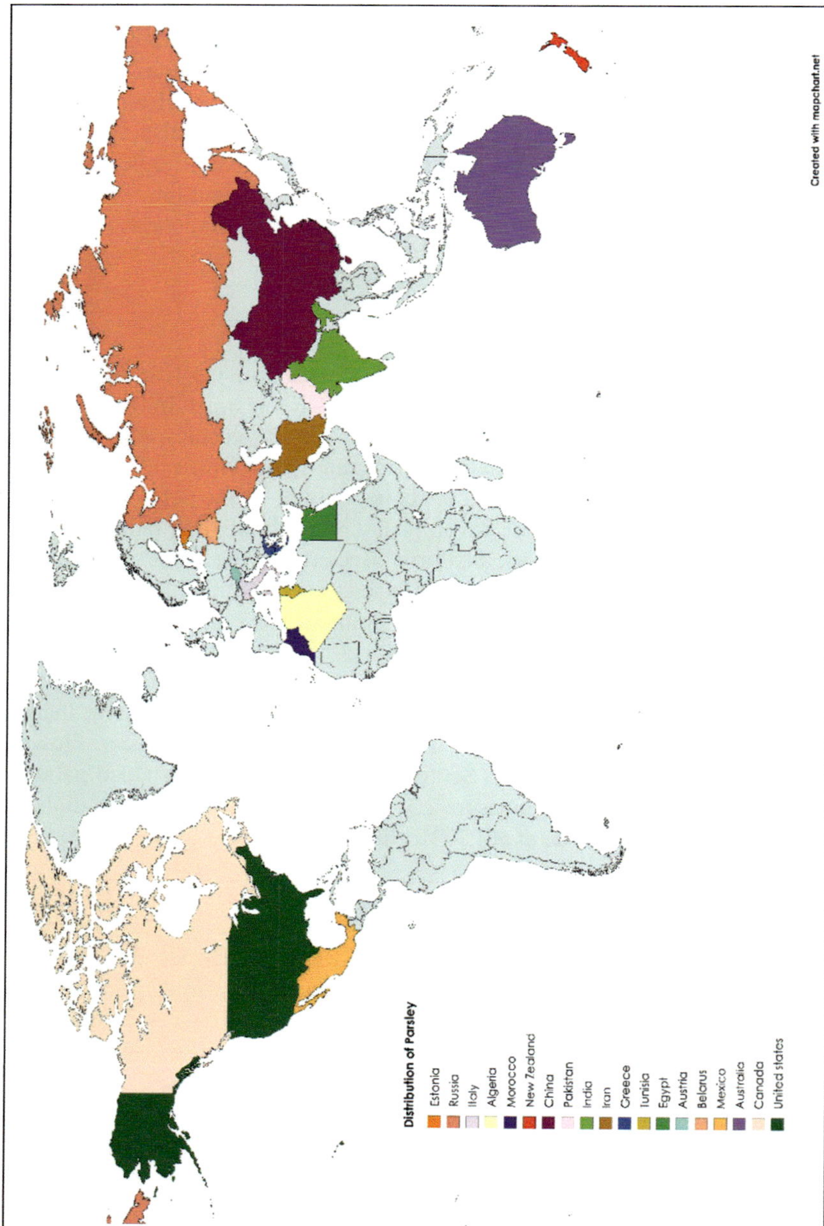

Fig. 1 Distribution of Parsley across the globe.

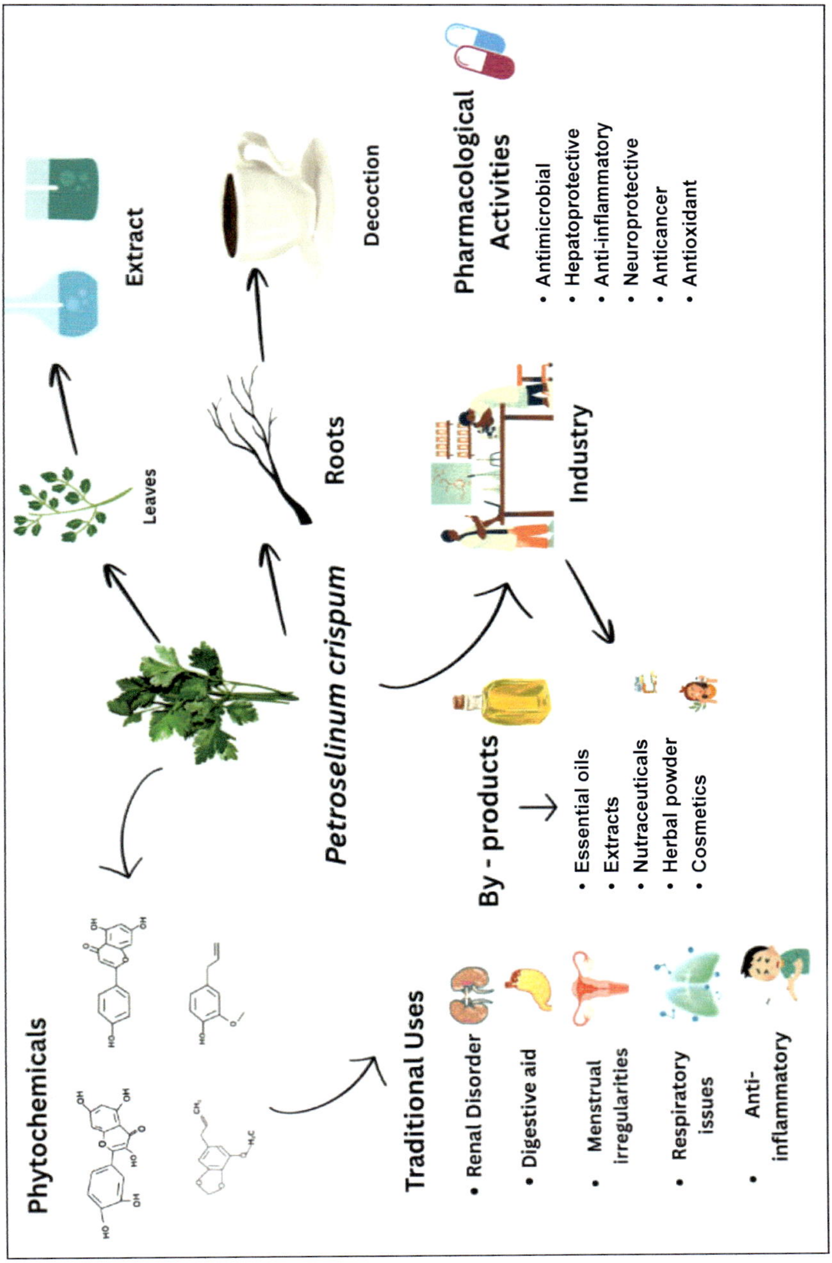

Fig. 2 Traditional and pharmacological uses of Parsley.

Traditional Uses

Parsley is known as Hazaa and Jafary in Persian traditional medicine and it is categorised as having a "Hot and Dry" temperament. Additionally, it was used to clear spleen and liver blockages and to reduce the inflammation caused by hemorrhoids, insect bites, and scars as depicted in Figure 2 and Table 1. In the Indian tradition, it was described as a warming agent and an astringent; other than the usual therapeutic treatment, it was also used to treat skin disorders like spots and vitiligo (Sarwar et al., 2016).

Table 1 Traditional uses of *Petroselinum crispum*.

Country	*Plant Part Utilised*	*Preparations*	*Treatment*	*References*
Algeria	Leaves	Decoction	Kidney and bladder diseases (diuretics)	Nouioura et al. (2024)
Morocco	Leaves	Infusion	Regulation of the menstrual cycle	Nouioura et al. (2024)
Egypt	Seeds	Powder or infusion	Treatment of flatulence	Marthe (2020)
Iran	Leaves	Aqueous extract-infusion/ decoction	Treatment of urinary retention and kidney disorders	Al-Hadi et al. (2013)
Mediterranean region	Leaves	Fresh extract	Dermal diseases	Akpaso (2023)
Europe	Leaves	Used as an herb	Smooth muscle relaxant, hepatoprotective	Abdu & Hauwa (2019)
Western Asia and Egypt	Seeds	Aqueous extract	Kidney stones, anti-hypertensive	Abu-Serie et al. (2019)
India	Leaves	Decoction	Urinary tract infections	Prabhat et al. (2024)
India	Seeds and whole plant	Powder or infusion and Juice or paste	Gastrointestinal problems and Treatment for insect bites and skin problems	Kurtaran et al. (2021)
Romania	Leaves	Infusion or decoction	Treatment for kidney and bladder disorders	Ancuceanu et al. (2018)
Germany	Leaves	Raw or infusion	Treatment for rheumatic conditions	Lamponi & Baratto (2020)
Italy	Leaves and seeds	Infusion or decoction	Treatment for menstrual disorders and urinary tract infections	Motti et al. (2019)

Contd...

Table 1 *Contd...*

Country	*Plant Part Utilised*	*Preparations*	*Treatment*	*References*
Catalonia	Root and aerial parts	Ointment	As an external antiseptic and antie-echymotic	Rigat et al. (2015)
Serbia	Leaves and roots	Infusion	Urinary tract infections and to relieve edema	Savikin et al. (2013)

Nutritional Composition of Parsley

Parsley is a nutrient-rich herb with distinct leaf, stem, and root characteristics. The leaves have high protein and the dietary fibre contents enhance digestion and metabolic wellness. In addition, they include significant levels of substantial minerals such as calcium, potassium, magnesium, and iron, which are necessary for oxygen delivery, cardiovascular function, and bone strength. Though the stem and root are quite low in their protein content, they still contribute meaningfully to the overall protein content of the plant. It is also rich in vitamins such as C and E, which play a crucial role in the mechanism of antioxidants and immune defence. Parsley also contains provitamin A and K, which are important for vision, blood clotting, and skin health. These vitamins are largely found in the leaves, making them the most efficient site for nutritional supplementation (Dobricevic et al., 2019). Its association with enhanced cellular metabolism and brain function is further characterised by the presence of substantial amounts of folates and vitamin B-complex. The herb's macronutrient profile shows moderate levels of protein and low fat. The study also shows that both dried or fresh parsley show significant nutritional contents, making them potent for a wide range of dietary applications (Fernandes et al., 2020).

Phytochemicals of *Petroselinum crispum*

The ethanolic extract of parsley leaves identified the presence of tannins, saponins, flavonoids, steroids, alkaloids, and cardiac glycosides, and these phytochemicals show therapeutic effects in various roles, such as anti-inflammation, antioxidant, and antimicrobial properties. Additional validation of parsley through GC-MS revealed six primary bioactive compounds in it such as cineole, 1-limonene, cyclohexane, phenol, neophytadiene and 9, 12, 15-octadecatrienoic acid. Both phenol and neophytadiene showed antibacterial and anti-inflammatory responses; cyclohexane and 1-limonene showed antimicrobial properties, while a wide range of activities like hepatoprotective, anticancer, anti-histamic, anti-arthritic, anti-asthma, antidiuretic were all shown by 9,12,15-octadecatrienoic acid (Abubakar et al., 2021). Parsley is a rich source of phenolic acid which includes p-coumaric acid, ferulic acid, and caffeic acid (Figure 3). These compounds help in controlling the oxidative equilibrium in cells, and also show anti-inflammatory and anti-

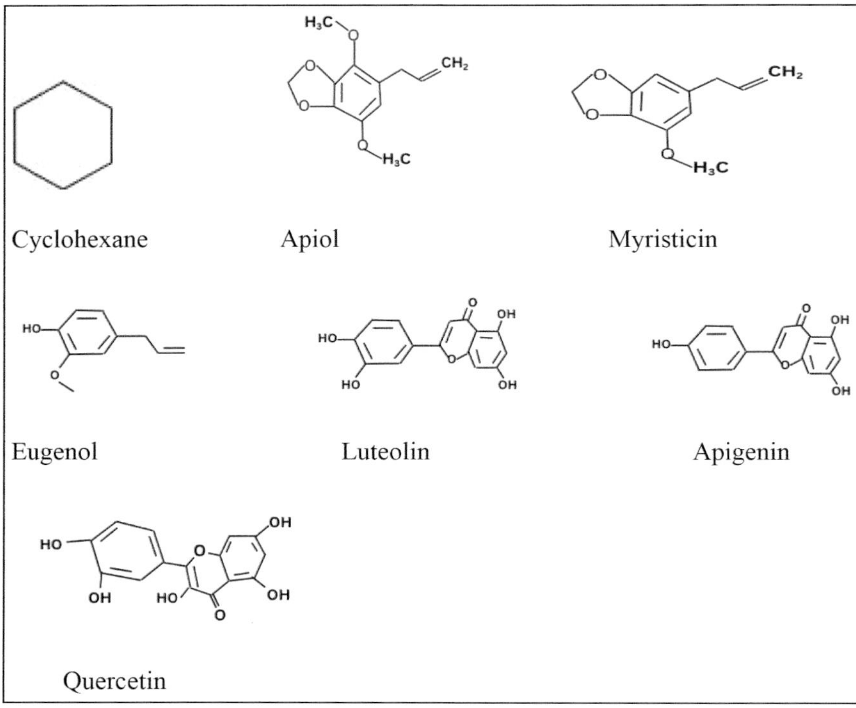

Fig. 3 Some important phytochemicals of *Petroselinum crispum*.

carcinogenic properties. Other phytochemicals like coumarins and carotenoids are also associated with medicinal properties like anticoagulant properties, while carotenoids boost immunological and vision function (Nouioura et al., 2024).

Flavonoids

Flavonoids like apigenin, luteolin, and kaempferol are rich in parsley. These compounds are mainly studied for their antioxidant properties and are mainly present in the leaves. Apigenin, which is abundant among these compounds, shows anti-inflammatory and anti-cancer activity by modulating certain key cellular pathways and is also found to show a high rate of free radical scavenging effects. Luteolin and kaempferol also aid the plant's pharmacokinetic profile by suppressing oxidative stress and helping in cellular protection (Tang et al., 2015). The petroleum ether, chloroform, and methanolic extracts of parsley showed the presence of apigenin, and it was verified by high-performance thin-layer chromatography (HPTLC). The concentrations of apigenin in all these three extracts were found to be 12.77 μg/mL, 14.18 μg/mL, and 27.63 μg/mL. The identification of apigenin confirms parsley as a useful dietary source of medicinal flavonoids

(Daimari et al., 2024). Through chromatographic and spectrographic analysis, apigenin-7-O-glucoside, apigenin-O-apiosylglucoside, two glycosylated forms of apigenin were also identified (Nielsen et al., 1999). Human investigations have verified the bioavailability of parsley's apigenin, suggesting possible biological effects (Meyer et al., 2006). Using methanolic extract, HPLC-DAD, and LCMS/MS, two different glycosides, 6-acetylapiin and petroside, were found among the flavonoids in parsley. 6-acetylapiin is an acetylated apiin derivative within apigenin backbone, which boosts the antioxidant activity. While petroside is responsible for anti-inflammatory, antioxidant and its chemotaxonomic significance with its Apiaceae family is a parsley-specific flavone glycoside with a disaccharide molecule (Yoshikawa et al., 2000). Apigenin has been shown to induce apoptosis and stunt the growth of cancer cells, including lung adenocarcinoma (A549) cells, which was abundant in methanol extract (Thangavelu et al., 2022).

Phenolic Acids

The phytochemical profile of parsley is closely related to its shikimate pathway. Shikimate dehydrogenase (SDH) is a key enzyme that regulates this pathway. This enzyme helps in the production of shikimic acid, which acts as a precursor for a wide range of secondary metabolites, namely polyphenols and phenylpropanoids, and it is also responsible for the production of phenolic acids (Hyskova et al., 2021). Several phenolic acids, such as caffeic acid, ferulic acid, and chlorogenic acid, are found in parsley. These substances are necessary for the plant's antioxidant properties. Lipid peroxidation is especially hindered by caffeic acid and helps in oxidative stress. Ferulic acid and chlorogenic acid additionally strengthen the plant's ability to neutralize reactive oxygen species, thereby contributing to its therapeutic potential in managing oxidative stress-related disorders (Tang et al., 2015). The total polyphenol content was measured using FC analysis and represented as gallic acid equivalents (GAE) was notably high. This high concentration of GAE was correlated with strong antioxidant activity, confirming the phenolic acid components' therapeutic value in inflammatory and cardiovascular diseases (Chaves et al., 2011).

Coumarins

Coumarins, including bergapten and umbelliferone, are found predominantly in seeds and roots of parsley. These compounds are recognised for their anticoagulant and vasodilatory effects. The presence of these coumarins supports the plant's traditional uses in treating inflammatory and circulatory problems, whereas bergapten is recognised for its photoprotective properties and umbelliferone exhibits antibacterial and anti-inflammatory activity (Tang et al., 2015). Components like Furanocoumarins were shown to have strong binding affinities with cancer-related receptors, enhancing the herb's anticancer potential. Other important constituents included quercetin, which mediates cell cycle arrest (Thangavelu et al., 2022).

Carotenoids

Parsley leaves contain a significant amount of carotenoids, especially beta-carotenoids and lutein. These pigments have certain biological roles in addition to the vibrant green coloration. As a provitamin-A molecule, β-carotene is necessary for immunological and visual function. While lutein also promotes eye health by blocking the harmful blue light and lowering the risk of age-related macular degeneration (Tang et al., 2015).

Essential Oils

Myristicin, apiol, and eugenol are the constituents that define the presence of essential oil in parsley. Myristicin possesses hepatoprotective, insecticidal, and antibacterial effects. Eugenol is recognised for its analgesic property, while apiol is characterised by its diuretic and antispasmodic benefits. These components are mainly extracted from the seeds of parsley, and due to their wide range of therapeutic effects, they are utilised in herbal medicine (Tang et al., 2015). Another significant compound found was the monoterpenoid alcohol and citronellol, which is well known for its anticancer, antibacterial, and anti-inflammatory actions. In addition, gamma linoleic acid was discovered, which is renowned for its antibacterial and anti-inflammatory effects and adds medicinal value for combating microbial infections (Daimari et al., 2024). The essential oils also provide a characteristic aroma besides their therapeutic applications (Meyer et al., 2006). The essential oils extracted through hydrodistillation and analysed by GC-MS were found to be abundant in monoterpenes and aromatic compounds. The main components were myristicin (8.3%), 1-limonene (32.4%), beta-phellandrene (14.6%), and allyl tetramethoxybenzene (11.2%). These substances contribute to parsley's aroma and also show antioxidant, antimicrobial, and therapeutic properties (Benmekhbi et al., 2022). The components of essential oils as myristicin and 1, 8-cineole, showed positive binding interactions with neurotransmitter receptors, indicating the possibility of neuroprotection (Thangavelu et al., 2022).

Hydrocarbons and Terpenoids

Hentriacontane, a hydrocarbon, was found to have antibacterial, anti-inflammatory, and antitumour properties. Diterpenes like neophytadiene and phytol also showed anti-inflammatory, antitumour activities. Along with this, neophytadiene showed antipyretic properties; the existence of these compounds implies that parsley has therapeutic and culinary uses (Daimari et al., 2024).

Fatty Acids

Various fatty acids were identified, such as oleic, undecanoic, and n-hexadecanoic acids, through GC-MS analysis of the methanolic extract of the plant. These substances exhibited anthelmintic, antioxidant, anti-viral, antifungal, and antibacterial effects. Together, they emphasize the plant's strong antimicrobial and protective functions (Daimari et al., 2024).

Biological Activities of *Petroselinum crispum*

Antimicrobial Activities

Parsley's essential oil, which is abundant in components like apiol and myristicin, has exhibited antibacterial action. It implies that the phytochemical profile of parsley, especially its phenolics and essential oils, is directly related to its microbial properties, allowing it to prevent the proliferation of microbial infections and also protect the cells from oxidative damage (Karimi et al., 2012). Gram-positive and Gram-negative bacteria were used to test the antibacterial qualities of parsley leaf extracts and essential oils. The most potent antibacterial activity was shown by the n-BuOH extract against *Bacillus subtilis* and *Escherichia coli,* while *Staphylococcus aureus* and *Pseudomonas aeruginosa* showed moderate activity. The essential oil also exhibited antibacterial properties, particularly against *Staphylococcus aureus*. These results indicate that parsley is a good antibacterial agent that works well with Gram-negative bacteria (Benmekhbi et al., 2022). The aqueous, ethanol, methanol, and isopropanol leaf extracts of parsley were used to assess against *E. coli* DH5α by both well diffusion and disc diffusion methods, among which the aqueous extract showed no zone of inhibition may be due to the poor diffusion caused by the precipitation of phytochemicals in water, indicating zero antibacterial activity. On the other hand, the methanolic extract showed the most prominent antibacterial effect, generating a 4 mm inhibition zone in the well diffusion method. While ethanol and isopropanol extracts showed weaker or inconsistent inhibition (Prabhat et al., 2024). Myristicin, which makes up 41.45% of the essential oil of *Petroselinum crispum* leaves (PCEO), showed strong antibacterial activity *in vitro*. The growth of *candida albicans* was inhibited by PCEO with a zone of inhibition of 19.4 ± 0.08 mm and a minimum inhibitory rate (MIC) of 1.25 mg/mL according to the agar diffusion assay, similarly, *Staphylococcus aureus* showed a zone of inhibition of 17.86 ± 0.09 mm and an MIC of 2.5 mg/mL, while *E. coli* and *Klebsiella pneumoniae* showed moderate inhibition of 8.67 ± 0.12 mm and 9.43 ± 0.09 mm. The high percentage of myristicin and other active volatiles, including beta-myrcene, sabinene, beta-caryophyllene, and beta-farnesene, which are all known to have antibacterial and antifungal activities, was thought to be responsible for antimicrobial activities (Foudah et al., 2022). Turnip rooted parsley's antibacterial effect against pathogenic strains such as *Staphylococcus aureus*, *Candida albicans,* and *Escherichia coli* was inhibited by the ethanolic extracts from the roots. *E. coli* had an inhibitory zone ranging from 8.3–11.2 mm, *S. aureus* showed the values between 9.1–12.4 mm, and *Candida albicans* showed an inhibitory zone of 7.5–10.6 mm. These data are consistent with the presence of bioactive substances that are known to disrupt the microbial membrane and also suppress the activity of certain enzymes, such as apigenin, luteolin, and essential oils like myristicin. The experiment also showed the MIC value as low as 1.25 mg/mL for *Staphylococcus aureus,* indicating the strong antimicrobial effect (Petropoulos et al., 2022).

Antihypertensive Activity

Through both *in vitro* and *in vivo* experiments, parsley's antihypertensive activity has been experimentally confirmed using the aqueous extract of its aerial parts (AEPC). Oral treatment of AEPC at 160 mg/Kg decreased the systolic, diastolic, and mean blood pressure in anesthetized L-NAME-induced hypersensitive rats. After a single dose and repeated administration for seven days, AEPC was more effective in hypersensitive rats than normotensive ones, indicating its selective potential in conditions of elevated blood pressure (Ajebli and Eddouks, 2019).

Anticancer Effects

The methanol extract of the aerial parts of parsley showed notable anticancer activity against human glioblastoma U87MG cells. It inhibited the adhesion of cells to proteins of the extracellular matrix, like fibronectin, fibrinogen, poly-L lysine etc, with the IC_{50} value between 19–24 μg/mL. The suppression of glioblastoma cell proliferation was observed after four days of incubation at a concentration of 1mg/mL. This indicates that the extract of parsley hinders integrin-mediated cell attachment, thereby decreasing the growth of tumour cells and their metastasis potential (Aissani et al., 2021).

Antioxidant Activity

The HPLC-DAD analysis of the methanol extract of parsley found five major phenolics such as quinic acid, gallic acid, acacetin, protocatechuic acid, and cirsilineol, which are recognised for their radical scavenging and redox modulating effects. Methanol extract showed a strong radical scavenging activity, with IC_{50} values of 105.82 μg/mL for DPPH, 19.95 μg/mL for ABTS, and 6.25 μg/mL for hydroxyl radicals. In addition, 88.21 mM $FeSO_4$ equivalents per gram of dry extracts indicated a significant ferric reducing antioxidant capacity (FRAP). The extract also exhibited a notable iron chelation capacity, further confirming its ability to counteract oxidative stress (Aissani et al., 2021). The biological activity of a fixed oil showed a weak antioxidant activity in the ABTS assay (EC_{50} = 0.4 mg/mL) and contained a low phenolic content (1.5 mg gallic acid equivalents/g) (Piras et al., 2022). The methanol extract showed a strong ability to neutralise radicals in the DPPH scavenging assay, with an IC_{50} value of 0.54 mg/mL, while the aqueous extract showed an IC_{50} value of 0.93 mg/mL. Similarly, in the ABTS assay methanolic extract exhibited an IC_{50} of 0.72 mg/mL compared to 1.24 mg/mL for the aqueous extract. These values significantly exhibited the higher antioxidant potential of methanol-extracted substances. The antioxidant activity of parsley is closely associated with quenching reactive oxygen species and also suppressing lipid peroxidation, thereby preventing oxidative damage to biomolecules like lipids, proteins, and DNA (Pajevic et al., 2021). The phenolic and flavonoid content of parsley leaf extracts was directly associated with their antioxidant capacity. The assessment of antioxidant activity using the DPPH scavenging assay confirmed the strong activity with the n-BuOH extract, resulting in the highest radical inhibition with an EC_{50} of 134.49 μg/mL when compared to other extracts like dichloromethane and aqueous extract. These findings demonstrated the potent free radical scavenging ability of polyphenols and flavonoids such as quercetin, apigenin, and kaempferol glycosides, which makes parsley an effective natural antioxidant source (Benmekhbi et al., 2022). Parsley shows a significant antioxidant activity that varies with its developmental stages, which is revealed in the comparative study of microgreens and baby greens. The baby greens showed a significantly higher hydrophilic antioxidant activity than microgreens, with 2.46 mmol of AA equivalents per 100g fresh weight, roughly 2.7 times more. A much higher concentration of total ascorbic acid of about 183.6% more than microgreens, and 64.2% higher amounts of total polyphenols were linked to baby greens, increased antioxidant capacity. Additionally, parsley's microgreens accumulated more carotenoids than baby greens, contributing 2.8 times more β-carotene and 1.8 times more lutein, both of which support antioxidant effects. The derivatives of apigenin, like apigenin-7-apiosyl-glucoside and apigenin-malonyl-apiosyl-glucoside, were together considered to be 96% of total phenolics, emphasizing their action in parsley's antioxidant profile (El-Nakhel et al., 2021). The antioxidant effects of parsley were associated with its potential to protect against liver and kidney damage caused by paracetamol. Overdosing on paracetamol produces excessive amounts of reactive oxygen species, which results in oxidative stress, endogenous antioxidant depletion, lipid peroxidation, and results in cellular damage. Parsley extract resists these effects by providing a rich environment of flavonoids and phenol compounds that function as strong radical scavengers (Nouioura et al., 2023). Assays for reducing power and DPPH radical scavenging were used to assess PCEO's antioxidant activity. At 5 mg/mL, PCEO showed a significant inhibition rate of 68.42% in the DPPH assay and also showed an absorbance of 0.517 in the ferric reducing assay. The antioxidant activity was mainly due to the presence of myristicin, apiol, beta-myrcene, sabinene, and beta-caryophyllene (Foudah et al., 2022). The parsley leaves grown in Benin were subjected to check for the antioxidant effect. The aqueous extract of the leaf was used to continue with the DPPH free radical scavenging activity, the extract showed a strong antiradical effect at the concentration of 63.66 μg/mL. Its strong antioxidant activity was associated with its high flavonoid and tannin content. These antioxidants can neutralize free radicals, resulting in the delay of lipid peroxidation and lower the risk of chronic disease and food deterioration (Gnintoungbe et al., 2023). When parsley was exposed to high concentrations of heavy metals like cadmium and zinc, it triggered a strong antioxidant defence mechanism. Increased activity was seen in the enzymes SOD, APX, CAT, and POD. SOD is necessary to convert superoxide radicals to hydrogen peroxide, which in turn is identified by APX and

CAT, which further detoxify it. POD also helps in the decomposition of hydrogen peroxide. The non-enzymatic antioxidants like glutathione and ascorbic acid scavenge the reactive oxygen species and also maintain the redox equilibrium. Parsley's resistance to heavy metals and potential for phytoremediation are shown by these reactions, which are correlated with decreased oxidative damage as seen by lower MDA and hydrogen peroxide levels (Nawaz et al., 2024).

Antidiabetic Activity

The fixed oil of parsley exhibited a strong alpha-glucosidase inhibitory activity of 91.2% inhibition at 20 mg/mL, showing its potential to treat Type 2 diabetes by delaying the absorption of glucose and the digestion of carbohydrates (Piras et al., 2022).

Anti-anaemic Effect

The aqueous extract of parsley was shown to prominently enhance the haematological indices that were said to decrease during anaemia, such as red blood cell count, haemoglobin, and haematocrit levels, when induced with phenylhydrazine. Increased serum and ferritin levels coincided with its therapeutic response, indicating that parsley not only helps in haemoglobin and red blood cell replacement but also efficiently replenishes iron stores. Decreased erythrocyte delicacy and hemolysis further supported the anti-anaemic effect. These findings show that the plant's antioxidant profile and rich mineral matrix work together to protect red blood cells, maintain iron metabolism, and defend against haemolytic damage caused by oxidative stress (Nouioura et al., 2024).

Immunomodulatory Activity

The sub-acute toxicity assessment of Wistar rats showed prominent immunomodulatory effects from the aqueous extract of parsley leaves. Significant haematological changes were seen after 28 days of oral treatment at dosages ranging from 50 to 200 mg/Kg body weight. The concentrations of haematocrit and haemoglobin, and red blood cell count, increased significantly in male rats, indicating a stimulatory effect on erythropoiesis. At 100 and 200 mg/Kg, the concentrations of white blood cells and lymphocyte count rose gradually, suggesting an improved immunological response. Female rats also showed an increase in haemoglobin concentration and lymphocyte count at 50 mg/Kg dose. These results suggest that the extract of parsley can enhance both innate and adaptive immunity (Jacques et al., 2021).

Anti-inflammatory Activity

The anti-inflammatory activity of PCEOs was evaluated using albumin and trypsin-induced protein denaturation assays, and the outcomes were compared to the standard drug ibuprofen. PCEO showed a good anti-inflammatory effect, with a decrease in albumin-induced anti-inflammation ranging from 22.2% to 90.4% at doses 5–1000 ppm, closely matching ibuprofen. Likewise, the trypsin-induced inflammation inhibition increased from 8.4% to 74.7% at 5–200 ppm. The blocking of pathways involving calcium flux, nitric oxide generation, and pro-inflammatory cytokines was shown by myristicin, which is responsible for its potent anti-inflammatory activity (Foudah et al., 2022).

Hepatoprotective and Nephroprotective Activity

The hepatoprotective and nephroprotective effects of parsley were clearly investigated by both biochemical and histopathological analysis. Regarding hepatic function, serum alanine aminotransferase (ALAT) and aspartate aminotransferase (ASAT) levels showed no significant elevation at any dose, with a slight decreasing trend, indicating the absence of hepatocellular damage. These findings were supported by histological sectioning of the same, which revealed no signs of necrosis, inflammatory infiltration, apoptosis, or structural problems. On the other hand, the treated animals showed a gradual increase in the total serum protein and also showed better hepatic function. Similarly, the nephroprotective activity was indicated by constant levels of renal biomarkers like urea, creatinine, and uric acid, which all remained within the normal physiological range (Jacques et al., 2021). According to the study conducted by Ismail and his teammates, aqueous extract of parsley has shown prominent hepatoprotective qualities. The rats exposed to carbon tetrachloride induced hepatotoxicity than in normal controls, indicating severe liver damage, and showed a significant increase in its hepatic enzymes of serum such as AST, ALP, and ALT. The rats treated with parsley extract, particularly at 500 mg/Kg, showed a reduced amount of these enzymes, indicating a regenerative and protective effect of hepatocytes. Histopathological assessment of liver tissues additionally supported these findings; the rats treated with CCL_4 alone exhibited necrosis and disorganization of hepatocytes. Whereas the co-treated rats with parsley extract showed a very minor necrosis and mild inflammatory changes (Ismail et al., 2022). The rats that were administered with parsley extract and exposed to paracetamol showed a protective effect against kidney and liver damage. According to biochemical research, the treatment of parsley reduced the serum levels of hepatotoxic indicators like lactate dehydrogenase (LDH), alanine aminotransferase (ALT), aspartate aminotransferase (AST), and alkaline phosphatase (ALP), all of which were raised by the administration of paracetamol. The normal range of albumin, total proteins, triglycerides, and indicators of renal functions such as creatinine and blood urea nitrogen, were restored by the nephroprotective effect of parsley's extract. Histopathological analysis of paracetamol overdose showed a severe alteration in the liver, such as congestion, degeneration, necrosis, and also renal lesions, including glomerular atrophy, tubular degeneration, and inflammatory infiltration. However, these alterations were improved when treated with parsley extract; this protective effect was contributed by its strong antioxidant properties (Nouioura et al., 2023).

Diuretic Activity

In Wistar rats, the study showed that the aqueous extract of parsley had a diuretic effect. Urinary excretion increased in a dose-dependent manner over a period of 24 hours, after increasing dosages of 500, 1000, and 1500 mg/Kg were administered. Without disturbing the normal renal function or negatively altering important kidney indicators like urea and creatinine. At the dosage of 1500 mg/Kg, the extract showed a gradual increase in urine output, which was equivalent to the impact of furosemide, which is a standard pharmaceutical diuretic. This indicates that the effect of the extract is potent and safe under experimental condition (Ismail et al., 2022). The five- group Wistar rats, which were treated with water overloaded with 0.9% NaCl saline solution and a dose-dependent extract, showed increased urinary excretion at 1500 mg/kg compared to the control group; this effect was statistically similar to the urinary excretion produced by furosemide at 40 mg/Kg. As a result of this increased excretion in the urine, both potassium (K^+) and sodium (Na^+) were eliminated. The Na^+/K^+ ratio for all the induced doses showed a value greater than 5, indicating the sparing effect of potassium (Atto et al., 2024).

Antiviral Effects

At a dosage of 3.3 mg/mL of dried parsley, the study found that the ethanol extract of parsley strongly induced the interaction between the human ACE2 receptor and the SARS-COV-2 spike protein, with an inhibition rate of 70%. Similarly, at 5 mg/mL, there was a reduction in the enzymatic activity of ACE2 by 86% indicating the disruption of the viral entry pathway. The bioactive substances found in the extract, such as oxypeucedanin and apigenin, which are known to interfere with the activities of viral proteins, are responsible for these effects. While oxypeucedanin may cause conformational changes in the spike protein, on the other hand, apigenin demonstrated a strong binding affinity to the SARS-COV-2 nucleocapsid protein. Additionally, parsley showed a strong scavenging effect of free radicals, particularly hydroxyl radicals (Lee et al., 2023).

Biosafety and Toxicity of Parsley

The phenolic content and safety parameters of a Moroccan aqueous polyherbal formulation that contains parsley along with *Coriandrum sativum* and *Apium graveolens*. A significant safety test indicated by acute toxicity experiments in mice, and it showed no death or any other symptoms or signs like nephrotoxicity and hepatotoxicity were observed, and MDA levels were also within the normal level, up to 2000 mg/Kg, and the LD_{50} was also found to be above the range of 2000 mg/Kg. Assessment of sub-acute toxicity showed no changes in physical body weight, food intake, or organ weights, and even the enzymes of the liver, like ALT, AST, showed normal ranges in the haematological and biochemical parameters test. Formulation's biosafety, as parsley was included as a key ingredient, was further characterized by histopathological analysis, which confirmed the absence of any tissue damage (Nouioura et al., 2024). No oral toxicity was observed in the characterisation of the aqueous extract of parsley leaves, liver, histology of the kidney, biochemical measures, or haematological parameters. It is proposed that the Wistar rats had an LD50 concentration more than 5000 mg/Kg (Gnintoungbe et al., 2023). Myristicin and apiol, found in parsley's seed oil, are toxic at higher doses. Myristicin can cause hallucination, liver damage, and neurotoxicity. While apiol is known for its abortifacient effects and hepatotoxicity (Akpaso, 2023).

Conclusion and Future Perspectives

Petroselinum crispum represents a multipurpose herb that combines medicinal relevance with nutritional richness. Phenolic acids, flavonoids, carotenoids, hydrocarbons, essential oils, terpenoids, and fatty acids are the phytochemicals that support the wide range of biological activity shown by the herb's leaves, roots, seeds, and extracts. Its antibacterial, antioxidant, anti-inflammatory, anticancer, antihypertensive, anti-anaemic, diuretic, immunomodulatory, hepatoprotective, and nephroprotective properties are supported by experimental research. Its effectiveness in preventing and treating chronic diseases is shown by its strong free radical scavenging, metal-chelating, and bioactive-mediated regulation of cellular pathways. Parsley's potential as a nutritional and therapeutic herb has been strengthened by its good biosafety profiles in toxicity analysis. Together, the results demonstrate that *Petroselinum crispum* has great potential for use in the pharmaceutical, nutritional, and food industries. It bridges traditional medicine and modern scientific validation.

Despite well-characterized nutritional and phytochemical composition of *Petroselinum crispum.* The precise molecular mechanisms, like receptor-level interactions and modulations of pathways, need further research, even though flavonoids, essential oils, and phenolic acids have been widely related to antioxidant, antibacterial, and anticancer activities. Apigenin and oxypeucedanin, for example, have been shown to disrupt the binding of the SARS-CoV-2 virus protein, but their therapeutic validation through clinical trials is still limited. Similarly, parsley's allelopathic bioactive compounds have not been fully identified, and this opens new opportunities for phytochemical discovery. Its bioactive profile can be improved by using standardized extraction procedures and advanced analytical techniques (HPLC-DAD, LC-MS/MS, GC-MS, NMR). Additional studies, comparing the antioxidant potencies of microgreens, baby greens, and mature plants, indicate that these variations vary depending on the stage, which needs more research for specific nutritional or medicinal uses.

Acknowledgements

The authors express appreciation for the support provided by the Deanship of Scientific Research, Vice Presidency for Graduate Studies and Scientific Research, King Faisal University, Saudi Arabia [Grant No. KFU254839].

References

Abdu, K. and Hauwa, D.G. 2019. *In vitro* evaluation of the antioxidant and antimicrobial activity of leaf extracts of *Petroselinum cripsum* (Parsley). *ChemSearch J.*, *10*(2): 25–29.

Abubakar, J., Edo, G. and Pasaoglulari Aydinlik, N. 2021. Phytochemical and GC-MS analysis on the ethanol extract of *Foeniculum vulgare* and *Petroselinum crispum* leaves. *Int. J. Chem. Technol.*, *5*(2): 117–124. https://doi: 10.32571/ijct.911711.

Abu-Serie, M.M., Habashy, N.H. and Maher, A.H. 2019. *In vitro* anti-nephrotoxic potential of *Ammi visnaga*, *Petroselinum crispum*, *Hordeum vulgare*, and *Cymbopogon schoenanthus* seed or leaf extracts by suppressing the necrotic mediators, oxidative stress, and inflammation. *BMC Complement. Altern. Med.*, *19*(1): 149. https://doi: 10.1186/s12906-019-2559-8.

Agyare, C., Appiah, T., Boakye, Y.D. and Apenteng, J.A. 2017. *Petroselinum crispum*: A review. In: Victor Kuete (Ed.), *Medicinal Spices and Vegetables from Africa*. Academic Press, pp. 527–547. https://doi: 10.1016/B978-0-12-809286-6.00025-X.

Aissani, N., Albouchi, F. and Sebai, H. 2021. Anticancer effect in human glioblastoma and antioxidant activity of *Petroselinum crispum* L. methanol extract. *Nutr. Cancer*, *73*(11): *12*: 2605–2613. https://doi: 10.1080/01635581.2020.1842894.

Ajebli, M. and Eddouks, M. 2019. Antihypertensive activity of *Petroselinum crispum* through inhibition of vascular calcium channels in rats. *J. Ethnopharmacol.*, *242*: 112039. https://doi: 10.1016/j.jep.2019.112039.

Akpaso, M.I. 2023. Improved fertility potential of ethanolic leaf extracts of *Petroselinum crispum*. *J. Med. Res.*, *2023*(1): 157–163. https://doi: 10.5281/zenodo.8312705.

Al-Hadi, Al-Marwa H., et al. 2013. Phytochemical screening, antibacterial, and cytotoxic activities of *Petroselinum crispum* leaves grown in Oman. *Iran. J. Pharm. Sci.*, *9*(1): 61–65.

Ancuceanu, R., Amgjel, A.I., Mariena, H. et al., 2018. Variation of iron contents, polyphenols, and flavonoids in *Petroselinum crispum* (Mill.) Fuss (Apiaceae). *Farm.*, *66*: 275–281.

Atto, V., Koko, K.B., Kamo, E.B.L.I., Konan, B.A. and Datte, J.Y. 2024. Assessment of the diuretic activity of *Petroselinum crispum* aqueous leaves extract in Wistar rats. *Azerbaijan Pharm. Pharmacother. J.*, *23*(1): 36–41. https://doi: 10.61336/appj/23-1-7.

Bahramsoltani, R., Ahmadian, R., Daglia, M. and Rahimi, R. 2024. *Petroselinum crispum* (Mill.) Fuss (Parsley): An updated review of the traditional uses, phytochemistry, and pharmacology, *J. Agric. Food Chem.*, *72*(2): 956–972. https://doi: 10.1021/acs.jafc.3c06429.

Benmekhbi, L., Mosbah, S., Laamraoui, H., et al., 2022. Evaluation of phytochemical properties and biological activities of leaf extracts and oil of *Petroselinum sativum* collected from Algeria. *Indones. J. Chem.*, *22*(6): 1566–1573. https://doi: 10.22146/ijc.74370.

Boutsika, A., Sarrau, E., Cook, C.M. et al., 2021. Evaluation of parsley (*Petroselinum crispum*) germplasm diversity from the Greek gene bank using morphological, molecular, and metabolic markers. *Ind. Crops Prod.*, *170*: 113767. https://doi: 10.1016/j.indcrop.2021.113767.

Chauhan, E.S. and Aishwarya, J. 2018. Nutraceuticals potential of *Petroselinum crispum*: A review. *J. Complement. Med. Alt. Healthcare*, *7*(2): 555707. https://doi: 10.19080/JCMAH.2018.07.555707.

Chaves, D.S.A., Fratttani, F.S., Assafim, M. et al., 2011. Phenolic chemical composition of *Petroselinum crispum* extract and its effect on haemostasis. *Nat. Prod. Commun.*, *6*(7): 961–964.

Daimari, E., Dutta, U. and Tochhawng, L. 2024. Phytochemistry of *Petroselinum crispum* (Mill.) Fuss, *Murraya koenigii* (L.) *Spreng.* and *Cinnamomum tamala* (Buch.-Ham.) T. Nees & C. H. Eberm. and *in silico* studies of the role of their bioactive components against cancer. *Plant Sci. Today*, *11*: 981–1003. https://doi: 10.14719/pst.3120.

Del Rio, D., Rodriguez-Mateos, A., Spencer, J.P.E., Tognolini, M., Borges, G., Crozier, A. 2013. Dietary (Poly)phenolics in human health: Structures, bioavailability, and evidence of protective effects against chronic diseases. *Antioxid. Redox Signal*, *18*(14): 1818–1892. https://doi: 10.1089/ars.2012.4581.

Dobricevic, N., Sic Zlabur, J., Voca, S. et al., 2019. Bioactive compounds content and nutritional potential of different parsley parts (*Petroselinum crispum* Mill.). *J. Cent. Eur. Agric.*, *20*(3): 900–910. https://doi: 10.5513/JCEA01/20.3.2417.

El-Nakhel, C., Pannico, A., Graziani, G. et al., 2021. Mineral and antioxidant attributes of *Petroselinum crispum* at different stages of ontogeny: Microgreens vs. Baby greens. *Agronomy*, *11*(5): 857. https://doi: 10.3390/agronomy11050857.

Fernandes A. et al., 2020. Phytochemical composition and nutritional value of pot-grown turnip-rooted and plain and curly-leafed parsley cultivars. *Agronomy, 10*(9): 1416. https://doi: 10.3390/agronomy10091416.

Foudah, A.I., Alqarni, M.H., Alam, A., Salkini, M.A., Ross, S.A. and Yusufoglu, H.S. 2022. Phytochemical screening, *in vitro* and *in silico* studies of volatile compounds from *Petroselinum crispum* (Mill) leaves grown in Saudi Arabia. *Molecules*, *27*(3): 934. https://doi: 10.3390/molecules27030934.

Gnintoungbe, G.S., Marc Medehouenou, T.C., Adounpke, F. et al., 2023. Phytochemical screening, antioxidant activity, and safety of *Petroselinum crispum* (Mill.) AW Hill Apiaceae leaves grown in Benin. *Open J. Appl. Sci.*, *13*: 36–50. https://doi: 10.4236/ojapps.2023.131004.

Hyskova, V., Belonoznikova, K., Smeringaiova, I., Kavan, D., Ingr, M. and Ryslava, H. 2021. How is the activity of shikimate dehydrogenase from the root of (parsley) regulated and which side reactions are catalyzed? *Phytochemistry*, *190*: 112881, https://doi: 10.1016/j.phytochem.2021.112881.

Ismail, M., Mohamed, M.E., Ismail, M.K.A. and Abbas, S. 2022. Effects of aqueous extract of *Petroselinum crispum* on liver and hematological parameters in carbon tetrachloride (CCl4) - induced hepatotoxicity in male Wistar rats. *Bayero J. Pure and Appl. Sci.*, *14*(2): 245–255. https://doi: 10.4314/bajopas.v14i2.30.

Jacques, K.K.J., Nadia, B.A., Roger, K.K., Seraphin, D.Y., Martin, S.Y. and Annick, T. 2021. Acute and subacute toxic aqueous extract of the leaves of *Petroselinum crispum* Mill. in male and female Wistar rats. *Eur. Sci. J.*, *17*(40): 178. https://doi: 10.19044/esj.2021.v17n40p178.

Karimi, M.H., Ebadi, P. and Amirghofran, Z. 2012. Parsley and immunomodulation. *Expert Rev. Clin. Immunol.*, *8*(4): 295–297. https://doi: 10.1586/eci.12.12.

Kurtaran, M., Koc, N.S., Aksun, M.S., Yildirim, T., Yilmaz, S.R. and Erdem, Y. 2021. *Petroselinum crispum*, a commonly consumed food, affects sirolimus level in a renal transplant recipient: A case report. *Ther. Adv. Drug Saf.*, *12*. https://doi: 10.1177/20420986211009358.

Lamponi, S. and Baratto, M.C. 2020. Bioactivity of hydro-alcoholic extract of *Petroselinum crispum*. *Longhua Chin, Med.*, *3*: 3(16). https://doi: 10.21037/lcm-20-47.

Lee, E.Y., Liu, Z., Li, Y., et al., 2023. Chemical constituents of parsley (*Petroselinum crispum*) leaf extract and its potential in mitigating the effects of viral infections. *ACS Food Sci. Technol.*, *3*(12): 2108–2116. https://doi: 10.1021/acsfoodscitech.3c00357.

Marthe, F. 2020. *Petroselinum crispum* (Mill.) Nyman (Parsley). In: Novak, J., Blüthner, W.D. (Eds.), *Medicinal, Aromatic, and Stimulant Plants. Handbook of Plant Breeding.* (Vol. 12). Springer, Cham. https://doi.org/10.1007/978-3-030-38792-1_13.

Meyer, H., Bolarinwa, A., Wolfram, G. and Linseisen, J. 2006. Bioavailability of apigenin from apiin-rich parsley in humans. *Ann. Nutr. Metab.*, *50*(3): 167–172. https://doi: 10.1159/000090736.

Mirmohammadmakki, F., Gharachorloo, M., Ghavami, M., Abdossi, V. and Azizinezhad, R. 2020. Quantitative changes in essential oils contents of Parsley *(Petroselinum crispum*) harvested in three consecutive months of spring. *Biotechnol. Bioprocess Eng.*, *6*: 48–55. https://doi: 10.22059/JFABE.2023.365275.1154.

Motti, R., Bonanomi, G., Emrick, S. and Lanzotti, V. 2019. Traditional herbal remedies used in women's health care in Italy: A review. *Hum. Ecol.*, *47*: 1–32. https://doi.org/10.1007/s10745-019-00125-4.

Nawaz, I., Mehboob, A., Khan, A.H.A. et al., 2024. Higher cadmium and zinc accumulation in parsley (*Petroselinum crispum*) roots activates its antioxidants defense system. *Biometals*, *37*(1): 87–100. https://doi: 10.1007/s10534-023-00529-2.

Nielsen, S.E., Young, J.F., Daneshvar, B. et al., 1999. Effect of parsley (*Petroselinum crispum*) intake on urinary apigenin excretion, blood antioxidant enzymes, and biomarkers for oxidative stress in human subjects. *Br. J. Nutr.*, *81*(6): 447–455. https://doi: 10.1017/S000711459900080X.

Nouioura, G., Kettani, T., Tourabi, M., et al., 2023. The protective potential of *Petroselinum crispum* (Mill.) Fuss. on paracetamol-induced hepatio-renal toxicity and antiproteinuric effect: A biochemical, hematological, and histopathological study. *Medicina (B Aires)*, *59*(10): 1814. https://doi: 10.3390/medicina59101814.

Nouioura, G., El Fadili, M., El Hachlafi, N. et al., 2024. *Petroselinum crispum* L., essential oil as promising source of bioactive compounds, antioxidant, antimicrobial activities: *In vitro* and *in silico* predictions. *Heliyon*, *10*(8): e29520. https://doi: 10.1016/j.heliyon.2024.e29520.

Nouioura, G., et al., 2024. Unveiling the antianemic activity, physicochemical aspects, antioxidant properties, and mineral profile of *Petroselinum crispum L. ACS Omega*, *9*(26): 28155–28166. https://doi: 10.1021/acsomega.4c01107.

Nouioura, G., Kettani, T., Elousrouti, L.T., Loukili, E.H., Lyoussi, B. and Derwich, E. 2024. Phenolic profile, and safety evaluation of the Moroccan aqueous polyherbal formulation containing *Petroselinum crispum* L., *Coriandrum sativum* L., *Apium graveolens* L.: Acute and sub-acute toxicity. *S. Afr. J. Bot.*, *170*: 308–317. https://doi: 10.1016/j.sajb.2024.05.049.

Pajevic, S., Mimica-Dukic, N., Nemes, I. et al., 2021. Arsenic content and phenolic compounds in parsley (‘*Petroselinum’ crispum* (Mill.) Fuss) and celery (‘*Apium graveolens*’ L.) cultivated in Vojvodina region, Serbia. *Food and Feed Res.*, *48*(2): 213–225. https://doi: 10.5937/ffr48-34625.

Petropoulos, S.A., Fernandes, A., Finimundy, T.C. et al., 2022. The bioactivities and chemical profile of turnip-rooted parsley germplasm. *Horticulturae*, *8*(7): 639. https://doi: 10.3390/horticulturae8070639.

Piras, A., Porcedda, S., Falconieri, D. et al., 2022. Supercritical extraction of volatile and fixed oils from *Petroselinum crispum* L. seeds: Chemical composition and biological activity. *Nat. Prod. Res.*, *36*(7): 1883–1888. https://doi: 10.1080/14786419.2020.1810031.

Prabhat, L.N.U. and Jagriti, L.N.U. 2024. Screening of the leaf extracts of culinary herbs (*Apium graveolens*, *Petroselinum crispum*, *Cichorium endivia*, and *Anethum graveolens*) for their antibacterial activity against *Escherichia coli. Cureus*, *16*: e54242. https://doi: 10.7759/cureus.54242.

Punosevac, M., Radovic, J., Lekovic, A. and Kundakovic-Vasovic, T. 2021. A review of botanical characteristics, chemical composition, pharmacological activity, and use of parsley. *Arh. Farm (Belgr)*., *71*(3): 177–196. https://doi: 10.5937/arhfarm71-30800.

Rigat, M., Valles, J., D'Ambrosio, U., Gras, A., Iglesias, J. and Garnatje, T. 2015. Plants with topical uses in the Ripolles district (Pyrenees, Catalonia, Iberian Peninsula): Ethnobotanical survey and pharmacological validation in the literature. *J. Ethnopharmacol.*, *164*: 162–179. https://doi: 10.1016/j.jep.2015.01.055.

Sarwar, S., Ayyub, M.A., Rezgui, M., Nisar, S. and Jilani, M.I. 2016. Parsley: A review of habitat, phytochemistry, ethnopharmacology, and biological activities. *J. Tradit. Chin. Med.*, *33*: 815–826.

Savikin, K., Zdunic, G., Menkovic, N. et al., 2013. Ethnobotanical study on traditional use of medicinal plants in South-Western Serbia, Zlatibor district. *J. Ethnopharmacol.*, *146*(3): 803–810. https://doi: 10.1016/j.jep.2013.02.006.

Sbai, H., Saad, I., Ghezal, N., Della Greca, M. and Haouala, R. 2016. Bioactive compounds isolated from *Petroselinum crispum* L. leaves using bioguided fractionation. *Ind. Crops Prod.*, *89*: 207–214. https://doi: 10.1016/j.indcrop.2016.05.020.

Subaş, T., Ozgen, U., Gokkaya, I. and Renda, G. 2024. *Petroselinum crispum* (Mill.) Fuss (parsley), A food and medicinally important plant: A review of recent studies between 2013–2023. *Ankara Universitesi Eczacilik Fakultesi Dergisi.*, *48*(2): 724–750. https://doi: 10.33483/jfpau.1362626.

Taheri, M., Hossein Mirzabe, A., Mansouri, A. and Reza Aliakbari, A. 2015. Determination of some engineering properties of parsley (*Petroselinum crispum* L.) seeds and modelling. *Agric. Eng. Int.*, *17*: 191–210.

Tang, E.L., Rajarajeswaran, J., Fung, S. and Kanthimathi, M. 2015. *Petroselinum crispum* has antioxidant properties, protects against DNA damage, and inhibits proliferation and migration of cancer cells. *J. Sci. Food Agric.*, *95*(13): 2763–2771. https://doi: 10.1002/jsfa.7078.

Thangavelu, S., Balasubramanian, B., Palanisamy, S. et al., 2022. Characterization and phytoconstituents of *Petroselinum crispum* (Mill) and *Coriandrum sativum* (Linn) and their impacts on inflammation: An *in vitro* analysis against human adenocarcinoma cells with molecular docking, *S. Afr. J. Bot.*, *146*: 776–788. https://doi: 10.1016/j.sajb.2021.12.024.

Yoshikawa, M., Uemura, T., Shimoda, H., Kishi, A., Kawahara, Y. and Matsuda, H. 2000. Medicinal foodstuffs. XVIII. Phytoestrogens from the aerial part of *Petroselinum crispum* MILL. (Parsley) and structures of 6"-Acetylapiin and a new monoterpene glycoside, Petroside. *Chem. Pharm. Bull.* (*Tokyo*), *48*(7): 1039–1044. https://doi: 10.1248/cpb.48.1039.

15 Extraction and Metabolic Profiling of Phenolic Acids, Flavonoids, and Carotenoids from Green Leafy Vegetables: A State of the Art of the Recent Techniques Used and Their Application

Bonta Ramesh Kumar[1*], Samsur Ali Dafadar[2],
Saikat Sen[3] and Srinivas Oruganti[4]

Introduction: Importance of Bioactive Components in Vegetables

Vegetables are a broad category of plants high in phytochemicals and minerals. The root, stem, leaf, and fruit are the most appetizing components of this food group. Leafy vegetables, also called "greens", "vegetable greens", or "leafy greens", contain a diverse range of bioactive compounds. Although present in small quantities, these metabolites serve a crucial function in plant secondary metabolism (Pateiro et al., 2023). Leafy vegetables contain vitamins, proteins, minerals, and antioxidants (Barba et al., 2014). Vegetable phytochemicals having antioxidant properties are predicted to promote health by decreasing the formation of free radicals, which are responsible for the onset of various chronic disorders (Kumar, 2017). Because of consumer awareness of health and nutraceutical benefits of green leafy vegetables (GLVs), there is an expanding body of studies on some of their active metabolites. Many of them are classed as bioactive compounds. Bioactive compounds present in plants are divided into many categories based on their molecular structure. They consist of dietary fiber, antioxidant vitamins, polyphenols, terpene derivatives, sulfur compounds, alkaloids, phytoestrogens, minerals, and polyunsaturated fatty acids (Barba et al., 2014). Figure 1 represents the most common bioactive components found in GLVs.

Phenolic compounds are distributed throughout the plant systems as secondary metabolites. They play significant roles in the plant's defensive systems against

[1-4] Center for Process Research and Innovation (CPRI), Dr.Reddy's Institute of Life Sciences, University of Hyderabad Campus, Gachibowli, Hyderabad 500046, Telangana, India.
* Corresponding author: bramesh.kmr@gmail.com; rameshkumarb@drils.org

DOI: 10.1201/9781003687832-15

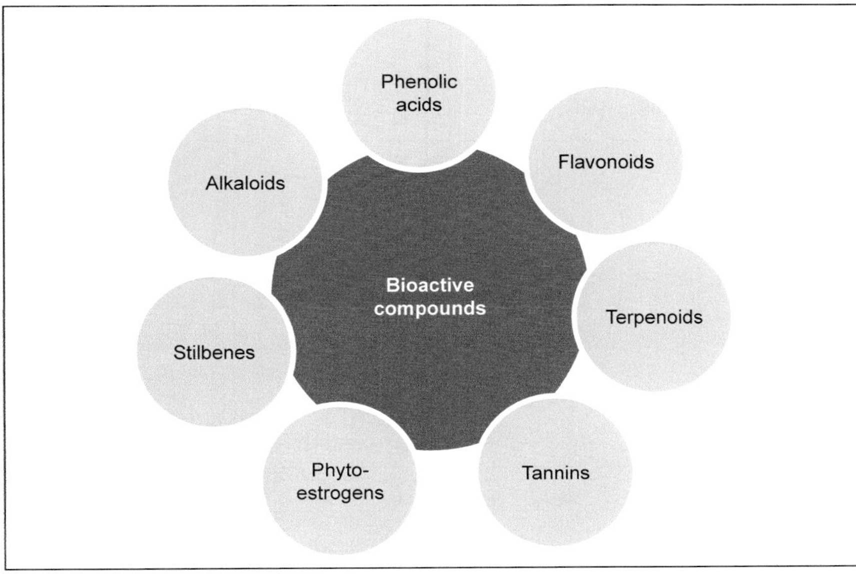

Fig. 1 Most common bioactives found in Green Leafy Vegetables.

stress, infections, and UV radiation (Bang et al., 2024). They are an enormous group of bioactive compounds (about 10,000), formed by the shikimic acid pathway. They are distinguished by at least one aromatic ring connected to one or more hydroxyl groups. Plant cell walls contain phenolic acids as a byproduct of the monolignol pathway and lignin breakdown products. They can be classified as hydroxy-cinnamic or hydroxy-benzoic acids based on their carbon structure. GLV, aerial portions (leaf, root, stem, and fruit) accumulate benzoic and cinnamic acid derivatives. While hydroxybenzoic acids are commonly found as glycosides, products derived from hydroxycinnamic acid are simple esters with a sugar moiety such as glucose. Hydroxycinnamic acids have a C6-C3 configuration, which consists of coumaric acid, sinapic acid, caffeic acid, and ferulic acid. Gallic acid, protocatechuic acid, syringic acid, and vanillic acid are the primary components of hydroxy-benzoic acids, which consist of a C6-C1 structure. Flavonoids' structural backbone consists of C6-C3-C6. They are composed of 15 carbon atoms linked by a heterocyclic C-ring to aromatic rings A and B. They are classified as flavonols, flavones, isoflavones, and anthocyanidins based on the C-ring substitution pattern. They are found in plant cell vacuoles and undergo a variety of chemical transformations, including hydroxylation, glycosylation, and methylation, to yield their corresponding derivatives (Kumar, 2017). Some of the important phenolic acids, including hydroxy-benzoic acids (Figure 2) and hydroxy-cinnamic acids (Figure 3), and flavonoids (Figure 4) found in GLVs are represented below.

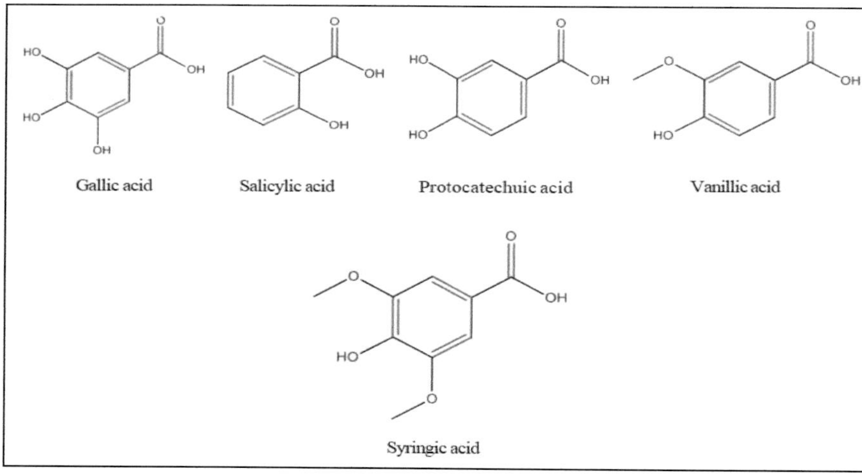

Fig. 2 Hydroxybenzoic acids.

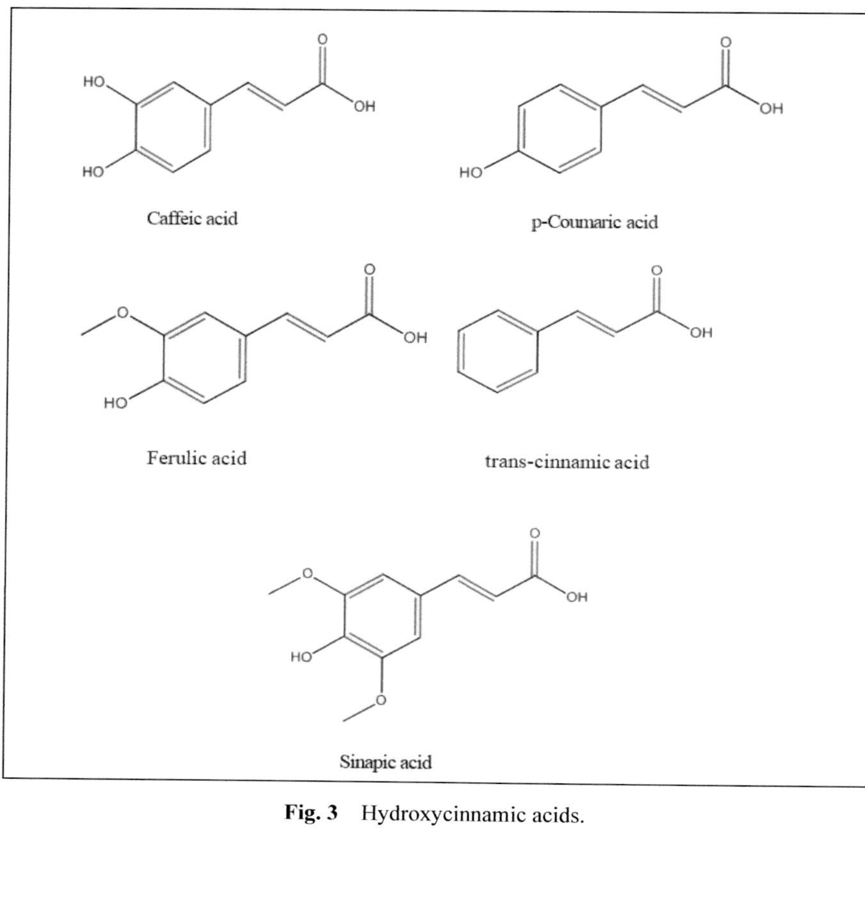

Fig. 3 Hydroxycinnamic acids.

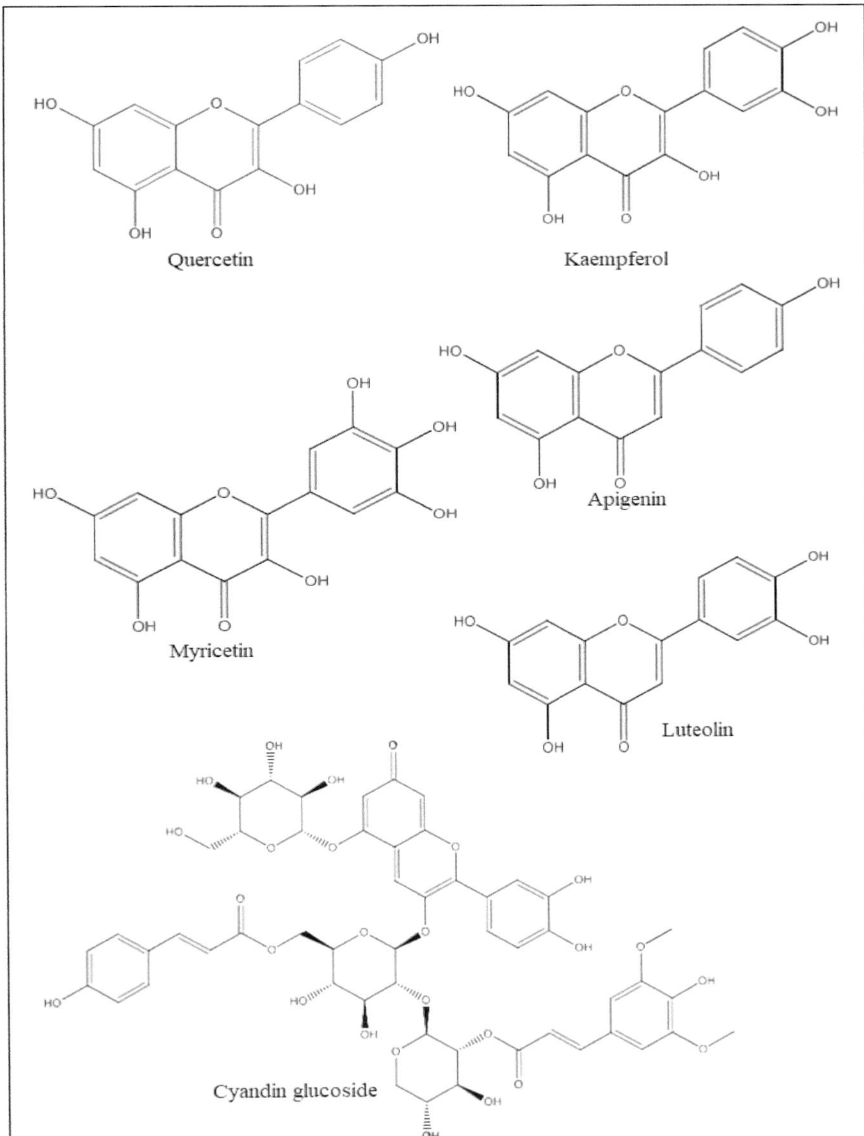

Fig. 4 Flavonoids.

Carotenoids are tetraterpene pigments that exist in yellow, orange, red, and purple shades. Carotenoids can be found in algae, plants, animals, some archaeal and fungal species, and photosynthetic bacteria. They are typically composed of a polyene chain with an end group at either end and nine conjugated double bonds

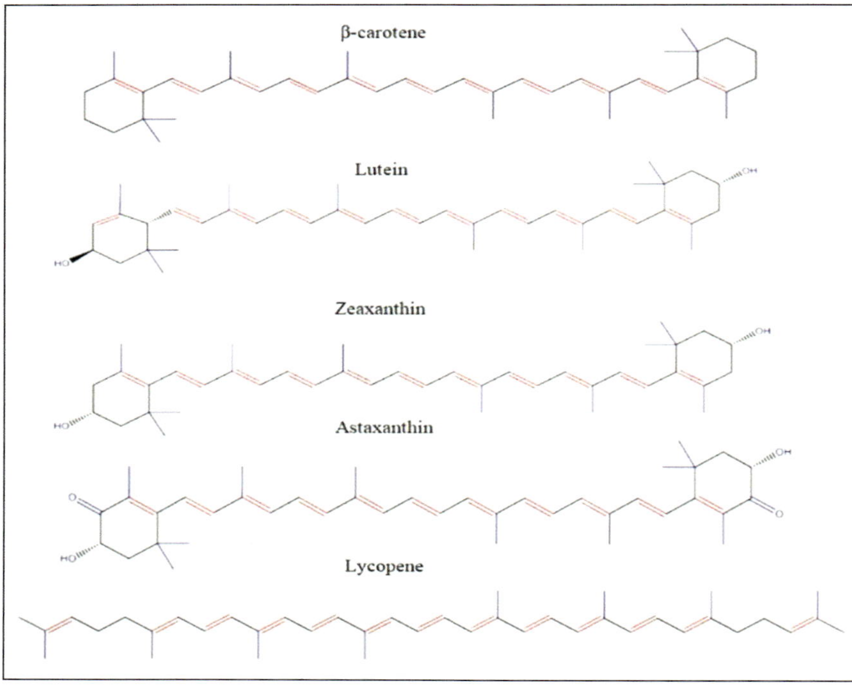

Fig. 5 Carotenoids.

(Maoka, 2023). They are classified into two types: xanthophylls and carotenes. These are known for pigmentation and aroma (Zeb, 2017). Carotenes are made up of hydrocarbon molecules. There are over 50 different types of carotenes in nature, including lycopene, α-carotene, and β-carotene. However, xanthophylls are carotenoids that contain oxygen atoms in the form of hydroxy, carbonyl, aldehyde, carboxylic, epoxide, and furanoxide groups. As a result, xanthophyll structure varies greatly. There are around 800 different types of xanthophylls in nature, including lutein, zeaxanthin, lactucaxanthin, astaxanthin, fucoxanthin, and β-cryptoxanthin (Maoka, 2023). Some of the important carotenoids are represented in Figure 5.

Plant based pharmaceuticals, which have a yearly market value of more than $100 billion, emerged as a result of research into plant based therapies. Despite the rise of synthetic drugs, herbal remedies are still popular due to their accessibility, tolerance, and absence of adverse effects. Mohd Zaid et al. (2022) found that 44% of new drugs use plants as lead ingredients. Phytochemicals are useful molecules found in a variety of industries such as agriculture, food and beverage, cosmetics, pharmaceuticals, and nutraceuticals. They are used in the production of pharmaceutical medications, nutritional supplements, functional foods, herbal therapies, and cosmetic formulations to fulfill the diverse demands and preferences of customers. The demand for a highly sensitive and selective analytical approach to identify polyphenols is increasing. Despite multiple investigations, the extraction and quantification of diverse polyphenols continues to be a challenge. The origin of vegetal bioactive compounds for novel therapeutic formulations is a continuing concern for the food and pharmaceutical industries, as their identification and dose determination is essential for gaining health benefits (Mungwari et al., 2025).

This chapter gives current information on the analytical determination of phenolic acids, flavonoids, and carotenoids from GLVs. It emphasizes chromatographic and spectroscopic methodologies, as well as contemporary extraction methods. The significance of the extraction and chromatographic processes described in this chapter provides a thorough grasp of the study of GLVs in terms of bioactive profile for health and industrial applications. Due to the numerous applications in this field, this chapter discusses a few key and relevant published studies and demonstrates the state of the art in analytical methodologies.

Leafy Vegetables: Significance, Distribution, and Nutritional Composition

Vegetables are one of the most important resources for preserving the body's alkaline reserve. The primary reason people value vegetables is because of their high vitamin, carbohydrate, and mineral content. Ascorbic acid, folic acid, carotenes, riboflavin, and minerals, including calcium, phosphorus, and iron are all abundant in vegetables (Randhawa et al., 2015). GLVs have been domesticated for hundreds or thousands of years due to the desirable traits of their leaves, such as morphology, quantity, and flavor. Many widely cultivated and consumed leafy crops, particularly from the Brassicaceae, Asteraceae, and Amaranthaceae families, belong to this group. Phenolic acids, which makes up 47.4% of vegetable polyphenols, are the primary source of polyphenols in non-Mediterranean parts of Europe, accounting for 41.2% of total intake (Masiala et al., 2024). A study by Pérez-Jiménez et al. (2010) stated that among the 452 foods in the Phenolic Explorer database, the 100 foods highest in polyphenols (average of the polyphenols determined by High Performance Liquid Chromatography (HPLC)) contain 17 foods made from plant leaves, nine of which are GLVs.

Many species of the Asteraceae family, which are widely cultivated worldwide, such as lettuce (*Lactuca sativa*), endive (*Cichorium endivia*), and red and green chicory (*Cichorium intybus* var. *foliosum*), are examples of GLVs that are frequently eaten raw in salads. The glucosinolates of GLV leaves, which belong to the Brassicaceae family, give them a unique peppery flavor and are especially well-liked when consumed raw in salads. Numerous health benefits have been demonstrated for these sulfur compounds, which are abundant in Brassicaceae (Bell and Wagstaff, 2014). Vegetables from the *Brassica* genus (Brassicaceae family) are some of the most consumed vegetables in the world

(Lin and Harnly, 2009). The Brassicaceae family have several species, including watercress (*Nasturtium officinale*) and cultivated rocket (*Eruca sativa*) (Masiala et al., 2024). Spinach (*Spinacia oleracea*), an important member of Amaranthaceae family is the most popular leafy vegetable consumed worldwide and valued for its high nutritional content. Spinach leaves have been consumed since ancient times after being domesticated in the Middle East more than 2,000 years ago. China was the primary producer (92.59%) of the 33.12 million tons produced worldwide in 2022 (FAO, 2022).

Accumulation and Distribution of Phenolic Acids, Flavonoids, and Carotenoids in GLVs

GLVs are an important source of phenolic compounds in vegetables. They are a large source of bioactive polyphenols; hence, there is a growing interest in studying their phenolic content and antioxidant properties. A study by Tresserra-Rimbau et al. (2013) on the population of Mallorca revealed that lettuce leaves are the fourth largest source of polyphenols among vegetables and account for 6% of flavonols among all foods. Green and red chicory leaves (*Chicorium intybus* var. *foliosum* L.) are widely consumed across the world. Their leaves are commonly used in human cuisine due to their characteristic bitterness caused by the presence of phenolic compounds and other useful nutrients. Red chicory contains the highest proportion of hydroxycinnamic acid (183 mg/100 g FW). A variety of flavonol glycosides, flavones, and other glycosylated anthocyanins, primarily found in red chicory types, have also been reported (Masiala et al., 2024). Kaempferol derivatives make up the most prevalent flavonoid category, with a range of 8.47–26.0 g per kg dw (77–88% of total phenolics). The primary flavonoid among them is kaempferol-3,4-diglucoside, which ranges from 8.07 to 23.68 g per kg dw. The second flavonoid, isorhamnetin-3,4-diglucoside, has a range of 1.07–4.75 g per kg dw (9–16.3% of total phenolics) (Pasini et al., 2012). In 2022, the globe produced 72.60 million tons of cabbage, making it one of the most popular green leafy vegetables in the world (FAO, 2022). It contains several subspecies of *Brassica oleracea* and *Brassica rapa* that have been chosen for their edible leaves, including kale (*B. oleracea* var. *sabellica*), Chinese cabbage or bok choy (*B. rapa* var. *chinensis*), Chinese white cabbage or Pe-tsai (*B. rapa* var. *pekinensis*), and head cabbage (*Brassica oleracea* var. *capitata*). Flavonoids, especially flavonols, are abundant in the leaves of most *Brassica* species. Kaempferol and quercetin glycosides, as well as their derivatives linked to hydroxycinnamic acids and sinapic acid, are among the many flavonoid glycosides that have been found. Sinapic acid (421.9 μg/g DM), caffeic acid (84.6 μg/g DM), p-coumaric acid (63.1 μg/g DM), and ferulic acid (25.7 μg/g DM) were the primary phenolic acids detected in green/white cabbage leaves cultivated outdoors in the spring (Park et al., 2014). Flavonols like quercetin (601.3 μg/g DM) and kaempferol (448.7 μg per/ g DM) are also found in high concentrations in the same plant matrix. For instance, kale accounts for 28% of flavonols in Brazil, making it the primary source (Carnauba et al., 2021). Flavonols, especially derivatives of kaempferol and quercetin, are mostly found in its leaves. Kaempferol-3-sinapoyl-diglucoside-7-diglucoside and quercetin-3-sinapoyl-diglucoside-7-diglucoside constituents, with 159 mg of rutin equivalent/100g of FW and 144 mg of rutin equivalent /100 g FW, respectively (Olsen et al., 2009). Bok choy leaves, like those of other cabbage cultivars, are high in flavonols, especially glycosylated Kaempferol. Petunidin-3-O-galactoside and petunidin-3-O-glucoside make up the majority of the purple leaves of the cultivar *B. rapa* var. *chinensis*, but there are also trace amounts of various glycosylated anthocyanins, including derivatives of delphinidin, cyanidin, peonidin, and malvidin (Zhang et al., 2014). *Gymnanthemum amygdalinum*, an Asteraceae member whose leaves are commonly known as "bitter leaf", is mainly consumed in several sub-Saharan African countries, notably Cameroon. They are used in the preparation of "Ndolé", in Cameroon and Nigerians prepare bitter leaf soup appellé "Onugbo", as a popular dish. Their leaves are well-known for accumulating derivatives of caffeoylquinic acid: 3,4-dicaffeoylquinic acid, 1,5-dicaffeoylquinic acid, 3,5-dicaffeoylquinic acid, and 4,5-dicaffeoylquinic acid (Johnson et al., 2011). *Ocimum gratissimum* leaves, often known as "false basil", are consumed in some Sub-Saharan African countries, particularly in the western region, as well as in India, Latin America, and the Caribbean. The plant's ethanolic leaf extract (EtOH 70%) contained the highest quantities of p-coumaric acid, followed by caffeic acid, myricetin, quercetin, and kaempferol (Ganguly et al., 2021). Vicenin-2, luteolin-7-O-glucoside, quercetin-3-O-glucoside, and quercetin-3-O-rutinoside were the predominant flavonoids in the leaves, corresponding to the principal flavonoid profile published by other authors (Ola et al., 2009).

Carotenoids are secondary metabolites that play crucial roles in plants and are nutritionally relevant. They have received a lot of interest due to their potential antioxidant properties. They are commonly referred to as provitamin A, because this vitamin is a result of carotenoid metabolism. Despite being overshadowed by chlorophyll, leafy vegetables contain the most β-carotene, a vitamin A-rich compound. There are also trace quantities of zeaxanthin, α-carotene, and antheraxanthin. Although roughly 600 carotenoids have been identified in nature, about 50 have provitamin A activity, and only about 40 are found in the average human diet (Kiokias et al., 2016). A study in Andhra Pradesh, India, measured the levels of lutein and β-carotene in 10 green vegetable species. The leafy vegetables used during the study were *Allmania nodiflora, Alternanthera sessilis, Amaranthus cruentus, Amaranthus tricolor, Amaranthus viridis, Celosia argentea, Chenopodium album, Dallocardia muricata, Hibiscus cannabinus,* and *Rumex vesicarius*. The reported consumption of the 10 selected GLV species provides 1489 μg of β-carotene, contributing over 40% of the daily intake recommended by WHO/FAO (Bélanger et al., 2010).

Extraction Methods: Widely Used Methods with Recent Advancements, Applications, and Limitations

Conventional Extraction Methods

The extraction method allows for the recovery of components from a solid mixture or solution. It is essential in sample preparation, experimental research, and the qualitative analysis of bioactive compounds. The initial step in the extraction process is known as pre-extraction. Pre-extraction can be performed on both fresh and dried materials (Zahran and Najafi, 2020). Fresh plant samples used in medicinal studies tend to be more sensitive and decay faster than dried samples (Kozłowska and Gruczyńska, 2018). For solid materials, pre-extraction may involve several steps such as screening, milling, grinding, and homogenization. In contrast, liquid samples can be filtered or centrifuged before extraction to ensure successful analysis.

The solubility of bioactive substances is influenced by the degree of phenolic polymerization and the type of solvent used, whether it is polar, nonpolar, or less polar. To effectively extract these substances from complex plant matrices, it is essential to use solvents with varying polarity. Generally, the complexity of the plant material determines the duration of the extraction process; however, longer extraction times may lead to the oxidation of phenolic compounds. Phenolic acids and their derivatives are typically extracted using polar and organic solvents, including acetone, methanol, ethanol, water, and ethyl acetate. Additionally, alkaline hydrolysis can be employed to remove phenolic compounds that are attached to cell walls. Furthermore, lipid fractions are typically present in extracts that have been vacuum concentrated. As a result, lipid extraction frequently uses solvents such as n-hexane, acetone, petroleum ether, and diethyl ether. One easy way to inhibit or prevent additional enzymatic activity in plant samples is to freeze them with liquid nitrogen. Frozen samples can then be extracted directly or crushed into a powder. This process helps isolate substances at the endocellular level. Polyphenols are best extracted from leafy vegetables using an aqueous solution of methanol, ethyl acetate, and acetone (Kumar, 2017). In dry samples, reducing particle size enhances the surface contact between the samples and extraction solvents. When samples are ground into smaller and more uniform particles, they exhibit improved interaction with the extraction solvents (Oubannin et al., 2024). The most widely used conventional extraction techniques include maceration, Soxhlet extraction, and hot water bath methods (Giacometti et al., 2018). These traditional techniques are often solvent-intensive and more suitable for small scale research.

Maceration involves soaking the plant material, whether ground or powdered, in an airtight container filled with a solvent, known as the menstruum. The mixture should be stirred frequently until the soluble components dissolve, and it can be stored at room temperature for up to three days (Manousi et al., 2019). Further, to remove contaminants, the mixture is filtered (Jovanović et al., 2017). Infusion and decoction are similar to maceration, involving the immersion of soluble materials in either cold or boiling water until they dissolve (Manousi et al., 2019). However, infusion typically requires a shorter steeping time. Percolation is another extraction method that shares similarities with decoction; it involves transferring powdered and dried material into hot water for a two hour maceration (Sultana et al., 2018). The Soxhlet extractor is the most common device used to extract bioactive compounds from plants (Oubannin et al., 2024). This equipment dissolves weakly soluble compounds from solid matrices through a continuous extraction process. The solvent repeatedly passes through the product, dissolving the desired compounds by heating, vaporizing, and condensing the solvent. This cycle continues until the extraction is complete. Despite their widespread use, conventional extraction methods have several drawbacks. They often require multiple extraction cycles to achieve good yields due to their low extraction efficiency. This inefficiency can lead to sustainability issues, resulting in longer processing times, higher energy costs, and greater solvent consumption (Mungwari et al., 2025).

Advanced Extraction Methods

Researchers are constantly seeking new and creative techniques to extract and concentrate bioactive substances, driven by the growing demand for natural products in cosmetics, nutraceuticals, and pharmaceuticals. To meet these needs, advanced extraction techniques have become essential, offering solutions that are both efficient and environment friendly. Several sophisticated extraction methods have been developed that effectively separate and concentrate these important components from their sources, surpassing the capabilities of traditional approaches. The value of modern techniques lies in their ability to increase yields of bioactive components by reducing extraction time, enhancing efficiency, and using fewer solvents. Additionally, these methods often lead to improved selectivity and help preserve sensitive molecules, enabling the extraction of powerful, high-quality bioactives. Importantly, they also minimize the use of hazardous solvents, contributing to more environmentally sustainable practices (Bhadange et al., 2024). Microwave-Assisted Extraction (MAE), Supercritical fluid extraction (SFE), Ultrasound-Assisted Extraction (UAE), Solid-Phase Microextraction (SPME), and Subcritical solvent extraction (SSE) are some of the most important advanced extraction methods being recently used in natural products extraction from plant matrices. Table 1 summarizes the principal mechanisms, advantages, and limitations of these extraction techniques.

While contemporary techniques have largely replaced traditional extraction methods in many areas, the latter are still used in small-scale, artisanal, or resource-limited contexts. Traditional extraction methods offer several advantages, including simplicity, cost-effectiveness, and low equipment requirements (Brewer et al., 2014). However, they often lack the efficiency of modern methods and typically produce smaller yields of the target compounds (Gori et al., 2021). Advanced extraction techniques usually require specialized tools and technology, which can be expensive and challenging for smaller businesses to access (Yu et al., 2024).

Table 1 Advanced extraction techniques used for bio-actives with advantages and limitations.

Method of Extraction	*Principle Mechanism*	*Advantages*	*Limitations*	*Main Target Class of Compounds*	*References*
Microwave Assisted Extraction	The analytes are separated from the sample matrix into the solvent by heating solvents in contact with the sample using microwave energy.	1. The solvent and sample are heated quickly, which speeds up the extraction kinetics and may increase the yields of bioactive chemicals. 2. Depending on the polarity and dielectric characteristics of a molecule, this method scan target analytes specifically. 3. It makes laboratory and industrial applications convenient by doing away with the requirement for intricate setups and lengthy extraction periods.	1. The depth to which microwave energy can penetrate the sample is limited. 2. Particularly in complex matrices, this may result in uneven heating and insufficient extraction of bioactive chemicals. 3. The quality of the extracted bioactive compounds may be impacted when interfering compounds are released as a result of microwave interaction with matrix components.	Phenolic acids, Flavonoids, Terpenoids, and Essential oils.	Bhadange et al. (2024)

Contd...

Table 1 *Contd...*

Method of Extraction	*Principle Mechanism*	*Advantages*	*Limitations*	*Main Target Class of Compounds*	*References*
Supercritical Fluid Extraction	It is a procedure that uses supercritical fluids, mainly CO_2, as the extraction solvent to separate components from a matrix.	1. It offers a high level of selectivity. 2. Because SFE is usually performed at comparatively moderate temperatures, there is less chance that heat-sensitive chemicals would degrade.	1. The limited solubility of supercritical CO_2 for polar compounds reduced extraction efficiency for specific molecule types. 2. The effectiveness of the extraction process and the quality of the final product may be impacted by this method's sensitivity to moisture.	Phenolic acids, Flavonoids, Tannins, Vitamins, Essential oils, and Carotenoids.	Bhadange et al. (2024)

Contd...

Table 1 *Contd...*

Method of Extraction	*Principle Mechanism*	*Advantages*	*Limitations*	*Main Target Class of Compounds*	*References*
Ultrasound Assisted Extraction	This technique uses ultrasonic waves to improve the extraction of bioactive components by causing cavitation, which breaks down the cell membrane and makes it easier for the components to be released into the solvent medium.	1. By improving mass transfer and dissolving cell walls and embraces, ultrasound increases the effectiveness of bioactive ingredient extraction from a variety of materials.	1. Energy consumption may be taken into account in large-scale applications since ultrasonic extraction uses energy to generate high-frequency waves. 2. Excessive ultrasonic conditions might cause sample degradation, especially for compounds that are heat-sensitive or fragile.	Phenolic acids, Flavonoids, Carotenoids, Saponins, Proteins, Polysaccharides.	Bhadange et al. (2024)
Subcritical Solvent Extraction	This technique uses hot water as a solvent to extract bioactive compounds from medicinal plants under atmospheric pressure or at higher temperatures (between 100°C and 374°C) and pressures (less than 5 MPa).	1. SSE usually employs water as a solvent, obviating the requirement for potentially environmentally hazardous organic solvents. By varying the temperature and pressure, SSE's selectivity can be changed.	1. Some types of compounds with restricted solubility may not be effectively extracted by them. At higher temperatures certain substances may be thermally sensitive and prone to degradation.	Phenolic acids, Flavonoids, Carotenoids, Polysaccharides.	Bhadange et al. (2024)

Contd...

Table 1 *Contd...*

Method of Extraction	*Principle Mechanism*	*Advantages*	*Limitations*	*Main Target Class of Compounds*	*References*
Solid-phase Micro Extraction	A solid phase fiber coated with a thin layer of suitable sorbent material is used to extract and concentrate target analytes from a variety of matrices.	1. The SPME method requires very little sample preparation. 2. Analytes can be quickly extracted and concentrated from the sample matrix. 3. Because SPME focuses analytes onto a limited volume of the stationary phase, it can achieve high sensitivity and enhance detection limits.	1. The extraction efficiency may be impacted by high molecular weight analyses of the SPME fiber. 2. The extraction efficiency and selectivity of SPME may be impacted by matrix interferences caused by complicated sample matrices.	Volatile and Semi-volatile organic compounds.	Bhadange et al. (2024)

Additionally, achieving selectivity for desired molecules can be difficult due to the wider range of phytochemicals that may be extracted (Zhou et al., 2024). Nonetheless, precise and focused extraction techniques can help minimize extensive purification processes and waste production (Li et al., 2024). Figure 6 represents the steps involved in obtaining the bioactives from crude plant samples via extraction, purification, and characterization.

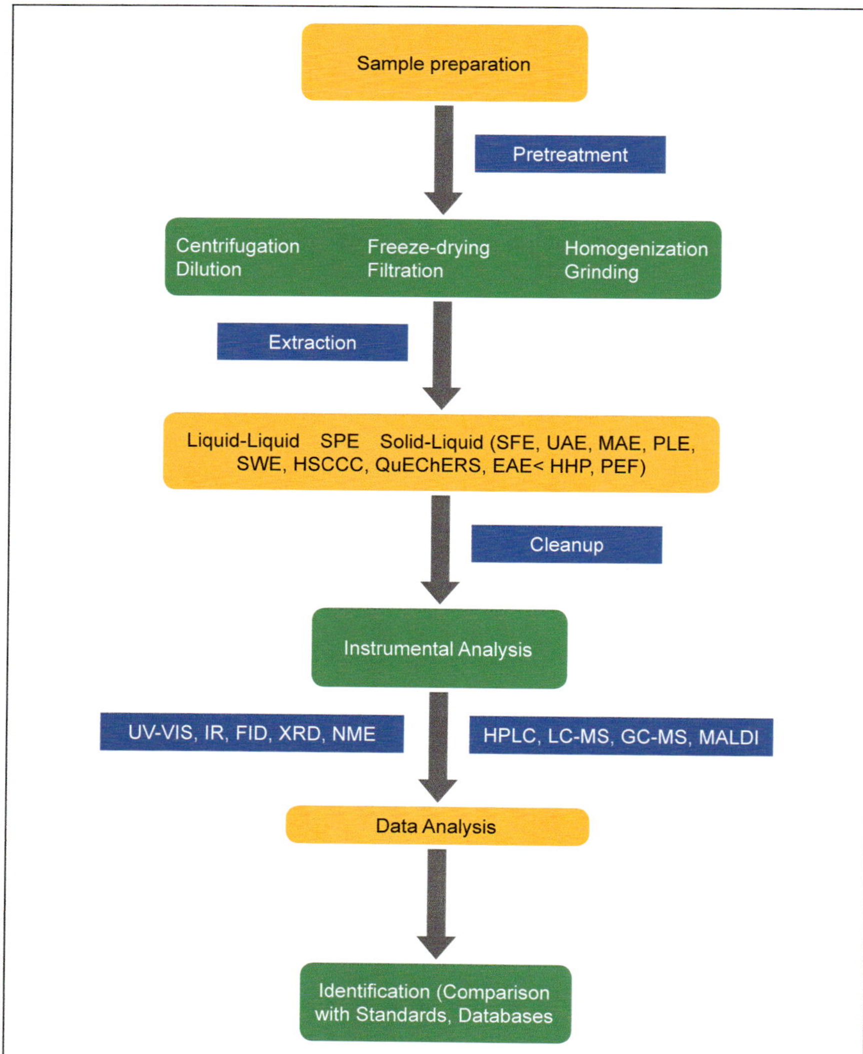

Fig. 6 Flowchart diagram of the steps involved in the extraction, purification, and characterization of bioactive compounds from plant matrices.

Role of Liquid Chromatography and Mass Spectrometry Techniques in the Metabolic Profiling of Phenolic Acids and Flavonoids from Green Leafy Vegetables

Liquid Chromatography-Mass Spectrometry (LC-MS) has become an essential tool in natural product research because it can identify and accurately determine the molecular weight of substances from small amount of plant extracts, even even with low concentrations. In standard HPLC procedures for phenolic compounds, reverse-phase (RP) octadecyl silica columns are typically employed. During RP-HPLC analysis, polar compounds, such as phenolic acids, elute earlier in the gradient than low-polarity molecules. Due to their significant structural diversity, there is a possibility of overlap among individual members of various classes. However, the general elution order is as follows: phenolic acids < cinnamic acids < flavonoids. The mobile phase and structural complexity significantly influence the elution time of compounds. Primary phenolic acids, such as ferulic acid, sinapic acid, and p-coumaric acid, can be separated using RP-HPLC within 40–50 minutes, although some analyses may extend to 70–90 minutes. The number of hydroxyl groups at the fourth position, followed by those at the third and second positions, increases the polarity of cinnamic and other phenolic acids. During HPLC analysis, compounds are separated in a single run despite their differing polarities, which requires gradient elution with higher concentrations of an organic modifier. Free phenolic acids, including caffeic acid, vanillic acid, syringic acid, ferulic acid, chlorogenic acid, protocatechuic acid, p-hydroxybenzoic acid, and p-coumaric acid, can be effectively separated using isocratic elution with a mixture of water, methanol, and acetic acid (Kumar, 2017). The phenolic profile of some of the most popular GLVs consumed in China was analyzed using the HPLC-MS technique. This study included vegetables from several families, such as Brassicaceae, Amaranthaceae, and Fabaceae (Zhang et al., 2019). To separate the components, three distinct extraction techniques were employed, targeting free phenolic acids, bound phenolic acids, and free esters of free phenolic acids. The analysis utilized a reversed-phase C18 column maintained at 45°C, and a diode array detector (DAD). The mobile phase consisted of acetonitrile and 0.1% aqueous formic acid. The HPLC system was coupled online with a mass spectrometer using an electrospray ionization (ESI) source in multiple reaction monitoring (MRM) mode. Seventeen primary phenolic acids were identified in these leafy vegetables, including gallic acid, 3,5-dihydroxybenzoic acid, chlorogenic acid, p-hydroxybenzoic acid, gentisic acid, 2,3,4-trihydroxybenzoic acid, vanillic acid, syringic acid, caffeic acid, protocatechuic acid, p-coumaric acid, sinapic acid, ferulic acid, 3-hydroxycinnamic acid, isoferulic acid, 2-hydroxycinnamic acid, and salicylic acid. Under optimal conditions, 15 of the identified phenolic compounds eluted within 15 minutes. In natural plants, most phenolic acids exist as trans-isomers. However, when exposed to ultraviolet (UV) light, they gradually convert into cis-isomers. To separate cis-isomers from trans-isomers, an RP-stationary phase, which includes an optically active molecule in the mobile phase, is typically used. The primary methods for detecting phenolic acids include fluorescence, DAD, and UV-visible detectors. Gallic acid and syringic acid have absorption maxima at 271 nm and 275 nm, respectively, while most benzoic acids have maximum absorption between 246–262 nm. Cinnamic acids exhibit their highest absorption in two wavelength ranges: 225–235 nm and 290–330 nm. For phenolic compounds and their derivatives, the detection range of 280 nm is most commonly used. The separation of flavonoids in RP-HPLC is based on hydrophobicity. Variations in selectivity depend on the polar groups attached, the carbon loading, and the characteristics of the stationary phase. The order in which various classes of flavonoids elute is as follows: flavanols < flavanones < flavonols < flavones. A higher number of methoxy groups on the flavonoid backbone increases the retention duration of the compounds, while a greater number of hydroxyl groups decreases it. Flavonoid analysis employs both binary and isocratic gradient elution techniques; however, isocratic gradient elution has certain limitations. During RP-HPLC analysis, acylated compounds tend to extend the elution duration, whereas glycosylated flavonoids slightly reduce the retention time. Additionally, a higher pH leads to the ionization of flavonoids, which shortens the retention period during RP separation. To prevent crystallization, small amounts of acetic acid (2%–5%) and trifluoroacetic acid (0.1%–1%) are added to the solvent mixture, enhancing the resolution and reproducibility of each separation (Kumar, 2017). The accurate molecular weights of the phenolic acids isolated via RP-HPLC can be determined using mass spectrometry (MS). This technique allows for the collection of MS data during the analysis, as well as the simultaneous monitoring of the retention time of the isolated chemical peak. As a result, LC-MS is now a recognized technique for phenolic acid identification and characterization (Kumar, 2017). Phenolic acids and flavonoids were extracted from the adult stems of *Basella alba* and *Basella rubra* using HPLC and electrospray ionization-mass spectrometry (ESI-MS) methods. Water and acetonitrile with 0.1% trifluoroacetic acid (TFA) were used as mobile phases to separate the flavonoids and phenolic acids on a reverse-phase C18 column. The identified phytochemicals included 4-hydroxybenzaldehyde, caffeine, ferulic acid, gallic acid, galloyl shikimic acid, p-coumaric acid, salicylic acid, trans-cinnamic acid, and diosmetin. Furthermore, the ESI-MS technique was employed for the provisional identification of these compounds. Notably, diosmetin—an O-methylated flavone—was reported in these plant species for the first time. The detected bioactives also demonstrated strong antioxidant, anti-angiogenic, and anti-cancer properties (Kumar et al., 2018). In a related investigation, the red *Basella* cultivar, *B. rubra*, was found to contain flavonoids and phenolic acids. These compounds were extracted from the fruits of *B. rubra* using a reverse-phase C18 column. The phenolic acids detected included gentisic acid, chlorogenic acid, sinapic acid, ferulic acid, and coumaric acid at 280 nm. Flavonoids such as myricetin, apigenin, kaempferol, luteolin, and quercetin were observed at 320 nm. Additionally, betacyanins were identified in the flesh of *B. rubra* fruits using HPLC-Quadrupole Time of Flight Mass Spectrometry (HPLC-QTOF-MS). The study also documented significant antioxidant and anticancer effects against human cervical cancer cell lines (Kumar et al., 2015). The detection of flavonoids is influenced by their chemical characteristics and the sensitivity of the analytes. To identify certain substances, researchers use the UV absorption spectra and retention times of reference standards. Even when two compounds are closely related and elute simultaneously, they can be differentiated using multiple wavelength detection methods that rely on variations in their absorption spectra. Acetonitrile and methanol, with low UV cut-off λmax values at 190 and 205 nm, do not interfere with the detection of UV-Vis absorption bands between 240–285 nm and 300–560 nm for the A and B rings of flavonoid aglycones. The wavelength range of these bands can vary slightly among flavones, depending on the substitution patterns of hydroxy and methoxy groups, as well as the types of glycosides (C-glycosides or O-glycosides). Metabolite profiling of phenolic acids, anthocyanidins, and flavonols has been conducted on *Brassica oleracea* var. *capitata* (Park et al., 2014). The study utilized LC-MS and HPLC methods to identify and quantify key compounds, including flavonols (quercetin and kaempferol), anthocyanidins (cyanidin and peonidin), and phenolic acids (caffeic, p-coumaric, ferulic, and sinapic acid). Phenolic compounds were separated using an HPLC-Photodiode Array (PDA) detector, featuring a C18 column along with a guard column. The mobile phase consisted of methanol and 0.1% formic acid in water. Measurements for anthocyanidins, phenolic acids, and flavonols were taken at wavelengths of 530 nm, 330 nm, and 360 nm, respectively. For phenolic acids and flavonols, ESI-MS was performed in negative ion mode, while for anthocyanidins, it was operated in positive-ion mode. Notably, the primary phenolic acids identified in *Brassica* vegetables were derivatives of sinapic acid. Through this combined technique, the study successfully identified and quantified phenolic acids, flavonoids, and anthocyanins across nine different *Brassica* species. The best ionization sources for the analysis of flavonoids are Atmospheric Pressure Chemical Ionization (APCI) and ESI. These ionization methods facilitate the coupling of HPLC with MS. Flavonoids in plants are predominantly found in their glycosidic forms, which can be categorized into two main types: O-glycosides and C-glycosides. Typically, the mass spectra of flavonoid glycosides exhibit similar fragmentation pathways, regardless of the ionization method used. However, in many cases, the desired fragment ions may not be generated due to contaminants in the sample or the compounds being analyzed. In such situations, Collision-Induced Dissociation (CID) MS/MS techniques can be beneficial as they produce the appropriate fragment ions. Identifying the structure of acylated flavonoid glycosides can be challenging because these compounds are often acylated with either aliphatic (e.g., acetic acid) or aromatic acids, which can attach directly to the glycosidic or aglycone portions. Although increasing the cone voltage may sometimes lead to their fragmentation, ESI spectra of glycosidic compounds typically show an aglycone ion, a pseudo-molecular ion such as ([M+H]), and ions associated with the solvent (Kumar, 2017). The phenolic, vitamin, and antioxidant profiles of several *Amaranthus* lines were examined by Sarker and Oba (2021). Phenolic compounds were identified using LC-ESI-MS and HPLC techniques. The

HPLC system consisted of an ODS-II column and a UV-Vis detector. The mobile phase was composed of 6% acetic acid in water as Solvent A and acetonitrile as Solvent B. The MS system was integrated with the HPLC and UV-Vis detector, with an online ESI ionization source. The detector was calibrated to continuously monitor flavonoids, cinnamic acids, and benzoic acids at wavelengths of 360 nm, 370 nm, 280 nm, and 254 nm. From the leaves, 25 phenolic acids and flavonoids were identified, including protocatechuic acid, salicylic acid, gentisic acid, gallic acid, vanillic acid, p-hydroxybenzoic acid, chlorogenic acid, ellagic acid, syringic acid, ferulic acid, coumaric acid, trans-cinnamic acid, caffeic acid, sinapic acid, quercetin, kaempferol, myricetin, apigenin, isoquercetin, rutin, hyperoside, and catechin. It was concluded that the hyphenated techniques employed in this study provided a comprehensive analysis of the flavonoids and phenolic acids from four *Amaranthus* genotypes. When analyzing flavonoids using ESI in positive mode, sodium and potassium adducts are commonly observed. These adducts form when the samples are kept in a glass solution and are particularly noted in the analysis of flavanol-3-O-glycosides and isoflavones (Kumar, 2017). Additionally, a combined loss of a methyl group and water, represented as $[M+H-33]^+$, is also detected. In the case of C-glycosylated flavonoids, complete removal of the sugar moiety is not observed. Instead, fragments or portions of the sugar ring are eliminated. The intensity of the protonated molecule ions $[M+H]^+$ seen in the mass spectrum depends on the number of sugar moieties attached to the aglycone. Specifically, the strength of $[M+H]^+$ ions decreases as the number of sugar rings increases. The bond that connects sugars to the flavonoid ring is a strong, acid resistant C-C bond. The results of these analyses may be improved by conducting collision experiments using appropriately high and low energy levels. Moreover, 6-C-glycosyl flavonoids tend to lose more water molecules compared to 8-C-glycosyl flavonoids for $[M+H]^+$ and $[M-H]^-$ ion fragmentation patterns at high collision energy (Kumar, 2017). The phenolic profiles of red and green oak-leaf lettuce cultivars were thoroughly characterized using Ultra-High Performance Liquid Chromatography (UHPLC) in combination with ESI, Quadrupole-Time of Flight Mass Spectrometry (QToF/MS), and DAD (Viacava et al., 2017). For this investigation, an ultra-performance liquid chromatography Ethylene Bridged Hybrid (UPLC-BEH) C18 VanGuard™ pre-column (1.7 μm) and a reverse-phase UPLC-BEH C18 column (100 mm × 2.1 mm; particle size 1.7 μm) were used. The mobile phases consisted of acetic acid in water (Solvent A) and methanol (Solvent B), both at 0.1% (v/v). Several classes of compounds were identified using the UV-Vis detector at different wavelengths: hydroxybenzoic acids at 254 nm, hydroxycinnamic acids and coumarins at 320 nm, flavanones at 280 nm, flavonols and flavones at 370 nm, and anthocyanins at 500 nm. A High Definition Mass Spectrometry (HDMS) system with an ESI source, operating in both positive and negative modes, was used for all MS data acquisitions. For the first time, 48 compounds were identified in lettuce with a reasonable degree of certainty. The identified phenolic compounds included caffeoyl quinic acids, p-coumaroyl quinic acids, caffeoyltartaric acids, p-coumaroyl tartaric acid, caffeoylmalic acid, dicaffeoylquinic acids, caffeoylquinic acid glycosides, p-coumaroylcaffeoylquinic acids, di-caffeoyltartaric acid, hydroxycinnamic acid derivatives, hydroxybenzoic acid derivatives, hydroxyphenylacetic acid derivatives, quercetin derivatives, kaempferol derivatives, luteolin derivatives, apigenin derivatives, eriodictyol-O-glucuronide, and cyanidin derivatives. The UHPLC-DAD-ESI-QToF/MS method proved to be a valuable and robust technique for effectively characterizing phenolic compounds. Tables 2 and 3 provide recent studies on profiling and characterizing phenolic acids and flavonoids from GLVs using HPLC and MS methods.

When used alongside ESI-MS, HPLC columns with a small diameter offer increased sensitivity because they elute the concentrated analyte in a smaller volume. This design also minimizes the introduction of impurities and buffers into the MS, which helps maintain the ionization source's sensitivity for a longer time. Guard columns are vital for enhancing the longevity and performance of analytical columns, as they protect these columns by removing strongly binding components like chlorophyll and other endogenous compounds commonly found during the analysis of crude plant samples. Advanced HPLC techniques, such as high-temperature liquid chromatography (HTLC), ultra-high-performance liquid chromatography (UHPLC), and two-dimensional liquid chromatography (LC × LC), utilize new types of columns to detect flavonoids and phenolic acids in plant samples. Recent advancements in instrumentation have made LC × LC and hydrophilic interaction liquid chromatography (HILIC) popular for analyzing plant samples. HILIC may be particularly effective when combined with MS due to its higher compatibility with the mobile phase used for separation.

In traditional Chinese medicine, polar and semi-polar fractions have been successfully detected using a combination of 2-D LC × HILIC and 2-D LC × RP-LC. The online coupling of HPLC and MS using ESI allows for the high-resolution characterization of various polar substances. Recent innovations, such as modifications to sprayer techniques (including pneumatically aided electrospray, ultrasonic nebulizer electrospray, and nanoelectrospray), have further optimized the application of ESI techniques (Kumar, 2017).

Role of Liquid Chromatography and Mass Spectrometry Techniques in the Metabolic Profiling of Carotenoids from Green Leafy Vegetables

The extraction of carotenoids from the sample matrix is essential for accurate analysis. Because carotenoids are inherently unstable, it is crucial to minimize degradation, oxidation, and isomerization during this process to ensure reliable results. Their high degree of conjugation makes carotenoids susceptible to damage from heat, light, oxygen, and metal ions. For example, in an aerobic environment, β-carotene rapidly breaks down when exposed to visible and UV light. Therefore, samples should be handled under low light or red light, and solvents should be evaporated using gases like helium, argon, or nitrogen. Additionally, it is important to store samples in a dark place at temperatures below –20°C (Kopec et al., 2012). Most carotenoids absorb light in the 400–500 nm range due to their conjugated double bonds; however, some precursors, such as phytoene and phytofluene,

Table 2 HPLC profiling of phenolic acids and flavonoids.

Plant Name	*Parts Used*	*Extraction Solvent*	*Method*	*Stationary Phase*	*Mobile Phase*	*Flow Rate (mL) and Wavelength (nm)*	*Phenolic Compounds Identified*	*References*
Amaranthus sps. (*A. hypochondriacus*, *A. cruentus*, *A. caudatus*, *A. tricolor*, *A. dubius*, *A. blitum*, *A. crispus, A. hybridus*, and *A. viridis*)	Leaves	75% methanol	UPLC-PDA	BEH C18 Column (2.1mm ×100 mm, 1.7 µm)	Water and Acetonitrile with 0.1% acid.	0.25 mL/min, 260 nm.	***Phenolic acids*:** Gallic acid,3,4-dihydroxybenzoic acid, 4-hydroxybenzoic acid, 2,4-dihydroxybenzic acid, vanillic acid, syringic acid, benzoic acid, caffeic acid, p-coumaric acid, ferulic acid, sinapic acid, and cinnamic acid. ***Flavonoids*:** Rutin, quercetin-3-β-D-glucoside, kaempferol-3-O-β-rutinoside, quercetin, and kaempferol.	Bang et al. (2024)
Basella alba and *Basella rubra*	Stem	Aqueous methanol	HPLC	RP-C18 column (250 x 4.6 mm)	Water and Acetonitrile with 0.1% TFA.	1.0 mL/min, 254 nm, 280 nm, and 330 nm.	***Phenolic acids*:** Gallic acid, p-Coumaric acid, Salicylic acid, Caffeic acid, Ferulic acid, Trans-cinnamic acid, Galloyl shikimic acid. ***Flavonoids:*** Diosmetin.	Kumar et al. (2018)

Contd...

Table 2 *Contd...*

Plant Name	*Parts Used*	*Extraction Solvent*	*Method*	*Stationary Phase*	*Mobile Phase*	*Flow Rate (mL) and Wavelength (nm)*	*Phenolic Compounds Identified*	*References*
Tinosporia cordifolia, Mentha arvensis.	Stem and Leaves	80% methanol	UHPLC	C18 column (150 x 2.0mm)	Formic acid in water (0.1%) and acetonitrile in water (0.1%)	0.2 mL/min	***Phenolic acids*:** Vanillic acid, Ferulic acid, Chlorogenic acid, p-coumaric acid, gallic acid, rosmarinic acid, sinapic acid, 3-feruloyl quinic acid, and neo-chlorogenic acid. ***Flavonoids*:** Luteolin-7-O-neohesperidoside, Quercetin-3,7-O-diglucoside, Luteolin-5-O-glucoside, Genkwanin-5-O-glucoside, Hesperidin, Diosmin, Didymin, Rutin, Quercetin dihydrate, Kaempferol, and Quercetin.	Mueed et al. (2023)

Contd...

Table 2 *Contd...*

Plant Name	*Parts Used*	*Extraction Solvent*	*Method*	*Stationary Phase*	*Mobile Phase*	*Flow Rate (mL) and Wavelength (nm)*	*Phenolic Compounds Identified*	*References*
Amaranthus (VA14, VA16, VA11, VA6)	Leaves	80% methanol with 1% acetic acid.	HPLC	STR-ODS-II (150 x 4.6 mm)	6% v/v acetic acid and Acetonitrile	1.0 mL/min, 254 nm, 280nm, 260 nm, and 380 nm.	***Phenolic acids*:** Protocatechuic acid, salicylic acid, gentisic acid, gallic acid, vanillic acid,p-hydroxy-benzoic acid, chlorogenic acid, ellagic acid, syringic acid, ferulic acid, coumaric acid, trans-cinnamic acid,p-coumaric acid, caffeic acid, and sinapic acid. ***Flavonoids*:** Kaempferol, quercetin, iso-isoquercetin, rutin, naringenin, myricetin, and hyperoside.	Sarker & Oba (2021)
Indian spinach (*Basella rubra*)	Fruits	Water	HPLC	C18 column (250mm × 4.6 mm)	Water with acetic acid (pH 2.6), and 80% acetonitrile	1.2 mL/min, 280 nm, and 320 nm.	***Phenolic acids*:** Chlorogenic acid, Sinapic acid, Ferulic acid, and Coumaric acid. ***Flavonoids*:** Apigenin, Kaempferol, Luteolin, Myricetin, and Quercetin.	Kumar et al. (2015)

Contd...

Table 2 *Contd...*

Plant Name	*Parts Used*	*Extraction Solvent*	*Method*	*Stationary Phase*	*Mobile Phase*	*Flow Rate (mL) and Wavelength (nm)*	*Phenolic Compounds Identified*	*References*
Green and Red Oak leaf lettuce (*Lactuca sativa*)	Whole plant	Methanol-water-acetic acid (30:65:5, v/v/v)	UHPLC-DAD	UPLC BEHC18 Column (100mm × 2.1 mm)	0.1% (v/v) acetic acid in water and 0.1% (v/v) acetic acid in methanol.	0.5 mL/min, 254 nm, 280 nm, 320 nm, 370 nm and 500 nm.	***Phenolic acids*:** Caffeoyl quinic acids, p-coumaroyl quinic acids, caffeoyltartaric acids,caffeoyltartaric acids, p-Coumaroyl tartaric acid, caffeoylmalic acid, dicaffeoylquinic acids and caffeoylquinic acid glycosides, p-Coumaroylcaffeoylquinic acids, di-caffeoyltartaric acid, hydroxycinnamic acid derivatives, hydroxybenzoic acid derivatives, hydroxyphenylacetic derivatives. ***Flavonoids*:** Quercetin derivatives, Kaempferol derivatives, Luteolinderivatives, Apigenin derivatives, Eriodictyo-O-glucuronide, and Cyanidin derivatives	Viacava et al. (2017)

Table 3 MS profiling of phenolic acids and flavonoids.

Plant Name	*Part Used*	*Ionization Method*	*Phenolic Compounds Identified*	*Retention Time (Minutes)*	*Observed [M+H]+*	*Observed [M-H]-*	*Scan Range (m/z)*	*Reference*
Basella alba and *Basella rubra*	Stems	ESI-MS	***Phenolic acids*** 4-hydroxybenzaldehyde	2.51	-	121.1	50–1000	Kumar et al. (2018)
			Caffeic acid	2.55	-	179.2		
			Ferulic acid	10.00	195.97	-		
			Gallic acid	3.43	-	169.10		
			Galloylshikimic acid	11.03	-	325.1		
			p-Coumaric acid	3.85	-	163.0		
			Salicylic acid	4.36	-	137.02		
			Trans-cinnamic acid	9.25	149.3	-		
			Flavanoids Diosmetin	10.86	-	301.4		
Amaranthus (VA14; VA16; VA11; VA6).	Leaves	ESI-MS	***Phenolic acids*** 2,4-dihydroxybenzoic acid	4.28	-	154.11	0–1000	Sarker & Oba (2021)
			3,4-dihydroxybenzoic acid	2.33	-	154.13		
			2,5-dihydroxybenzoic acid	3.82	-	154.13		
			2-hydroxybenzoic acid	48.23	-	137.21		
			4-hydroxybenzoic acid	31.34	-	137.02		
			3,4,5-trihydroxybenzoic acid	9.24	-	169.12		
			4-Hydroxy-3-methoxybenzoic acid	30.13	-	167.17		

Contd...

Table 3 *Contd...*

Plant Name	*Part Used*	*Ionization Method*	*Phenolic Compounds Identified*	*Retention Time (Minutes)*	*Observed [M+H]+*	*Observed [M-H]-*	*Scan Range (m/z)*	*Reference*
			3,5-Dimethoxy-4-Hydroxybenzoic acid	34.57	-	197.12		
			3,4-dihydroxycinnamate	32.12	-	179.07		
			3-(3,4-Dihydroxycinnamoyl) Quinic acid	31.13	-	353.14		
			4-Hydroxy-cinnamic acid	42.21	-	163.05		
			3-Methoxy-4-hydroxy Cinnamic acid	47.88	-	193.16		
			3-Hydroxycinnamic acid	49.59	-	163.24		
			4-Hydroxy-3,5-dimethoxy Cinnamic acid	49.03	-	223.16		
			3-Phenylacrylicacid	67.26	-	147.12		
			Flavanoids Naringenin	26.71	-	271.05		
			Quercetin-3-O-glucoside	54.33	-	463.27		
			Quercetin-3-O-galactoside	53.29	-	463.44		
			Quercetin-3-O-rutinoside	53.14	-	609.28		
			Myricetin-3-O-rutinoside	4.58	-	626.31		
			Kaempferol-3-O-rutinoside	17.77	-	593.43		

Contd...

Table 3 *Contd...*

Plant Name	*Part Used*	*Ionization Method*	*Phenolic Compounds Identified*	*Retention Time (Minutes)*	*Observed [M+H]+*	*Observed [M-H]-*	*Scan Range (m/z)*	*Reference*
			4,5,7-Trihydroxyflavone, 5,7- Dihydroxy-2-(4-hydroxyphenyl)-4-benzopyrone	15.55	-	270.24		
			2-(3,4-dihydroxy phenyl)-3,5,7-trihydroxychromene-4-one	7.54	-	301.12		
Green radish; Purple cabbage; Celery; Chinese cabbage, Baby cabbage; Amaranth, White radish, Asparagus lettuce; Radish, Spinach, Cauliflower, Broccoli, Phaseolus Vulgaris.	Aerial parts	ESI-MS	***Phenolic acids*** Gallic acid	2.10	-	169.0		Zhang et al. (2019)

Contd...

Table 3 *Contd...*

Plant Name	*Part Used*	*Ionization Method*	*Phenolic Compounds Identified*	*Retention Time (Minutes)*	*Observed [M+H]+*	*Observed [M-H]-*	*Scan Range (m/z)*	*Reference*
			3,5-Dihydroxybenzoic acid	3.48	-	153.1		
			Chlorogenic acid	4.33	-	353.0		
			p-Hydroxybenzoic acid	4.47	-	137.0		
			Gentisic acid	4.54	-	152.9		
			2,3,4-Trihydroxybenzoic acid	3.56	-	141.0		
			Vanillic acid	5.07	-	167.0		
			Caffeic acid	5.21	-	179.0		
			Syringic acid	5.37	-	197.0		
			Protocatechuic acid	5.42	-	153.2		
			p-Coumaric acid	7.56	-	163.0		
			Sinapic acid	8.63	-	223.0		
			Ferulic acid	8.57	-	193.0		
			3-Hydroxycinnamic acid	9.00	-	163.1		
			Isoferulic acid	8.98	-	193.1		
			2-Hydroxycinnamic acid	10.26	-	163.1		
			Salicylic acid	10.43	-	137.1		

primarily absorb in the UV range. Different carotenoid species can be distinguished and identified through spectral analysis, but it is important to note that the UV-visible spectra for common carotenoids only reveal the features of the carotenoid's chromophore.

A distinctive characteristic of carotenoids is that, unlike many compounds that have a single absorption band, most carotenoids exhibit three roughly separate absorption peaks. The wavelength of maximum absorption and the fine structural details can vary significantly between different carotenoids. The most common method for separating carotenoids is RP-HPLC. However, several normal-phase methods have been developed for separating specific combinations of xanthophylls or carotenoid ketones, such as astaxanthin, which have a low affinity for C18 or C30 columns designed specifically for carotenoids. While C30 columns generally provide better separation, especially for closely related molecules like lutein and zeaxanthin, C18 columns can sometimes suffice for distinguishing different carotenoids. Carotenoids have also been analyzed using UHPLC, primarily in conjunction with other fat soluble vitamins. However, it is uncommon to find UHPLC techniques specifically developed for quantifying a diverse range of carotenoids (Kopec et al., 2012). The carotenoid profile of many leafy vegetables, including *Basella rubra* L., *Peucedanum sowa* Roxb., *Moringa oleifera* Lam., *Trigonella foenum-graecum* L., *Spinacia oleracea* L., *Sesbania grandiflora* (L.) Poir., and *Raphanus sativus* L., was assessed. Open column chromatography (OCC) on a neutral alumina column was used to purify the recovered carotenoids, allowing for their identification based on distinctive UV-visible absorption spectra. The separated carotenoids were analyzed using RP-HPLC on a C18 column with UV-visible photodiode array (PDA) detection under isocratic conditions. The mobile phase consisted of a 60:20:20 v/v/v mixture of acetonitrile, methanol, and dichloromethane, with 0.1% ammonium acetate. Both methods identified lutein, α-carotene, violaxanthin, neoxanthin, and zeaxanthin as the main carotenoids (Lakshminarayana et al., 2005). RP-LC is a popular technique that achieves great resolution by utilizing C18 and C30 columns with non-polar stationary phases (Xu et al., 2023). While C30 phases, including YMC C30 and Prontosil C30, provide greater separation of structurally identical molecules and geometric isomers, C18 phases, like Spherisorb ODS-2 and BEHC18, are appropriate for hydrophobic carotenoids (Delpino-Rius et al., 2014; Pop et al., 2014; Gebregziabher et al., 2021; Londoño-Giraldo et al., 2021; Feng et al., 2022; Grujić et al., 2022). The polarity and solubility of carotenoids influence the selection of mobile phases in HPLC. Commonly used solvents include methanol (MeOH), acetonitrile (ACN), methyl tert-butyl ether (MTBE), tetrahydrofuran (THF), acetone, and water. Ammonium acetate (NH4OAc) is often used as a buffer modifier, while ethyl acetate (EtoAc) is used in some procedures to improve solubility (Lourenço-Lopes et al., 2022). HPLC was utilized to examine the presence of β-carotene in the leaves of spinach, cabbage, carrot, brinjal, tomato, and bitter gourd (Zahra et al., 2022). The HPLC system operated in isocratic flow mode and utilized a UV-Vis detector along with a reversed-phase C18 column. The separation was achieved using a mobile phase composed of acetonitrile, dichloromethane, and methanol in a ratio of 70:20:10. This study demonstrated that HPLC is a reliable method for analyzing carotenoids.

Most C30 techniques utilize a gradient to achieve optimal separation of various carotenoid species and their isomers. These methods often involve a mixture of MTBE, methanol, and a small quantity of water. Some C30 techniques can also simultaneously separate tocopherols and chlorophylls, along with carotenoids and their isomers. Others may use a combination of THF, acetonitrile, methanol, isopropyl alcohol, ethyl acetate, and dichloromethane. Generally, a temperature of 23°C ± 1°C is sufficient for separating various carotenoid species, while 30°C has been identified as the optimal temperature for isomer separation. It is important to note that cis isomers typically exhibit reduced fine structure and absorb maximally at wavelengths that are about 2–6 nm lower than those of the all trans carotenoid (Kopec et al., 2012). Another study identified carotenoids from the leaves of *Brassica oleracea* L. var. *botrytis*, which were cultivated in four different regions of Pakistan's Malakand division: Buner, Swat, Malakand, and Lower Dir. The researchers used an RP-HPLC with a DAD system to separate and identify the carotenoids at 450 nm. The mobile phase consisted of a combination of solvents: Solvent A was a mixture of methanol and deionized water in a 92:8 ratio, v/v, containing 10 mM ammonium acetate; Solvent B included 0.1 mM ammonium acetate and deionized water; and Solvent C was pure MTBE. The study identified several compounds, including phytoene, all-E-neoxanthin, 9-Z-neoxanthin, all-E-lutein, 9-Z-lutein, 9′-Z-lutein, α-carotene, 9′-Z-neoxanthin, γ-carotene, and all-E-β-carotene. The major carotenoids found across all varieties were β-carotene, lutein, fucoxanthin, δ-carotene, neoxanthin, γ-carotene, and phytoene (Zeb et al., 2022).

Carotenoids, which are compounds with parent ions typically ranging from 536 to 600 Daltons, can be accurately identified and characterized using HPLC-MS and HPLC-MS/MS. Most applications commonly utilize ESI and APCI as ionization sources in mass spectrometry. APCI is particularly effective for ionizing lipophilic substances and is often used for nonpolar carotenoids such as β-carotene, α-carotene, and lycopene (Pintea et al., 2020). This method operates at higher temperatures, which reduces fragmentation and enhances the ionization of volatile compounds; this often results in a dominant $[M+H]^+$ molecular ion. In contrast, oxygenated carotenoids, known as xanthophylls, such as lutein, zeaxanthin, and violaxanthin, are better suited for ESI. This technique provides gentler ionization, minimizing fragmentation and yielding intact molecular ions in both positive ($[M+H]^+$) and negative ($[M-H]^-$) modes (Kurek et al., 2025). The majority of M^+ and $[M+H]^+$ ions are found in hydroxy carotenoids, including zeaxanthin and β-cryptoxanthin. The sodium adduct ions $[M+Na]^+$ and $[M+H]^+$ are mostly provided by keto-carotenoids like astaxanthin. The protonated molecule $[M+H]^+$ and the alkaline metal adductions $[M+Na]^+$ and $[M+K]^+$ are the main products of Violaxanthin's ESI-MS. Both carotenes and xanthophylls exhibit the protonated molecule $[M+H]^+$ as a molecular mass ion in APCI-MS. Xanthophylls exhibit the same fragment ions, including $[M+H-H_2O]^+$ and $[M+H-AcOH]^+$. Furthermore, the structural properties of the hydroxylated end group in carotenoids are reflected in the intensities of these dehydrated fragment ions. Despite having the identical chemical formula ($C_{40}H_56O_2$), lutein (β,ε-carotene- 3,3′-diol), lactucaxanthin (ε,ε-carotene-3,3′-diol), and zeaxanthin (β,β-carotene-3,3′-diol) displayed significantly different ion peak intensities at m/z 569 $[M+H]^+$, m/z 551 $[M+H-H_2O]^+$, and m/z 533 $[M+H-2H_2O]^+$. Zeaxanthin displayed a dehydrated ion at m/z 551 $[M+H-H2O]^+$ and a base peak at m/z 569 $[M+H]^+$. On the other hand, zeaxanthin with a 3-hydroxy-β- end group scarcely showed a fragment ion at m/z 533 $[M+H-2H_2O]^+$, but both lutein and lactucaxanthin with a 3-hydroxy-ε-end group displayed m/z 551 $[M+H-H_2O]^+$ as the most prevalent ion (Kaczor and Baranska, 2016). Carotenoids containing hydroxyl or epoxy groups benefit significantly from ESI, which enhances detection sensitivity in complex food matrices. Although both ESI and APCI methods provide valuable information about the composition of carotenoid pigments, ESI is more effective for analyzing polar carotenoids, while APCI is better suited for nonpolar species (Wu et al., 2024). In contrast, triple quadrupole (QqQ) and ion trap (IT) mass spectrometers offer more informative data, enabling the execution of tandem mass spectrometry (MS/MS) and multi stage mass spectrometry (MSn) experiments. Specific MRM studies allow for both qualitative data collection and individual quantification of carotenoids present at very low levels. Tandem setups, such as Quadrupole-Ion Trap (Q-IT) and Quadrupole-Time of Flight (Q/ToF) mass spectrometry (IT-TOF-MS), are also employed. Furthermore, some LC-MS systems utilize Q/Orbitrap and ion mobility (IM)/MS for carotenoid investigations. Various LC-MS configurations exist, each offering unique features that enable data collection at different levels of linearity, sensitivity, mass resolution, and accuracy (Giuffrida et al., 2020).

A total of seven carotenoids were simultaneously identified in 12 Brassicaceae plants, including broccoli, Brussels sprouts, cabbage, cauliflower, Chinese cabbage, pak choi, choy sum, kailan, daikon radish, red radish, rocket salad, and watercress. LC-APCI-MS/MS was used, and the carotenoids were separated on a YMC Carotenoid C30 column connected to a triple quadrupole mass spectrometer in positive mode with an APCI source. The mobile phases used were methanol: water (9:1) as solvent A and isopropanol: methanol (7:3) as solvent B. The target analytes were identified using optimized transitions in MRM mode. The identified carotenoids were violaxanthin, neoxanthin, antheraxanthin, lutein, zeaxanthin, β-cryptoxanthin, and β-carotene. The authors reported that the LC-MS approach was user-friendly and demonstrated to be a quick and efficient method for simultaneously determining carotenoids, tocopherols, and phylloquinone in *Brassica* species (Lee et al., 2020). In a related study, the HPLC-APCI-MS/MS technique identified the presence of eight carotenoids from 26 types of GLVs that are primarily consumed in Southeast Asia (Lee et al., 2022). A YMC Carotenoid C30 column, along with a guard column, was used as the stationary phase. The mobile phase consisted of 90% methanol (solvent A) and a mixture of isopropyl alcohol (IPA), and methanol in a 70:30 (v/v) ratio (solvent B). To confirm the precursor and product ions, full scan and product ion scans were conducted using separate standard solutions. The analytes were quantified in MRM mode with optimized transitions. The identified carotenoids included violaxanthin, neoxanthin, antheraxanthin, lutein, zeaxanthin, β-cryptoxanthin, α-carotene, and β-carotene. Among the 26 GLVs studied, lutein and β-carotene were the predominant carotenoids, followed by violaxanthin and neoxanthin. This hyphenated approach successfully and accurately differentiated between the two isomers, α-carotene and β-carotene. A recent study utilized HPLC and LC-MS methods to analyze the carotenoid profiles in the leaves of four different accessions of *Amaranthus gangeticus* (L3, L5, L7, and L9) (Sarker and Oba 2021). The HPLC process involved an STR ODS-II column and a UV-Vis detector. The mobile phase used for separating the carotenoids consisted of acetonitrile, methanol, and dichloromethane in a ratio of 60:20:20 (v/v/v), with 0.1% ammonium acetate, under isocratic flow conditions. For the LC-MS analysis, a TOF mass spectrometer was employed, which was connected online to an HPLC system equipped with an ESI source operating in negative ion mode and measuring in the m/z range of 0–1000. The carotenoids identified in this study included neoxanthin, violaxanthin, lutein, zeaxanthin, and β-carotene. The hyphenated techniques used in this investigation proved to be robust and reliable for the accurate identification of carotenoids. A study on 18 conventional GLVs compared the profiles of carotenoids and tocopherols using a single ion monitoring LC-MS method to identify potential GLVs for healthy diet formulations (Eun-Young et al., 2023). The GLVs used for the study were *Amaranthus lividus, Angelica gigas, Glehnia littoralis, Heracleum moellendorffii, Peucedanum japonicum, Pimpinella brachycarpa, Aralia continentalis, Kalopanax septemlobus, Artemisia princeps, Cirsium setidens, Ligularia fischeri, Petasites japonicas, Rudbeckia laciniata, Solidago virgaurea* var. *asiatica, Taraxacum mongolicum, Adenophora triphylla* var. *japonica, Allium victorialis* var. *platyphyllum,* and *Toona sinensis*. A single ion monitoring (SIM) method combined with LC-MS was used to analyze the carotenoids. The study utilized a Q-TOF mass spectrometer with APCI in positive ion mode as the ionization source. For the LC separation, a YMC C30 carotenoid column was used. The mobile phases consisted of methanol and water in a ratio of 95:5 with 5 mM ammonium formate, and a mixture of methyl tertiary butyl ether, methanol, and water in a ratio of 90:7:3 with 5 mM ammonium formate. The carotenoids detected included β-carotene, zeaxanthin, lutein, luteoxanthin, neoxanthin, and violaxanthin. Among the conventional GLVs analyzed, (all-E)-lutein was found to be the most abundant carotenoid, comprising between 44.4–52.1% of the total. It was followed by (all-E)-violaxanthin and (all-E)-β-carotene. These findings demonstrate the reliability and validity of the LC-MS/SIM approach utilized in this study.

Research clearly shows that LC-DAD-MS and MS/MS systems are effective tools for analyzing natural carotenoids. The development of these systems is progressing steadily. As the LC-DAD-MS and MS/MS technologies advance, we can expect to gain more insights into the distribution and roles of small natural carotenoids. Table 4 presents studies on the separation and identification of carotenoids from GLVs using HPLC and LC-MS techniques.

Table 4 Profiling of Carotenoids.

Plant Name	*Parts Used*	*Extraction Solvent*	*Compounds Identified*	*Method Used*	*Retention Time (Min)*	*Stationary Phase*	*Mobile Phase*	*Wavelength (nm) / Flowrate (mL/min)*	*MS Condition*	*Observed V-alue*	*Reference*
Brassica oleracea var. *italica; B. oleracea* var. *gemmifera; B. oleracea* var. *capitata; B. oleracea* var. *botrytis; B. oleracea* var. *alboglabra; B. rapa* var. *pekinensis; B. rapa* var. *chinensis; B. rapa* var. *parachinensis; Raphanus sativus; Nasturtium officinale*	Leaves	Ethanol and hexane: Ethyl acetate (1:1) and ethanol containing BHT	Violaxanthin	LC- AP CI-MS/ MS	6.88	YMC Carotenoid C30 Column, (150 mm× 2.1 mm)	Methanol: water (9:1) and IPA:methanol (7:3)	0.3 mL/min	Triple quadrupole MS with positive APCI source and gas temperature –200°C. Vaporizer temperature –350°C, drying gas flow rate of 2.33×10^{-4} m3/s, nebulizer pressure of 1.38 ×105 Pa and capillary voltage of 4500 V.	601.4	Lee et al. (2020)
			Neoxanthin		7.23					601.4	
			Antheraxanthin		10.39					585.5	
			Lutein		13.58					569.4	
			Zeaxanthin		15.73					569.4	
			β-Cryptoxanthin		19.19					553.5	
			β-Carotene		21.44					537.4	

Contd...

Table 4 *Contd...*

Plant Name	*Parts Used*	*Extraction Solvent*	*Compounds Identified*	*Method Used*	*Retention Time (Min)*	*Stationary Phase*	*Mobile Phase*	*Wavelength (nm) / Flowrate (mL/min)*	*MS Condition*	*Observed V-alue*	*Reference*
Chayote leaves; Celery; Garland chrysanthemum; Green amaranth; Iceberg lettuce; Kale; Leeks; Pea shoots; Spinach; Sword leaf lettuce; Sweet potato leaves; Wolfberry leaves; Cassava leaves; Kangkong; Melinjau; Pegaga; Sweet leaf bush; Turmeric leaves; Ulam raja; Agathi leaves; Ceylon spinach; Fenugreek leaves; Gongura leaves; Moringa leaves; Ponnaganni	Leaves and Aerial Parts.	Ethanol containing BHT	Violaxanthin	LC- AP CI-MS/ MS	9.73	YMC CarotenoidC30 column (150 mm × 2.1 mm) with a guard column	90% methanol and IPA: methanol (70:30)	0.3 mL/min	Triple quadrupole MS with positive APCI source and gas temperature –200°C. Nebulising gas flow –2.5L/min, Interface temperature –350°C, desolvation temperature –150°C and drying gas flow rate - 3L/min.	601.45	Lee et al. (2022)
			Neoxanthin		10.26					601.50	
			Antheraxanthin		15.01					585.45	
			Lutein		18.67					551.40	
			Zeaxanthin		19.50					569.45	
			β-Cryptoxanthin		21.36					553.45	
			α-Carotene		22.50					537.45	
			β-Carotene		23.69					537.45	

Contd...

Table 4 *Contd...*

Plant Name	*Parts Used*	*Extraction Solvent*	*Compounds Identified*	*Method Used*	*Retention Time (Min)*	*Stationary Phase*	*Mobile Phase*	*Wavelength (nm) / Flowrate (mL/min)*	*MS Condition*	*Observed V-alue*	*Reference*
Amaranthus lividus; Angelica gigas; Glehnia littoralis; Heracleum moellendorffii; Peucedanum japonicum; Pimpinella brachycarpa; Aralia continentalis; Kalopanax septemlobus; Artemisia princeps; Cirsium setidens; Ligularia fischeri; Petasites japonicas; Rudbeckia laciniata; Solidago virgaurea var. *asiatica; Taraxacum mongolicum; Adenophora triphylla* var. *japonica; Allium victorialis* var. *platyphyllum; Toona sinensis*	Foliage	Acetone: ethanol: cyclohexane (1:1:2) containing 0.1% BHT	β-carotene	LC- AP CI-QT OF-MS/ SI M	~ 32.5	YMC CarotenoidC30 column (150 mm × 4.6 mm)	Methanol: water (95:5) containing 5mM of ammonium formate and methyltertiarybut-yl; ether:methanol: water (90:7:3) containing 5mM ammonium formate	0.5 mL/min.	Quadrupole TOF-MS with APCI positive mode. Drying gas flow - 10L/ min, Nebulizing gas flow - 3 L/ min, Interface temperature –400°C, DL temperature –300°C, and the Heat block temperature –300°C.	537.44	Eun-Young Ko et al. (2023)
			Zeaxanthin		~ 21.5					569.43	
			Lutein		~ 20.0					551.42	
			Luteoxanthin		~ 17.0					601.42	
			Neoxanthin		~ 15.2					601.42	
			Violaxanthin		~ 14.8					601.42	

Contd...

Table 4 *Contd...*

Plant Name	*Parts Used*	*Extraction Solvent*	*Compounds Identified*	*Method Used*	*Retention Time (Min)*	*Stationary Phase*	*Mobile Phase*	*Wavelength (nm) / Flowrate (mL/min)*	*MS Condition*	*Observed V-value*	*Reference*
Amaranthus gangeticus (L3,L5,L7 and L9)	Leaves	Ice cold acetone	Neoxanthin	HPLC and LC- ESI - MS	2.52	CTO-10 AC STR ODS-II column, (150×4.6 mm)	Acetonitrile/ methanol/ dichloromethane (60:20:20, v/v/v) containing 0.1% ammonium acetate	450 nm, 1 mL/min.	TOF-MS with UV-Vis-Visdetector and ESI source. Scan range: 0–1000m/z. Needle voltage: –2000V	438.54	Sarker & Oba (2021)
			Violaxanthin		2.63					446.43	
			Lutein		3.84					445.22	
			Zeaxanthin		4.28					452.63	
			β-carotene		20.16					449.52	
Brasicca capitates; Spinacia oleracea	Leaves	Acetone	β-carotene	HP LC	~10.0	RP-C18 column	Acetonitrile dichloro- dichloro-methane, and methanol (70:20:10)	452 nm, 1 mL/min.	NA	NA	Zahra et al. (2022)

Contd...

Table 4 *Contd...*

Plant Name	*Parts Used*	*Extraction Solvent*	*Compounds Identified*	*Method Used*	*Retention Time (Min)*	*Stationary Phase*	*Mobile Phase*	*Wavelength (nm) / Flowrate (mL/min)*	*MS Condition*	*Observed V-alue*	*Reference*
Brassica compestris; Brassica rapa; Brassica juncea; Malva neglecta; Spinacia oleracea	Leaves	Ice-cold acetone and ethanol	All trans anhydrolutein-III	HP LC-DA D	5.2	RP-C18 Column (4.6 × 100 mm)	Methanol: deionized water (92:8) with 10mM ammonium acetate, deionized water containing 0.1mM Ammonium acetate and MTBE.	450 nm, 1 mL/min.	NA	NA	Zeb et al. (2017)
			β-carotene-5,6-epoxide		8.5						
			All trans Neoxanthin		10.0						
			All trans violaxanthin		10.7						
			13-ciszeaxanthin		12.7						
			9-cis zeaxanthin		13.0						
			Lutein		15.3						
			9-cis-Lutein		17.6						
			9'-cis-Lutein		18.2						
			Apo-caroten-8-al		23.4						
			B-cryptoxanthin ester		29.4						
			Alltrans-β-carotene		33.0						
			13-cis-β-carotene		33.4						

Contd...

Table 4 *Contd...*

Plant Name	*Parts Used*	*Extraction Solvent*	*Compounds Identified*	*Method Used*	*Retention Time (Min)*	*Stationary Phase*	*Mobile Phase*	*Wavelength (nm) / Flowrate (mL/min)*	*MS Condition*	*Observed V-alue*	*Reference*
Brassica oleracea *L.* var. *botrytis*	Leaves	Ice-cold acetone and ethanol	Phytoene All-E-neoxanthin 9-Z-neoxanthin All-E-lutein 9-Z-lutein 9′-Z-lutein αcarotene 9′-Z-neoxanthin γ carotene All-E-βcarotene	HP LC-DA D	-	C18 column (4.6 × 100 mm)	Methanol:deionized water (92:8) with 10mM ammonium acetate, ammonium acetate (0.1mM) with- ionized water and MTBE.	450 nm, 1.0 mL/min.	NA	NA	Zeb et al. (2022)

Conclusion and Future Perspectives

Natural product research is gaining popularity around the world. Medicinal plants have attracted significant attention due to their rich sources of bioactive components, such as phenolic acids, flavonoids, carotenoids, vitamins, and minerals (Mueed et al., 2023). To understand the biochemical and pharmacological activities of medicinal plants, functional foods, and other nutraceuticals, researchers need to identify detailed chemical profiles. This information is crucial for evaluating their therapeutic potential. GLVs are a vital part of the human diet, and have become increasingly popular due to their rich source of phytochemicals. These compounds serve as important antioxidants and help prevent chronic diseases. To maximize the health benefits of these phytocompounds from GLVs, they must be extracted with high purity and thoroughly characterized. A comprehensive review of various extraction methods—including solid-phase microextraction, subcritical solvent extraction, MAE,, UAE, and SSE—provides valuable insights into their applications and advantages. Together, these advanced methodologies enhance the field of extraction techniques by offering specialized solutions for a wide range of bioactive compounds and natural matrices. Incorporating these methods into research and industrial processes improves efficiency, reduces environmental impact, and enhances selectivity (Bhadange et al., 2024). Numerous studies have investigated the phenolic profiles of various plant species using RP-HPLC. However, recent advancements in MS have significantly enhanced the analysis of complex biological samples. MS techniques require less sample dilution due to reduced noise, improved detection sensitivity, and greater specificity. As a result, several limitations of LC-UV—such as high detection and quantification limits, as well as interference from complex samples—have been addressed (Chiriac et al., 2021). LC-MS has become an essential tool in natural product analysis, particularly for comprehensive profiling of secondary metabolites due to its unique advantages, including effective chemical identification, high resolution, and sensitivity at a relatively low cost. The rapid development of LC-MS technology can be attributed to its ability to characterize molecules based on C-ring cleavages and to provide precised data. It is important to note that HPLC parameters and ionization techniques can influence fragmentation patterns. When investigating conjugated derivatives of phenolic compounds, especially flavonoids, careful attention is necessary, as even minor errors can lead to inaccurate identification of the target molecules (Kumar, 2017). One of the significant advancements in carotenoid analysis is the design and development of novel packing materials for chromatographic columns. Unlike fully porous particles, core-shell technology utilizes superficially porous particles to enhance separation efficiency by reducing band broadening and increasing resolution. Additionally, C30 stationary phases have been improved to enhance isomer separation, which is crucial for accurately distinguishing between carotenoids with similar structures. By combining sophisticated chromatographic techniques with high-resolution mass spectrometry, methods such as UHPLC-

QTOF-MS, supercritical fluid chromatography-mass spectrometry (SFC-MS), and bi dimensional chromatography coupled with MS facilitate greater structural characterization and quantification. These approaches allow for detailed profiling of carotenoids and the discovery of new isomeric and derivative forms. The classification of closely related carotenoid compounds based on collision cross-section values has been improved by integrating ion mobility spectrometry (IMS) with MS (Giuffrida et al., 2020). Several research studies have employed LC-MS techniques to investigate these bioactive compounds from GLVs. However, most of the papers cited in this chapter represent only modest attempts at this juncture. There is a pressing need for more studies on the distribution patterns of phytocompounds in GLVs using these methodologies to better understand the structural-functional relationship and their roles in the food and pharmaceutical industries. Future advancements in the industry will undoubtedly rely on a sophisticated understanding of the relationship between extraction parameters and compound properties. This approach will lead to sustainable and optimized systems for extracting bioactive compounds from natural sources. Researchers and practitioners are encouraged to adopt these cutting-edge hyphenated techniques, as they represent the future of extracting, identifying, and characterizing natural products from plant sources.

References

Bang, Jun-Hyoung, Ick-Hyun, J., Raveendar, S., Won Tea, J., Sangtaek, O., Tae-Young, H, et al. 2024. Comparative analysis of polyphenolic compounds in different *Amaranthus* species: Influence of genotypes and harvesting year. *Antioxidants, 13*(4): 501.

Barba, F.J., Esteve, M.J. and Frigola, A. 2014. *Studies in Natural Products Chemistry: Bioactive Components from Leaf Vegetable Products*. Elsevier B.V.

Bélanger, J., Balakrishna, M., Latha, P., Katumalla, S. and Johns, T. 2010. Contribution of selected wild and cultivated leafy vegetables from South India to lutein and beta-carotene intake. *Asia Pac. J. Cli. Nutr., 19*(3): 417–424.

Bell, L. and Wagstaff, C. 2014. Glucosinolates, myrosinase hydrolysis products, and flavonols found in rocket (*Eruca sativa* and *Diplotaxis tenuifolia*). *J. Agric. Food Chem., 62*(20): 4481–4492.

Bhadange, Y.A., Carpenter, J. and Saharan, V.K. 2024. A comprehensive review on advanced extraction techniques for retrieving bioactive components from natural sources. *ACS Omega, 9*(29): 31274–31297.

Brewer, L.R., Kubola, J., Siriamornpun, S., Herald, T.J. and Yong Cheng, S. 2014. Wheat bran particle size influence on phytochemical extractability and antioxidant properties. *Food Chem., 152*: 483–490.

Carnauba, R.A., Neuza, M.A.H. and Lajolo, F.M. 2021. Estimated dietary polyphenol intake and major food sources of the Brazilian population. *Br. J. Nutr., 126* (3): 441–448.

Chiriac, E.R., Chitescu, C.L., Geana, E.I., Gird, C.E., Socoteanu, R.P., and Boscencu, R. 2021. Advanced analytical approaches for the analysis of polyphenols in plant matrices: A review. *Separations*, *8*(5): 65.

Delpino-Rius, A., Eras, J., Marsol-Vall, A., Vilaró, F., Balcells, M. and Canela-Garayoa, R. 2014. Ultra performance liquid chromatography analysis to study the changes in the carotenoid profile of commercial monovarietal fruit juices. *J. Chromatogr. A., 1331*: 90–99.

Eun-Young, K., Lee, Ji-Ho, Sivanesan, I., Mi-Jung, C., Young-Soo, K. and Saini, R.K. 2023. Carotenoid and tocopherol profiling in 18 Korean traditional green leafy vegetables by LC-SIM-MS. *Foods, 12* (6): 1312.

FAO. 2022. *World Food and Agriculture: Statistical Yearbook*. Rome.

Feng, X., Yu, Q., Li, B. and Kan, J. 2022. Comparative analysis of carotenoids and metabolite characteristics in discolored red pepper and normal red pepper based on non-targeted metabolomics. *LWT, 153*: 112398.

Ganguly, S., Kumar, J. and Seal, T. 2021. Characterization of secondary metabolites in different parts of *Ocimum gratissimum* L. by *in vitro* Antioxidant activity and high-performance liquid chromatography–diode-array detector analysis. *Pharmacogn. Mag., 17* (74): 209–215.

Gebregziabher, B.S., Zhang, S., Qi, J., Azam, M., Ghosh, S., Feng, Y., et al., 2021. Simultaneous determination of carotenoids and chlorophylls by the HPLC-UV-VIS method in soybean seeds. *Agronomy, 11*(4): 758.

Giacometti, J., Kovačević, D.B., Putnik, P., Gabrić, D., Bilušić, T., Krešić, G., et al., 2018. Extraction of bioactive compounds and essential oils from Mediterranean herbs by conventional and green innovative techniques: A review. *Food Res. Int., 113*: 245–262.

Giuffrida, D., Zoccali, M. and Mondello, L. 2020. Recent developments in the carotenoid and carotenoid derivatives chromatography-mass spectrometry analysis in food matrices. *TrAC Trends Anal. Chem., 132*: 116047.

Gori, A., Boucherle, B., Rey, A., Rome, M., Fuzzati, N. and Peuchmaur, M. 2021. Development of an Innovative maceration technique to optimize extraction and phase partition of natural products. *Fitoterapia,148*: 104798.

Grujić, V.J., Todorović, B., Kranvogl, R., Ciringer, T. and Ambrožič-Dolinšek, J. 2022. Diversity and content of carotenoids and other pigments in the transition from the green to the red stage of *Haematococcus pluvialis* microalgae identified by HPLC- DAD and LC-QTOF-MS. *Plants, 11*(8): 1026.

Johnson, C.E., Long-Ze, L., Harnly, J.M., Oladeinde, F., Kinyua, A.M., Michelin, R., et al., 2011. Identification of the phenolic components of *Vernonia amygdalina* and *Russelia equisetiformis*. *J. Nat. Prod., 4*: 57–64.

Jovanović, A.A., Đorđević, V.B., Zdunić, G.M., Pljevljakušić, D.S., Šavikin, K.P., Gođevac, D.M., et al., 2017. Optimization of the extraction process of polyphenols from *Thymus serpyllum* L. herb using maceration, heat- and ultrasound- assisted techniques. *Sep. Purif. Technol., 179*: 369–380.

Kaczor, A. and Baranska, M. 2016. *Carotenoids: Nutrition, Analysis, and Technology: Structural Studies of Carotenoids in Plants, Animals, and Food Products*. John Wiley and Sons, Ltd.

Kiokias, S., Proestos, C. and Varzakas, T. 2016. A review of the structure, biosynthesis, absorption of carotenoids-analysism, and properties of their common natural extracts. *Curr. Res. Nutr. Food Sci., 4*: 2227.

Kopec, R.E., Cooperstone, J.L., Cichon, M.J. and Schwartz, S.J. 2012. *Analysis of Antioxidant-rich Phytochemicals: Analysis Methods of Carotenoids*. John Wiley and Sons Ltd.

Kozłowska, M. and Gruczyńska, E. 2018. Comparison of the oxidative stability of soybean and sunflower oils enriched with herbal plant extracts. *Chem. Pap., 72* (10): 2607–2615.

Kumar, B.R. 2017. Application of HPLC and ESI-MS techniques in the analysis of phenolic acids and flavonoids from green leafy vegetables (GLVs). *J. Pharm. Anal., 7*(6): 349–364.

Kumar, B.R., Anupam, A., Manchikanti, P., Rameshbabu, A.P., Dasgupta, S. and Dhara, S. 2018. Identification and characterization of bioactive phenolic constituents, anti-proliferative, and anti-angiogenic activity of stem extracts of *Basella alba* and *rubra*. *J. Food. Sci. Technol., 55* (5): 1675–1684.

Kumar, S.S., Manoj, P., Giridhar, P., Shrivastava, R. and Bharadwaj, M. 2015. Fruit extracts of *Basella rubra* that are rich in bioactives and betalains exhibit antioxidant activity and cytotoxicity against human cervical carcinoma cells. *J. Funct. Foods., 15*: 509–515.

Kurek, M.A., Aktaş H., Pokorski, P., Pogorzelska-Nowicka, E. and Custodio-Mendoza, J.A. 2025. A Comprehensive Review of Analytical Approaches for Carotenoids Assessment in Plant-Based Foods: Advances, Applications, and Future Directions. *Appl. Sci., 15*(7): 3506.

Lakshminarayana, R., Raju, M., Krishnakantha, T.P. and Baskaran, V. 2005. Determination of major carotenoids in a few indian leafy vegetables by high-performance liquid chromatography. *J. Agric. Food Chem., 53* (8): 2838–2842.

Lee, H.W., Zhang, H., Liang, X. and Ong, C.N. 2020. Simultaneous determination of carotenoids, tocopherols, and phylloquinone in 12 Brassicaceae vegetables. *LWT, 130*: 109649.

Lee, H.W., Bi, X. and Jeyakumar Henry, C. 2022. Carotenoids, tocopherols, and phylloquinone content of 26 green leafy vegetables commonly consumed in Southeast Asia. *Food Chem., 15*(385): 132729.

Li, Y., Yang, Z., Jin, C., Liu, Y., Zhang, Y., Ma, X., et al., 2024. Akebia fruit: A review on its extraction, phytochemicals, bioactivities, and applications. *eFood, 5*(2): e142.

Lin, L. and Harnly, J.M. 2009. Identification of the phenolic components of collard greens, kale, and Chinese broccoli. *J. Agric. Food Chem., 57* (16): 7401–7408.

Londoño-Giraldo, L.M., Bueno, M., Corpas-Iguarán, E., Taborda-Ocampo, G. and Cifuentes, A. 2021. HPLC-DAD-APCI-MS as a tool for carotenoid assessment of wild and cultivated cherry tomatoes. *Horticulturae, 7* (9): 272.

Lourenço-Lopes, C., Fraga-Corral, M., Garcia-Perez, P., Carreira-Casais, A., Silva, A., Gandara, J., Prieto, M. A. 2022. A HPLC-DAD method for identifying and estimating the content of fucoxanthin, β-carotene and chlorophyll a in brown algal extracts. Food Chem. Adv., 1:100095.

Manousi, N., Sarakatsianos, I. and Samanidou, V. 2019. *Engineering Tools in the Beverage Industry: Extraction Techniques of Phenolic Compounds and Other Bioactive Compounds from Medicinal and Aromatic Plants*. Woodhead Publishing.

Maoka, T. 2023. Carotenoids: Distribution, function in nature, and analysis using LC- Photodiode Array Detector (DAD)-MS and MS/MS system. *Mass Spectrom., 12*(1): A0133.

Masiala, A., Vingadassalon, A. and Aurore, G. 2024. Polyphenols in edible plant leaves: An overview of their occurrence and health properties. *Food Funct., 15*: 6847–6882.

Mohd Zaid, N.A., Sekar, M., Bonam, S.R., Gan, S.H., Lum, P.T., Begum, M.Y., et al., 2022. Promising natural products in new drug design, development, and therapy for skin disorders: An overview of scientific evidence and understanding their mechanism of action. *Drug Des. Devel. Ther., 16*: 23–66.

Mueed, A., Shibli, S., Al-Quwaie, D.A., Ashkan, M.F., Alharbi, M., Alanazi, H. et al., 2023. Extraction, characterization of polyphenols from certain medicinal plants and evaluation of their antioxidant, antitumor, antidiabetic, antimicrobial properties, and potential use in human nutrition. *Front. Nutr., 21*(10): 1125106.

Mungwari, C.P., King'ondu, C.K., Sigauke, P. and Abiodun Obadele, B.A. 2025. Conventional and modern techniques for bioactive compounds recovery from plants: Review. *Sci. Afr., 27*: e02509.

Ola, S.S., Catia, G., Marzia, I., Francesco, V.F., Afolabi, A.A. and Nadia, M. 2009. HPLC/DAD/MS characterisation and analysis of flavonoids and cynnamoil derivatives in four Nigerian green-leafy vegetables. *Food Chem., 115* (4): 1568–1574.

Olsen, H., Aaby, K. and Borge, G.I. 2009. Characterization and quantification of flavonoids and hydroxycinnamic acids in curly kale (*Brassica oleracea* L. convar. *Acephala* var. *sabellica*) by HPLC-DAD-ESI-MSn. *J. Agric. Food Chem., 57* (7): 2816–2825.

Oubannin, S., Bijla, L., Ahmed, M.N., Ibourki, M., El Kharrassi, Y., Devkota, K., et al., 2024. Recent advances in the extraction of bioactive compounds from plant matrices and their use as potential antioxidants for vegetable oils enrichment. *J. Food Compos. Anal., 128*: 105995.

Park, S., Arasu, M.V., Jiang, N., Seung-Hyun, C., Yong Pyo, L., Jong-Tae, P. et al., 2014. Metabolite profiling of phenolics, anthocyanins, and flavonols in cabbage (*Brassica oleracea* var. *capitata*). *Ind. Crops Prod., 60*: 8–14.

Pasini, F., Verardo, V., Caboni, M.F. and Luigi Filippo D'Antuono, L.F. 2012. Determination of glucosinolates and phenolic compounds in rocket salad by HPLC- DAD-MS: Evaluation of *Eruca sativa* Mill. and *Diplotaxis tenuifolia* L. genetic resources. *Food Chem., 133* (3): 1025–1033.

Pateiro, M., Domínguez, R., Munekata, P.E.S., Nieto, G., Bangar, S.P., Dhama, K., et al., 2023. Bioactive compounds from leaf vegetables as preservatives. *Foods, 12*(3): 637.

Pérez-Jiménez, J., Neveu, V., Vos, F. and Scalbert, A. 2010. Systematic analysis of the content of 502 polyphenols in 452 foods and beverages: An application of the phenol- explorer database. *J. Agric. Food Chem., 58* (8): 4959–4969.

Pintea, A., Dulf, F.V., Bunea, A., Socaci, S.A., Pop, E.A., Opriță, V.A. et al., 2020. Carotenoids, fatty acids, and volatile compounds in apricot cultivars from Romania: A chemometric approach. *Antioxidants, 9* (7): 1–27: 562.

Pop, R.M., Weesepoel, Y., Socaciu, C., Pintea, A., Vincken, J.P. and Gruppen, H. 2014. Carotenoid composition of berries and leaves from six Romanian sea buckthorn (*Hippophae rhamnoides* L.) varieties. *Food Chem., 147*: 1–9.

Randhawa, M.A., Khan, A.A., Javed, M.S. and Wasim Sajid, M.W. 2015. *Handbook of Fertility: Nutrition, Diet, Lifestyle, and Reproductive Health: Green Leafy Vegetables: A Health Promoting Source*. Academic Press.

Sarker, U. and Oba, S. 2021. Color attributes, betacyanin, and carotenoid profiles, bioactive components, and radical quenching capacity in selected *Amaranthus gangeticus* leafy vegetables. *Sci. Rep.,11*: 11559.

Sultana, S., Akram, M., Muhammad Ali Shah, S., Ayaz, S., Riaz Ur Rehman, M., Ahmad, H. et al., 2018. *Moringa oleifera* Lam. a miraculous medicinal plant: Review. *Pak. J. Biol. Sci., 3*(1): 81–84.

Tresserra-Rimbau, A., Medina-Remón, A., Pérez-Jiménez, J., Martínez-González, M.A., Covas, M.I., Corella, D., et al. 2013. Dietary intake and major food sources of polyphenols in a Spanish population at high cardiovascular risk: The PREDIMED Study. *Nutr. Metab. Cardiovasc. Dis., 23* (10): 953–959.

Viacava, G.E., Roura, S.I., Berrueta, L.A., Iriondo, C., Gallo, B. and Alonso-Salces, R.M. 2017. Characterization of phenolic compounds in green and red oak-leaf lettuce cultivars by UHPLC-DAD-ESI-QToF/MS using MSE scan mode. *J. Mass Spectrom., 52* (12): 873–902.

Wu, J., Chen, C., Li, Y., Cao, K., Wang, X., Fang, W. et al., 2024. Integrated ESI-MS/MS and APCI-MS/MS based metabolomics reveal the effects of canning and storage on peach fruits. *Food Chem., 430*: 137087.

Xu, J., Lin, J., Peng, S., Zhao, H., Wang, Y., Rao, L., et al., 2023. Development of an HPLC-PDA Method for the determination of capsanthin zeaxanthin, lutein, β-cryptoxanthin, and β-carotene simultaneously in chili peppers and products. *Molecules, 28* (5): 2362.

Yu, M., Gouvinhas, I., Chen, J., Zhu, Y., Deng, J., Xiang, Z., et al., 2024. Unlocking the therapeutic treasure of pomegranate leaf: A Comprehensive review on phytochemical compounds, health benefits, and future prospects. *Food Chem. X., 23*: 101587.

Zahra, N., Saeed, M.K., Shahzad, K., Firdous, S., Ahmad, I., Ashraf, M., et al., 2022. Beta carotene determination in different vegetables by high performance liquid chromatography. *Pak. J. Anal. Environ. Chem., 23* (1): 160–167.

Zahran, H. and Najafi, Z. 2020. Enhanced Stability of refined soybean oil enriched with phenolic compounds of olive leaves. *Egypt. J. Chem., 63* (1): 215–224.

Zeb, A. 2017. A simple, sensitive HPLC-DAD method for simultaneous determination of carotenoids, chlorophylls, and α-tocopherol in leafy vegetables. *J. Food Meas. Charac., 11*: 979–986.

Zeb, A., Khan, S. and Ercişli, S. 2022. Characterization of carotenoids, chlorophylls, total phenolic compounds, and antioxidant activity of *Brassica oleracea* L. var. *botrytis* leaves from Pakistan. *Biologia, 77*(2): 315–324.

Zhang, Y., Chen, G., Dong, T., Pan, Y., Zhao, Z., Tian, S. and Hu, Z. 2014. Anthocyanin accumulation and transcriptional regulation of anthocyanin biosynthesis in purple bok choy (*Brassica rapa* var. *chinensis*). *J. Agric. Food. Chem.*, *62*(51): 12366–12376.

Zhang, L., Li, Y., Liang, Y., Liang, K., Zhang, F., Xu, T. et al., 2019. Determination of phenolic acid profiles by HPLC-MS in vegetables commonly consumed in China. *Food Chem., 276*: 538–546.

Zhou, Y., Chu, M., Ahmadi, F., Agar, O.T., Barrow, C.J., Dunshea, F.R., et al., 2024. A comprehensive review on phytochemical profiling in mushrooms: Occurrence, biological activities, applications, and future prospective. *Food Rev. Int., 40*(3): 9.

Index

About the Editor

Hosakatte Niranjana Murthy is currently a Professor at the Department of Life Sciences, Christ University, Main Campus, Bengaluru, India. Before this, he served at the Department of Botany, Karnatak University, Dharwad, India, from 1986 to 2022. Apart from teaching experience in plant biology, Prof. Murthy has post-doctoral and collaborative research experience in many international research institutes. He worked at Biotechnology Division of Tata Energy Research Institute, New Delhi, India (1992); Crop Science Department, University of Guelph, Guelph, Canada (1993); Research Centre for the Development of Horticultural Technology, Chungbuk National University, Cheongju, Republic of Korea several times, the last being in 2022-2024; and Department of Biological Sciences, University of Nottingham, Nottingham, United Kingdom (2005-2006) as a post-doctoral fellow/ visiting scientist.

Prof. Murthy has completed more than 15 research projects funded by various agencies and supervised 22 PhD students. Prof. Murthy has published 335 research articles in international peer-reviewed journals with high impact factors. His research has been cited more than 7700 times by fellow researchers, and his H-index (Hirsch Index) of 46 is recorded by Scopus.